ICE BLINK

Canadian History and Environment Series
Alan MacEachern, Series Editor
ISSN 1925-3702 (Print) ISSN 1925-3710 (Online)

The Canadian History & Environment series brings together scholars from across the academy and beyond to explore the relationships between people and nature in Canada's past.

ALAN MACEACHERN, FOUNDING DIRECTOR
NiCHE: Network in Canadian History & Environment
Nouvelle initiative canadienne en histoire de l'environnement
http://niche-canada.org

UNIVERSITY OF CALGARY
Press

ICE

Navigating Northern Environmental History

BLINK

EDITED BY

Stephen Bocking and Brad Martin

Canadian History and Environment Series
ISSN 1925-3702 (Print) ISSN 1925-3710 (Online)

© 2017 Stephen Bocking and Brad Martin

University of Calgary Press
2500 University Drive NW
Calgary, Alberta
Canada T2N 1N4
press.ucalgary.ca

LIBRARY AND ARCHIVES CANADA CATALOGUING IN PUBLICATION

Ice blink : navigating northern environmental history / edited by Stephen Bocking and Brad Martin.

(Canadian history and environment series, 1925-3710 ; 7)
Includes bibliographical references and index.
Issued in print and electronic formats.
ISBN 978-1-55238-854-9 (paperback).—ISBN 978-1-55238-855-6 (open access PDF).—ISBN 978-1-55238-856-3 (PDF).—
ISBN 978-1-55238-857-0 (EPUB).—ISBN 978-1-55238-858-7 (MOBI)

1. Human ecology—History. 2. Canada, Northern—History.
3. Canada, Northern—Environmental conditions. 4. Canada—Boundaries—Arctic regions. 5. Native peoples—Canada, Northern—Government relations.
6. Native peoples—Canada, Northern—Politics and government. 7. Arctic regions—International status. I. Bocking, Stephen Alexander, 1959-, editor
I. Martin, Brad, 1972-, editor III. Series: Canadian history and environment series ; 7

GF512.N6I23 2016   971.9   C2016-907175-8
                                                             C2016-907176-6

The University of Calgary Press acknowledges the support of the Government of Alberta through the Alberta Media Fund for our publications. We acknowledge the financial support of the Government of Canada. We acknowledge the financial support of the Canada Council for the Arts for our publishing program.

Copyediting by Edwin Janzen
Cover photo: Coloubox image #1886094
Cover design, page design, and typesetting by Melina Cusano

Table of Contents

INTRODUCTION

Navigating Northern
Environmental History

Stephen Bocking

Ice blinks—white glare under clouds, indicating light-reflecting ice beyond the horizon—are a distinctive feature of the northern environment.
Long used by travellers for navigation, they are one way in which people
have known the north and made their way within it. As natural features
that gain their meaning through their use by humans living and travelling
in the north, they also exemplify the links between people and nature in
this place.

Today, however, the dominant image of northern Canada is not that
of the reliable relation between ice and sky represented by ice blinks, but
of turmoil: a place being transformed by larger forces of environmental
change. Models predict, satellites track, and northerners testify to the consequences of rapid warming, transforming ice and polar bears into icons
of global change. Expectations of an ice-free ocean are encouraging exploitation of oil, gas, and other resources, and exploration of new shipping
routes. Territorial claims have gained urgency, as nations map the seabed.
Indigenous peoples seek a sustainable economy, a secure environment, and
a say in the region's future. Throughout, a changing environment—ice,

water, tundra, forests, minerals, wildlife—remains central to northern plans, hopes, and anxieties.

This image of unprecedented change has tended to encourage the assumption that northern change stems mainly from a changing climate—implying that before the current era of global warming the north was a static landscape, almost outside history. But a longer view can challenge this image by demonstrating that we are witnessing today only the latest episode in a century of historic change, and that this history has encompassed much more than the consequences of a changing global climate. Consider what we see if we extend our gaze ever deeper into twentieth-century northern history. In recent decades we have seen the evolution of Indigenous self-government, with land claims and regional governments reconfiguring northern power and authority. Before that, in the late 1970s, came expectations of transformation through a new northern oil-and-gas economy. The decade before, Prime Minister John Diefenbaker sketched in his "northern vision" his ambition to integrate the region into the Canadian state and economy—supplanting the Cold War view of the north as a military frontier. Even earlier, dreams of a pastoral economy based on reindeer, followed by airplanes and mineral discoveries, inspired diverse visions of the future of the north. Before that, the Klondike Gold Rush briefly promised instant wealth. Change, often viewed at the time as unprecedented, has been a constant during this century.

This volume presents novel perspectives on this century of transformation. It is the product of a new generation of northern scholars collaborating on the study of the environmental history of northern Canada. Mainly based outside the region—in southern Canada and the United States—they, like northern travellers perceiving an ice blink, are seeking an understanding of conditions at a distance. The stories they tell concern the evolving relations between people and the northern environment: how this environment has changed over time, how human activities have affected this environment while being themselves shaped by it, and how culture, knowledge, and interests have been tied to these relations. As we will see, these stories, while akin to those of environmental historians elsewhere, take distinctive forms in the north. Some of them are of newcomers: railway promoters capitalizing on Klondike fever, surveyors seeking mineral deposits, pilots tracing transportation routes, technicians installing surveillance systems, miners exploiting landscapes, scientists tracking contaminants. Several chapters consider the Canadian state: its efforts to impose a pastoral economy, supermarket food in place of fresh meat, or community economic development in place of traditional ways of life. Others concern Indigenous people: their identities, ways of life, and evolving relations with the land, the state, scientists, and the wider world. Our authors have pursued these stories across the north: from Quebec and British Columbia to the territories and the High Arctic, while paying careful attention to the links between these places and the rest of the world. They (and the volume itself) therefore express an inclusive understanding of the north, encompassing not just the Canadian Arctic, or an arbitrary administrative region such as the territories above latitude 60° north, but those places customarily imagined as northern Canada. (We discuss in more detail below the varied definitions of the north as a physical and an imagined place.) They also demonstrate the interdisciplinary character of environmental history, linking environmental change with social and political history, geography and anthropology, and the history of science and technology. To help us navigate this century of northern environmental history, we have arranged these chapters into three eras: the colonial, modern, and contemporary north.

Forming Northern Colonial Environments

Even while drawing on universalizing discourses of imperialism and progress, colonization nevertheless took a distinctive form in the north, which became evident in human-environment relations. In the western Arctic, commercial harvesting depleted bowhead whale populations, as well as caribou, polar bears, and musk oxen. In the east, whalers, traders, and missionaries brought new technology, while the fur trade and the Hudson's Bay Company reshaped Inuit environmental and economic relations. Novel biota from the south included disease, with often devastating impacts on communities. With agricultural expansion encountering the limits imposed by soil, climate, and distance, other ways of colonizing northern environments became more prominent. Visions of a domesticated north focused on caribou and musk oxen. A royal commission examined the potential, and in 1922 Wood Buffalo National Park was established, providing space for experiments in transferring and managing bison. Conservation became an instrument of sovereign authority, exercised in a dispute with Denmark involving musk ox hunting on Ellesmere Island. Game reserves, the Northwest Game Act of 1917 and 1929, and the Migratory Birds Convention Treaty infringed on Indigenous hunting activities. Ideas from elsewhere guided these initiatives: ecological theory, sport hunting narratives, and the notion of Indigenous people as irrationally wasteful. The government also initiated a series of scientific surveys.[1]

Economic interests took much of the initiative in reshaping relations between people and the northern environment. In 1920, the discovery of oil at Fort Norman revived interest in northern minerals, two decades after the Klondike Gold Rush. Airplanes promised speed and access (displacing dog teams, canoes, and feet), and a new view of and control over the landscape. Industry sponsored surveys, the federal government provided maps and aerial photos, and discoveries followed: silver and pitchblende east of Great Bear Lake in 1932, and the following year, gold north of Great Slave Lake. Northern colonization gained a distinctive form: scattered sites of exploitation, lasting only as long as the market required. Industrialization (although constrained by geography and ecology) remade northern environments and communities, while drawing the region into global economic networks: exporting commodities, importing capital, machinery, and expertise.[2]

Four of our authors examine this era in northern environmental history. Jonathan Peyton presents a novel perspective on the Klondike, from the point of view of aspiring capitalists and miners in the Stikine region of northern British Columbia. Competing interests proposed railroads to the gold fields, and governments granted concessions, imagining an all-Canadian route. But these plans failed, as did those of most travellers. Yet these schemes—too readily dismissed as irrelevant—had historical significance. Surveys and practical experience eventually catalyzed new ways of linking the Stikine to the world. These failed railroads therefore had consequences, influencing perceptions of the region and its future.

As Andrew Stuhl explains, domesticating northern wildlife implied not only a new view of northern landscapes but a novel interest in scientific advice. In 1926, Robert and Alf Erling Porsild arrived from Denmark to begin the Canadian Reindeer Project. They travelled to Alaska to learn about its reindeer industry, and then surveyed the north to identify a suitable range. Their project itself became an experiment in both applied botany and a new role for the state: managing the relations between north-erners and their landscape. National policy was now expressed through an animal peculiarly suited to both the environment and the state's priorities. Their work illustrates how science could become the basis not only for sur-veying but manipulating resources, a task rendered feasible by reducing the landscape to just a few variables. Yet this conjunction of biology and policy would quickly pass, as changing northern state and scientific prior-ities rendered the reindeer project a relic of an earlier era.

Tina Adcock examines another aspect of the sense of change that overtook the north during the interwar era. Guy Blanchet and George Douglas—two seasoned northern travellers—worked as prospectors and geological surveyors, helping to build the new resource economy. Yet they regretted the passing of another north: remote from the modern world, where hard travel on land and water could preserve one's vital spirit. Their thoughts and experiences illustrate the complexity of responses to techno-logical and environmental change: while many welcomed faster and eas-ier travel and new economic opportunities, this also provoked disquiet, doubt, and a sense of loss.

Even as airplanes became essential to northern travel and transform-ation, aviators and their machines had to adapt to environmental realities. Marionne Cronin examines how pilots and other employees of Canadian Airways translated experience in the northern environment into techno-logical change. Her view of northern aviation focuses on the technology itself, as material articulations of values, ideas, and power. Northern geog-raphy, including rivers and lakes, determined flying routes and landing sites, and weather and other challenges required airplanes to be modified if they were to work properly. The north did not simply receive, but active-ly reshaped technology from elsewhere.

Transformations and the Modern North

The Second World War and the postwar era transformed relations between humans and northern nature. Wartime imperatives remade the north into a military zone defined by access and mobility, evident in airfields, the CANOL pipeline, and the Alaska Highway. During the Cold War, the north remained a strategic environment, now as a bulwark between the superpowers. By 1957 three radar systems, including the Distant Early Warning Line, had brought the region within the North American defence system. Canada also asserted its own view of the north as national territory through aerial photography, mapping, weather observations, and military exercises, and the northern environment defined the Canadian military's vocation as the winter warfare specialist within the Western alliance. Science became a strategic necessity: Cold War requirements accelerated collection of climate and terrain data, and encouraged the study of human responses to the northern environment. The Arctic Institute of North America promoted and coordinated research on both strategic and scientific priorities.[3]

Modernization and development soon joined strategic defence as northern imperatives. The Department of Northern Affairs and National Resources, created in 1953, focused federal efforts, as new highways and other links, as well as an expanding mining economy, tightened the region's ties to global markets while degrading local environments. The Geological Survey, the Fisheries Research Board, and other agencies redefined the northern landscape in terms of the resources demanded by these markets. Strategic and economic initiatives typically bypassed Indigenous communities, perpetuating a view of the north as an unoccupied space to be transformed in response to priorities formed elsewhere. Nevertheless, these communities experienced profound environmental and social change. Low prices disrupted the fur trade economy, and reports of hardship, starvation, and disease captured national attention. For the federal government, intervention became an imperative: replacing hunting and trapping with integration within the Canadian state and economy. Social, educational, and health services were expanded, communities relocated or consolidated into larger settlements, wildlife became subject to management, and Indigenous peoples were encouraged to take up wage employment. These initiatives marginalized Indigenous interests, knowledge, and values, and provoked reaction: anger and frustration over hunting regulations, and assertions of resource rights. By the 1960s, this program of social and environmental engineering had begun to unravel.[4] Three of our chapters provide insights into this period in northern history.

Amidst these transformations, the hunting and sharing of country foods remained central to Indigenous ways of life. So did uncertainty: wildlife migrations, and variations in climate and other aspects of the environment, led to hunger, particularly when accompanied by disease. As Liza Piper explains, food has also been essential to relations between Indigenous communities and newcomers. Depletion of musk ox, caribou, and walrus by explorers, whalers, trappers, and traders justified conservation initiatives that complicated and often criminalized food gathering. Motivated by belief in the inherent deficiency of a non-agricultural diet, the mid-century state engaged in "food colonialism." Health initiatives guided by nutritional science, social programs (such as family allowances), and education encouraged northerners to adopt southern food ways: gardening rather than trapping, canned food in place of fresh meat, residential school cafeterias instead of family hunting trips. Piper demonstrates how food links bodies to environments, and people to the politics of colonialism.

Tina Loo begins her chapter with a witness to starvation. In the early 1950s, reports of desperation, including Richard Harington's photos of dying Inuit and Farley Mowat's *People of the Deer*, provoked the Canadian government to take responsibility for northern social conditions. Its mandate expanded to include transformation of the north into a modern society, populated by citizens amenable to administration. Concepts of sustainable community development that had gained currency elsewhere linked northern economic priorities to issues of power and poverty, with expert knowledge applied to improving peoples' lives. Loo presents this history of northern development as a history of hope. Yet these efforts also fell short of their promise: defining development as a technical matter, they failed to challenge the political and economic forces and structures that defined northern existence.

The DEW Line was the most ambitious element of the postwar militarization of the north. It was an unprecedented exercise: both materially, with construction equipment and workers transported to sites across the region; and technically, as devices and practices perfected in southern laboratories were installed in the north. As Matthew Farish and Whitney Lackenbauer explain, this fusion of science and security demanded a new form of expertise, embodied in the Western Electric engineers and technicians who conceived, assembled, and tested this complex system in an unfamiliar and challenging environment. Like dams and other projects, the DEW Line epitomized the ideology of high modernism, in which military and corporate forms of power are embodied in technology. But while this system reconfigured the north as part of continental defence, the region imposed its own requirements: getting southern technology to work in northern environments demanded improvisation and local knowledge.

Environmental History and the Contemporary North

By the late 1960s, the future of the northern environment had become a matter of active debate. Some foresaw rapid development, especially of its energy resources, and continued integration into global markets. In 1961, the Canada Oil and Gas Land Regulations set the terms for petroleum exploration. Activity expanded after 1968, when oil was found in Alaska, raising hopes for geologically similar regions in Canada, including the Mackenzie Delta/Beaufort Sea and High Arctic regions. Hydroelectric projects in British Columbia, Manitoba, Quebec, and Labrador, and oil sands development in northeastern Alberta, illustrated the provinces' belated interest in their own "forgotten" norths. They also testified to how many now viewed the north as an energy-rich hinterland—a prospect welcomed in Ottawa and the provincial capitals, but one that aroused concern and controversy elsewhere. Protection of the northern environment became a national issue, encouraged by both the wider emergence of environmentalism in Canada and elsewhere and by specific concerns about threats to "fragile" northern ecosystems. These concerns also testified to the influence of science on views of the northern environment. Although this notion of a fragile north soon faded, by the 1990s global change, including contaminants, depletion of stratospheric ozone, and especially climate change had begun to dominate scientific perceptions of this environment.[5]

For many, however, the assertion of Indigenous authority over land, water, and wildlife became central to their vision of the future of the north. Land claim negotiations and self-government agreements provided opportunities for Aboriginal people to regain authority over land and resources, restoring ties between place and livelihood ruptured by state interventions. By 1973 Dene, Métis, Inuvialuit, and Inuit had formed representative organizations, initiating a dramatic political evolution rooted in a longer history of resistance to the state. New resource management institutions drew on both scientific and Indigenous knowledge, including a view of northern landscapes not as uninhabited wildernesses but as social and cultural systems.[6] Aboriginal environmental politics have continued to evolve, through activism, legal affirmation of treaty rights, development of co-management arrangements, and the assertion of Indigenous knowledge. Evolving views of the relations between people, economies, and the northern environment have also been important, including affirmation of the central role of hunting and food gathering in Indigenous communities, and the place of Indigenous people in circumpolar environmental affairs.

Five chapters examine this era in northern environmental history. We begin in northern Quebec, where a lengthy history of exploitation of furs, forests, minerals, and rivers has reflected the imperatives of both global resource demands, and, more recently, Quebec nationalism. Exploring the consequences of this history for people and the land, Hans Carlson takes us on his travels with the Cree. His companions' stories record the ties between place, livelihood, and culture, and their efforts to negotiate their past, present, and uncertain future. In 1975, the James Bay and Northern Quebec Agreement provided the legal basis for the Cree to pursue their way of life in terms of their own practices and knowledge. Development has nevertheless transformed their relations with the land, creating a new politics in which they have both adapted to and challenged change—in part by forming links with other places, as in 1990, when they brought their opposition to the Great Whale hydroelectric project to New York City, travelling via Odeyak (a canoe-kayak hybrid). These travels are also, as Carlson explains, an opportunity to reflect on the ties between personal history and the history of a place, stories of the past and present events, and memory, meaning, and the land.

From Quebec, we travel west to the Yukon. Land claims and self-government agreements under the 1993 Yukon Umbrella Final Agreement established Indigenous authority over much of the territory and its resources. As Paul Nadasdy explains, however, these agreements have also reshaped how people relate to the land, to animals, and to each other. Because they are based on territorial jurisdiction—the foundation of modern states—they have required Aboriginal people to think territorially: to become managers and to create bureaucracies framed in the language of maps. But in doing so they risk neglecting the social relations between people and animals that once formed the basis of land-use practices, in effect securing their future at the expense of their past.

Mines have a finite lifespan, and scores of abandoned sites now lie scattered across the north—sixty-four in the Yellowknife region alone. Many still release contaminants, forming a toxic "landscape of exposure" with lasting environmental and social impacts. As Arn Keeling and John Sandlos explain, understanding this legacy requires linking the histories of labour and of landscape to form a perspective on mining more elaborate than the simple cycles of boom and bust. Such a view also accommodates distinctive features of the industry in the north, including remote locations and the presence of Indigenous people, whose ways of life render them more vulnerable to contaminants and to damage to local living resources. Today, rising mineral prices are encouraging efforts to reopen or remediate some mines. Keeling and Sandlos identify in these efforts the continuing influence of the past on the future of northern mining: when these "zombie" mines are brought back to life, so too are memories of the conflicts they once provoked.

Northern contaminants come not only from mines and other local sources, but from more distant places. In my chapter, I explain how these substances—including radioactive fallout, particulates, and pesticides and other persistent organic pollutants—have encouraged new scientific perspectives and methods, novel environmental and health initiatives, and a new relationship between Indigenous peoples, scientists, and governments. This is partly a story of surprise; the discovery of contaminants in northern ecosystems, animals, and food confounded assumptions about what "belongs" in the northern environment. While contaminants gave scientists an opportunity to extend their historical role as interpreters of the northern environment, Indigenous communities and institutions asserted their own perspectives on these substances and their significance to food, health, and knowledge. Contaminants thus provide an opportunity to examine the distinctive social dynamics and structures of northern environmental knowledge, and the material relations between the north and the rest of the world.

Images of melting sea ice have made climate change the most obvious link between the north and the rest of the planet. Emilie Cameron considers its implications for northerners, their livelihoods, and their "right to be cold." Climate science implies certain ideas about time, space, and action: it focuses on the future, seeks prediction and adaptation, and assumes that local places are self-evident. It also implies that, however useful Indigenous knowledge may be in relation to these places, it has little to add to global perspectives. However, a more critical perspective on scale and knowledge can open up other ways of understanding northern climate change. One way is by encouraging an awareness of the history of climate science itself, including its formation in the context of colonization, and the local character of its "global" perspective. As Cameron explains, this awareness can also provide a basis for enabling northerners to contribute to a more inclusive understanding of the consequences of global change for the north.

These chapters examine episodes drawn from a century of northern environmental history. Lacking features familiar elsewhere—agriculture and settlement, industrialization and urbanization, roads and railroads—this can be challenging terrain for historians. Certain enduring questions have guided them—questions that relate to themes of interest to environmental historians elsewhere, but also raise issues distinctive to this region. For example, how has the northern environment changed over time as a result of natural forces or human activities? What has been the place of Indigenous people in human-environment relations in the north? What distinctive roles has the state played in northern environmental relations? How has technology influenced how people relate to the northern environment? What role has human experience played in these relations? How has knowledge both shaped and been influenced by these relations? How has movement—of nature, people, their products and ideas—been a factor in northern history? And finally, how have the identities of northern places and spaces themselves been formed?

Environmental Change

Changing northern environments today are commonly interpreted in terms of human impacts on the global environment, especially climate change. The place in climate models of ice, the Arctic Ocean, and methane released from permafrost, the role of ice cores in studies of the changing atmosphere, and assessments of impacts on endangered wildlife testify to how scientific interpretations of the northern environment are contributing to our understanding of global environmental change. Its geopolitical and economic consequences—new sea routes and resource development opportunities, a "contest" for northern territory, impacts on local uses of sea ice and other features—testify to its social implications.

This novel image of an open and vulnerable north stems, in part, from the perception that the region has been insulated by its distance from industry and human populations—that it is, in effect, the last relic of a once pristine planet. However, while current global changes may be unprecedented, the north has known change across a range of scales and time periods since long before the current era. Dramatic fluctuations in animal populations—including caribou, as well as, famously, lemmings and the predators they support—are a distinctive feature of northern ecosystems, even in the absence of human activities. Environmental change itself has thus often been ambiguous, sparking debate as to whether specific instances are "natural" features of northern landscapes and wildlife or the result of human activity. In the 1950s, for example, apparent declines in caribou populations captured attention. At the time, many attributed this to hunting; yet, before and since, other explanations have been considered, including population cycles, changes in migration routes, and other human impacts, such as harassment by aircraft, fire, or mineral exploration near calving areas. Similar ambiguities have been evident in debates regarding fire: its origins as a natural phenomenon, in Indigenous firing practices, or in the intervention of outsiders—and whether fire inflicts damage on northern landscapes, or is a natural and normal aspect of northern environmental change, a distinction that has often had political implications.[7]

Indigenous people have also accumulated a history of change: hunting caribou and other species, with potentially significant impacts, or harvesting wood and manipulating fire to create optimum habitat for wildlife, including bison and moose. Nonetheless, the arrival of newcomers provoked unprecedented change. By the early 1900s, walrus and bowhead whale populations in the western Arctic had collapsed, with serious consequences for local human populations; other species also felt the impacts of the whaling industry, including caribou, polar bears, and musk oxen. Robert Peary and other explorers, as well as hunters supplying the musk ox robe trade, also depleted populations. These impacts reflected both economic imperatives and different views of nature. While Indigenous peoples recognized ecological variability and distributed hunting activities accordingly, newcomers responding to the demands of southern markets tended to focus their harvesting in specific areas, depleting local populations. Diverse views of nature were also expressed in conservation initiatives, which, in seeking to manage or minimize change, often provoked it. The 1916 Migratory Birds Convention disrupted seasonal Indigenous hunting activities, while Kluane and other wildlife reserves excluded hunters from traditional territories. Experiments in stocking game—including bison in Wood Buffalo National Park and reindeer in the northwestern Arctic—affected landscapes, wildlife, and those who had traditionally relied on them.[8]

The war and postwar era witnessed a new order and scale of environmental change. The construction of the Alaska Highway left a disrupted and polluted landscape exposed to disease, sport hunting, fire, and development. Cold War activities, from building the DEW Line in the 1950s to training flights over Labrador in the 1980s, imposed additional impacts on local sites and regions. Mining development had diverse and often devastating consequences, transforming northern environments from the Klondike to the Yellowknife region to northern Ontario. These impacts were felt throughout the cycle of prospecting, exploitation, and abandonment, exhibiting the implications of changing mining practices. Prospectors, road crews, and trappers burned northern forests, often intentionally to expose the rock. In 1940, C. H. D. Clarke commented that "Fire is the thing to fear"—less because of concerns regarding forests, as in temperate regions, than because of its impacts on caribou. Mining wastes were dumped in lakes and their fish populations were depleted. More recently, seismic lines, oil spills, and tundra defaced with tire tracks have accompanied petroleum exploration, while dams have drowned rivers and forests, disrupted flow regimes, and released mercury from the soil. These consequences together testify to the distinctive environmental changes imposed by energy developments across the north.[9]

These developments also had indirect consequences, including new geographies of exploitation, production, and control. Newcomers affected certain regions more than others: coastal areas attracted whalers, and mineral-rich regions attracted prospectors. Other areas, including much of the eastern Arctic, tended to be bypassed. Development catalyzed regional transformation, as roads and aviation routes opened up new areas. In the Klondike, and ever since, northern miners have displaced wildlife, which, combined with shifting patterns of subsistence, has affected the wellbeing of Indigenous people. The movement of agricultural species and diseases together formed a "broken frontier" of ecological imperialism, with consequences for both ecosystems and Indigenous communities. Efforts to adapt food plants and animals from elsewhere to northern conditions, while often unsuccessful, have sometimes had significant local impacts—as seen, for example, in the clearing for grazing of winter caribou habitat in northern Saskatchewan, with lasting consequences for the local Indigenous economy. Diseases have formed their own history of environmental and social change. A history of disruption of northern communities by pathogens, often in combination with hunger and other hardships, culminated in epidemics of influenza, measles, and other afflictions in the twentieth century, often brought north by military or industrial operations. In many communities, epidemics catalyzed the transition into the modern world, along with new economic activities, health, and social services.[10] This history has blurred the boundaries between nature and humans: environmental change has affected humans as much as other species; hybrid landscapes have formed in which people and their consequences touch every part of nature, and nature is present in every aspect of human activities; and northerners and the northern environment have shaped the effects of interventions such as community relocations and hunting regulations.

The Indigenous North

Northern Canada is an Indigenous landscape. Historians have described the relations between Indigenous ways of life and the northern environment: how people travelled, hunted, shared food, organized their communities, and formed knowledge about the world. Much of this study has been framed in the context of colonization: treaties, resource exploitation, expansion of education, health, and social services, community relocation and other aspects of the extension of southern authority into the north. Colonization has taken a distinctive form in the north compared to elsewhere in North America. Throughout this history, Indigenous people and newcomers have formed relations mediated by or with consequences for nature. Early in the twentieth century, explorers like Peary relied on Indigenous technology and survival strategies. Fur traders formed economic relations with Inuit trappers, tying their wellbeing to factors beyond their control, including fluctuations in fox populations and foreign demands for furs. Official views on Indigenous people evolved over the decades: from encouraging traditional subsistence activities, to, by the 1950s, more firmly interventionist policies, motivated by the ambitions of an activist state, a declining fur trade, signs of destitution, and a tendency to view Indigenous people as wasteful and irrational. Wildlife conservation reworked hunting, food sharing, and other aspects of community life. Relocation, whether motivated by insecurities about sovereignty, conservation concerns, or the desire to avoid welfare dependency, disrupted relations with the environment. So did residential schools and resettlement of Inuit from camps to larger communities, illustrating how colonization merged environmental and social change.[11]

However, Indigenous people have not been passive recipients of colonization, but active participants in shaping the north and their place within it. Land claims and treaty negotiations, devolution of authority to territorial governments, legal decisions, co-management arrangements, and the planning of national parks and reserves have provided the basis for asserting authority over environmental relations, including hunting, fishing, and the regulation of development. To assert their claims, Aboriginal groups have had to demonstrate their indigeneity in particular places. In the 1970s, the Inuit Land Use and Occupancy Project showed how their hunting and travel experience could be translated into cartography, and Thomas Berger's Mackenzie Valley Pipeline Inquiry assembled testimony regarding the experience and meaning of landscapes. Energy projects have prompted changes in northern governance to be more consistent with Indigenous ways of life. For example, the James Bay and Northern Quebec Agreement was intended to ensure that Cree culture would continue, even while opening the region to hydroelectric development. This has also encompassed a reconsideration of the environmental impacts of development: not in terms of an imagined, pristine nature, but in relation to how people use and understand the environment, and equitable ownership and access to benefits. This acknowledges that country foods—caribou, seals, walrus, fish—remain essential in many northern communities, reinforcing the relevance of Indigenous knowledge and ways of life.[12]

Indigenous people have also had to work out relations with interests beyond their control, including industry, governments, and scientists. Their implications have been evident in, for example, land claims, co-management agreements, and other aspects of the long-term process of transferring political power to Aboriginal people. But while ensuring that hunting and other traditional activities can continue, these have also been reframed in terms of science, property, and bureaucracy, potentially undermining the ways of life they were intended to preserve. Indigenous knowledge still often carries less weight within management systems accustomed to quantitative models, particularly when their combination is seen as merely a technical task, neglecting the power relations that accompany knowledge.[13] Animal rights activism has undermined the sealskin and fur trades, fuelling distrust between Inuit communities and environmental groups. Other tensions have arisen regarding polar bears and claims regarding their status as an endangered species. Indigenous people have also asserted their interests internationally, including, as already noted, their opposition to the Great Whale Project; more recently, they have used circumpolar institutions to express their views regarding contaminants, climate change, and sustainable development.

The State

Throughout the twentieth century, the evolving state in northern Canada—its capabilities, roles, and objectives—has had consequences for people's relations with the environment. This became evident even in one of the state's primary roles: asserting territorial sovereignty. It has often had environmental dimensions: hunting regulations in the 1910s, military activities during and after the Second World War, mapping and aerial surveys, scientific activities (with scientists asserting a Canadian presence in the north), and the Arctic Waters Pollution Prevention Act of 1970. State efforts to reshape the northern economy have also had environmental consequences. In the 1920s, an emerging federal wildlife bureaucracy attempted to conserve and domesticate musk oxen and reindeer. In subsequent decades, support for resource development (through surveys, transportation facilities, and financial incentives) redefined the northern landscape as part of the national economy. Economic development became part of colonization, as educational, health, and social services transformed communities, ways of life, and relations with the landscape and wildlife. These interventions, often justified in scientific terms, marginalized local customs and attitudes, advancing a view of species, especially caribou, as production units enabling efficient use of the northern landscape. In the 1970s, as environmental concerns became prominent, the state responded by extending the mechanisms of administrative rationalism into the north through regulations and environmental impact assessment.

The activities of the state in the north paralleled those elsewhere in Canada (albeit with, in the territories, the federal government acting in the place of a provincial government): partnerships with economic interests, formation of educational and social services, management of wildlife and other resources. All were aspects of the wider expansion of the state since the 1940s. In the north, as elsewhere, the state has also imposed boundaries on nature, and on particular ways of knowing, managing, and regulating. But the northern state has also exhibited distinctive features—the product of local history and geography: poor agricultural prospects, concern with territorial authority, dispersed settlements requiring transportation networks, and the presence of other agencies, such as the Hudson's Bay Company, the military, and the resource industries. Today, the state's relations with Indigenous communities through land claims, co-management arrangements, and novel approaches to managing national parks have exemplified the status of the north as a terrain of experimentation in governance.

Technology

As generations of historians have shown, people everywhere have used technology to live within, exploit, and transform nature. In the north, it has enabled survival, with Indigenous clothing, shelter, sleds, komatiks, and other objects and practices together forming a distinct technological tradition. This tradition's displacement by devices from elsewhere, even as the north has been colonized, illustrates how technological change has been tied to social and environmental transformation. Yet new technology could also support older traditions: snowmobiles have enabled hunters to continue their practices after moving to larger communities.[14]

But northern technology is more than tools. Indigenous fire-making and other practices, and, more recently, industrial equipment—airplanes, dams, buildings, and research instruments—have created new environments. This became especially evident in transportation technology; within little more than a generation, feet, canoes, and dogsleds co-existed with and then were displaced by airplanes. Airplanes and communication technologies distributed resource extraction, military operations, and other activities (and their impacts) more widely across the landscape, making distance itself essential to experiencing the north. Airplanes, aerial photography, and mapping also encouraged many to view the north as a resource-rich hinterland, lacking people and history, but legible and amenable to management—linking northern history to the global role of fossil fuels in the conversion of the natural world into resource commodities. Photography presented the north as an object of knowledge and as Canadian territory, overseen by an always-present state.[15] Technology has also enabled some to be almost indifferent toward the north, with adaptation replaced by self-contained environments equipped with "southern" amenities.

Technological change has not been an autonomous force, but one shaped by institutions, ideas, and the environment. Mining companies had adopted airplanes by the 1920s, but the Geological Survey only embraced their potential in the 1950s. A shift on the part of prospectors from using aircraft mainly for transport to developing aerial survey techniques required not only new instruments but new ideas about what counted as useful knowledge.[16] Throughout this history, the environment—rivers to follow, lakes to land on, frigid cold requiring special engine care—demanded adaptation and influenced the design of aviation networks. Technology generated demand for itself: as flying routes spread throughout the north, aviators and prospectors required topographic and geological maps based on aerial surveys. It also reinforced social and racial distinctions: for several decades, it was mainly white men who flew, installed radar facilities, and operated scientific equipment.

Experience

As environmental historians elsewhere have noted, nature is also known and formed through experience: what people see, sense, and feel through their bodies.[17] In the north, Indigenous and newcomers' ways of life and the consequences of colonization, technological change, and landscape transformation have been shaped by how people have experienced this environment. Indigenous people have done so through activities defined by the seasons: movement, hunting and sharing food, and raising families; through their relations with animals; and through their knowledge of the landscape. Colonization has been experienced through residential schools, wildlife management, and other interventions. More recently, experience with climate change has inspired novel perceptions of cold and ice that can no longer be taken for granted, together with nervousness about the prospect of unstoppable change. Links between experience and landscape have also been evident in the lives of newcomers. Some found extreme hardship; others, a sense of ease—contrasting experiences that often stemmed from intimacy with or distance from the northern environment. Experience has inspired diverse perceptions of the north: as a harsh, unforgiving terrain, a place inspiring feelings of wonder and a sense of the sublime, a pristine but fragile ecosystem exhibiting dramatic variations in abundance and productivity, or a stockpile of resources. This experience has been framed in terms of distance, time, or winter and other seasons, forming views of the north as a place of extremes that regulate human activities, including arrivals and departures.[18]

Historical change has been experienced in many ways. Through their experiences, northern miners established new connections to nature, even underground, as well as novel patterns of production and consumption, importing most of what they ate and producing only for export. Technological change was also experienced in many ways. Graham Rowley recalled his experience in the late 1930s in the British Canadian Arctic Expedition: "We had to live in the ways the Inuit had evolved, and to travel by dog team in land that was still unexplored. There was the excitement of the unknown and of finding what lay over the next hill."[19] In contrast, flying was described as a "profound leap into a new dimension," transforming time and space.[20] (But when this experience became routine, pilots, once the focus of romantic imagining and excitement, became mere cogs in a large flying machine.) The significance of experience has also been evident in the history of northern science, including methods of travel and fieldwork. And, as noted below, experience has been tied to the evolving authority of knowledge. Even in the postwar era, northern scientific experience has been viewed in terms of adventure and heroic masculinity—of bodies and machines wresting knowledge from a challenging environment.[21]

Knowledge

Throughout northern history, knowledge has shaped how people and institutions understand and act in the northern environment. For much of the twentieth century, it has been a terrain for scientists—usually from elsewhere—accumulating knowledge across a range of disciplines. This history includes evolving scientific practices: surveys, experiments, and aerial photography. Northern scientists have accumulated an extensive record of field research, which we can interpret in terms of the history of field sciences—ecology, geology, wildlife science, oceanography, and climatology—in the north and elsewhere. These practices have implied novel ways of seeing and knowing. For example, aerial surveys demonstrated how technology combined with knowledge to impose a more distant, less intimate view of the north, emphasizing some landscape features while obscuring others.[22] The northern environment has itself influenced research topics, methods, and results, demonstrating the importance of place to scientific practices and knowledge. Some habitats, such as the Mackenzie Delta and "polynyas" (areas of ocean that remain free of ice), have attracted particular attention from scientists, remaking these places and their phenomena as objects of research. They have also tended to define the arctic environment itself as anomalous compared to temperate regions.

Ideas about what counts as knowledge in the north have also changed over time. One persistent question has been whether the north is distinctive: if knowledge and techniques developed elsewhere can be applied there, and whether knowledge from the north is valid outside the region.[23] A related question has concerned who can provide authoritative knowledge about the north. Claims to speak as an expert on the north were once based on ample northern experience, and on exhibiting self-reliance, endurance, and the ability to use Inuit technology and ways of travel and survival. For much of the twentieth century, this knowledge was produced by RCMP officers (the "government's eyes and ears"), trappers, traders, and others who submitted reports regarding the abundance of wildlife and other matters of concern to the state, the Hudson's Bay Company, or other interests. Authoritative knowledge was defined as the product of individuals who had demonstrated endurance and self-reliance in the north— people like W. H. B. Hoare, who between 1924 and 1926 conducted the first government-sponsored study of caribou. But this relation between experience and authority was eventually undermined by new technologies and forms of knowledge: airplanes, scientific equipment, and theories framed in terms of scientific disciplines. Advanced training elsewhere— not arduous experience in the north—became the basis upon which one could speak with authority on northern matters. By the 1950s, northern knowledge had become the province of experts trained in the south, applying new theories and techniques: cosmopolitan knowledge triumphing over (albeit still drawing on) local knowledge. The aerial perspective itself became defined as objective and authoritative: a disembodied, disinterested view from above. Knowledge has linked northern Canada to the rest of the world. Imperial, continental, and international research networks, and disciplines that aspire to global relevance, have shaped scientists' questions, methods, and identities (as northern scientists, or as merely, say, ecologists or entomologists who happen to work in the north).[24] With their colleagues elsewhere, northern scientists, too, shared this postwar confidence in expertise as a source of rational and efficient solutions to social challenges.

In practice this has often meant that science has served as the sharp edge of southern intervention, imposing legibility, linking control over knowledge to control over territory. Science has been closely tied to the evolution of northern resource exploitation—as in, for example, the transition from managing the fur trade in the 1940s to managing wildlife in the 1950s, and the accompanying shift from Elton's Oxford ecology to North American wildlife biology. Other aspects of this evolution have included the shift from soil surveys in the interwar era to postwar geological surveys, and the emerging role in the 1970s of environmental science in administering resource development through surveys, impact assessments, regulations, and public hearings. Even more recently, expertise in ice engineering, pipeline construction, and other activities specific to the region have given the energy industry the ability to operate in ever-more-difficult places, illustrating how science may not just implement but expand northern economic ambitions.

Close ties developed between science and northern administration, as federal agencies, including the Defence Research Board, the Canadian Wildlife Service, and the Fisheries Research Board, adopted a firmly scientific identity, with research embodying not only knowledge but policy. Other areas of expertise, including anthropology and nutrition, have further supported government initiatives. Science has also supported claims to the north as national territory. The presence of scientists asserted Canada's occupation of the Arctic, while during the Cold War "survival science" enabled the military to operate throughout the region.[25] In the early postwar era, ecologists claimed that northern animal populations provided an opportunity for Canada to make a distinctive contribution to science, while physicists made a similar assertion regarding ionospheric research—each group capitalizing on Canada's "natural advantage" in northern research.[26] Through such means, scientists contributed to asserting Canada's cognitive sovereignty over its north, reinforcing the principle that knowing it meant owning it. Yet, more recently, ice core studies and analysis of the role of the Arctic in global climate have redefined the north as a site for constructing not only national, but global knowledge. Knowledge has also often been contested. Scientists have acted as policy advocates, challenging dominant interests and assumptions, asserting the urgency of protecting "fragile" northern ecosystems, debating forest fire control, or experimenting with community-based research.[27] During the Mackenzie Valley Pipeline Inquiry in the 1970s, knowledge helped define opposing views regarding northern resource development, exemplifying how struggles over northern futures have often turned on the politics of knowledge.

Indigenous knowledge and practices have been central to these issues of northern knowledge and expert authority. The relation between science and Indigenous knowledge has been one of the more contentious issues in the politics of northern knowledge. Scientists have asserted evolving and sometimes contradictory perspectives on Indigenous knowledge. Early in the century, explorers relied on Inuit knowledge and technology to survive; surveyors and mapmakers "discovered" the north by translating what Indigenous people already knew; and conservation initiatives, like those of the Hudson's Bay Company, drew upon Cree knowledge. But as scientists began to serve as agents of colonialism and modernization, they often marginalized Indigenous knowledge while reframing northern homelands as territories ready for exploitation. For example, on northern lakes such knowledge—though essential to mining and transportation—was dismissed, while science justified an industrial fishery that depleted fish populations and displaced local fishers.[28] Scientific attitudes have often been at odds with Indigenous values, with wildlife conservation influenced by racial stereotyping, the "sportsmen's code" of ethical hunting, and the view of wildlife as a "crop" to be managed. Scientific techniques have been similarly contentious: Dene and Inuit have viewed counting, tagging, radio collars, and the handling of live animals as disrespectful.[29] Indigenous knowledge has itself been an object of research, with the north serving as a laboratory for anthropologists. However, scientists and holders of Indigenous knowledge have also created opportunities to collaborate respectfully—in land claims research, land and wildlife management, and other fields.

Indigenous knowledge has also provided a new way of understanding northern knowledge by enabling a post-colonial history of science, in which Indigenous perspectives provide not just data but a new way of understanding history. This encompasses distinctive ways of understanding the environment and history through community-based research, storytelling, and oral traditions, and collaboration between academics and Indigenous peoples. This has required reconsidering ideas considered fundamental to Western society: of humans as uniquely rational, sentient, and distinct from nature, and of knowledge as the product of specialized inquiry. Instead, humans are thought to share the landscape with other sentient species, linked by relationships of respect and reciprocity, with knowledge gained through experience and passed from generation to generation.[30]

Mobility

Movement has been essential to northern places and lives. It is intrinsic to northern nature: migrations of birds, whales, caribou, and other species; currents of air and water that link the north to global climate systems; substances put in motion by humans, including radioactive fallout and other contaminants. Movement has been similarly essential to northern human history: long-distance Indigenous trading relationships, and voyages of exploration, colonization, and exploitation, as people and ships travelled to the north and resources were shipped out.[31] Northern history thus echoes the significance of mobility in environmental history: the flows of people, species, materials, capital, knowledge, and influence that have formed the basis for the relations between cities and regions, imperial networks, resource economies, and global institutions.

Throughout the twentieth century, markets and other economic institutions have compelled movement. The Klondike Gold Rush was founded on the mobility of ideas about monetary policy and wealth, risk, and opportunity; the networks of transport and mercantile exchanges that carried miners north and supplied them also linked the region to more distant natures and economies. A decade after the rush, the Mackenzie region was busy with whalers and fur traders responding to global demand, with dramatic impacts on local species and people. As the north became embedded within the global economy, transportation, commercial, and administrative networks formed an "industrial circuitry" across the north. During the Second World War, the military mobilized the north with roads and airfields, and radium was shipped from Great Bear Lake to Manhattan Project facilities. In the postwar era, modernity itself became mobility and technology was celebrated for eliminating perceptions of the north as remote and inaccessible—that is, as immobile. Networks of transport and communication have been linked to the spatial extension of political and economic power, with airplanes, radio, even the Alouette satellite essential to incorporating the north into the nation. Power became defined in terms of mobility, exercised by relocating Indigenous people, or by detecting and restricting the movement of those defined as outsiders—hunters from Greenland in the 1910s and Soviet bombers in the 1950s. Mobile ideas have had other consequences: the concept of citizenship motivated interventions in the lives of Indigenous people; and, more recently, environmental and animal rights activism has affected an Inuit economy already tied to global fur and sealskin markets.[32] Knowledge from elsewhere, its mobility underpinned by the institutions and disciplines of modern science, has displaced less mobile, "local" forms of knowledge, but the emerging status of Indigenous knowledge as both local and mobile has also become evident in circumpolar environmental negotiations.

Making Northern Places

Definitions of the north are many and varied. The ideas they represent about this place—where it is, who is there, its past and future—can help guide us as we navigate northern environmental history. Canadian governments have long defined the north merely in terms of degrees of latitude: the latitude 60° north that marks the southern boundary of the territories, or the mapping of "provincial norths" as the regions above 55° north. Of course, these administrative conveniences do not correspond to physical conditions; geographers have, accordingly, proposed classifications of regions within the North, such as the Arctic and Subarctic, or Louis-Edmond Hamelin's mapping of extreme, far, middle and near norths, and his construction of an index of nordicity. Other definitions of the North in terms of its "essential" characteristics have often reflected perspectives prevalent during particular historical episodes. Some have been framed in relation to geography and ecology, or social interests and identities, or in terms of the north as an imagined space in literary, artistic, and cartographic works.[33]

Many observers have described the north in terms of its physical features and species: boreal forest, permafrost, tundra, ice; caribou, polar bears, narwhals, walruses. Seasonal cycles—dark, cold winters, brief but brilliant summers—define Inuit rhythms of moving, harvesting, and socializing. Extreme conditions and great distances have inspired images of the north as difficult and dangerous, challenging explorers, and, more recently, the energy industry. Scientists have drawn a variety of boundaries between north and south, defining it in terms of climate, the tree line, permafrost—or even, in ionospheric study, in terms of a line running through Ottawa, making nearly all of Canada part of the north. They have also noted distinctively fragile, relatively unproductive ecosystems that are slow to recover from disturbance, animals that exhibit dramatic cycles in abundance, and mass caribou migrations. These observations reflected their view of the north as extreme, an anomaly in relation to "normal" temperate environments. Some of these generalizations, like those regarding "fragile" ecosystems, have also become matters of debate.[34]

The north is not just a physical but an imagined space onto which people have projected ideas and ambitions, often in the service of particular interests in the south. Many have traced national identity itself to the northern climate and environment, calling on Canadians to embrace their nordicity. It has also often been defined as the region exempted from national development: a marginal space beyond the frontier of agricultural settlement and the one remaining region in which Indigenous people constitute a majority. But above all, ideas about the north have always been subject to change and challenge, exhibiting the shifting nature of colonialism itself. Some imagined here a pastoral economy based on those species most suited to this landscape, reindeer and musk oxen—that is, until prospectors redefined the north in terms of its geology. Others found freedom and opportunities for strenuous, masculine adventure—an antidote to southern civilization, or even (inspired by beliefs regarding the environment, health, and racial and cultural superiority) its rejuvenation. Some perceived a "mysterious north," the site of inexplicable phenomena such as mass migrations by lemmings and uncountable herds of caribou. By the 1950s, other ambitions had come to the forefront: to survey the north and make it legible, and to control, rather than adapt to, this "fickle" environment. Cold War strategists saw the north not as a place but an exposed flank, its vast spaces safeguarding southern cities. Economic interests saw a resource hinterland, viewing northern water and minerals only in terms of the materials themselves, and not their social context. Environmentalists considered the north the last Canadian wilderness, a pristine space requiring protection. The distinctive relationship between Quebec and its north, with the development of James Bay becoming a nationalistic imperative, illustrates the significance of provincial contexts to northern places. Different views of the north have implied contrasting economic and political interests and preferences, as Thomas Berger illustrated by invoking visions of the north as a homeland and as a resource frontier. Contradictory perspectives have been evident even within particular groups, such as wildlife managers, who combined anti-modernist sentiments with faith in bureaucratic management. Perceptions of the relations between the north and elsewhere have also been important: once an outpost of the British Empire, then integrated within North America and dominated by the United States, and most recently part of a circumpolar region defined by the Arctic Council and other agencies.[35]

While it is helpful to distinguish these distinct themes in northern environmental history, it must also be remembered that they have always been closely related. Environmental change has been tied to the activities of Indigenous people and the state; technology has shaped experience in the north; and knowledge has had diverse material consequences, while contributing to ideas about northern places. Mobility of many kinds has been a consistent presence throughout. The following chapters demonstrate these relations in particular times and places. In our conclusion, we will also return to these themes, to consider how these chapters advance our understanding of the environmental history of northern Canada.

Notes

1   Mary-Ellen Kelm, "Change, Continuity, Renewal: Lessons from a Decade of Historiography on the First Nations of the Territorial North," in *Northern Visions: New Perspectives on the North in Canadian History*, ed. Kerry Abel and Ken S. Coates (Peterborough: Broadview Press, 2001), 77–90; Liza Piper and John Sandlos, "A Broken Frontier: Ecological Imperialism in the Canadian North," *Environmental History* 12, no. 4 (2007): 759–95; Liza Piper, "The Arctic and Subarctic in Global Environmental History," in *A Companion to Global Environmental History*, ed. J. R. McNeill and Erin Stewart Mauldin (Blackwell Publishing, 2012), 153–66; John Sandlos, *Hunters at the Margin: Native People and Wildlife Conservation in the Northwest Territories* (Vancouver: UBC Press, 2007).

2   Marionne Cronin, "Northern Visions: Aerial Surveying and the Canadian Mining Industry, 1919–1928," *Technology and Culture* 48 (2007): 303–30; Mark O. Dickerson, *Whose North? Political Change, Political Development, and Self-Government in the Northwest Territories* (Vancouver: UBC Press, 1992); Liza Piper, *The Industrial Transformation of Subarctic Canada* (Vancouver: UBC Press, 2009); Morris Zaslow, *The Northward Expansion of Canada, 1914–1967* (Toronto: McClelland and Stewart, 1988).

3   Kenneth S. Coates and William R. Morrison, "The New North in Canadian History and Historiography," *History Compass* 6, no. 2 (2008): 639–58; Matthew Farish, "Frontier Engineering: From the Globe to the Body in the Cold War

Arctic," *Canadian Geographer* 50, no. 2 (2006): 177–96; P. Whitney Lackenbauer and Matthew Farish, "The Cold War on Canadian Soil: Militarizing a Northern Environment," *Environmental History* 12, no. 4 (2007): 920–50.

4   Lyle Dick, *Muskox Land: Ellesmere Island in the Age of Contact* (Calgary: University of Calgary Press, 2001); R. Quinn Duffy, *The Road to Nunavut: The Progress of the Eastern Arctic Inuit since the Second World War* (Montreal: McGill-Queen's University Press, 1988); Peter Kulchyski and Frank James Tester, *Kiumajut (Talking Back): Game Management and Inuit Rights, 1900–70,* (Vancouver: UBC Press, 2007); Sandlos, *Hunters at the Margin*; Peter J. Usher, "Caribou Crisis or Administrative Crisis? Wildlife and Aboriginal Policies on the Barren Grounds of Canada, 1947–60," in *Cultivating Arctic Landscapes: Knowing and Managing Animals in the Circumpolar North*, ed. David G. Anderson and Mark Nuttall (New York: Berghahn Books, 2004), 172–99.

5   Annika E. Nilsson, "A Changing Arctic Climate: More than Just Weather," in *Legacies and Change in Polar Sciences: Historical, Legal and Political Reflections on the International Polar Year*, ed. Jessica M. Shadian and Monica Tennberg (Farnham: Ashgate, 2009), 9–33; Robert Page, *Northern Development: The Canadian Dilemma* (Toronto: McClelland and Stewart, 1986).

6   Anderson and Nuttall, eds., *Cultivating Arctic Landscapes*; Kelm, "Change, Continuity, Renewal."

7   Kulchyski and Tester, *Kiumajut*; Sandlos, *Hunters at the Margin*;

Usher, "Caribou Crisis"; on fire: Stephen J. Pyne, *Awful Splendour: A Fire History of Canada* (Vancouver: UBC Press, 2007).

8   Henry Lewis, "Maskuta: The Ecology of Indian Fires in Northern Alberta," *Western Canadian Journal of Anthropology* 7, no. 1 (1977): 15–52; Piper, "The Arctic and Subarctic"; Piper and Sandlos, "A Broken Frontier"; Sandlos, *Hunters at the Margin*.

9   Lackenbauer and Farish, "The Cold War on Canadian Soil"; Pyne, *Awful Splendour*; Piper, *Industrial Transformation*; Clarke quoted in: C. H. D. Clarke, *A Biological Investigation of the Thelon Game Sanctuary*, National Museum of Canada Bulletin No. 96 (1940), 111.

10   Kathryn Morse, *The Nature of Gold: An Environmental History of the Klondike Gold Rush* (Seattle: University of Washington Press, 2003); Peter R. Mulvihill, Douglas C. Baker, and William R. Morrison, "A Conceptual Framework for Environmental History in Canada's North," *Environmental History* 6, no. 4 (2001): 611–26; Piper and Sandlos, "Broken Frontier."

11   David Damas, *Arctic Migrants, Arctic Villagers: The Transformation of Inuit Settlement in the Central Arctic* (Montreal: McGill-Queen's University Press, 2002); Dick, *Muskox Land*; Tina Loo, *States of Nature: Conserving Canada's Wildlife in the Twentieth Century* (Vancouver: UBC Press, 2006); Sandlos, *Hunters at the Margin*.

12   Fikret Berkes, *Sacred Ecology: Traditional Ecological Knowledge and Resource Management* (Philadelphia: Taylor and Francis, 1999); Hans Carlson, *Home is the Hunter: The James Bay Cree and Their Land*

(Vancouver: UBC Press, 2008); Milton M. R. Freeman, "Looking Back—and Looking Ahead—35 Years after the Inuit Land Use and Occupancy Project," *Canadian Geographer* 55, no. 1 (2011): 20–31; Claudia Notzke, *Aboriginal Peoples and Natural Resources in Canada* (Captus Press, 1994); Paul Sabin, "Voices from the Hydrocarbon Frontier: Canada's Mackenzie Valley Pipeline Inquiry (1974–1977)," *Environmental History Review* 19, no. 1 (1995): 17–48.

13   Paul Nadasdy, *Hunters and Bureaucrats: Power, Knowledge, and Aboriginal-State Relations in the Southwest Yukon* (Vancouver: UBC Press, 2003).

14   Nadasdy, *Hunters and Bureaucrats*, 35.

15   Stephen Bocking, "A Disciplined Geography: Aviation, Science, and the Cold War in Northern Canada, 1945–1960," *Technology and Culture* 50, no. 2 (2009): 265–90; Peter Geller, *Northern Exposures: Photographing and Filming the Canadian North, 1920–45* (Vancouver: UBC Press, 2004).

16   Cronin, "Northern Visions."

17   Joy Parr, *Sensing Changes: Technologies, Environments, and the Everyday, 1953–2003* (Seattle: University of Washington Press, 2010).

18   Dick, *Muskox Land*; Bill Waiser, "A Very Long Journey: Distance and Northern History," in *Northern Visions: New Perspectives on the North in Canadian History*, ed. Kerry Abel and Ken S. Coates (Peterborough: Broadview Press, 2001), 37–44.

19   Graham Rowley, *Cold Comfort: My Love Affair with the Arctic*

(Montreal: McGill-Queen's University Press, 2007), xi.

20   Prentice G. Downes, *Sleeping Island: The Story of One Man's Travels in the Great Barren Lands of the Canadian North* (Western Producer Prairie Books, 1988 [1943]), 27.

21   Michael Bravo and Sverker Sorlin, eds., *Narrating the Arctic: A Cultural History of Nordic Scientific Practices* (Canton, MA: Science History Publications, 2002); Richard C. Powell, "'The Rigours of an Arctic Experiment': The Precarious Authority of Field Practices in the Canadian High Arctic, 1958–1970," *Environment and Planning A* 39 (2007): 1794–811.

22   Bocking, "Disciplined Geography."

23   Ronald E. Doel, Urban Wråkberg, and Suzanne Zeller, "Science, Environment, and the New Arctic," *Journal of Historical Geography* 44 (2014): 2–14.

24   Stephen Bocking, "Science and Spaces in the Northern Environment," *Environmental History* 12 (2007): 868–95; Cronin, "Northern Visions."

25   Farish, "Frontier Engineering."

26   Edward Jones-Imhotep, "Communicating the North: Scientific Practice and Canadian Postwar Identity," *Osiris* 24, no. 1 (2009): 144–64.

27   Michael Bravo, "Science for the People: Northern Field Stations and Governmentality," *British Journal of Canadian Studies* 19, no. 2 (2006): 221–45.

28   Piper, *Industrial Transformation.*

29   Sandlos, *Hunters at the Margin*; Usher, "Caribou Crisis."

30 John Bennett and Susan Rowley,
    eds., *Uqalurait: An Oral History
    of Nunavut* (Montreal: Mc-
    Gill-Queen's University Press,
    2004); Julie Cruikshank, *Do
    Glaciers Listen? Local Knowledge,
    Colonial Encounters, and Social
    Imagination* (Vancouver: UBC
    Press, 2005); Dick, *Muskox Land*;
    Kelm, "Change, Continuity,
    Renewal."

31 Dolly Jørgensen and Sverker Sörlin,
    *Northscapes: History, Technology,
    and the Making of Northern Envi-
    ronments* (Vancouver: UBC Press,
    2013); Robert McGhee, *The Last
    Imaginary Place: A Human His-
    tory of the Arctic World* (Chicago:
    University of Chicago Press, 2005);
    Piper, "The Arctic and Subarctic."

32 Jones-Imhotep, "Communicating
    the North"; Morse, *Nature of Gold*;
    Andrew Stuhl, "The Politics of the
    'New North': Putting History and
    Geography at Stake in Arctic Fu-
    tures," *Polar Journal* 3, no. 1 (2013):
    94–119; George Wenzel, *Animal
    Rights, Human Rights: Ecology,
    Economy and Ideology in the Cana-
    dian Arctic* (Toronto: University of
    Toronto Press, 1991).

33 Sherrill E. Grace, *Canada and
    the Idea of North* (Montreal:
    McGill-Queen's University Press,
    2002); Louis-Edmond Hamelin,
    *Canadian Nordicity: It's Your
    North, Too* (Montreal: Harvest
    House, 1978, 1979).

34 Bennett and Rowley, *Uqalurait*;
    Bocking, "Science and Spaces";
    Jones-Imhotep, "Communicat-
    ing the North"; Mulvihill et al.,
    "Conceptual Framework"; Waiser,
    "Long Journey."

35 Kenneth Coates, "The Discovery
    of the North: Towards a Con-
    ceptual Framework for the Study
    of Northern/Remote Regions,"
    Northern Review 12/13 (1993–94):
    15–43; Caroline Desbiens, *Power
    from the North: Territory, Identity,
    and the Culture of Hydroelectricity
    in Quebec* (Vancouver: UBC Press,
    2013); Farish, "Frontier Engineer-
    ing"; E. C. H. Keskitalo, *Negotiating
    the Arctic: The Construction of an
    International Region* (New York:
    Routledge, 2004); Sandlos, *Hunters
    at the Margin*.

PART 1

Forming Northern Colonial Environments

2

Moving through the Margins: The "All-Canadian" Route to the Klondike and the Strange Experience of the Teslin Trail

Jonathan Peyton

This is the grave the poor man fills,
After he died from fever and chills,
Caught while tramping the Stikine Hills,
Leaving his wife to pay the bills.

—Pierre Berton[1]

And of all the mad, senseless, unreasoning, and hopeless rushes I doubt if the world has ever seen the equal. Day after day crowds of men of all classes and conditions, hauling their sleighs, struggling, cursing, and sweating, thrashing their horses mules and dogs, all filled with the mad hopeless idea that if they could get as far as Telegraph Creek they would be in good shape for the Klondike. ... Some gave up on the river, sold their outfits and went back. Thousands arrived at Glenora and

Telegraph Creek and started over the Teslin Trail but by this
time it was April or May, and the snow was beginning to go off
the trail leaving pools of water and swamps through which it
was almost impossible to transport their outfits. ... Hundreds
stopped at Glenora until the river steamers started to run in the
spring, and they went home poorer and wiser.

—George Kirkendale [2]

The history of the Klondike Gold Rush is well documented. In August
1896, placer gold was discovered on Rabbit Creek (later renamed Bonan-
za) in the Yukon River Valley. By midsummer of the following year, gold
fever was in full bloom in the United States, and, to a lesser extent, in
Canada. This fever was exacerbated by a series of financial crises in the
United States related to anxieties about the gold-centric monetary policy. [3]
The winter of 1897–98 saw an estimated forty thousand miners en route to
the area around Dawson, the booming new town at the centre of the mining activity. Wealthy participants took the all-water route via steamship
around Alaska and south to Dawson along the Yukon River. Most, however, took steamships north to Dyea or Skagway, where they embarked
on the harrowing trek over the Chilkoot and White passes, before sailing
across Bennett Lake and up the Yukon River. The treacherous conditions
of the trails and unrelenting force of the environment have become the
stuff of legend. There were other, less publicized, less popular routes: the
Edmonton route with its long portages, the Dalton route with plenty of
grazing land in summer, the Taku River route, the ultimately impassable
Copper River route, and the Stikine route. [4] Information on all routes was
sketchy and largely compromised by the self-interest of promoters. Facilities were haphazard, and, at the outset at least, travellers were forced to rely
on their own ingenuity to survive and succeed. Many abandoned the effort before reaching Dawson, many died while trying, and many returned
home hopeless and bankrupt. But because a few did succeed, people of all
backgrounds kept venturing north. [5]

This chapter deals with the motivations and effects of a particular
historical episode: the attempt by the federal Liberal government, contracted construction firms, and a state-sanctioned network of engineers
and bureaucrats to construct an "all-Canadian" railway from the Stikine

River to Teslin Lake during the Klondike Gold Rush in the late nineteenth century.[6] The first section details the political economy of several conflicting attempts to build the railway, as well as the material accounting of the natural environment that accompanied these commercial adventures. The second section tracks the men who followed in the wake of these railway schemes, who engaged directly with animals and nature in the Stikine, and, in the process, formulated a new understanding of the Stikine environment, both locally and in metropolitan centres. This was not development as "project of rule," but rather as an incremental practice that could inculcate new ways of knowing nature in a peripheral landscape.[7] In this instance, it involves casting our historiographical gaze both north and south. This chapter builds on literature that places territorial and provincial norths within the historiography of the Canadian north.[8] Many recent advances in northern scholarship have built upon the methodological and analytical impetus provided by this "provincial norths" literature.[9] More recently, an emerging group of scholars has examined the industrialization of the north and the exploitation of northern mineral and energy resources more generally.[10] Northern scholars are increasingly attuned to the vagaries of development and the place of technology in its mobilization.[11] Indeed, this collection includes several contributions that build on this northern resource development literature.[12] My view is that by emphasizing connections to a broader scale of northern development, but also to the intimacies and particularities of uniquely northern environmental encounters, northern historians can unpack the relational narratives that characterize encounters like the attempts to traverse the Teslin Trail.

Construction of the railway and accompanying wagon road began, but was quickly undercut by a combination of politics, bureaucratic failure, and engineering success on competing routes. The project—and the feasibility of the Teslin Trail route to the Klondike—ultimately failed, and the events of the railway failure provide context for the environmental changes set in motion in the Stikine. While the railway scheme remained unrealized, it was an important catalyst for new valuations and understandings of nature on the plateaus north of Glenora and Telegraph Creek. This inquiry into a fledgling transportation network, along with changing human-nature relations, therefore illuminates the complex politics of

Fig. 2.1: "Sketch map shewing the different routes to the Yukon Gold Fields." Canada, High Commissioner, Yukon District of Canada (London: McQuorodale, 1897).

nature that emerged as colonial and metropolitan ideas, peoples, and hardware first entered the region.

The Stikine route had several legs. The first was a simple though increasingly expensive steamboat journey from the metropolitan centres of the Pacific Northwest (Seattle or Tacoma) or from Vancouver or Victoria to Wrangell, Alaska, a mostly Tlingit community located on Wrangell Island in the silt-laden Stikine Delta. From there, miners had several options. Those who reached Wrangell in the summer and fall could wait to secure passage on one of the intermittent and sometimes reliable steamboats brought north to the Stikine. Wrangell had a new sawmill, and men wishing to avoid the substantial steerage and goods transport cost of the steamboat often opted to build their own rudimentary rafts. Most of these men fared poorly, as the notoriously fickle tidal shallows of the delta rarely let inexperienced pilots through unscathed. A better option was to hire Tlingit guides and their canoes, though many judged the escalating price of expert knowledge and marine technology too dear. Choices

FIG. 2.2: "On the Road to Klondyke: Mounting the Summit of the Divide above Telegraph Creek," illustration by C. E. Fripp. Yukon Archives, *The Graphic*, 27 August 1898, vol. 58, no. 1500, front page.

dwindled when the ice froze in winter. The only real option then was to haul goods and grub on the ice and snow approximately 300 kilometres to Glenora and Telegraph Creek. Still others, perhaps even worse off, eschewed the river completely, preferring the overland route over the derelict Ashcroft Trail, cut thirty years earlier during an abandoned attempt to construct a telegraph cable across the Bering Strait. From Glenora and Telegraph Creek, aspiring miners and their animals packed their loads directly north to Teslin Lake, two hundred kilometres distant, where they met the Hootalinqua River and the Yukon River system. Fiercely cold and exposed in winter, and a soupy mess in summer, the Teslin Trail was given many other names: Telegraph Trail, the Bughouse Trail, the Devil's Trail, the Cold March...

The need to order this new northern landscape encouraged state institutions to press their imprint on land that had formerly been managed by the Tahltan and other Indigenous peoples of the area.[13] The Royal Canadian Mounted Police arrived to manage the transient population and to collect duty on goods crossing the international border, missionaries came to minister to wayward souls, a Gold Commissioner organized the traffic in minerals, and a small army of merchants supplied the material needs of miners. Through their initiatives, southern agencies established a more concrete and pervasive presence in the Stikine. This was an ideological as much as an institutional change. Land and nature were reordered within new conceptions promoted by state enterprises, their agents and technologies, and the movements of capital they enabled.[14]

The massive increase in population numbers and the formalization of state conceptions of land in the Stikine had an immediate impact on human relations with the natural world. Increased exploitation resulting from demographic pressures is one element of this tension, but this chapter is equally concerned with how itinerant miners and locals experienced nature differently in the wake of the rush. Interactions between uses and perceptions of nature informed and altered natural-cultural interrelationships during and after the rushers had swept through the Stikine. In her work on the commodification of nature during the Klondike Gold Rush, Kathryn Morse claims that the proper way to interrogate the complex natural and social history of the Klondike is by looking at how participants came to know nature and how they forged connections to others through labour and experience. Morse sees gold as an abstraction which allowed the commodification of "knowledge, experience, and connections to nature" of the rushers and the area's Indigenous peoples.[15] The imposition of an international commodity market linked to cities and markets elsewhere encouraged nature and wildlife to be consumed in different ways. Valuable for more than subsistence, animals and their habitat were consumed not only on dinner tables, but as art on walls, through gun sights, in photographs, and as fodder for the great narratives of hunting and "frontier" travel.[16] In the wake of the Gold Rush, northern nature became a space to consume as much as a place to sustain. This, in turn, altered Indigenous peoples' relationships to the environment and animals alike: "The gold rush affected how, where, and why Native peoples hunted, fished, and marketed their catch, but it also changed the Indians' own connection to nature through the foods they themselves consumed."[17]

Morse has little to say about the Stikine, as her work focuses on American events; however, her more general comments about the nature of the gold rush are instructive.[18] Like Morse, I see resources as cultural concepts, imbued with meaning by miners and others moving into the north.[19] I also draw from a growing body of research in environmental history and geography that interrogates the commodification of nature by framing resources as social relations, understood as much in terms of their changing materialities as by their biophysical realities.[20] In this instance, the focus is less on gold and its extraction than on the narratives, mobilities, and relationships enabled by the rush itself. The Gold Rush was part of a much wider political economy of resource development in the Canadian north that was largely formed later in the twentieth century.[21] The rush of miners through the Stikine was the major material catalyst for these new expressions of value and meaning. As one scholar recently put it, the narrative experiences of the Klondike were (and continue to be) frequently engaged as "technologies in the making of this environment."[22] I suggest that both narrative and the materialities of socio-natural experience were fundamental to understanding and knowledge of the Teslin Trail and the erstwhile all-Canadian route.

The Failure of the Railway and the Stikine Route to the Klondike

The Stikine route was advertised with a certain nationalist bombast as the beginning of an "all-Canadian route" to the Klondike. It was promoted as a simpler route, longer but far less taxing. Promoters in Vancouver and Victoria (and throughout the Empire) exaggerated the emerging transportation infrastructure: they wrote of an armada of steamboats to transport people and increasingly precious goods up the Stikine River from Wrangell, Alaska; or, in winter, dog-teams to haul brand-new outfits to the boom towns of Glenora and Telegraph Creek. From the north banks of the Stikine, promoters claimed it was a quick 150-mile overland journey on a good trail to Teslin Lake and the headwaters of the Yukon River system.[23]

But the Stikine route depended on a development initiative conceived by the federal government and its potential commercial partners. The federal Minister of the Interior, Clifford Sifton, seeking to appease both coastal merchants eager for trade and a public concerned about lost revenue and territorial influence, toured the Stikine and Taku river deltas in October 1897 with a view to establishing a route to the Klondike through Canadian territory. Impressed with the possibilities he saw in the Stikine, Sifton eventually signed an untendered bid with the engineering firm of Mackenzie and Mann in January of the following year to build a wagon road and short-gauge railway between Glenora and Teslin Lake. The contract stipulated that the road would be finished in six weeks, while the railway would be operational by September. William Mackenzie and Donald Mann, experienced railway entrepreneurs, enjoyed a favoured business relationship with Sifton, having already constructed several railways in the Prairie Provinces (which together would become the Canadian Northern Railway).[24]

This simple explanation obscures the complicated politicking by multiple interests, particularly within British Columbia, that preceded the announcement of the Mackenzie and Mann contract. Several groups sought the contract directly through the provincial government. In December 1896, prominent Victoria journalist Alexander Begg wrote to BC Premier John Herbert Turner on behalf of the "MERCHANTS and BUSINESS MEN" of Victoria, asking that a survey be undertaken in the Stikine with a view to having a "convenient and practicable route" constructed to win back the Yukon trade from "UNITED STATES DEALERS." The country had been neither surveyed nor explored, he assured Turner, although he claimed that he had been informed by "Dr. G. W. Dawson, Head of the Geological Survey of Canada, that a very favourable route, entirely within British territory, can ... be made available."[25] In April, Begg, acting as president of a newly formed Stickeen and Teslin Railway Company, amended the petition to request an exclusive charter (and potential assistance) for the construction of a line between Glenora and Teslin Lake.[26] By May, Begg was anxious to begin work while the Railway Bill was being debated in the Legislative Assembly. Undoubtedly aware of competition and the fleeting nature of the opportunity, he pressed Turner to grant the charter because he was "very anxious to proceed to Ottawa, to deal with the Dominion Government as to further aid for the Construction of the Railway."[27] Begg's company won the charter from the province and later from the federal government, but his Stickeen and Teslin Railway never began construction, suffering a series of financial and organizational difficulties before sputtering to a halt.[28] As often happened in an age of railway speculation, Begg's charter was eventually purchased by Mackenzie and Mann for $50,000, forming the legal basis for their construction plans.[29]

Other groups filled the competitive void when Begg's efforts lagged. The Victoria Board of Trade began to lobby, insisting on information and access that Turner was holding close to his vest.[30] In spite of his economic and political interest in the railway scheme, Turner was careful not to overextend provincial interest and jurisdiction. Sifton and Turner maintained contact after Sifton's tour of the Stikine, and by mid-November Turner was pressing for a firm commitment from the Minister of the Interior: "Very numerous and earnest representations are being made from all quarters ... [but] it would be manifestly a waste of energy for the Dominion and Provincial Governments to proceed on independent lines of action to secure what is really a common object, and therefore it is highly desirable that there should be unison and co-operation of effort."[31]

Three other groups emerged with serious intentions in the lead-up to the announcement of the Mackenzie and Mann contract. J. T. Bethune, a local real estate agent in Victoria, wrote to Turner in November on behalf of "a strong London Company" to announce the group's intention to build a "sleigh road" between the Stikine River and Teslin Lake, and to inquire about the possibility of public outlay in that enterprise. Turner's reply was apparently unsatisfactory for Bethune's backers: the government intended to build the road itself anyway, but would consider supplying one-third of the cost to a maximum of $3,000 if the company would guarantee that the finished route would be made public upon completion.[32] A body of correspondence between Turner and the H. Maitland Kersey Syndicate, also of London, detailed the attempt by the group to secure funding and/or permission to build on the still-theoretical Teslin Lake route. Represented alternately by Lord Charles Montague, R. T. Elliott, and F. M. Yorke, the syndicate had more substantial financial backing than the Bethune group. The syndicate's proposal ultimately failed because of commitments that both levels of government had made to Mackenzie and Mann and because it had not demonstrated sufficient financial means.[33] It is unclear how much overlap existed between Mackenzie and Mann's separate negotiations with the federal and provincial governments. The firm had secured significant enticements from Turner's government by the signing of the contract with Sifton in late January 1898, including a subsidy of $2,250 per mile of railway constructed and a grant of free lands required for the right of way and terminal facilities.[34]

But the provincial concessions were minor compared to those granted by the federal government. In Clifford Sifton's northern vision, the railway would lead to Canadian control of the Klondike and was therefore necessary at any cost.[35] In exchange for construction of the Yukon-Canadian Railway, as the line came to be known, which was estimated to cost $22,000 per mile of track, Mackenzie and Mann received lucrative concessions: a limited monopoly over further railway construction (possibly south to the Portland Canal at Stewart, on the central coast of BC, north to Dawson, and east to Edmonton), a land grant of 3,750,000 acres (based on 25,000 per mile of track), with all mineral rights tax-free for five years. Gold produced from any of this land would be taxed at one percent, while the toll on the railway would be fixed by the government for seven years.[36] These allowances would prove too generous for Sifton's political opponents, especially Sir Charles Tupper, leader of the opposition Conservatives, who opposed the terms of the railway contract.[37]

Meanwhile, Warburton Pike led a secondary railway scheme in the Stikine. Pike was a well known itinerant writer, sportsman, and entrepreneur whose wilderness acumen had furnished the basis for two popular accounts of northern adventuring exploits, but who would later become known for a series of business failures in the Stikine and elsewhere.[38] Curiously, Pike was not keen to build north, but rather to head northeast from Glenora to somewhere near the head of Dease Lake. He eschewed the gold of the Klondike, preferring to focus on what he felt were longer-term prospects of mineral wealth in the remote Cassiar district. Pike operated as a kind of lobbyist/contractor for his Cassiar Central Railway Company, both in the provincial capital, Victoria, and in Ottawa. Pike's efforts did not conflict with the operations of Mackenzie and Mann, as they were intent on building in different directions with different purposes. Indeed, there may have been some professional overlap: there is some evidence that Pike was also involved in Mackenzie and Mann's planning and promoting operations.

Once the Cassiar Central Railway construction contract was secured with the provincial government, Pike organized the transport of railway building materials and workers from Victoria and Vancouver to the Stikine. He secured the cooperation of the provincial government through his connections to Premier Turner.[39] Pike's letters and requests foregrounded the potential of "opening up the Cassiar [Stikine] District" for capital investment, exploration, and settlement (and its accompanying revenue) through infrastructure initiatives, because, in his words, "that part of the country is at present a deserted waste."[40] He needed no formal monetary outlay from the government, but would be willing to conduct all business through BC merchants and suppliers, use local construction materials where they could, and hire only BC labour.

Pike won the charter in May of 1897, but he sold it days later to a consortium of British financiers associated with the Transvaal Goldfields Company. They quickly formed a subsidiary, the Africa British Columbia Company, which hired Pike to oversee operations in the Stikine. Large warehouses were built at Glenora, surveys were conducted, mining lots were chosen along the corridor, and animals, equipment, and supply materials were brought north. Construction for the Railway would start at both ends, Glenora and Dease Lake. Employing two thousand men, Pike planned to have the line finished in a year, and had already begun surveying and planning for a proposed wagon road between Dease and Teslin lakes. But construction suffered a series of setbacks, including the sinking of a tractor and a shipment of rails after the transport scows ran aground on a sandbar. Pike and his associates persevered until late spring of 1899, but were forced to give up the project after the Boer War weakened the financial capabilities and interest of the parent company.[41]

The railway dreams were built on surveys completed in the years before the Gold Rush by the Geological Survey of Canada (GSC).[42] William Ogilvie had been sent briefly to survey the area during the winter of 1894–95 as part of his duties as Yukon Commissioner. Writing in November 1897 (though his findings were not published until 1898), Ogilvie claimed that with suitable infrastructure investment, the northern portion of the Stikine plateau could develop into "the richest gold field the world has ever seen."[43] Ogilvie had originally accompanied George M. Dawson on his northern survey in 1887, an endeavour that included a reconnaissance of the Stikine watershed and which first pointed to the region's mining potential.[44] Another party led by GSC civil engineer William Tyndale Jennings was in the plateau north of the river when gold fever hit. Jennings, along with assistant surveyors A. S. Ross, Arthur St. Cyr, Edmund J. Duchesnay, and Morley Ogilvie, were already looking for the most practicable route for the railway. Failing that, they sought the best route for a "highway," an ambitious term for the wagon road. Jennings' report, released in February 1897, was positive about construction prospects. His ideal route began at Little Canyon, some fifty miles downstream from Glenora, for an estimated cost of almost $4 million dollars, slightly higher than the estimate eventually provided by Mackenzie and Mann.[45]

More importantly, Jennings and company began to construct an inventory of the landscape with construction hazards and natural assets in mind. St. Cyr and Ogilvie reported more specifically on the possible rail and road route north to Teslin Lake, mapping out potential avenues and proposing construction methods to overcome the perceived climatic and geologic obstructions. These state-based surveys began a process of institutional data creation that would have far-reaching consequences for the way residents and newcomers related to the Stikine environment.[46] Jennings and company mapped the watershed topographically, but he also commented directly on the "resources" that could be useful in railway construction (i.e., timber, rock, animals).[47] The simple act of data creation presaged a new valuation of nature. Members of Jennings' party also commented directly on anticipated costs. Edmund Duchesnay, a junior member of the party, believed the railway could be constructed for $1,575,925, while Jennings estimated the cost at $2,850,000, warning that construction would be difficult through terrain that was "covered in moss and occasionally mire and unpleasant to travel over in unseasonal weather."[48] Even building the wagon road would be expensive and would require unconventional construction methods. The moss would have to be cut out and the tree canopy removed and the soil underneath left to dry. Ditches would have to be dug the whole length of the road, and coarse gravel would have to be laid to protect the intractably mushy ground. This would take two months and cost between 100 and 250 dollars per mile. He recognized that there was little forage for pack animals and recommended government caches to lessen the hardship of the many anticipated travellers. His recommendations were not taken up; little was done to improve the trail, and Jennings' comments on forage proved true.[49]

In spite of political will, economic rationale, and geological incentives, only ten kilometres of track and thirty kilometres of wagon road were built before party politics scuppered the Stikine route. Railway construction was abandoned when the Conservative Party-controlled Senate balked at the terms and abruptly voted down the contract just as the large second wave of miners poured into Glenora. Aside from the usual partisan bickering and self-interest that often accompanied massive government expenditure, the Tories were concerned about the lack of competition in the award of a possible trade, land, and transport monopoly for Mackenzie and Mann, and possible taxation difficulties at the transshipment point at Wrangel.[50]

In the end, none of the grand railway plans materialized. In their stead, a "wagon road" was cut. At least, that was the official position. In actuality, some corduroy was laid north of Telegraph Creek, but miners were largely left to fend for themselves on a sodden, muddy, poorly marked trail to Teslin. The promises of easy access to the gold fields rang hollow. An informal infrastructure of caches, grazing areas, campsites, and one Hudson's Bay Company outpost provided some semblance of infrastructure, but in reality, miners went only as far as their own luck and perseverance took them. Many of the men who came through Glenora turned back without attempting the run to Dawson; others (and many of their animals) died on the ice of the river or on the exposed plateau that bounded the Teslin Trail. The overwhelming experience of travel on the trail constituted a unique interaction with a new northern nature.

The Experience of the Teslin Trail

Aspiring miners who made it up the river had little option but to follow the trail as envisioned by the GSC surveyors and railway speculators. But the actual Teslin Trail that they faced was an awful mess of muddy confusion and disorder. Accounts left by men on the trail demonstrate its tenuous nature and the alienating experience of travel upon it. As they battled against the trail, the Klondike seemed further and further away. Thomas Frederick Seldon offered increasingly vivid descriptions of the deteriorating trail in his diary, in a style emblematic of what other chroniclers also saw and represented. In the beginning of July, after a winter of waiting in Glenora and a profitable dip into the horse market, Seldon's group began the trek north where they encountered "a decent trail for some miles and then it was a terror. Horses were sinking down in mud between trees and roots of trees. It makes one shake expecting a broken leg every minute." The next days brought no respite: "the trail has been dreadful, swamps and fallen timber, had several [horses] down but no limbs broken—passed several dead horses. … People have no idea what it is like & people who have spoken so highly of it ought to be made to pack a train of mules and then be hauled up for cruelty to animals."[51] Travel in winter was easier but posed its own set of problems. Diarist O. T. Switzer noted that, "the extreme cold has also caused us a lot of trouble in travelling along these creeks. The ice freezes so thick that it does not leave water way sufficient, and it forces the water out over the ice, along the edge of the stream and it overflows."[52] This was made even more slippery when an early thaw and subsequent freeze reformed ice on the creeks and the trail itself. After two months on the trail in winter, Hunter Fitzhugh told his sister in a letter, "I wish I could tell you all of my strange and ridiculous experiences, but it would take acres of paper, oceans of ink, and horse powers of work to do it up in style."[53] In general, the Teslin Trail was passable in winter, though most miners attempted it in spring and summer when it was essentially impassable even with healthy animals, an adequate outfit, and a sterling constitution. A particularly difficult section of the trail encountered during a second trip, in the summer, inspired Fitzhugh to complain, "The strain on our minds and bodies during the five days it took to get through that water was maddening. Our lives and possessions were both in the greatest danger, and the work was fearful, for we had to put one sled on top

Fig. 2.3: "Freighting by Wheelbarrow, Teslin Lake Trail," ca. 1898. BC Archives, D-02068.

of the other in order to keep our stuff dry."[54] Everybody who passed over the trail had similar tales of wetness and woe.

Travel on the river, either in a boat or in a sled over the ice, was equally dangerous and presented its own hardships. Steamboats and hired Tlingit canoes made the trip simply and without much difficulty, but self-constructed boats and sleds faced considerable natural obstacles. Scows and rafts built at Wrangel or on Cottonwood Island in the Stikine delta often sank because they were overloaded, poorly built, or overcome by the currents and migrating sandbars of the river. Sleds could travel quickly, but, at least for the leader of the impromptu convoys, they posed considerable risk from patchy ice. Dog teams often plunged through the ice, leaving the driver to scramble first for his own safety, then for the safety of his cargo, and then to rescue his freezing dogs. The everyday labour of sledding (packing the sleds correctly, righting tipped sleds, tending to dogs and equipment, lighting fires, and sleeping on top of snow and ice, keeping relatively dry) provided routine but hardly reduced the risk encountered on the ice.

Glenora was the staging ground for the Teslin Trail. Five thousand miners wintered there in 1897–98, waiting for the opportunity to head north.[55] It was an ephemeral city of tents and improvised shacks, built hastily to house miners, working animals, and the goods and services required by a transient population. The town had a favourable reputation as a lawful and vice-averse place, probably a result of the presence of civil authority in the Gold Commissioner, postmaster, and police. New markets developed for meat, timber, knowledge, and assistance of various kinds. There were new opportunities for both Indigenous peoples and neophyte miners open or resigned to the possibility of trading or working in a trade. A solid trail of fourteen kilometres connected Glenora to Telegraph Creek, nominally the head of navigation on the river. Telegraph Creek was a smaller settlement with a larger permanent population and the site of a considerable Tahltan camp still located at the site. Many preferred "Telegraph" because its reputation for hospitality extended to nervous newcomers.

Miners who reached Glenora and Telegraph Creek had already passed through Wrangel, the Alaskan/Tlingit town in the Stikine delta. Wrangel was widely derided as an unscrupulous place full of potential hustles and the iniquity that often flourished in itinerant rush towns. Indeed, there was clearly a market for the kind of behaviour that made most chroniclers blush. Apart from the constant danger of swindles, Wrangel was the only transshipment point, posing many economic difficulties for British and Canadian miners intending to bring goods purchased elsewhere into the transboundary waters of the Stikine. For a time, American customs inspectors ignored the guarantee of free shipment and navigation established in the International Waters Act of 1871, charging duty on goods coming through town on steamers from the south. This provided Clifford Sifton with one of his main motivations for an "all-Canadian" route. In the middle of the delta was Cottonwood Island, which served the multiple functions of campsite, boatbuilding site, and staging ground for the assault on the river itself. Mackenzie and Mann used the island to establish a camp for workers and to assemble machinery and materials for their construction projects. The island proved to be a telling metaphor for the Stikine route to the Klondike. It was flooded and essentially disappeared during the spring break-up of 1898. The rising waters consumed miners' camps, destroyed half-finished rafts, and drowned animals not evacuated by the residents of Wrangel.[56]

At the end of the Teslin Trail was Teslin Lake, an itinerant community adjacent to a First Nations settlement, and the lake itself, a long, narrow body and the beginning of the water-bound portion of the journey to Dawson. In the imaginations of many miners, Teslin became a focal point, the anticipated end of the difficult part of the journey. But by the summer of 1898, Teslin was no more than a transit stop for the few who made it over the trail: "Things are dead in Teslin. No railroad made it so some log huts partly built are stopped. Men clear out as soon as they build their boats."[57] O. T. Switzer arrived in Teslin in February 1898 and saw it change from a promising business outpost and supply centre to a place essentially devoid of economic activity in less than six months. It was an exceptionally expensive place to live, largely because of the exorbitant packing rates from Glenora, which reached $800 per tonne by June.[58] By then, "living in Teslin Lake [was] an expensive luxury."[59]

The men who used the Teslin Trail were often confronted with the stark contrast between the rumour and the reality of travel on the trail itself. News of the railway's construction added to excitement about the Stikine route. Guy Lawrence and his father, who sailed to the Pacific from Blackpool, England, chose the Stikine route because they had been told by promoters in London that the Yukon-Canadian Railway was virtually ready.[60] Merchants and boards of trade in Vancouver and Victoria claimed the railway was as good as finished, and, with its "invaluable help" in transporting goods, the trip to the goldfields could be completed in six weeks. Promoters in eastern Canada promised a similarly easy, though invigorating, journey.[61] The rhetoric used by the "Dunsmores" [sic: read Dunsmuir] was typical: "The only easy way to the Klondike. Five hundred dollars from Victoria to Dawson, with 500 pounds of baggage; first-class steamer to Wrangel; newly-equipped handsomely furnished river steamer up the Stickeen to Telegraph Creek; there pack trains will be waiting to take you to Teslin lake over a fine grass country beautiful scenery, beautiful lakes and fine trout fishing; on arriving at [Teslin] lake the steamer Dunsmore, just finished at tremendous cost, will carry you down the lake to the Hootlinqua river; passengers can lead anywhere, take out a million or two and go home before wintering-down in the Yukon, on which river there will be plenty of steamers."[62] It was to be a lovely holiday with the added bonus of a pot of gold at the end.

Others came north to the Stikine via the Ashcroft Trail, which began in the Chilcotin country and followed an overgrown route originally and rapidly laid out by the Collins Overland Telegraph Expedition sponsored by Western Union, as they were rushing to construct a transpacific cable across the Bering Strait.[63] Rancher Norman Lee was actually impressed by the Ashcroft Trail for the first weeks of his journey. But that happiness quickly faded. Soon, he wrote, "we were in the thick of the misery—as regards mud and shortness of feed. ... Every day one of our horses had to be left beside the trail, and not ours alone, as it was scarce possible to travel a hundred yards without finding dead or abandoned horses; I have seen in one place, two dead horses on each side of the trail." The trail was littered with jettisoned goods, and eventually became "one succession of mud and swamps from one end to the other."[64] The trail was even a disappointment from a literary point of view. American essayist Hamlin Garland, seeking the sublime landscape of wilderness adventure, was disappointed, "not because it was long and crossed mountains, but because it ran through a barren, monotonous, silent, gloomy, and rainy country, It ceased to interest me. It had almost no animal life, which I love to hear and see. Its lakes and rivers were for the most part cold and sullen, and its forests sombre and depressing." It was a foolish route to the goldfields of the Yukon, Garland claimed, and unless mining was developed in the Stikine region, the Ashcroft Trail should be "given back to the Indians and their dogs."[65]

The experience of nature in the Stikine was complicated by the vagaries of weather and climate. Rain, snow, and cold made the difficult experience of travel a profoundly visceral one.[66] But it was the capriciousness of climate that most affected the movements of miners, the distances they were able to cover, and the caloric and emotional energy that was required of them. An unseasonably mild winter in the Stikine in 1897–98 caused variations in the seasonality of the river, the surrounding plateau, and the new trail that transected it. The river froze and thawed quickly, often forcing dogsleds to travel through smaller creeks of slush and ice. The ice was unstable, forcing the first sled in the team, and its driver and animal engines, into dangerous positions. The Teslin Trail changed almost instantly from a passable, snow-covered route to a sloppy, sticky mess of thawed mud and snow. Storms and blizzards imperiled miners, particularly when they were travelling. Hunter Fitzhugh recounted a story of a Japanese chemist and his group, who, "when about three miles from Wrangel … were struck by a squall and to save their lives had to throw nearly everything over board."[67] Pervasive rain and wetness compounded the difficulty of the journey, but it was the unexpected perils that yielded the more acute psychological implications.

In order to demonstrate the Stikine route's viability, the Royal Canadian Dragoons were sent through the Stikine in May 1898 on their way to the Klondike. Over two hundred scarlet-clad soldiers came through Glenora with a hundred tonnes of stores, just as rumours surrounding the failure of the Yukon-Canadian Railway were reaching the town. This increased confusion and speculation. But the sudden appearance of the Yukon Field Force (YFF), as the unit was known, had a more immediate and tangible effect; it further reduced the already small supply of animals that could be hired for packing. Likely because of scarcity and value, their local contractor had failed to provide the agreed-upon number of animals. Lt.-Col. T. D. B. Evans, the YFF commanding officer, was forced to delay troop movement so that he could raise an adequate animal convoy to take supplies and food north on the Teslin Trail. The YFF eventually bought or commandeered the majority of serviceable horses in the area, over three hundred in total, including many employed on existing pack trains. This drove up packing rates— already exorbitant—and further demoralized camp residents.[68] The institutional presence of the state, exemplified by the army's regimented conduct, was meant to reassure the populace of the mining camps. Instead, the army destabilized existing economies and transportation networks, raised the price of animals, and unwittingly fomented the anti-government feeling that was taking hold in camp and gaining traction through the failure of the railway scheme.

While journeying to the unexpected and unknown, miners were confronted with hardship and toil. The exertion forced a new reckoning with the nature around them. Many expressed profound ambivalence at the contrast between the modernity of the urban environment in metropolitan centres and the virility and immediacy of nature in the Stikine. The transportation of goods through the Stikine forced miners to contemplate the contradiction embedded in the consumption and production of a new nature that was at once familiar yet profoundly discomfiting. Henry Franklin wrote to his father in New York about the absurdity of taking

"20 tons of machinery and 36 horses [in an] outfit that weighs about 100 tons," all for the purpose of acquiring some "yellow dust." He could not believe he was about to "climb over 4 ft of snow for 300 miles then build boats and steam for another 1000 miles even the hay and oats for the horses we have to carry with us it is a tough proposition."[69] Miners often expressed an uneasy perception of the unreality of the Klondike experience. For many, the practice and toil of trudging tonnes of grub and supplies through an ostensibly barren landscape was akin to dreaming. It was otherworldly. Not only were they confronted with an entirely new nature, but they were also faced with the intimate changes that their involvement with nature produced.

This was the first major incursion of modernity into the Stikine; an unprecedented influx of people and money passed through the Stikine and a new state authority was established.[70] Inevitably, vestiges were left behind and incorporated by locals through the many new interactions that took place after the Klondike Gold Rush. The arrival of miners, their capital, goods, and ideas affected interactions with nature. They burned wood, used it for construction purposes, managed the river as a transportation conduit, and dug into the earth in search of mineral wealth. Above all, they consumed animals and fish to sustain the physical exertions required to propel themselves northward. The relations between humans and their environment were also affected by the attempt to build a railroad between the Stikine River and Teslin Lake to facilitate the movement of goods and people. The perception of the Stikine in the metropolis had been that it was economic and perhaps socially peripheral and certainly geographically remote. The activities around the Gold Rush changed that.

This story of railroad construction schemes and their ecological impacts is relevant to more general discussions regarding the side effects of unsuccessful developments, and notions of marginality and scale in environmental histories of remote regions.[71] Several scholars have used the work of James Scott to great effect to explore the unintended and often pernicious consequences of development in Canada.[72] Scott's insistence on the ordering, simplification, enumeration and ultimate control of nature as a tool of statecraft is particularly revealing in this chapter. The railway schemes were, on the face of it, failures. But embodied in those failures was an inchoate awareness of the Stikine's physical features, which in turn became the hard evidence that the Stikine could and should be developed. As miners, tradesmen, surveyors, engineers, government officials, and shopkeepers moved into and through the Stikine, it became, as Scott would have it, "legible."[73] This performed the dual function of making the Stikine less "marginal" and less "remote" to the state and to others captivated by the emerging possibility of development. Places like the Stikine watershed, so long at the margins, now began to be connected to other places through expanding markets and enterprises, as state actors and development dreamers began to enact new visions and uses of nature and resources.

A new ordering of nature emerged out of the Gold Rush and the failed railway. The interactions between miners and nature in the Stikine commodified bodies and brought about ambitious plans to move people and goods across the watershed. The Stikine watershed would remain largely undeveloped in the coming decades, but these early efforts were fundamental because they created the prospect of improvement and allowed the Stikine to be considered as a place where progress could be achieved. The failed railway initiatives fostered political debate and necessitated lobbying and negotiation of economic concessions that turned elements of the Stikine environment into new commodities. The rail schemes and the construction of the Teslin Trail also required collection of geological and natural data that described landscape characteristics in ways designed to facilitate profit and progress. Furthermore, prospective miners, as they fought against the strange experience of the trail, helped to bring the Stikine into the wider cultural fabric, as they wrote home about their exertions and frustrations. The promise of the railway and easy passage to the Klondike goldfields enabled the embodied experiences of the miners and forced state actors to reckon with nature and resources that had to be deciphered and ultimately controlled. Together, railway schemes and miners' experiences show how the conditions of possibility for development were created in the Stikine, as it was slowly moved to within the reach of the state and of metropolitan entrepreneurship, becoming a place that could be improved. The railway projects—the first major development projects brought to the Stikine—and the experiences they made possible are historically important because they changed perceptions of the Stikine as a place where development could happen. This shift stemmed from the identification and cataloguing of the region's characteristics and resources undertaken to attempt to prove the feasibility of the railways and wagon roads. In the coming decades, a number of relatively small economies and projects would emerge in the region: a flourishing big-game hunting and guiding industry, the construction of the Yukon Telegraph, placer mining operations throughout the Stikine watershed, and large-scale exploration and surveying programs, largely under the auspices of the Geological Survey of Canada. These received their major impetus in the waning years of the nineteenth century as the Stikine entered into the vernacular as a place of opportunity.

Notes

1 Cited in Pierre Berton, *Klondike: The Last Great Gold Rush, 1986–1899* (Toronto: McClelland and Stewart, 2001), 226.

2 City of Vancouver Archives (hereafter cited as CVA), PAM 1943–46, George Kirkendale, *The Stikine Trail, 1898: A Narrative of Glenora, Telegraph Creek, Cassiar Central Railway, Teslin and Dease Lakes,* 1943.

3 Kathryn Morse, *The Nature of Gold: An Environmental History of the Klondike Gold Rush* (Vancouver and Seattle: UBC Press and University of Washington Press, 2003). See ch. 1, "The Culture of Gold," 16–39.

4 *The Klondike News* (Dawson City), 1 April 1898 (vol. 1, no. 1).

5 For a general, popular discussion, see Pierre Berton, *Klondike: The Last Great Gold Rush, 1986–1899* (Toronto: McClelland and Stewart, 2001). For more nuanced, scholarly analyses, see Julie Cruikshank, *Reading Voices: Oral and Written Interpretations of Yukon's Past* (Vancouver and Toronto: Douglas and McIntyre, 1991); Morse, *The Nature of Gold*; Colin M. Coates and Ken Coates, eds., "The Klondike Gold Rush in International

Perspective," *Northern Review* 19 (Winter 1998); Charlene Porsild, *Gamblers and Dreamers: Women, Men, and Community in the Klondike* (Vancouver: UBC Press, 1998).

6 There was a competing "All-Canadian" route heading northwest from Edmonton, which received significant institutional and legislative support from the local MP, Frank Oliver, though the term was primarily a promotional designation appealing to patriotic and/or imperial sensibilities.

7 The phrase is borrowed from Bernard Cohn, *Colonialism and Its Forms of Knowledge: The British in India* (Princeton: Princeton University Press, 1996). See also, Ian McKay, "The Liberal Order Framework: A Prospectus for a Reconnaissance of Canadian History," *Canadian Historical Review* 81, no .4 (2000), 617–45. On the immanence of northern development in Canada more generally, see the chapter by Loo in this volume.

8 For an outline, see Ken Coates and William Morrison, *The Forgotten North: A History of Canada's Provincial Norths* (Toronto: James Lorimer, 1992).

9 Some prominent recent examples include Hans Carlson, *Home is the Hunter: The James Bay Cree and Their Land* (Vancouver: UBC Press, 2008); Caroline Desbiens, *Power from the North: Territory, Identity, and the Culture of Hydroelectricity in Quebec* (Vancouver: UBC Press, 2013); Arn Keeling, "'Born in an Atomic Test Tube': Landscapes of Cyclonic Development at Uranium City, Saskatchewan," *Canadian Geographer* 54, no. 2 (2010): 228–42; Frank Tough, "*As Their Resources Fail*": Native Peoples and the Economic History of Northern Manitoba, 1870–1930 (Vancouver: UBC Press, 1996); Donald G. Wetherell and Irene R. A. Kmet, *Alberta's North: A History, 1890–1950* (Edmonton: University of Alberta Press).

10 Liza Piper, "Subterranean Bodies: Mining the Large Lakes of North-west Canada, 1921–1960," *Environment and History* 13 (2007): 155–86; Liza Piper, *The Industrial Transformation of Subarctic Canada* (Vancouver: UBC Press, 2009); John Sandlos and Arn Keeling, "Claiming the New North: Development and Colonialism at the Pine Point Mine, Northwest Territories, Canada," *Environment and History* 18, no. 1 (2012): 5–34.

11 See, in particular, Dolly Jorgenson and Sverker Sorlin, *Northscapes: History, Technology and the Making of Northern Environments* (Vancouver: UBC Press, 2013). See also the chapters by Cronin, Farish and Lackenbauer, and Piper in this volume.

12 See the chapters by Farish and Lackenbauer, Loo, and Keeling and Sandlos in this volume.

13 Information on Tahltan resource management practices is found primarily in the ethnographic literature: Thomas McIlwraith, "*We Are Still Didene*": Stories of Hunting and History From Northern British Columbia (Toronto: University of Toronto Press, 2012); Sylvia L. Albright, *Tahltan Ethnoarchaeology* (Burnaby: Department of Archeology, Simon Fraser University, 1984); Judy Thompson, *Recording Their Story: James Teit and the Tahltan* (Vancouver: Douglas and McIntyre, 2007); G. T. Emmons, *The Tahltan Indians* (Philadelphia: University of Pennsylvania Press, 1911); Janice Sheppard, *The History and Values of a Northern Athapaskan Indian Village* (PhD diss., University of Wisconsin, 1983); Canadian Museum of Civilization, box 121, folder 3, 1210.4b, VI-O-8M, 1912–15, James A. Teit, "Report on Tahltan Fieldwork Among the Tahltan, Kaska, and Bear Lake Indians," 1912–15.

14 For an elaboration on these encounters in colonial British Columbia, see Cole Harris, "How Did Colonialism Dispossess? Comments from an Edge of Empire," *Annals of the Association of American Geographers* 94, no. 1 (March 2004): 165–82.

15 Kathryn Morse, *The Nature of Gold*, 68. See also William Cronon, "Kennecott Journey: The Paths Out of Town," in *Under an Open Sky: Rethinking America's Western Past*, ed. William Cronon, George Miles, and Jay Gitlin (New York: W. W. Norton, 1992), and Gunther Peck, "The Nature of Labor: Fault Lines and Common Ground in Environmental and Labor History," *Environmental History* 11, no. 2 (2006): 212–38.

16   Finis Dunaway, *Natural Visions: The Power of Images in American Environmental Reform* (Chicago: University of Chicago Press, 2005); Peter Geller, *Northern Exposures: Photographing and Filming the Canadian North, 1920–45* (Vancouver: UBC Press, 2006); Greg Gillespie, *Hunting for Empire, Narratives of Sport in Rupert's Land, 1840–1870* (Vancouver: UBC Press, 2007); Gregg Mitman, *Reel Nature: America's Romance with Wildlife on Film* (Cambridge: Harvard University Press, 1999).

17   Morse, *The Nature of Gold*, 162; for the Yukon, see Ken Coates, *Best Left as Indians: Native-White Relations in the Yukon Territory, 1840–1950* (Vancouver: UBC Press, 1984); see also Robert G. McCandless, *Yukon Wildlife: A Social History* (Edmonton: University of Alberta Press, 1985). For an environmental history of the Caribou Gold Rush, see Megan Prins, "Seasons of Gold: An Environmental History of the Caribou Gold Rush" (master's thesis, Simon Fraser University, 2007).

18   Graeme Wynn, review of Kathryn Morse, *The Nature of Gold: An Environmental History of the Klondike Gold Rush*, *Agricultural History* 79, no. 2 (Spring 2005): 243–46.

19   Although contemporary in nature, the most prominent Yukon-based work in this vein comes from Paul Nadasdy. See in particular *Hunters and Bureaucrats: Power, Knowledge, and Aboriginal-State Relations in the Southwest Yukon* (Vancouver: UBC Press, 2003); "The Politics of TEK: Power and the 'Integration' of Knowledge," *Arctic Anthropology* 36, nos. 1–2 (1999): 1–18; and "Wildlife as

Renewable Resource: Competing Conceptions of Wildlife, Time, and Management in the Yukon," in *Timely Assets: The Politics of Resources and their Temporalities*, ed. E. Ferry and M. Limbert (Santa Fe, NM: School of Advanced Research Press, 2008): 75–106. See also Nadasdy's chapter in this volume.

20   For an overview, see Karen Bakker and Gavin Bridge, "Material Worlds? Resource Geographies and the 'Matter of Nature,'" *Progress in Human Geography* 30, no. 1 (February 2006): 5–27.

21   Liza Piper, *The Industrial Transformation of Subarctic Canada* (Vancouver: UBC Press, 2010); Arn Keeling, "'Born in an Atomic Test Tube.'"

22   Lisa Cooke, "North Takes Place in Dawson City, Yukon, Canada," in *Northscapes: History, Technology, and the Making of Northern Environments*, ed. Dolly Jorgenson and Sverker Sorlin (Vancouver: UBC Press, 2013): 223–46. See also Emilie Cameron, "Copper Stories: Imaginative Geographies and Material Orderings of the Central Canadian Arctic," in *Rethinking the Great White North*, ed. Andrew Baldwin, Laura Cameron, and Audrey Kobayashi (Vancouver: UBC Press, 2011).

23   Dianne Newell MacDougall, "Canada's Share of the Klondyke: The Character of Gold Rush Publicity, 1897–98" (master's thesis, Carleton University, 1974).

24   H. V. Nelles and Christopher Armstrong, *Monopoly's Moment: The Organization and Regulations of Canadian Utilities, 1880–1930* (Philadelphia: Temple University Press, 1986); Duncan McDowall, *The Light: Brazilian Traction, Light*

and Power Company, 1899–1945 (Toronto: University of Toronto Press, 1988); R. B. Fleming, *The Railway King of Canada: Sir William MacKenzie, 1849–1923* (Vancouver: UBC Press, 1991).

25 Provincial Archives of British Columbia (hereafter cited as PABC), GR 0441, Premier's Records, box 5, file 1, Alexander Begg, *Memorial To the Honourable the Executive Council of the Province of British Columbia*, Victoria, December, 1896. Begg claimed that Vancouver and Nanaimo merchants had submitted similar memorials.

26 PABC, GR 0441, Premier's Records, box 5 file 1, Alexander Begg to J. H. Turner, 10 April 1897.

27 PABC, GR 0441, Premier's Records, box 5, file 1, Begg to Turner, 4 May 1897.

28 PABC, GR 0444, British Columbia, Executive Council Records, Alexander Begg to The Honourable the Executive of British Columbia, July 28, 1897.

29 Peter Murray, *Home From the Hill: Three Gentlemen Adventurers* (Victoria: Horsdal and Schubart, 1994), 37.

30 PABC, GR 0441, Premier's Records, box 5, file 3, F. Elworthy to J. H. Turner, 18 November 1897.

31 PABC, GR 0441, Premier's Records, box 5, file 3, J. H. Turner to Clifford Sifton, 20 November 1897.

32 PABC, GR 0441, Premier's Records, box 5, file 3, J. T. Bethune to J. H. Turner, 12 November 1897; PABC, GR 0441, Premier's Records, box 5, file 3, Bethune and Charles Lugrin to Turner, 26 November 1897; PABC, GR 0441, Premier's Records, box 5, file 3, Turner to Bethune, 30 November 1897.

33 PABC, GR 0441, Premier's Records, box 7, file 3, J. H. Turner to R. L. Elliott, 7 January 1898.

34 PABC, GR 0441, Premier's Records, box 7, file 5, Mackenzie, Mann & Co. to J. H. Turner, 28 December 1897; PABC, GR 0441, Premier's Records, box 7, file 5, J. H. Turner to Donald Mann, 30 December 1897.

35 Library and Archives Canada (hereafter cited as LAC), Clifford Sifton Papers, vol. 224, Clifford Sifton to William Van Horne, 11 January 1898.

36 D. J. Hall, *Clifford Sifton, Volume One: The Young Napoleon, 1861–1900* (Vancouver: UBC Press, 1981), 179–80.

37 The debate about the Yukon-Canadian occupied Parliament between 22 February and 17 March 1898, with Sifton and Tupper each playing prominent roles. See Canada, "Official report of the debates of the House of Commons of the Dominion of Canada: third session, eighth Parliament … comprising the period from the third day of February to the twenty-first day of April inclusive" (Ottawa: S. E. Dawson, 1898).

38 Warburton Pike, *Through the Subarctic Forest: Down the Yukon by Canoe in 1887* (New York: E. Arnold, 1896); Warburton Pike, *The Barren Grounds of Northern Canada* (London: MacMillan, 1892).

39 PABC, GR 0441 Premier's Records, box 5, file 3, 1883–1933, Warburton Pike to J. H. Turner, 13 January 1897; PABC, GR 0441, Premier's Records, box 5, file 3, Pike to Turner, 5 April 1897. The term "Cassiar" is often synonymous with "Stikine," though it often applies to the

larger section of northwest British Columbia.

40 PABC, GR 0441, Premier's Records, box 5, file 3, Pike to Turner, 5 April 1897.

41 Murray, *Home From the Hill*, 31–42.

42 Morris Zaslow, *Reading the Rocks: The Story of the Geological Survey of Canada, 1842–1972* (Toronto: Mac-Millan, and Ottawa: Department of Energy, Mines and Resources and Information Canada, 1975).

43 William Ogilvie, "Extracts From the Report of an Exploration Made in 1896–1897," in *The Yukon Territory* (London: Downer, 1898), 383–423. Ogilvie suggested that there was great mining potential in the Teslin-Hootalinqua area. He estimated the prospects for the northern part of the Stikine plateau as even greater: "from Telegraph Creek northward to the boundary line we have in the Dominion and in this province an area of 550 to 600 miles in length, and from 100 to 150 miles in width, over the whole of which rich prospects have been found. We must have from 90,000 to 100,000 square miles, which, with proper care, judicious handling, and better facilities for the transportation of food and utensils, will be the largest, as it is the richest, goldfield the world has ever seen" (p. 417).

44 George Mercer Dawson, *Report on an Exploration made in the Yukon District, N.W.T. and Adjacent Portion of British Columbia, 1887* (Ottawa: Geological Survey of Canada, 1898). A useful account of a previous surveying trip Dawson made to British Columbia can be found in Jason Grek-Martin, "Vanishing the Haida: George Dawson's Ethnographic Vision and the Making of Settler Space on the Queen Charlotte Islands in the Late Nineteenth Century," *Canadian Geographer* 51, no. 2 (Fall 2007): 373–98.

45 William Jennings, "Report on Routes to the Yukon" (Ottawa: Queen's Printer, 1898).

46 For more general discussions of this point, see Morris Zaslow, *The Opening of the Canadian North, 1870–1914* (Toronto: McClelland and Stewart, 1971); Alan Cooke and Clive Holland, *The Exploration of Northern Canada 500–1920: A Chronology* (Toronto: Arctic History Press, 1976); Marionne Cronin, "Northern Visions: Aerial Surveying and the Canadian Mining Industry, 1919–1928," *Technology and Culture* 48, no. 2 (2007): 303–30.

47 Jennings, "Report on Routes to the Yukon."

48 Jennings, "Report on Routes to the Yukon"; PABC, MS-0051, Edmund Duchesnay, "Field Notes of E. J. Duchesnay on survey trip to Telegraph Creek, Stikine River, Sept. 1897."

49 Jennings, "Report on Routes to the Yukon."

50 Hall, *Clifford Sifton, Volume One: The Young Napoleon, 1861–1900*, 177–82.

51 Yukon Archives (hereafter cited as YA), MSS361 (2006/163), Thomas Frederick Seldon fonds, 1898–99.

52 YA, MSS207 (90/48), O. T. Switzer fonds, 1897–1900, 20 February 1898. Switzer's diary and letters home to his parents were published serially in the *Philipsburg Ledger*. It appears this arrangement was made before Switzer came north.

53 YA, MSS169 (81/101), Robert Hunter Fitzhugh fonds, 1897–99, Fitzhugh to his sister, December 1897. For more on Fitzhugh, see Joanne Hook, "He Never Returned: Robert Hunter Fitzhugh in Alaska," *Alaska Journal* 15 (Spring 1985): 33–38.

54 YA, MSS169 (81/101), Robert Hunter Fitzhugh fonds, 1897–99, Fitzhugh to his sister, 5 May 1898.

55 Estimates vary from five thousand to over ten thousand people, which is likely too large. The plateau at Glenora is small, and miners' shacks and tents were squeezed into a very confined space. Diarists describe seeing tents and shacks along the banks of the river for several miles each way from Glenora.

56 Larry Pynn, *The Forgotten Trail: One Man's Adventures on the Canadian Route to the Klondike* (Toronto: Doubleday Canada, 1996).

57 YA, MSS361 (2006/163), Thomas Frederick Seldon.

58 YA, MSS207 (90/48), O. T. Switzer Diary, 23 June 1898.

59 YA, MSS207 (90/48), O. T. Switzer Diary, 16 July 1898.

60 Guy Lawrence, *Forty Years on the Yukon Telegraph* (Quesnel, BC: Caryall Books, 1965). The line was also sometimes referred to as the Yukon-Teslin.

61 Dianne Newell, "The Importance of Information and Misinformation in the Making of the Klondike Gold Rush," *Journal of Canadian Studies* 21, no. 4 (Winter 1986–87): 95–111.

62 YA, MSS207 (90/48), O. T. Switzer Diary, 6 April 1898.

63 The Collins Overland Telegraph was an ambitious scheme to connect San Francisco to Moscow by telegraph. The route came directly north through Oregon, Washington, and British Columbia, then skirted west through the Yukon Territory and Alaska. Another party from the company was working in Siberia. The line and trail were abandoned in 1867 (with $3 million already spent) after a competing company successfully laid an underground cable between Newfoundland and Ireland. It was also known as the Russian-American Telegraph and the Western Union Overland Telegraph. See John Dwyer, *To Wire the World: Perry M. Collins and the North Pacific Telegraph Expedition* (Westport, CT: Praeger, 2001); Rosemary Neering, *Continental Dash: The Russian-American Telegraph* (Ganges, BC: Horsdal and Schubart, 2000).

64 Norman Lee, *Klondike Cattle Drive: The Journal of Norman Lee* (Vancouver: Mitchell Press, 1960), 20.

65 Hamlin Garland, *On the Trail of the Goldseekers: A Record of Travel in Verse and in Prose* (New York: Macmillan, 1906), 180–81; Keith Newlin, "Prospecting For Health: Hamlin Garland's Klondike Adventures," *American Literary Realism* 35, no. 1 (Fall 2002): 72–92.

66 See Ken S. Coates and William R. Morrison, "Winter and the Shaping of Northern History: Reflections from the Canadian North," in *Northern Visions: Perspectives on the North in Canadian History*, ed. Kerry Abel and Ken S. Coates (Peterborough, ON: Broadview Press, 2001): 23–36.

67 YA, MSS169 (81/101), Hunter Fitzhugh to his mother, 7 November 1897.

68 Arthur L. Disher, "The Long March of the Yukon Field Force," *The Beaver* (August 1962): 4–15.

69 YA, MSS169 (81/145), Henry W. Franklin to his father, n.d.

70 For an argument that suggests that European modernity and the life-ways of northern Indigenous peoples were not mutually exclusive, see Cruikshank, *Do Glaciers Listen? Local Knowledge, Colonial Encounters, and Social Imagination* (Vancouver: UBC Press, 2005), in particular ch. 2, "Constructing Life Stories: Glaciers as Social Spaces."

71 James C. Scott, *Seeing Like A State: How Certain Schemes to Improve the Human Condition Have Failed* (New Haven: Yale University Press, 1999). See also Tania Murray Li, *The Will to Improve: Governmentality, Development and the Practice of Politics* (Chapel Hill: Duke University Press, 2007); and Anna Tsing, *Friction: An Ethnography of Global Connection* (Princeton: Princeton University Press, 2005). See also the chapter by Cameron in this volume.

72 Tina Loo, "People in the Way: Modernity, Environment, and Society on the Arrow Lakes," *BC Studies* 142, no. 33 (2004): 161–96; Tina Loo and Meg Stanley, "An Environmental History of Progress: Damming the Peace and Columbia Rivers," *Canadian Historical Review* 92, no. 3 (2011): 399–427; Philip Van Huizen, "Building a Green Dam: Environmental Modernism and the Canadian-American Libby Dam Project," *Pacific Historical Review* 79, no. 3 (2010): 418–53.

73 Scott describes this "as a central problem of statecraft." Scott, *Seeing Like A State*, 2.

3

The Experimental State of Nature: Science and the Canadian Reindeer Project in the Interwar North

Andrew Stuhl

Like a black mass of some fluid the herd slowly approached the edge of the plateau—began to flow down first slowly—a few deer at a time but soon gathering impetus and speed and ending in a wild rush. ... It was a grand sight that I will never forget. ... The drive is on its way to Canada.

—Alf Erling Porsild, 1929[1]

In 1929, on the tattered pages of a field journal, botanist Alf Erling Porsild recorded a "grand sight." Before his eyes, a herd of nearly three thousand reindeer set off for northern Canada from the Seward Peninsula in Alaska. The scientist had good reason to commemorate this moment: it marked his success in transforming the Arctic into reindeer country.

The Canadian Department of Interior initiated the Canadian Reindeer Project in 1926 with hopes of bringing industry and civilization to the nation's northernmost frontier. Reindeer, a species foreign to North America but common in other parts of the circumpolar world, seemed perfectly adapted to both the physical environments of the region and the state's priorities. Through the consumption of tundra plants, these domesticated animals could create readily accessible meat, hides, and bones. These items were critical to Inuit livelihoods, but had become scarce as caribou populations dropped over the turn of the twentieth century. More than a "natural" form of social support or a possible commercial good for export, reindeer were a tool of northern colonization. State officials hoped herding units would organize dispersed and semi-nomadic Inuit hunters and thus facilitate regulation of people and land uses in the Arctic.[2]

Between 1926 and 1944, Porsild laboured tirelessly to make these colonial dreams come true. The Canadian government employed him to investigate existing reindeer industries in Alaska and determine the conditions underpinning their success. This investigation was a precursor to another, in which he surveyed thousands of square miles of the north to identify a suitable home for a Canadian herd. His official report of these travels provided the basis for the government's choice of the Mackenzie Delta as a home for its "experiment."[3] Pulling double duty as a colonial official, Porsild oversaw the construction of a new town on the Mackenzie River to house the reindeer—Reindeer Station—and ventured to Norway to find Saami herders to instruct Inuit apprentices. Working together with other researchers and civil servants in Ottawa, Porsild erected a regulatory apparatus to supervise people, animals, and the land in the Arctic.[4] For the scientist, the view of the herd entering Canada for the first time in 1929 was indeed a vision. Reindeer ushered in a new era in the north.

In this essay, I position the arrival of reindeer in Canada as a watershed in northern history. Attention to the practice of science in the Reindeer Project reveals shifts in relations between the Canadian state and the Arctic, as well as in human relationships with nature in the region itself. Before reindeer, the Canadian government had mobilized scientific knowledge about the north as a means of enumerating the resources of these remote territories.[5] With the employment of Porsild, however, the state mobilized science not to document entire landscapes or bring samples of them back to museums, but to reduce complex human and physical environments to a few key variables. State agencies could then manipulate these variables to serve particular political and economic ends, like developing markets for meat in urban centres or demarcating public domain in the far north. Reindeer projects, then, were experimentalist both in the sense of the scientific method and insofar as the Canadian government had never before managed arctic resources directly.

As they did for Porsild, reindeer may afford Canadian historians new perspectives on the north. For the purposes of this essay, my focus is on the decades between the two world wars. The interwar era has received less scholarly attention than the more iconic northern episodes that precede and follow it, like the Klondike Gold Rush or the infrastructure megaprojects implemented after Second World War. Some scholars have characterized Ottawa in the 1920s and 1930s as neglectful of northern territories, citing an isolationist and reactive posture to issues of sovereignty and welfare. But recent analyses by environmental historians have demonstrated that, in other arenas, the government was more involved and exacting. By focusing on nature in northern history—the restriction of Aboriginal hunting, the establishment of northern parks, and the expansion of fisheries and mineral industries—these historians have revealed how central the remote north was to national politics and world economies. The control of northern natural resources became a vehicle for extending federal jurisdiction over northern peoples and linking marginal environments to global markets. By reshaping the north, Canada redefined its dominion at home and abroad.[6]

A close look at reindeer in this essay builds on these interpretations of northern history by incorporating the central role of science in mediating human relationships with the natural world. As detailed in the first section of this chapter, the physical and human landscapes above the Arctic Circle presented the Department of Interior with challenges not found in other northern areas. Copper deposits had been identified near the lower reaches of the Coppermine River, but there were minimal food resources available locally to sustain a labour pool. Moreover, Inuit occupied the arctic coastline across the northern edge of western Canada, and were reluctant to sign over rights to their land through treaty arrangements. Government officials looked to scientists, including Porsild, when crafting solutions to these problems.

As Porsild conducted field research and instituted a plan for herding, his science fundamentally changed the Canadian state and the arctic environment. As we will see in the second half of this essay, managing animal husbandry economies required institutional arrangements and approaches not found in the development of the Canadian north elsewhere.

As scholars in this volume convincingly show, a multi-national mining industry and the machinery of expertise associated with it were crucial to the exploitation of many regions north of sixty from the late 1800s through the 1940s.[7] Private mineral companies, alongside governmental engineers, geologists, and surveyors, rendered distant northern lands knowable and therefore more easily subject to corporate and federal power. Government agents quickly realized, however, that these institutions and forms of knowledge were not suitable for a development regime based on reindeer. In learning how to manage a foreign species, the Canadian government reimagined criteria for scientific authority, forged novel partnerships with the United States, collected unprecedented scientific data about the tundra, and generated new instruments for regulating northern land uses. In the conclusion, I suggest these results of the reindeer experiment comprise part of the continuing legacy of the interwar north, even as governments and corporations jettisoned herding programs in the second half of the twentieth century.

Taming the Arctic: The Impulses behind the Canadian Reindeer Project

Experience has proven that there are periodical fluctuations in the number of fur-bearing animals and that caribou cannot be depended upon to follow the same migration routes each year. With the object of broadening the basis of subsistence of the natives, especially in view of the rapid advance of mining in the Northwest Territories, the Department of the Interior has for a considerable time been looking into the possibilities of increasing the numbers of the larger ruminants.

—Canada's Reindeer Experiment, 1936[8]

Reindeer first came to North America in the late 1800s under the direction of missionaries. These men hoped to alleviate starvation among Native northerners in Alaska whose subsistence base of caribou and marine mammals had been destroyed by commercial fisheries industries.[9] Missionaries believed the reindeer could turn non-arable hinterlands into productive grazing lands and so-called primitive Inuit hunters into so-phisticated herders.[10] Sheldon Jackson brought a herd across the Bering Strait from Siberia to Port Clarence, Alaska, in 1892. In Canada, Dr. Wilfred Grenfell spearheaded the introduction of reindeer to Newfoundland in 1908 with motivations that mirrored Jackson's.[11]

These initial introductions had differing fates. Jackson's Alaskan herd swelled in the early 1900s. Smaller herds were spun off the main group and driven to Inuit settlements along the Bering Strait and Beaufort Sea coasts. Congress grew interested in the possibility of pairing the civilizing mission behind missionary-led herding with education, and dispatched US Bureau of Education staff and Saami herders from Scandinavia to each village to teach the would-be Inuit herders. Meanwhile, countless obstacles plagued the Canadian herds. Predacious wolves, pestering flies, straying animals, and poor grazing lands were all problems reindeer officials could not solve. In addition, as the federal bureaucracy expanded to incorporate new departments for the north and for wildlife in the late 1910s, Ottawa bureaucrats disappointed in reindeer shuffled the responsibility for herding programs. After many of the animals died, the Parks Branch took over those that remained, only to transfer them to the Anticosti Island Administration in 1923. This herd remained small and isolated compared to what became of the Canadian Reindeer Project in the 1930s and 1940s.[12]

After 1918, Canadian government officials gained new motivation and partners to develop reindeer industries. Reindeer garnered attention following the completion of a series of scientific expeditions to the western Arctic in 1918. Inspired by their travels in the north, expedition members championed reindeer as a vehicle of economic development and game management. Like missionaries before them, biologists, geographers, geologists, and anthropologists leaned on their own experience to deploy a complicated rhetoric about protection and exploitation. Along with federal officials, northern police, and whaling captains, scientists imagined the north as meat-producing factory and game sanctuary. There, domesticated musk ox and reindeer met the needs of Canadians through the commodification of northern prairies and the conservation of large native mammals.[13]

These attempts by expedition scientists to redefine the Arctic through reindeer mediated post-First World War concerns about food scarcity and industrialization in North America. As western ranches gave way to settlement, European agricultural fields recovered from the wounds of battle, and urban populations exploded, the Canadian north appeared as both a promising frontier for livestock and a landscape on the verge of repeating the west's environmental history. Historian John Sandlos has argued the discursive practices relating to conservation in the north created an "Arctic Pastoral," in which bureaucrats, sportsmen, scientists, and other conservationists portrayed the Native hunter as "irrational and destructive" and the arctic tundra as an environment ripe for government-sponsored development. In combination with other measures, such as regulating hunting and creating national parks, taming the musk ox and introducing reindeer sought to stem the supposed "wanton slaughter" of certain game species, like caribou. In so doing, bureaucrats in Ottawa sought to establish northern lands and animals as national resources and southern bureaucrats as the logical managers of these assets.[14]

As we attend to the elaboration of the Canadian Reindeer Project in the 1920s and 1930s, it is important to distinguish between two kinds of northern nature at stake in the Arctic Pastoral: wildlife and tame-life. Administrators articulated the impulses behind reindeer herding as distinct from—though related to—the conservation of game. These impulses responded in part to the particular challenges of engaging Inuit and the arctic environment in the project of development.

The distinctions between wild and tame can be brought into focus by attending to the legal relationship between governmental agents and Inuit in the early twentieth century. The federal government was unable to secure a treaty with Inuit (as they had with Dene living along the Mackenzie River), as the Inuit did not sign Treaty 11, the comprehensive agreement of 1921. According to one Inuit scholar, "Our people had the necessary tools for surviving and there was enough game around to meet their needs, so they didn't see the need to sign any treaty."[15] While the Royal Canadian Mounted Police and missionaries had enforced legal and moral codes at whaling and fur-trading outposts in Inuit territory, the formal designation of Inuit as wards of the state did not occur until the Indian Act was amended in 1924.[16] Still, Inuit had never agreed to the terms of this amendment nor dissolved any rights to the land, and thus federal agents must have been eager to find some additional apparatus to bring Inuit under the purview of national law. Indeed, in the contemporary case of musk ox conservation, consultants to the Advisory Board on Wildlife

Protection suggested the federal government practice diplomacy with Inuit to enlist northerners in the project of protecting nature. This seemed a more effective alternative to doubling-down on hunting regulations that could not be adequately enforced.[17]

Accounts from Inuvialuit residents suggest that reindeer were living forms of bureaucracy in the Canadian Arctic. According to Randall Pokiak, Inuit living in the Mackenzie Delta and along the arctic coast in the early 1900s were troubled by the recent influx of Alaskan Inupiat into the area, as they deemed these foreign Natives responsible for the recent decline in caribou populations. Alaskan Inupiat had travelled eastward since the 1880s, first with commercial whalers who had over-harvested caribou in the Bering Strait and north slope regions, and later to avoid an epidemic of Spanish Influenza after 1918. Calling on a local shaman, Inuvialuit leaders hoped to alter the migration patterns of caribou to force the Inupiat to return to Alaska. The caribou did go away, but did not return, and the Inuvialuit thus became amenable to new means of procuring food. When government agents approached the Inuvialuit in the 1920s, Aboriginal leaders negotiated for the delivery of reindeer from Alaska, having "heard stories from the Inupiat that reindeer had the same diet as caribou."[18] Oral histories indicate that one Inuvialuit man, Mangilaluk, negotiated with the government on behalf of local communities, suggesting to treaty officers that, "if they brought reindeer from Alaska to Tuk area they would think about signing an agreement."[19] In Canada, then, reindeer created possibilities for making Inuit into colonial subjects, whether through religion, commerce, or law.

The control of nature also played out differently in introducing tame-life than it did with conserving wildlife. On the ground, corrals brought widely dispersed animals and herders to one geographic location at two distinct points in the year, allowing for counting, branding, slaughter, and evaluation. On a broader scale, legislation in Ottawa also enclosed people and resources in the Arctic. The creation of a six-thousand-acre Reindeer Grazing Preserve in the Mackenzie Delta and a federal protection ordinance for reindeer made northern nature a federal responsibility. In the context of an introduced reindeer industry, protected areas would have to be administered according to the demands of recruiting Inuit to herding animals. This was quite different than preventing the hunting of wild species. In a 1935 meeting of the Interdepartmental Reindeer Committee, a body organized in 1932 to help guide the Project's evolution, biologist Rudolph Anderson and anthropologist Diamond Jenness contrasted the functions of national parks with those of the Reindeer Grazing Preserve. The scientists agreed that the national parks were designed to protect all wildlife in perpetuity. But in the preserve, hunting and trapping must be carried out by herders, as they required a certain amount of fur for winter clothing. The scientists noted it would be nearly impossible to attract Natives to herding if trapping privileges were denied. In making this distinction, Jenness and Anderson helped construe the Reindeer Preserve as an experimental space for managing Inuit and nature. With aims to domesticate, study, and develop, reindeer projects were more like colonial botanical gardens than hunting preserves.[20]

Most importantly, reindeer herding became a key mechanism in plans for arctic economic development in ways that wildlife and national parks did not. Reconnaissance work in the Coronation Gulf during the Canadian Arctic Expedition of 1913–18 returned with the promise of extensive copper deposits. Scientists argued that, in order to capitalize upon these resources, a local food source would need to be established, since populations of migrating caribou had been decimated. Many southern Canadians also believed that white men were unlikely to want to live in the north and might be physically unable to do so. Reindeer and Inuit offered solutions to these problems. Inuit could be responsible for maintaining reindeer herds, the meat from which could be shipped to the Coronation Gulf, reducing overhead costs for privately or federally sponsored mineral extraction. Drawing connections between labour needs, environmental changes, and the possibilities of reindeer and mineral economies, promoters of northern development often articulated Inuit as the Arctic's most valuable asset. Without them, the government would "spend millions" to get people to live and work there. Jenness, who had recently returned from three years of study among the Copper Inuit of the Coronation Gulf, distilled the situation for his audience at a 1923 lecture at the Victoria Memorial Museum. "Unless we use the Eskimos," he argued, "we can never develop the Northland."[21]

As Canadian scientists and bureaucrats began to see the value of reindeer for arctic development, they were forced to think differently about the existing northern fur trade. Especially after the stock market crash in 1929, the discourse around reindeer in North America asserted the value of stability found in husbandry economies while denouncing the volatility of dealing in fur and its associated markets. Herding and harvesting reindeer appeared to state agents as more stable than the fluctuations inherent in animal populations and the fur trade, which was rapidly expanding across the Beaufort Sea coast after the Great War. Regulating hunting of native species would not necessarily address the unpredictability of markets and nature, but building up reindeer as a subsistence base might. Such a "native-run industry," the kind government agents advocated for in the early 1930s, gradually replaced the visions of a grand northern meat industry.[22]

Rhetoric about reindeer reflected the complicated project of administering the Arctic during the interwar period. Emerging from missionary-sponsored civilizing missions, reindeer projects found new impulses after 1918. Industrial boosterism, the limits of legal relationships with Inuit, concerns about the volatility of wild nature and markets, and desires to build a northern mineral industry all infused the conversation on herding programs. After the Great War, many Canadians believed that taming the Arctic was the key to the region's future. Over the 1920s, a series of trials and errors would test this optimism and catalyze new relationships between science and the federal government.

Exploratory Engineer or Botanist? Testing Definitions of Arctic Expertise

It was not clear what kind of expert would lead the Canadian Reindeer Project. In the winter of 1919–20, members of a royal commission on musk ox and reindeer sat down to a series of meetings in Ottawa. This body was brought together by explorer-anthropologist Vilhjalmur Stefansson to "investigate the possibilities of the reindeer and musk-ox industries in the Arctic and Sub-arctic regions of Canada."[23] The Commission called thirty-five witnesses to testify on the opportunities and obstacles facing a reindeer industry. That no trained botanist had spent enough time in reindeer country to give evidence before the Commission did not appear to be an issue, though it became one in 1926.

The Commission heard from whalers, missionaries, northern police, elected officials, explorers, and self-identified scientists.[24] Despite their divergent training, all had spent considerable time living or traveling in the north. Importantly, eight had recently returned from the Canadian Arctic Expedition of 1913–18.[25] Commission members queried witnesses on issues suspected to be indicators of failure or success of reindeer introductions, including "vegetation, soil, climate, topography, and surface conditions."[26] Testimony detailed the extent and distribution of vegetation in certain geographical districts; the presence or former presence of caribou, which was assumed to denote the potential for reindeer; and the existence of mosquitoes and prevailing winds (to account for a troublesome reindeer pest, the mosquito).[27] Excitement for the reindeer industry grew with each meeting. The Commission outlined vast swaths of the north as Canadian reindeer country, including several islands in Hudson Bay, the entire Ungava and Mackenzie Districts, the interior of the Yukon, and the arctic coast from the international boundary to Kent Peninsula. Given northern Canada's similarities with Alaska, Siberia, and northern Europe—all areas with thriving reindeer industries—witnesses believed animal husbandry would finally capitalize upon "vast tracts of country that are not utilized."[28]

Yet there were also many concerns with existing knowledge about northern lands. Commissioners admitted that there was much "conflicting evidence" about whether Inuit would take to herding, how reindeer managed pests, and how much time plants needed to recover after grazing. Commissioners underlined the importance of continued governmental presence "to remove the elements of doubt and uncertainty, and so tend to encourage private enterprise and investment." This could be accomplished through "careful study," of individual localities, so as to "utilize to the best possible advantage, as means of control, any suitable valleys or other special topographical features, which may be available." Participants agreed that the Canadian government should lead the initial reindeer trials, beginning with a small, manageable herd, working out any kinks in logistics, and paving the way for future investment by private groups.[29]

Despite this faith in government-led development, the first attempt to cash in on reindeer following the Royal Commission came from the north's biggest corporation and biggest personality. Resigning from the body, Vilhjalmur Stefansson introduced reindeer to Baffin Island in 1921, in conjunction with a new subsidiary of the Hudson's Bay Company, the Hudson's Bay Reindeer Company. This project was a public disaster, as the $200,000 spent to transport six hundred reindeer and six Saami herders from Norway was squandered in six years. In 1921, the herd was delivered to Baffin Island, and, by 1927, most of the reindeer had died or disappeared, prompting the government to cancel the Company's grazing permit.[30] Reports of this debacle—and the controversy they inspired— eventually catalyzed the hiring of botanist Alf Erling Porsild and major changes in relations between science and the state.

The Hudson's Bay Company hired Alaskan reindeer specialist W. T. Lopp in 1925 to assess the problems with Baffin Island and to survey the island for potential areas for continued experimentation. Lopp's report concluded that the Company herd failed because of the choice of location, calling the plot "virtually worthless as range for reindeer." Lopp's insistence on this root problem seemed to call the whole enterprise into question. While the Company could buy more animals, replace its manager, Storker Storkerson, or invite more Saami families to assist them, Lopp wrote that the tundra itself—the "handicap" of its operations—could be neither "remedied nor removed." His surveys of the remainder of Baffin Island showed little hope for future industries there.[31]

Newspapers across the United States and Canada covered Lopp's report, sparking a controversy with Stefansson. Stefansson interpreted the report as inflicting severe damage on his reputation as an expert on northern matters. During one of the anthropologist-explorer's high-profile lectures in Ottawa, Edward Sapir, the director of anthropology at the Geological Survey of Canada, challenged Stefansson on the Baffin Island ordeal, asking for some explanation for the "Reindeer experiment on Baffinland." Stefansson blamed the failure on issues of management, suggesting that the Hudson's Bay Company did not heed his advice and also had mistreated Storkerson. Sensing that he was losing favour with Canadian audiences—whether in that lecture hall, or in broader commercial, intellectual, or political circles—Stefansson sent a private letter to Carl Lomen, the head of the Lomen Reindeer Corporation in Alaska. Lomen's business had boomed since the mid-1910s, with herds dotting much of the Alaskan coastline and markets for reindeer meat popping up across the United States. The explorer-anthropologist admitted to Lomen that Lopp's report might result in the Hudson's Bay Company backing out of the reindeer business, and that Lomen should consider working with Stefansson in buying up land on Baffin Island. Lomen did not take up the offer, signaling Stefansson's increasingly marginal role in reindeer industries after 1925.[32]

Stefansson was known for his contentious nature, but this case was as much about changing requirements for knowledge about the north as about his knack for the spotlight.[33] Before 1921 and the formation of the Hudson's Bay Reindeer Company, members of the Canadian Department of Interior relied on substantial northern experience—like the kind Stefansson had amassed in his ten years in the Arctic, or that embodied by the witnesses to the Royal Commission on Muskox and Reindeer—over pure academic scientific training when appraising the needs of a possible reindeer industry. Those with both academic expertise and northern experience, like many of the scientists invited to the Commission, seemed especially useful sources.[34] In early 1926, after Lopp's report was made public, the Department of Interior retained its emphasis on northern experience and academic expertise, but refined its interest in a particular type of knowledge and know-how: applied botany.

These shifting definitions of arctic expert authority materialized in correspondence among Canadian bureaucrats trying to decide on a suitable manager for the Canadian Reindeer Project. In January 1926, before the publication of Lopp's report, the head of the Northwest Territories and Yukon Branch could see little value in a botanist. O. S. Finnie wrote the Deputy Minister of the Interior, making a plea for a man with practical skills to lead a government reindeer project. "I do not think the qualifications as a Botanist is sufficient," he wrote, because "I believe we would get better results if we could get a practical reindeer man who knows the kind of feed that the reindeer live on, and one who is a good traveller and could go through the country and size up the situation accurately and quickly."[35] Finnie wanted to hire Lopp, but the Alaskan was unavailable due to his contract with the Hudson's Bay Company. In addition, the stress placed on the ability to travel raised concerns about his age—Lopp was nearly seventy years old.[36]

By 1927, though, Finnie expressed a firm commitment to applied botanical science as a way of knowing and managing reindeer. When a second private venture, the Dominion Reindeer Company of Vancouver, inquired in 1927 about leasing land in the Northwest Territories, Finnie responded with caution. He was unable to recommend any location "until the different districts in the North West Territories had been thoroughly cruised with a view to determining their value as feeding grounds for the reindeer." Finnie admitted to the director of the National Herbarium, M. O. Malte, that his hesitance with the Dominion Reindeer Company emerged from the Baffin Island incident. A lease was granted to the Dominion Reindeer Company in the eastern Arctic in 1928, contingent on a scientific survey of the region. This survey was never completed, the Dominion Reindeer Company never introduced reindeer, and the government terminated the lease in 1931. As Finnie noted, the recent history with commercial enterprise had "served as a lesson" for governmental managers of reindeer experiments.[37]

What happened in the interim to change Finnie's mind? Beyond the report issued by Lopp, Finnie and other reindeer enthusiasts in Canada were convinced of the value of applied botanical science by their counterparts in Alaska. In March of 1926, two months after Finnie voiced skepticism about botanists, high-ranking Canadian official W. W. Cory visited New York City and Washington, DC to consult with US officials on best practices for a Canadian reindeer industry. While in the American capital, Cory met with Dr. E. W. Nelson, the chief of the United States Bureau of Biological Survey, an agency that assisted both the Lomen Corporation and the US Bureau of Education with reindeer operations in Alaska. In 1920, Nelson had dispatched two scientists to Fairbanks, Alaska, where they began surveys and experiments on reindeer, including their principal movements, feeding habits, and major predators, pests, and diseases. Nelson impressed upon Cory that a single man could not handle the duties of getting the Canadian Reindeer Project off the ground. They also required surveying Canada for suitable forage and building the systems of reindeer management, like the supervisory hierarchies, corrals, and storage facilities needed to round up, slaughter, and process reindeer.

Moreover, Nelson advocated for trained ecologists to fill these roles, as the Bureau's grazing scientist, Lawrence Palmer, had made clear the importance of scientific studies of reindeer feed. Nelson attributed the steady growth of reindeer populations in Alaska to Palmer's ability to translate his studies to the management of grazing lands. In 1901, one thousand animals roamed the coasts of Alaska; by the mid-1920s, that number had exploded to over two hundred thousand. Palmer had also argued that, when his research was fully applied, reindeer country in Alaska could support three million livestock. Nelson suggested that the Department of Interior hire two botanists and have them apprentice with Palmer for six months, learning the particulars of reindeer ecology and the reindeer business. Cory relayed this news to Finnie, and with both men sold on the model of the Alaskan industry, they began to see botanical expertise, both academic and applied, as fundamental to reindeer management.[38]

These conversations among Stefansson, Lopp, Finnie, Malte, Cory, and Nelson redefined the terms of state power and science in the Arctic. Scholars have described relationships among the Canadian state and explorers during the interwar era as tumultuous, especially as the north became a site of economic and political development and as bureaucrats endured debates provoked by Vilhjalmur Stefansson.[39] In the case of reindeer and the Arctic, fields of expertise were similarly unstable. Since the Canadian government was experimenting with taming the Arctic for the first time, it needed new experimenters. Naturalists, explorers, geographers, geologists, topographers, biologists, and anthropologists had been instrumental in documenting and discovering the north before the Great War. But none of these specialists appeared as "qualified men" for the duties necessary in starting a government herd.[40]

In 1925, Finnie and Cory referred to the person capable of guiding the Project as an "Exploratory Engineer." But by May 1926, Finnie had hired both a "Botanist" and an "Assistant Botanist." These titles captured the shifting position of science relative to the state, as well as the place of the north in Canada during the interwar period. The jobs required an expert traveller who could make use of the north's existing transportation networks and yet "size up the situation" from the dogsled when necessary. He was a botanizer, who systemically collected data about vegetation patterns from landscapes in Alaska, Yukon, the Northwest Territories, and the Keewatin District. Finally, he was a project manager, who could apply extensive and intensive studies in selecting appropriate pastures and best management practices, the foundations of a new industry in the north.[41]

Finding the Men for the Job: Alf Erling Porsild, Robert Porsild, and a Transnational Reindeer Network

With the help of Dr. Malte of the National Herbarium, Finnie prepared a memo soliciting Canadian universities and governmental departments for trained botanists. This memo stated clear requirements for knowledge of systematic botany, with specific competence in the flora of the Canadian north. They wanted candidates who could work independently in a harsh, remote, and difficult terrain. They had to have common sense and a proven ability to apply knowledge to economic benefit. Yet queries to schools in Ontario, Alberta, Saskatchewan, and Quebec failed to turn up a single nomination. No government men applied for the job, either. But by a stroke of luck, Malte knew of two Danish brothers who fit the Department of Interior's bill.[42]

Malte had recently been contacted by Morten Porsild, the director of the Arctic's first biological station at Disko Island (Greenland). Morten's sons, Alf Erling and Robert Porsild, grew up in the shadow of the station, within a transient community that offered useful training in northern botany and arctic travel. The two men spent their youth building elaborate plant collections, competing with one another to win their father's approval. They met government officials and arctic scientists from around the world and cut their teeth on dog sledding while accompanying research parties. When Morten Porsild contacted Malte to inquire about employment possibilities for his sons, Malte was ecstatic to learn of trained botanists with arctic experience—even though the men knew little about reindeer.[43]

The decision to hire the Porsild brothers again made use of the United States Bureau of Biological Survey's director, E. W. Nelson. W. W. Cory first contacted Alf Erling and Robert Porsild, meeting the two brothers while they visited Chicago. Cory was impressed by Alf Erling's knowledge of Inuit culture and language and of northern vegetation. In April 1926, Finnie, Cory, and Alf Erling Porsild went to Washington to meet with Nelson. Here, Nelson facilitated what would become the brothers' indoctrination to reindeer: a half-year spent visiting the Alaskan operations and studying the work of Bureau of Biological Survey biologist Lawrence Palmer. Finnie wrote the brothers in May 1926, detailing the orders to be completed.[44]

The Porsilds' recruitment makes visible a network that bureaucrats used to manage the Canadian Reindeer Project in the 1920s and 1930s. We have briefly visited several nodes in this network: Disko Island; the Bureau of Biological Survey's headquarters in Washington, DC; and, in Ottawa, the Royal Commission's hearings and Finnie's reindeer team. As we will see, between 1926 and 1936 this network expanded to include Fairbanks, Alaska; Kautekeino, Norway, where Porsild hired Saami herders hired to train Inuit apprentices; the Norlite Building in Ottawa, where the Interdepartmental Reindeer Committee met to discuss the Project's progress; and the Mackenzie Delta, the eventual home for Canadian reindeer. This network comprised the intellectual, political, and physical space of the Project, and continued attention to it helps us to comprehend further the relationship between science and the state in the interwar north. At these sites, and via the knowledge produced therein, the Canadian government was able to design and implement the Project and direct the day-to-day operations of Saami herders, Inuit apprentices, and imported reindeer in the Arctic. This network was also responsible for realizing the long-imagined dream of northern reindeer herds in the persons of arctic vegetation specialists Robert and Alf Erling Porsild.

Their abilities to speak an Inuit language and travel in harsh northern conditions, combined with their studies at a pioneering institution for arctic science, met the expectations of both the Canadian Department of Interior and the US Bureau of Biological Survey.[45] While Finnie had been initially resistant to the value of a botanist, other northern promoters bristled against the shifting domains of credibility surrounding reindeer. The Porsild brothers' lack of practical experience with reindeer was not lost on Vilhjalmur Stefansson, who lobbied his peers to reconsider their hiring.[46] However, these deficiencies did not bother Finnie, Cory, and Nelson, who came to believe in the Porsilds' talents and skills, and were convinced that time spent in Alaska gaining hands-on experience with reindeer and grazing science would fill in any remaining gaps. While his brother Robert eventually left the reindeer business, Canadian bureaucrats and scientists soon identified Alf Erling Porsild as a leading authority on arctic vegetation and reindeer.[47]

A Regime for Reindeer: Lawrence Palmer, Lichens, and Legibility in Reindeer Country

In May 1926, the Porsilds headed for Fairbanks, where they began their studies with Lawrence Palmer. They carried a note from the director of the US Bureau of Biological Survey that served as instructions for the Alaskan ecologist. Palmer was to offer the Porsilds his "fund of information" on reindeer.[48] This fund had been generated by his quadrat studies on tundra re-growth and carrying capacity at the Fairbanks experimental station, and his collaboration with the US Bureau of Animal Industry on the nutritive quality of various types of forage.[49] E. W. Nelson also recommended that the Porsild brothers be introduced to the practical workings of the herds, trying their hands at corralling, capturing, marking, castrating, and branding.

Discerning the Porsilds' apprenticeship with Palmer is crucial to our understanding of the Canadian Reindeer Project and how it unfolded in the remainder of the twentieth century. This partnership guided the Porsilds in siting the Project and crafting its inner workings. Palmer emphasized the importance of a particular kind of knowledge in first selecting and subsequently managing a reindeer grazing area. The Porsild brothers, Nelson wrote, "should be taught as much as possible concerning the forage plants used by these animals, with a special view to the differences between the summer and winter forage and the need of safeguarding the winter forage areas from use in summer in order that the range may be perpetuated. ..."[50] Recognizing and protecting forage were foundational to managerial decisions in the Canadian Reindeer Project. As such, these twinned convictions were inscribed onto the physical landscape of the Mackenzie Delta and the social interactions of developers, Saami, Inuit, and governmental supervisors.

To understand how this could be so, we must first gather the details of what the Porsilds learned in Alaska, and thus become familiar with the work of Lawrence Palmer. Palmer had studied forestry and botany at the University of Nebraska between 1911 and 1915 before becoming a grazing assistant with the US Forest Service. Hired in 1919 as an assistant biologist at the Bureau of Biological Survey, Palmer considered himself a botanist, biologist, ecologist, and range manager—suggestive of the kinds of relations between plant studies, agricultural development, and state power in place at the time.[51] He applied his knowledge of grazing relationships in the American west to the study of reindeer. His first five years in Alaska were taken up with reconnaissance surveys of the herds along Alaska's meandering coastline. These surveys supplied Palmer with a sense of the reindeer industry in Alaska, and the seasonal movements of people and animals across the land. As with range management in the west, Palmer concluded that the bases of the industry were the major species of plants that provided nutrition for reindeer. He arrived at this conclusion after careful study of these plants in the field and at the experimental station in Fairbanks.

Nelson's instructions to impress upon the Porsilds the significance of winter forage likely did not surprise Palmer. After all, it was the Fairbanks biologist who had first articulated the significance of this component of the reindeer industry. Palmer developed an elaborate system of experimental pastures and quadrat studies in Fairbanks. These he explored in several lines of research, including the conditions governing forage and range management, the various relations of lichens to grazing, the relative carrying capacity of lichen and non-lichen ranges, and the methods of feeding and their effects.[52] Palmer parcelled out eleven pastures, each with slightly different vegetation based on its position on the slope of the hill on which the farm sat. He brought reindeer to graze within these pastures, learning about how the animals ate, what plants they selected in different seasons, how they dealt with snow, and how the plants responded in spring. He established quadrats within these different pastures and performed his own tests, cutting plants and picking them by hand. These experiments convinced Palmer that winter forage, comprised mostly of the genus *Cladonia*, was essential to a modern, successful reindeer industry.[53] Beyond supplying the local industry with valuable data, Palmer was certain that the study of lichens would also open an entire field of inquiry for the Alaskan and broader scientific communities.[54]

By 1926, Palmer had made a case for organizing the entire industry around *Cladonia*. He noticed that winter ranges were patchier than summer ranges, and so winter resources had to be protected—especially given the observation that reindeer bunched up in colder temperatures, potentially overgrazing their food source. A closer look at the nutritive quality of winter forage plants and their reaction in quadrat studies to mowing, picking, and feeding showed surprising results.[55] Even after only a few years of observation, Palmer noticed that it might take winter lichen ranges ten to fifteen years to "come back to a normal height growth of four to five inches," and thus "proper management of the winter range presents an exceptionally important problem."[56] Palmer had initially estimated that each reindeer required thirty acres of land per year. But this number was based on the supposition that tundra could recover from grazing within five to ten years. After allotting time for recovery, especially the winter range, he asserted that carrying capacity must be on the order of forty to sixty acres per head. Extrapolating to the available land in Alaska suitable for grazing, he estimated that the territory could support three million reindeer, three times as many as the fully stocked industry had in 1926.[57]

Palmer's conclusions about *Cladonia* and carrying capacity fit into the Bureau of Biological Survey's larger scheme of modernizing the reindeer industry. Palmer lamented that reindeer handling in Alaska suffered "from lack of application of improved modern methods."[58] What he meant was modern science, and more specifically, the concept of rotational grazing. This concept Palmer imported to Alaska through the US Department of Agriculture from sheep and cattle range science in the west. In theory, this approach made maximum use of available forage by moving herds between a series of summer and winter pastures, and prevented overgrazing by allowing some tracts of land to go fallow each year. In order to make this kind of grazing possible, Palmer noted, the industry's management and infrastructure would need careful overhaul and supervision. The territory must be divided into grazing units; fences should be erected to keep herds separate and prevent strays or mixing; corrals should be constructed to facilitate round-ups and slaughter; and permanent winter cabins needed to be built to ease herd management in winter, the most important phase for the protection of *Cladonia*. But most importantly, rotational grazing depended on open herding, where animals were free to select food on their own. This approach contrasted with the Saami tradition of close herding, where herders and animals stayed together as they moved over the land. Both Palmer and Nelson agreed that rotational grazing replaced the "crude methods of the original herders" and instilled in the industry "definite scientific investigations [and] oversight."[59]

As we consider the relationship of the Porsilds and Palmer—and the connections among scientists, the Canadian state, and the Canadian Reindeer Project—we must remember that the concepts of winter forage and

Section	Composition					Density	Palatability	Forage value
	Lichen	Browse	Sedge	Weeds	Moss			
Coast summer range:								
St. Lawrence Island	0	5	91	3	1	90	65	58.5
Kivalina	5	15	47	31	2	79	68	53.7
Kotzebue Sound	10	26	51	5	8	93	64	59.5
Seward Peninsula	7	15	53	24	1	68	60	40.8
Norton Sound	11	22	50	5	12	92	51	46.9
Yukon-Nunivak Island	9	15	57	15	4	90	60	54.0
Kuskokwim	6	40	34	17	3	70	67	46.9
Average	7	20	55	14	4	83	62	51.4
Interior summer range:								
Broad Pass	18	28	27	12	15	96	70	67.2
Gulkana-Tangle Lakes	16	34	29	10	11	88	68	59.8
Average	17	31	28	11	13	92	69	63.5
Coast winter range:								
St. Lawrence Island	65	12	2	11	10	40	80	32.0
Kotzebue Sound	50	25	15	10	0	60	70	42.0
Seward Peninsula	50	15	30	5	0	70	75	52.5
Norton Sound	50	10	30	4	6	87	67	58.3
Yukon-Nunivak Island	50	10	30	2	8	99	66	65.3
Kuskokwim	47	30	10	3	10	70	70	49.0
Average	52	17	20	6	6	71	71	50.0
Interior winter range:								
Broad Pass	50	20	8	4	18	85	76	64.6
Gulkana-Tangle Lakes	53	23	11	6	7	85	83	70.5
Average	52	22	10	5	13	85	80	67.5

1 Forage value derived by multiplying the percentage of density of forage stand by the percentage of palatability.

FIG. 3.1: Lawrence Palmer quantifies the potential of the arctic landscape based on the seasonal forage needs of reindeer and the types of vegetation along coastal and interior Alaska. Table by Lawrence Palmer, "Progress of Reindeer Grazing Investigations in Alaska," United States Department of Agriculture Bulletin 1423 (Washington, DC, 1927), 20.

carrying capacity hinged on the application of a scientific management regime. This regime made room for the expertise of scientists to guide the activities of Saami herders and Native apprentices. To visualize the linkages between scientific knowledge, state supervision, and the reindeer industry, consider the tables and maps Palmer presented to his readers in his 1926 US Department of Agriculture publication (Figs. 3.1–3.2). Through reference to the chemistry of various tundra plants and the spatial distribution of what he called "tundra types," Palmer argued for the merits of a rational, scientific manager to preside over people and nature in the north. Such a person could consider the particular nutritive value of *Cladonia* and the landscape mosaic of topography, vegetation, and climate, while directing the right number of herders and reindeer to the right places at

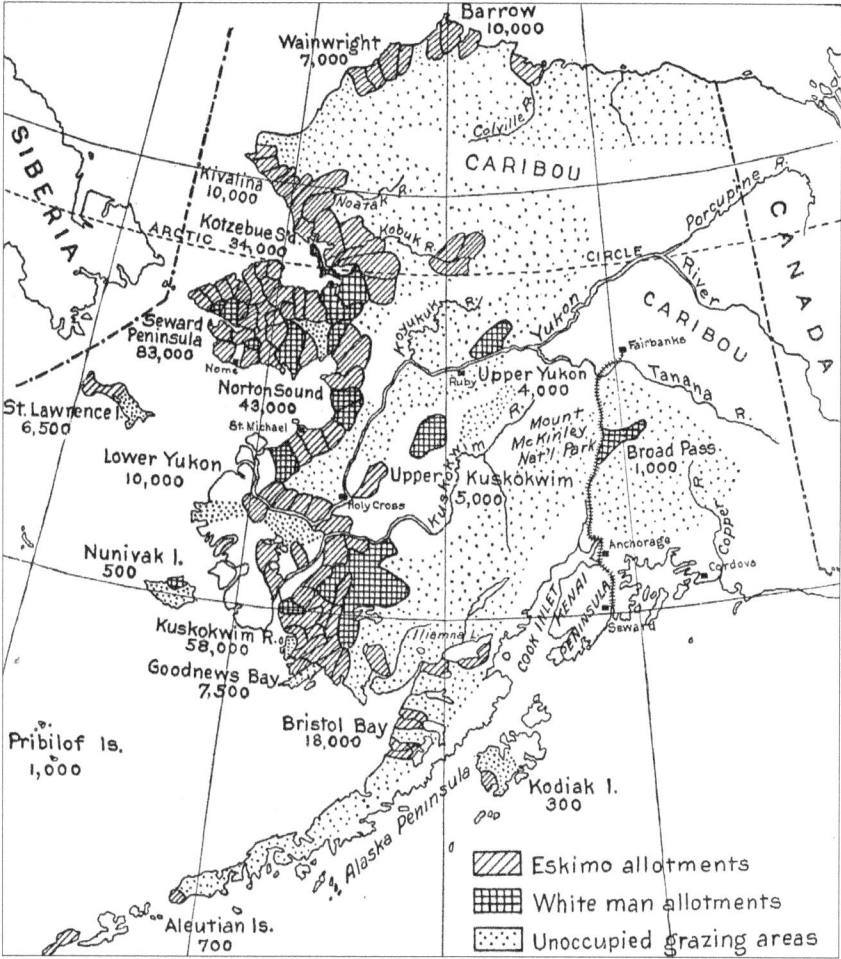

FIG. 3.2: Lawrence Palmer converts the arctic landscape into the terms of reindeer ecology. Map by Lawrence Palmer, "Progress of Reindeer Grazing Investigations in Alaska," United States Department of Agriculture Bulletin 1423 (Washington, DC, 1927), 2.

the right times. Winter forage, carrying capacity, and grazing units were thus mechanisms for what scholars have called *legibility*, the capacity of governments to represent the resources of particular territories so as to exploit them. Palmer's charts and maps provide telling examples of "the radical reorganization and simplification of flora to meet man's goals."[60]

The Porsilds' studies with Palmer brought science and the state a long way from the Royal Commission on Muskox and Reindeer. Enclosures and quadrat studies produced new knowledge about the tundra, which highlighted a set of problems, solutions, and problem solvers unique to a style of reindeer management founded on grazing ecology. Lichens and "reindeer mosses" were known to Canadian bureaucrats before the Porsilds' visit with Palmer, but *Cladonia*, "winter forage," "carrying capacity," and "tundra types" had not yet been quantified or made intelligible. Similarly, the creation of a scientific grazing manager reordered the positions of Saami, native Inuit, government teachers, and federal administrators relative to one another. Armed with charts, maps, specimens, and observations, the scientist-manager abstracted himself from the day-to-day operations of the industry, even as he governed them. Perhaps paradoxically, this scientific and managerial ethos meant that Saami could remain authorities on tacit knowledge about reindeer in ways that no longer threatened researchers or bureaucrats. The novelty of Palmer's ecology and its applications might be why some Alaskans considered him not a practical reindeer man, but a man with a briefcase, issuing figures pulled from thin air.[61] It may also account for renowned ecologist Frederic Clements' interest and support of Palmer's research, which he called "exceedingly important and helpful" in the development of ecological science.[62]

When the Porsild brothers were given orders to learn what Palmer had to teach them about reindeer, a passage was opened between the Canadian Reindeer Project and scientific ideas emerging from the Fairbanks station. The Porsilds visited extensively with Palmer, touring his experimental pastures and travelling with him around Alaska to observe herds. Palmer walked the brothers through the practices of marking, corralling, and butchering, and shared "all his reindeer files" with Alf Erling Porsild. The Bureau biologist also conveyed his views about the advantages of open herding, and, by association, the superiority of "modern" methods for handling reindeer over Native Alaskan and Saami ways of knowing the animal.[63]

The Porsilds left Nome, Alaska, in December 1926, completing a trek to the Mackenzie Delta to test a possible route for the delivery of the herd to Canada. Upon arriving in Aklavik, Alf Erling Porsild wrote O. S. Finnie to proclaim the reconnaissance mission with Palmer a success. Porsild developed his observations of the northern tundra and Inuit culture through the lens of Palmer's science. Noting the plant cover in the Mackenzie Delta flats, Porsild characterized them as one of many tundra "types," which "entirely conform[ed] with similar deltas of Buckland, Kubuk, or Noataq in Alaska." While on his way to the International Boundary, he observed the herds owned by Inupiat in the vicinity of Point Barrow. Porsild lamented that the "lack of white initiative and of adequate supervision" had resulted in poor management and even a notable decrease in the size of reindeer.[64] To tame the Arctic, one first needed to recognize its wild ways.

Home on the Range: Surveying the Canadian North, Building the Canadian Reindeer Project

Between 1926 and 1931, Alf Erling Porsild visited Alaska twice (once to study with Palmer and a second time in 1929 to select the animals to comprise the Canadian herd), scoured the Canadian north for a home for reindeer, and also visited Kautekeino, Norway, to hire three Saami families to teach Inuit how to herd.[65] Ultimately, Porsild recommended that only two districts, the Mackenzie Delta and the Dease River valley, were suitable for a governmental reindeer herd.[66] Over the next few years, Porsild helped build Reindeer Station, oriented the Saami families to the place, and waited for the herd to arrive.[67] When viewed together, these activities and the reports Porsild wrote about them reveal how Palmer's regime of reindeer configured Porsild's observations and conclusions, and the final construction of the Canadian Reindeer Project.

Between April and August 1927, A. E. Porsild and his brother completed a survey of the "Husky Lakes" region between present-day Inuvik and Tuktoyaktuk. Porsild was hopeful this landscape could house the Canadian Reindeer Project. "Magnificent lichen cover over vast areas," he scribbled in his journal. "50 to 80% [of which are] pure lichen." With excitement, he pictured the region with Palmer's lichen ecology in mind: "*Cladonia rangiferina* and *Cl. silv.* and *Cl. Uncinalis, Cl. alpestris, Cetr. aiv.* and many others. *Cladonia rangiferina* and *Cl. silv* probably covers more ground than all the rest together. ... This lake would be ideal location for winter reindeer camp."[68] Based on his observations of forage in the area, Porsild estimated the country could support up to 250,000 reindeer.[69]

The carrying capacity of the Husky Lakes region did not dwarf the other areas that Finnie asked the Porsilds to study, which included the shores of Great Bear Lake and the Keewatin District. In 1928, the brothers inspected the valley of the Dease River, which extends northeast of Great Bear Lake, and the "northern plains" running south and west of the lake. Alf Erling Porsild described the region as a "natural grazing unit," as it was "closed in from all sides" and afforded abundant vegetation. But he increased the number of acres there to be allotted per reindeer. These grazing units presented a "tundra type" different from the Mackenzie Delta, and Porsild found it difficult to estimate grazing potential in this "unmapped country." Still, the botanist suggested that the twenty-five million acres of the Great Bear Lake basin could support a total of three hundred thousand reindeer.[70]

The Great Bear Lake basin presented other ecological problems particular to reindeer. The southern shores of the lake were "too heavily timbered to make herding and control of tame reindeer practicable."[71] But more importantly, both the Dease Valley and the northern plains grazing units offered little protection from mosquitoes, the ubiquitous, though temporary, pest of reindeer and reindeer industries. A. E. Porsild's diaries are peppered with comments about how annoying the mosquitoes could be, as well as how troublesome they were to effective reindeer management. When visiting Palmer in 1929, Porsild learned that nearly fifty head of Palmer's stock at Fairbanks had been killed by mosquitoes in the previous year. In 1936, Porsild spelled out the consequences of mosquitoes for the potential expansion of reindeer industries. "Nowhere in the area under consideration are the hills high enough to permit reindeer to escape flies during the summer," he concluded. For this reason, Porsild surmised, reindeer ranching would be "limited to the sea-coast and adjacent hinterland."[72]

Alf Erling Porsild's thoughts on *Cladonia*, carrying capacity, and mosquitoes make clear how an enriched awareness of ecology was at play in siting the Canadian Reindeer Project. This ecological knowledge found its clearest expression in descriptions of vegetation in the Mackenzie Delta, the eventual home for Canadian reindeer. "Reindeer ranching under a system such as has been evolved in Alaska," he argued, "requires summer and winter pastures."[73] With rotational grazing in mind, Porsild admired the patchwork of tundra plants evident in the Mackenzie Delta and the arctic coast. He employed Palmer's models of tundra types and the Alaskan scientist's ecological counting methods to determine the exact proportion, distribution, and nutritive values of sedges, grasses, and lichens.[74] In this grazing unit, Porsild pointed out that the highest parts of the interior were covered by a "hard and fairly dry type of tundra," while low-lying areas were comprised of brackish lakes and lagoons. "Although not so rich in succulent grasses and herbs as the Alaska tundra," Porsild wrote, "this type of pasture is nevertheless more valuable as summer pasture for reindeer, as it is not so susceptible to damage by the trampling of grazing herds." Ranking the "Husky Lakes" region as the best winter grazing land in Canada, Porsild commented on its "high percentage of palatable species," and the possibility for its "maximum development." In one turn of phrase, Porsild even pictured reindeer in this winter pasture "put[ting] on their back fat," directly linking the growth of plants with the growth of a northern reindeer industry.[75]

Porsild's observations did not seem to favour the obstacles or opportunities of reindeer in Canada, but rather the logical consequences of ecological data for the institution of animal husbandry. His attention to mosquitoes and forage makes this clear, as does his impression of the Hudson Bay coast. In 1929–30, Porsild teamed with the Royal Canadian Air Force to perform aerial surveys of vegetation on the shores of Hudson Bay. While exhilarated by the plane's ability to ease the rigours of fieldwork, Porsild admitted that the Keewatin district was poor country for reindeer. He later wrote that the flights proved this area was "entirely unsuited to reindeer"; viewed from a plane, "the almost total absence of soil and closed plant cover is most striking."[76] Ironically, then, ecology made reindeer possible in the Mackenzie Delta even as it circumscribed its possibilities within Canada. As historian P. Wendy Dathan has noted, just as the Royal Commission's grand plans for reindeer came into being in the Delta through scientific research on lichens, the application of this knowledge to other potential reindeer landscapes confirmed that Canada's north would never boast a vast industry.[77]

Porsild did not only apply his new knowledge to natural conditions, he also used it to affirm ideas about the social organization of the Project. In his journals and his reports to the Department of Interior and the Royal Geographical Society, Porsild found evidence to support a hierarchical regime of supervision over grazing units, Saami instructors, Inuit herders, and the tundra itself. While crossing from Barrow, Alaska, to Aklavik in the winter of 1926–27, Porsild met Tarpoq, an Inupiat man and owner of a reindeer herd. Porsild found that Tarpoq was "a good reindeer man under the supervision of a white man," but when he had been left unsupervised by the US Bureau of Education, he started to "neglect his herd when his increase and profits is [sic] not up to his expectations."[78]

Financial concerns and the need for governmental oversight animated Porsild's engagement with local Inuit in the Mackenzie Delta region. While at Atkinson Point, Porsild noted that the Inuvialuit had "too much easy cash" and had not yet learned the value of caring for their possessions. Moreover, he found that in the region between the international boundary and the Mackenzie Delta, Inuit had given up their customary seal hunt in favour of trapping fox, as the latter activity afforded them enough money to buy dog food (rather than hunt seal for it) and purchase other goods, like flour, tobacco, rifles, and ammunition. The botanist also worried about the future of Inuit in a fur economy, which was more volatile than one based on herding.[79] Later, in 1929, Finnie echoed Porsild's sentiments, suggesting that the "natives … might be seriously affected by the periodic fluctuations in the numbers of fur-bearers and by changes in the fur markets."[80] For Porsild and Finnie, reindeer helped subdue these wild elements of the north.

Scientists and state officials drew clear boundaries between white society and Inuit culture even as reindeer projects meant to erase them. But they sometimes got their lines crossed on the roles of scientists and Saami herders. While the reindeer were being driven from Alaska to the Mackenzie Delta, they faced incredible delays: a trip that was estimated to take eighteen months was completed in just under five years. These delays inflamed relationships between the Saami, the Canadian government, and Porsild. Anxious for the herd to arrive in Canada, bureaucrats in Ottawa suggested that Porsild relieve the current supervisor of the drive, Saami Andy Bahr, and guide the animals to their destination. Porsild was infuriated, both because he had advocated for "white men" to lead the drive originally and because he interpreted this order as a demotion from his position as scientist. On the verge of losing both Porsild and the reindeer, a representative of the Lomen Corporation of Alaska stepped in, paying the botanist a handsome $2,500 to manage the delivery of the herd.[81]

Porsild's conclusions, themselves a result of Palmer's teachings, directly informed the creation of a series of instruments to further guide the Canadian Reindeer Project. In 1931, he travelled to Kautekeino, Norway, as the Project's ambassador, identifying and selecting three Saami families to relocate to the Mackenzie Delta to train Inuit in reindeer herding. In 1931 and 1932, Porsild chose the site for and helped build Reindeer Station, the government's first town in the Arctic.[82] In 1933, following Porsild's recommendations, Parliament established federal ordinances to protect the reindeer as a national resource and created a six-thousand-acre Reindeer Grazing Preserve to contain and control northern pastures, as well as who trapped or hunted in them. In that year, the Inter-Departmental Reindeer Committee, having formed to consult Department of Mines and Resources staff on best practices for the reindeer industry, nominated Porsild to become the Canadian Reindeer Project's first superintendent.[83] In October 1935, after supervising the herd's arrival and the first six months of operations at Reindeer Station, Porsild left the north for Ottawa. After spending ten summers and seven winters in the Arctic, "getting Canada's first Government-owned reindeer off to a good start," he took up new roles as chief botanist at the National Herbarium and as a consultant for the Interdepartmental Reindeer Committee.[84] His continued investment in the Canadian Reindeer Project throughout the early 1930s is remarkable, especially given the retrenchment of the civil service in Canada and the reorganization of northern bureaucracies following the Great Depression.[85]

By 1940, the reindeer inhabited a landscape that looked quite different from that which Porsild had surveyed in the late 1920s. The image of thousands of reindeer the botanist had projected onto the landscape had been replaced by regular, seasonal movements of people and animals. In the spring, Saami herder Mikkel Pulk, together with Inuit apprentices, pushed the main herd from its winter range to the coastal area, where fawning commenced in early April and lasted until June. In the summer, the reindeer were driven to Richards Island, where consistent winds dispersed mosquitoes. Before the annual roundup, which took place at the summer corral near Kittigaruit, herders caught fish and harvested whales, and prepared this meat for the long winter.[86] Reindeer supervisors, hired through the Department of Mines and Resources, directed the schedule of the main herd and supervised the nascent, Native-owned herds.[87] In addition, they kept supplies and equipment on hand, maintained communication via radio with government agents in Aklavik, and arranged for the training in reindeer husbandry of as many young boys as possible.[88] Finally, supervisors issued regular reports to Ottawa to be reviewed and evaluated by the Interdepartmental Reindeer Committee. That these movements of supervisors, herders, apprentices, and reindeer had become routine by 1940 belies the dramatic transformations in the scientific understanding of arctic nature and government capacity that had taken place in the previous two decades.

Conclusion: The Experimental State of Nature

Despite these foundations, the Canadian Reindeer Project fell apart in a matter of twenty years. While six teams of Inuit became owners of herds after 1938, all of these operations had collapsed by 1959. In that year, the Canadian government handed the project to private developers, having little to show for its million-dollar investment. It continued to be passed back and forth between private and public hands throughout the 1960s and 1970s. Today, the small extant herd in the Mackenzie Delta is owned in part by a private individual and in part by the Inuvialuit Regional Corporation.[89]

As reindeer herding in the western Arctic fell apart, activities in the late 1940s and early 1950s reinforced the project's status as a product of governments testing out science and development in a so-called wild north. Inspired by Cold War geopolitics—which identified the western Arctic as a vulnerable border zone with the Soviet Union—Ottawa went north with renewed vigour in the 1940s and 1950s. Bureaucrats placed high priority on defense and modernization initiatives, which relegated the Canadian Reindeer Project to an antiquated status. Kittigaruit, where the reindeer were corralled in the summer, became home to a radar station in the mid-1940s. By the mid-1950s, the animals grazed the Yukon North Slope's coastal vegetation in the shadows of Distant Early Warning Line stations.[90] As Ottawa planned the relocation of Aklavik and Reindeer Station residents to the new "East-Three" site (Inuvik), the feasibility of maintaining the reindeer program was brought into question.[91] By 1958, much of Reindeer Station's labour pool had been channelled toward construction projects related to "East Three" and defense.[92] The government transferred the herd to private developers in that year, maintaining a staff person at Reindeer Station for oversight.[93]

The demise of government-sponsored reindeer herding coincided with the rise of high modernism in the north. High-modernist ideals and priorities helped reinforce reindeer herding as an outmoded form of development. In the 1950s, Canadian officials, Interdepartmental Reindeer Committee members, and university researchers began wondering why this project—seemingly destined to succeed—never lived up to its potential. Drawing on ecological and sociological analyses of the Alaskan industry, they turned the Canadian Reindeer Project into a case study of historic attempts to develop the north. Analysts hung the Project's troubles on Inuit culture, immature science, and poor planning.[94] But beyond the particulars was a broad conclusion about the past. Both government and science had moved on from interwar ways into a new era of commanding a strategic yet vulnerable environment.

The Canadian Reindeer Project makes clear how science and state priorities for the Arctic call into being new relationships between humans and nature. By the end of the First World War, Ottawa held little institutional knowledge for understanding the north as reindeer country, despite having introduced reindeer to several parts of Canada over the turn of the century. This kind of information—maps, statistics, and archives of research papers—had been instrumental in contemporary cases of mining and commercial fishing elsewhere in the north. There, bureaucrats turned to scientists at the Geological Survey of Canada and the Department of Fisheries to assist private industry in manipulating human and natural resources.[95] Legibility in the Subarctic was a product of well-established ideas, communities, and stereotypes being transported to terrestrial and aquatic environments. The reindeer, it turned out, was a whole other animal.

There was no university system in place to provide a pool of students trained in animal husbandry science, northern botany, or arctic ecology. There were no archives, maps, or statistics upon which bureaucrats could rely to plot their reindeer schemes. The Royal Commission on Muskox and Reindeer was one attempt to create this database and to define the characters and characteristics of reindeer expertise; but this led to false starts, including Vilhjalmur Stefansson's spectacular failure on Baffin Island.

In response, the government turned to Alaska, which by 1926 had begun to amass a wealth of research through biologist Lawrence Palmer, and to Denmark, which had already conducted arctic science and empire work in Greenland. The government also imagined Inuit as central to Canada's success in the north, not just as a population that could be supplanted by fishermen or prospectors.

Just as state interests forged new commitments to science, international partners, and local residents, so too did the production of scientific knowledge about arctic life alter how state agents understood the north and their capacity there. Visions of vast herds had preceded Palmer and the Porsilds, but the biologists created the instruments and concepts by which the dream of a domesticated Arctic could become a reality. Through notions of lichens, tundra types, and carrying capacity, these scientists helped bureaucrats to quantify northern terrain, to see it not as barren or backward, or even as an unending prairie, but as a set of districts with varying potential for people and reindeer. At the same time, ecologists and bureaucrats played on widely held fears of wildness—in markets, Native cultures, and game populations—to underscore the stability that animal husbandry economies would bring to people and nature in the Arctic. The arrival of reindeer in Canada, then, can help scholars think carefully about the nature of power and the power of nature. Interventions with domesticated species required nuanced knowledge of the arctic environment, while ecological science showed both the opportunities for government action and the limits northern nature imposed upon southern ambition.

The Canadian Reindeer Project, and especially Alf Erling Porsild's involvement in it, suggests the value of approaching northern history not only from the perspective of environmental history, but of the history of science, as well.[96] As a sparsely populated region distant from North American metropolitan centres, the Arctic did not enter the orbits of public consciousness and national identity via traditional pathways. Southerners have not consumed the far north through personal encounters with physical landscapes or goods that originate from it, but rather by subscribing to ideas produced about the place. As Emilie Cameron documents in her chapter in this volume, modern ecologists recapitulate colonial power dynamics through their research on arctic climate change.[97] Yet the place of science in the north is not guaranteed—it is created, contested, and sustained in time. The Reindeer Project was Canada's first experiment with scientific resource development in the Arctic, one founded not on mining and its disciplines, but on other forms of nature and knowledge. And it was not the last experiment of its kind. As other chapters in this volume make clear, ecological field research, state power, and manipulations of northern nature intensified and further intertwined following Second World War. These post-war episodes, then, amplify the Reindeer Project's importance and the legacy of the interwar period in Canadian history.

This point can be crystallized by returning to the ways historical actors referred to the Canadian Reindeer Project. In 1936, a crowd of military officials, academicians, interested citizens, and Department of Interior bureaucrats gathered at the Royal Geographical Society to hear Alf Erling Porsild speak on "The Reindeer Industry and the Canadian Eskimo." In the discussion that followed the presentation, Albert Charles Seward, a professor of Botany at Cambridge University, offered his support for Porsild's work and the Canadian government's initiative: "I feel that you will agree with me when I congratulate Mr. Porsild on having most successfully carried out this great experiment," Seward announced. "It was an experiment which I think there is no doubt will yield very valuable results, not only as regards value to the Dominion of Canada but particularly in improving conditions under which the Canadian Eskimo are living in those far northern regions."[98]

At the Royal Geographical Society event, Professor Seward chose his words carefully. The Canadian Reindeer Project was a "great experiment"—a study of how the Canadian government could administer the Arctic and its resources more effectively. It was *experimental*, relying upon an unorthodox technology of a foreign domesticated species, creating innovative alliances among ecological botanists, private corporations, and new governing bodies, and building new spaces for development in research stations, grazing preserves, and herding villages. It was also *experimentalist*, as the project employed professional ecological scientists whose quadrat studies and surveys yielded crucial data to guide the implementation of a government-run animal husbandry economy and civilizing program. Without exaggerating or downplaying it, the Canadian Reindeer Project was an attempt to remake the north and Canada's relationship to it. Reindeer Station and the reindeer country that surrounded it became a natural laboratory for state power and scientific knowledge, the best expression of an experimental state of nature.

Notes

1 Library and Archives Canada (hereafter cited as LAC), microfilm reel M-1958, A. E. Porsild, "Trip to Alaska to Select Reindeer to be Purchase for Delivery to the Mackenzie Delta, NWT, Autumn and Winter, 1929–1930. Daily Journal of A. E. Porsild," p. 39.

2 A note about terminology: in this essay, I refer to Native northerners living in the western Arctic. The region of the western Arctic includes portions of Alaska, Yukon, and the Northwest Territories lying above the Arctic Circle, roughly corresponding to the area between Port Hope, Alaska, and Kugluktuk, Nunavut, on today's map. "Native northerners" is a broad category, which I specify by referring to Inuit (as opposed to Dene). Within the category of Inuit are Inupiat, (Alaskan Inuit) and Inuvialuit (Canadian Inuit living in the Mackenzie Delta and on the Yukon coast). I also use the term Alaskan Natives to refer to Inupiat, Aleut, and Native Americans in Alaska. Historical actors often referred to Inuit as Eskimos. Such a reference continues in modern-day Alaska, but has been replaced with other terms in Canada.

3 A short history of Porsild's involvement with the Canadian Reindeer Project is given in Alf Erling Porsild, *Reindeer Grazing in Northwest Canada: Report of an Investigation of Pastoral Possibilities in the Area from the Alaska-Yukon Boundary to Coppermine River* (Ottawa: F. A. Acland, 1929), 6–14, 29. The language of "experiment" was explicitly used in reference to a governmental herd as early as January 1920. See LAC, RG 33 105,

"Royal Commission: Reindeer and Muskox Industry, Vol. 1: Hearings of Royal Commission on Muskox and Reindeer." The "Canadian Reindeer Project" was also likely to be called "Canada's Reindeer Experiment," as it was in 1936 in the publication R. H. G. Bonnycastle, "Canada's Reindeer Experiment," *Proceedings of the North American Wildlife Conference*, 3–7 February 1936, Senate Committee Print, 74th Congress, 2nd Session (Washington, DC, 1936), 424–27. Numerous newspaper articles also referred to the project as an experiment. For one example, see "Canadian Reindeer are doing very well," *Globe and Mail*, 1 October 1939.

4 For a very useful and comprehensive treatment of Porsild's research and correspondence during his "reindeer years," see Patricia Wendy Dathan, "The Reindeer Years: Contribution of A. Erling Porsild to the Continental Northwest" (master's thesis, McGill University, 1988). Dathan's thesis is not explicitly concerned with the relationships between science and the state, but rather Porsild's contributions to an as-yet inchoate discipline of circumpolar Arctic botany between the 1920s and 1940s. As I was finishing this chapter, Dathan published this monograph from her research: Wendy Dathan, *The Reindeer Botanist: Alf Erling Porsild, 1901–1977* (Calgary: University of Calgary Press, 2012), an extensive treatment of Porsild from a biographical perspective.

5 Trevor H. Levere, *Science and the Canadian Arctic: A Century of Exploration, 1818–1918* (New York: Cambridge University Press, 1993).

See also Stuart E. Jenness, *Stefansson, Dr. Anderson, and the Canadian Arctic Expedition, 1913-1918* (Gatineau, QC: Canadian Museum of Civilization, 2011).

6   For an example of historians viewing the Canadian government as neglectful of the north during the interwar period, see Kenneth Coates, P. Whitney Lackenbauer, William R. Morrison, and Greg Poelzer, *Arctic Front: Defending Canada in the Far North* (Toronto: Thomas Allen Publishers, 2008), 54–55. For recent interpretations of the interwar north by environmental historians, see Janice Cavell and Jeff Noakes, *Acts of Occupation: Canada and Arctic Sovereignty, 1918-1925* (Vancouver: UBC Press, 2010); Liza Piper, *The Industrial Transformation of Subarctic Canada* (Vancouver: UBC Press, 2009); John Sandlos, *Hunters on the Margin: Native People and Wildlife Conservation in the Northwest Territories* (Vancouver: UBC Press, 2007); and Morris Zaslow, *The Northward Expansion of Canada, 1914-1967* (Toronto: McClelland and Stewart, 1988), especially 141–45. See also the chapters by Tina Adcock and Marionne Cronin in this volume.

7   See the chapters by Tina Adcock and Jonathan Peyton in this volume.

8   Bonnycastle, "Canada's Reindeer Experiment," 424.

9   In the western Arctic, hunting pressure on the caribou intensified through the expansion of commercial whaling industries on the Beaufort Sea. Two good sources on this phenomenon have been provided by John Bockstoce. See John R. Bockstoce, *Whales, Ice,*

and Men: The History of Whaling in the Western Arctic (Seattle: University of Washington Press, 1986); and John R. Bockstoce and Daniel B. Botkin, "The Historical Status and Reduction of the Western Arctic Bowhead Whale (*Balaena mysticetus*) Population by the Pelagic Whaling Industry, 1848-1914," *Scientific Reports of the International Whaling Commission*, Special Issue, no. 5, 107–41.

10   There are many historical records that suggest the various cultural and economic benefits of reindeer in the interwar period. For one example, consider C. L. Andrews, *The Eskimo and his Reindeer in Alaska* (Caxton Printers, 1939), 30.

11   Gilles Seguin, "Reindeer for the Inuit: The Canadian Reindeer Project, 1929-1960," *Muskox* 6, no. 38 (1991): 1–10. John Sandlos, "Where the Reindeer and Inuit Should Play: Animal Husbandry and Ecological Imperialism in Canada's North" (unpublished manuscript). My thanks to the author for allowing me to review this piece.

12   C. L. Andrews, *The Eskimo and his Reindeer in Alaska* (Caldwell, ID: Caxton Printers, 1939), 30–37. Sandlos, "Where the Reindeer and Inuit Should Play," 4–7.

13   I return to these themes with a more specific treatment below. See notes 28–33.

14   For an overview of the "Arctic Pastoral" concept, see Sandlos, *Hunters at the Margins,* 161–70.

15   Randal Pokiak, *Inuvialuit History* (Inuvik: Inuvialuit Cultural Resource Centre, n.d.), 58 (accessed 26 January 2011). I would like to thank the Inuvialuit Cultural Resource Centre for granting me access to its materials, as well as for

offering gracious support and patience with my questions of them.

16 This amendment appears to have been an attempt to legitimize federal appropriations for forms of relief that were distributed to Inuit through missions and fur posts. See John Leonard Taylor, *Canadian Indian Policy During the Inter-War Years, 1918–1939* (Ottawa: Department of Indian Affairs and Northern Development, 1984), 87–88.

17 See Sandlos, *Hunters at the Margins*, 126.

18 Pokiak, 44–45.

19 This quotation is Pokiak's, from *Inuvialuit History*, 58. Elisa Hart records other accounts of Mangilaluk in *Reindeer Days Remembered* (Inuvik: Inuvialuit Cultural Resource Centre, 2001), 14.

20 For a comparison of national parks and the ordinances used to establish the Reindeer Grazing Preserve, consider the minutes of the Interdepartmental Reindeer Committee of 18 June 1935. LAC, microfilm reel T-1332, vol. 82, file 7128, "Lapp herders, 1929–1938." For a discussion of the relationships between ideas of nature in relation to laboratories and other scientific spaces, see Robert Kohler, *Landscapes and Labscapes: Exploring the Lab-Field Border in Biology* (University of Chicago Press, 2002).

21 See LAC, RG 33 105, "Royal Commission: Reindeer and Muskox Industry, Vol. 1." The quotation in the above paragraph is from George Comer (p. 32). Notes from Jenness's lecture are found in LAC, Rudolph Martin Anderson fonds, MG 30 40, vol. 14, file 1–eskimos, "A Lecture delivered at the Arts and Letters Club, by Diamond Jenness,

Victoria Memorial Museum, Jan 9, 1923: 'Our Eskimo Problem.'"

22 Porsild, *The Reindeer Industry and the Canadian Eskimo*, 4.

23 This was the subtitle to the 1922 report of the Department of the Interior, *Reindeer and Muskox: Report of the Royal Commission upon the possibilities of the Reindeer and Musk-Ox Industries in the Arctic and Sub-Arctic Regions* (Ottawa, 1922). See the first page of the report for the full subtitle.

24 *Reindeer and Muskox*, 18. Porsild documented the existing knowledge on flora in the western Arctic in *Canada's Western Northland: Its History, Resources, Population, and Administration* (Ottawa: Department of Mines and Resources, Lands, Parks, and Forests Branch, 1937), 130–41.

25 The list of witnesses can be found in Department of Interior, *Reindeer and Muskox*, 9–11.

26 *Reindeer and Muskox*, 12.

27 *Reindeer and Muskox*, 18.

28 The published report lists these witnesses, but only prints excerpts of their testimony. The records of the hearings are found in LAC, RG 33 105, under the titles "Royal Commission: Reindeer and Muskox Industry, Vol. 1" and "Royal Commission: Reindeer and Muskox Industry, Vol. 2." The quotation is from Bishop Isaac Stringer at the 4 February 1920 meeting (pp. 197–98 of the Vol. 1 hearings).

29 *Reindeer and Muskox*, 19–24.

30 Seguin and Sandlos both discuss the Baffin Island situation. See Seguin, "Reindeer for the Inuit," 1–10, and Sandlos, "Where the Reindeer and Inuit Should Play," 9.

31 The details of Lopp's report were found in several sources. For Lopp's thoughts as he pieced together his report, see Alaska and Polar Regions Department Archives, Elmer E. Rasmuson Library, University of Alaska Fairbanks (hereafter cited as APR-UAF), Kathleen Lopp Smith Family Papers, ser. 4, box 2, folder 4, Lopp to R. H. Parsons, 7 September 1925. For his final thoughts, see APR-UAF, Kathleen Lopp Smith Family Papers, ser. 4, box 2, folder 5, Lopp to R. H. Parsons, 1 July 1926.

32 For the interaction between Sapir and Stefansson, see LAC, Rudolph Martin Anderson Fonds, MG 30 40, vol. 11, file 8, "The Friendly Arctic." This file contains notes on the lecture as recorded by Anderson's wife, not a complete transcript. For correspondence between Stefansson and Carl Lomen, see Letter from Stefansson to Carl Lomen, June 20, 1927 Lomen Family Papers, ser. 2: Carl Lomen, Box 10, Folder 232, APR-UAF.; Stefansson thanks Lomen for submitting a rebuttal to Lopp's report in the *Ottawa Morning Journal* in Letter from Stefansson to Carl Lomen, Apr 17, 1928, Lomen Family Papers, ser. 2: Carl Lomen, Box 10, Folder 232, APR-UAF. The details of Stefansson's bid for the Lomens to purchase land on Baffin Island are covered in Letter from Stefansson to Leonard Baldwin, July 20, 1925, Lomen Family Papers, ser. 2: Carl Lomen, Box 10, Folder 232, APR-UAF.

33 On Stefansson's controversial nature, see Richard Diubaldo, *Stefansson and the Canadian Arctic* (Montreal: McGill-Queen's University Press, 1978).

34 I deduce that northern experience and scientific training, when combined, were especially useful from the following quotation in the Royal Commission's report: "Your commissioners were fortunate in being able to secure much valuable information as to the potentialities, from a reindeer point of view, of the territory lying within about thirty miles of the Arctic coast, between the International Boundary on the west, and Kent peninsula on the east." *Reindeer and Muskox*, 26. This area was the exact area covered by the recent Canadian Arctic Expedition of 1913–18, and the International Boundary Survey of 1908–12.

35 O. S. Finnie to Gibson, January 1926, as quoted in Dathan, 32.

36 Much of the Arctic region at the time offered little inland transportation infrastructure for would-be surveyors. In order for a fieldworker to travel and "size up the situation" efficiently, as Finnie wished, he or she would have to master dog-sledding techniques, as there was no feasible alternative for covering ground after the open-water season.

37 For quotation at end of paragraph, see LAC, W. T. Lopp, 1925–1939, microfilm reel T-13267, vol. 759, file 4824, O. S. Finnie to Rev. Canon C. W. Vernon, 10 June 1929. See also LAC, RG 132, vol. 23, file 364, O. S. Finnie to W. H. Collins, 14 May 1928; LAC, RG 132, vol. 23, file 364, O. S. Finnie to M. O. Malter, 20 February 1928. See also Sandlos, "Where the Reindeer and Inuit Should Play," 10.

38 The details of this meeting between U.S. and Canadian officials are drawn from several sources. See

LAC, RG 132, vol. 31, file 4492, Letter to A. E. Porsild, 23 March 1926. See also LAC, microfilm reel T13273, vol. 765, file 5095—Porsild, 1926–36, O. S. Finnie to Colonel Starnes, 19 May 1926. For more general descriptions of the hiring of the Porsilds, see North, *Exodus*, 27–34, and Dathan, 33–34.

39   See Sandlos, *Hunters at the Margins*; and Cavell and Noakes, *Acts of Occupation*.

40   In addition to these broader social and environmental patterns, the interactions of Finnie, Cory, and the U.S. reindeer men echo the complexities of the emergence of grazing sciences in America and Canada over the turn of the twentieth century. See John Sandlos, "Where the Scientists Roam: Ecology, Management, and Bison in Northern Canada," *Journal of Canadian Studies* 37, no. 2 (2002): 93–129; Morgan Sherwood, *Big Game in Alaska* (New Haven: Yale University Press, 1981); and Christian C. Young, "Defining the Range: The Development of Carrying Capacity in Management Practice," *Journal of the History of Biology* 31 (1998): 61–83. On Palmer, see Sherwood, 66–67. Of the wildlife specialists with the Bureau of Biological Survey in Alaska, there were Ernest P. Walker, W. H. Osgood, Alfred M. Bailey, Seymour Hadwen, and Lawrence J. Palmer. Only Walker was a "permanent employee."

41   See LAC, microfilm reel T-13267, vol. 759, file 4824—W. T. Lopp, 1925–1939, O. S. Finnie to Moran, 4 December 1925. "Exploratory Engineer" seems to be an established class or rank within the governmental payroll system. On

hiring a "Botanist" and "Assistant Botanist," see LAC, microfilm reel T13273, vol. 765, File 5095—Porsild, 1926–36, vol. 765, O. S. Finnie to Moran, 20 April 1926.

42   LAC, microfilm reel T13273, vol. 765, File 5095—Porsild, 1926–36, vol 765, Memo from R. A. Gibson, 19 April 1926.

43   See Dathan, 2–10, for a rich description of the Porsilds' upbringing.

44   See LAC, microfilm reel T13273, vol. 765, file 5095—Porsild, 1926–36, O. S. Finnie to Messrs. A. E. and R. T. Porsild, 19 May 1926. See also Dathan, 48, and North, 28–35.

45   See Porsild, *Reindeer Grazing in Northwest Canada*, 5–6.

46   APR-UAF, Lomen Family Papers, ser. 2, box 10, folder 232, Vilhjalmur Stefansson to Carl J. Lomen, 20 June 1927. Stefansson wrote, "I think it would also be well to point out that the Porsild brothers coming from Greenland are from a country where there is no reindeer industry and where the wild reindeer (caribou) are practically extinct." This is ironic, as Stefansson, who was born in Canada, was also from a country with no reindeer industry.

47   Finnie, *Canada Moves North*, 16. See also Percy Cox, Professor Seward, and Colonel Vanier, "The Reindeer Industry and the Canadian Eskimo: Discussion," *Geographical Journal* 88, no. 1 (1936): 17–19. This paper was a published version of a discussion with A. E. Porsild after he presented his paper, "A Four Year Trail from Alaska to the Mackenzie Delta," to the Royal Geographical Society.

48   Quoted in Dathan, 49.

49   N. R. Ellis, L. J. Palmer, and G. L. Barnum, "The Vitamin Content of Lichens," *Journal of Nutrition* 6, no. 5 (1933): 443–54. This paper references another: L. J. Palmer and G. H. Kennedy, "A Report on Digestibility in Reindeer. The Digestibility and Nutritive Properties of a Mixed Ration of Lichens and Oats, of Lichens Alone and of Alfalfa Hay Alone." The latter was, at the time, in the process of publication. Below, I cite sources that predate the Porsilds' arrival to demonstrate that Palmer was working on these studies in the early 1920s, and not just in the early 1930s.

50   Quoted in Dathan, 49.

51   APR-UAF, L. J. Palmer Collection (hereafter cited as LJP), box 2, folder 1, "Record of Members: Society of American Foresters."

52   L. J. Palmer and Seymour Hadwen, "Reindeer In Alaska," in *United States Department of Agriculture Bulletin no. 1089, Sept. 22, 1922* (Washington, DC: Government Printing Office, 1926), 19–23.

53   L. J. Palmer, "Progress of Reindeer Grazing Investigations in Alaska," in *United States Department of Agriculture Bulletin no. 1423, Oct., 1926* (Washington, DC: Government Printing Office, 1926), 1–11.

54   APR-UAF, LJP, L. J. Palmer to E. W. Nelson, 22 March 1923.

55   L. J. Palmer, "Progress of Reindeer Grazing Investigations in Alaska," 30. Palmer determined the nutritive quality of lichens by sending samples to the Bureau of Chemistry and the Bureau of Plant and Industry in Washington, DC. See National Archives Pacific Alaska Region, Anchorage, Alaska

(hereafter cited as NARA), RG 75, Bureau of Indian Affairs, Lawrence J. Palmer Correspondence, box 2, Digestion Studies, 1920–34, L. J. Palmer to Chief of Bureau, Biological Survey, 6 November 1929.

56   NARA, RG 75, box 4, "Progress Report: Quadrat Studies," 1 November 1923."

57   L. J. Palmer, "Progress of Reindeer Grazing Investigations in Alaska," 31.

58   L. J. Palmer, "Progress of Reindeer Grazing Investigations in Alaska," 33.

59   L. J. Palmer, "Progress of Reindeer Grazing Investigations in Alaska," 3. For an interesting comparison between Palmer's science and the Lomen Corporation's business practices, see Carl J. Lomen, *Fifty Years in Alaska* (New York: D. McKay, 1954).

60   The best source on legibility remains James C. Scott, *Seeing Like a State: How Certain Schemes to Improve the Human Condition Have Failed,* (New Haven: Yale University Press, 1999). See, in particular, pp. 1–9.

61   "Extracts from Report of W. B. Miller on Kuskokwim Reindeer Investigations, Dated Jan 17, 1931," an addendum within LAC, RG 33 105, "Royal Commission: Reindeer and Muskox Industry, Vol. 1: Hearings of Royal Commission on Muskox and Reindeer" (Washington, DC, 1931). Palmer was described in the addendum as follows: "The new man comes here with his brief case full of formulas which he intends to apply. He publishes and describes their use as an improvement of methods, etc."

62 NARA, RG 75, box 4, Quadrat Studies, 1922–35, Frederic Clements to L. J. Palmer, 6 November 1926.

63 Quote about files in LAC, microfilm reel M-1958, A. E. Porsild, "Field Journal of an Expedition through Alaska Yukon and the Mackenzie District being a botanical reconnaissance with special reference to the suitability of the country for domesticated reindeer. Also many notes on the physiography of the country, its inhabitants, wild life and general economic conditions. 1926–1928, by AE Porsild," 8. See also Dathan, 53–62; and APR-UAF, LJP, box 3, folder 4, L. J. Palmer to Chief of Bureau, Biological Survey, 27 August 1926.

64 LAC, microfilm reel M-1958, A. E. Porsild, "Field Journal of an Expedition through Alaska Yukon and the Mackenzie District…," 57. Porsild, *Reindeer Grazing in Northwest Canada*, 14.

65 See note 7.

66 He later noted that the government preferred the Mackenzie Delta because its proximity to Alaska would ease the delivery of reindeer. Indeed, Finnie had already suspected that the Mackenzie District—a region much larger than the Mackenzie Delta—would be the best home for reindeer. See LAC, microfilm reel T13273, vol. 765, file 5095—Porsild, 1926–36, O. S. Finnie to Messrs. A. E. and R. T. Porsild, 19 May 1926.

67 Porsild, "The Reindeer Industry and the Canadian Eskimo," 6–10.

68 As cited in Dathan, 118.

69 Porsild, *Reindeer Grazing in Northwest Canada*, 40.

70 Porsild, *Reindeer Grazing in Northwest Canada*, 19–39.

71 Porsild, *Reindeer Grazing in Northwest Canada*, 34.

72 Porsild, "The Reindeer Industry and the Canadian Eskimo," 16.

73 Porsild, "The Reindeer Industry and the Canadian Eskimo," 6.

74 LAC, microfilm reel M-1958, A. E. Porsild, "Field Journal of an Expedition through Alaska Yukon and the Mackenzie District…", 70.

75 Porsild, *Reindeer Grazing in Northwest Canada*, 29–32. See also LAC, microfilm reel M-1958, letter from A. E. Porsild [no addressee indicated], 25 June 1927.

76 Porsild, "The Reindeer and the Canadian Eskimo," 6–7.

77 See Dathan, 134–84. Porsild's aerial surveys also suggest the complicated legacy of the airplane for scientific knowledge and governmental priorities in regard to the north. For more on this theme, see Marionne Cronin's and Tina Adcock's chapters in this volume.

78 LAC, microfilm reel M-1958, letter from A. E. Porsild [no addressee indicated], 16 June 1927. Porsild wrote, "The case of Tarpoq is not unique. Similar cases have been met with in other places in Alaska, when an eskimo, though a good reindeer man under the supervision of a white man, when left to himself, soon starts to neglect his herd when his increase and profits is not up to his expectations."

79 LAC, microfilm reel M-1958, A. E. Porsild, "Field Journal of an Expedition through Alaska Yukon and the Mackenzie District…", 73–78.

80 Porsild, *Reindeer Grazing in Northwest Canada*, 5–6.

81 For a summary of this controversy, see Dick North, *Exodus*, 195–97, 217.

82 There were other towns in the western Arctic at the time, but these had emerged in response to private industry. Herschel Island was a product of the commercial whaling industry, just as Aklavik, Fort McPherson, and Tuktoyaktuk were created as outposts of the fur trade. Reindeer Station, in contrast, was built from the ground up to service the government's reindeer program. Families were relocated from these outlying towns to the new settlement. This history prefigured the attempt in the late 1950s to relocated residents of Aklavik to the government hub of Inuvik.

83 NWT Archives (Yellowknife) (hereafter cited as NWTA), G79-069, file 1-3, "Minutes from a meeting of the Interdepartmental Reindeer committee," 18 January 1933.

84 LAC, microfilm reel M-1958, "To the Mackenzie Delta to inspect and report on Reindeer Experiment, June-August, 1947, Daily Journal of A. E. Porsild," 3.

85 On retrenchment, see North, *Exodus*, 170–72; and Richard Finnie, *Canada Moves North* (New York: Hurst and Blackett, 1942), 65–70. Finnie and Cory resigned at the end of 1931. The Northwest Territories and Yukon Branch was abolished, and its territories came under jurisdiction of the Dominion Lands Board.

86 Richard Finnie, *Canada Moves North*, 141. See also LAC, RG 109, vol. 491, file 111, A. E. Porsild, 1947 Report, 1-7. Porsild notes that between 1936 and 1947, the Canadian Reindeer Project sent Aklavik five thousand carcasses, amounting to 500,000 pounds of meat.

87 LAC, microfilm reel, T13323, vol. 822, file 7128—Lapp Herders, 1929–38, "List for File: Staff at Reindeer Station," 15 November 1935. Inuit who signed up as apprentices earned $25 a month and a "ration" valued at $200, along with meat and skins.

88 Canada, Northwest Territories Administration, Lands, Parks, and Forests Branch, *Canada's Reindeer* (Ottawa, 1940), 5-7.

89 R. M. Hill, *Mackenzie Reindeer Operations*, Northern Coordination and Research Centre, Department of Indian Affairs and Northern Development, August 1967.

90 Pokiak, 80. Hart and Cockney, 1. On the Bar-1 DEW Line site, see David Neufeld, "Commemorating the Cold War in Canada: Considering the DEW Line," *Public Historian* 20, no. 1 (1998): 9–19.

91 For a compelling history of Inuvik's construction in the 1950s and its effect on local labour and land use, see Dick Hill, *Inuvik: A History, 1958-2008: The Planning, Construction and Growth of an Arctic Community* (Victoria: Trafford Publishing, 2008), 38–59.

92 R. M. Hill, *Mackenzie Reindeer Operations*, Northern Coordination and Research Centre, Department of Indian Affairs and Northern Development, August 1967.

93 Hart, 84. On labour pool problems in conjunction with the DEW Line and Inuvik's construction, see NWTA, G1979-003-70-1, "Annual Report, Reindeer Station, April 1, 1958–March 31, 1959." The Canadian Reindeer Project did not end with Second World War—or even in 1959, as is implied here. It continued to be passed back and forth between private and public hands

throughout the 1960s and 1970s. Today, the small extant herd in the Mackenzie Delta is owned partly by a private individual and partly by the Inuvialuit Regional Corporation.

94 On published analyses of the Canadian Reindeer Project, see Joseph Sonnenfeld, "An Arctic Reindeer Industry: Growth and Decline," *Geographical Review* 49, no. 1 (1959): 77–94; George W. Scotter, "Reindeer Husbandry as a Land Use in Northwestern Canada," in *Proceedings of the Productivity and Conservation in Northern Circumpolar Lands Conference*, ed. W. A. Fuller and P. G. Kevan, IUCN new ser., 16 (1969): 159–69; and Scotter, "Reindeer Ranching in Canada," 1972. Unpublished memos and annual reports from Reindeer Station post managers between 1952 and 1959 detail many of these "problems" with reindeer, and make gestures toward conducting research on the Alaskan industry's failure. See, for example: NWTA, G1979-003-70-1, Laco Hunt, "Some Observations on the Reindeer Industry, 1952"; and

NWTA, G1979-003-70-1, R. C. Robertson to "Mr. Steele," 27 May 1960. Of course, these documents gave little consideration to the ways in which Inuit residents consciously avoided the program.

95 On fishing and biological research, see Piper 192–223; on mining and geology: Piper, 114–39.

96 Stephen Bocking has framed northern environments as spaces wherein scholars can work out the integration of environmental history and the history of science. Stephen Bocking, "Science and Spaces in the Northern Environment," *Environmental History* 12, no. 4 (2007): 867–94.

97 See Emilie Cameron's chapter, "Climate Anti-Politics: Scale, Locality, and Arctic Climate Change," in this volume.

98 Percy Cox, Professor Seward, and Colonel Vanier, "The Reindeer Industry and the Canadian Eskimo: Discussion," *Geographical Journal* 88, no. 1 (1936): 18.

4

Shaped by the Land:
An Envirotechnical History
of a Canadian Bush Plane

Marionne Cronin

Walter Gilbert celebrated New Year, 1930, perched on the edge of his seat in an airplane rattling northward from Edmonton to Fort McMurray, Alberta. As McMurray came into view through the swirling snow of a growing storm, Gilbert excitedly turned to his wife to point out their new home. That Jeanne shared his excitement as she slumped down into the upturned collar of her winter coat, trying to keep warm in the plane's increasingly chilly cabin, was doubtful. Gilbert, however, was revelling in the excitement of achieving his long-held ambition of working as a pilot in the Canadian north.[1]

Walter, Jeanne, and the aircraft were all participants in a wider set of developments taking place across Subarctic Canada in the interwar years, as the northward expansion of resource industries transformed the region's social, economic, political, and environmental structures.[2] In the 1920s, industries such as forestry and mining had become increasingly interested in exploiting the resources of the Canadian Shield. The Shield's rocky, boggy terrain, however, when combined with the region's intricate waterways, made road and rail construction time- and resource-intensive, and thus made it challenging for these industries to access forests or minerals located beyond the Shield's edge. Early Canadian aviators, however, soon identified a commercial opportunity in this quandary. Capitalizing on the ability of war-surplus flying boats and later float- and ski-equipped aircraft to make use of northern rivers and lakes as ready-made runways, they began to market their services as a means of transcending the region's obstacles. The resulting style of aviation, which came to be known as bush flying, emerged out of the symbiotic relationship that developed between early Canadian aviation and the northern resource industries.

This relationship profoundly shaped twentieth-century histories of northern Canada and undoubtedly had important consequences for northern environments and peoples. But what if we shifted our focus for a moment from how the introduction of a new technology affected the North, to ask instead what histories of technology might emerge if we were to take up environmental history's commitment to foregrounding the role of the non-human environment and to examine the history of northern aircraft through this lens? How might it change our understanding of these technologies as being generated and produced outside the north? Such questions also suggest fruitful methods for approaching aviation histories more generally. How might such an analysis contribute to our understanding of aviation, encouraging us to treat it not as a transcendent practice disconnected from the ground over which it moves, or the air through which it passes, but as an envirotechnical system produced through the intermingling of the social and material?

In doing so with a focus on northern Canadian aviation, we can begin to see that while aviation was a significant ingredient in remaking northern interwar environments, the very material encounter between the aircraft and the environment simultaneously refashioned the technology. In other words, Canadian bush planes were produced not only in southern factories, but also through their interactions with northern Canadian environments. Taking the history of Canadian Airways' Mackenzie District operations in the 1920s and 1930s as a case study exposes the way these northern landscapes, waterscapes, and airscapes left their traces in the material bodies of these machines. By considering aircraft as envirotechnical objects that combine the material and social in their very construction and use, this analysis reveals how Mackenzie District bush flying produced an envirotechnical aviation system that sculpted aircraft use, performance, and design in specific ways.[3] This history also illustrates how, when combined with the history of technology's interest in the contingencies of technological change, environmental history can enrich our understanding of the multi-directional, multi-faceted, complex, and sometimes messy nature of the interaction between northern technologies and environments.

"Envirotech" Histories

A growing body of "envirotech" studies seeks to explore just these sorts of interactions by, in Sara Pritchard's terms, simultaneously opening the technological and environmental black boxes.[4] These works begin with a rejection of the traditional separation of technology (culture) and environment (nature) to argue that, far from being disconnected, technologies and environments are co-constitutive.[5] By blurring the division between technology and environment, envirotech analyses explore the capacity of technologies to modify environments, while simultaneously highlighting the "nature" of technology itself.[6]

Applying an envirotech-inspired analysis to the history of northern Canadian aviation—and more specifically to the history of the aircraft used along these routes—brings to light the braided natures of this history. For instance, not only did the interaction between these machines and their environments mould ideas about what sort of place the North was and what made for a "good" bush plane, the immersion of the planes' material bodies in the northern environment also produced physical changes in their constituent materials and affected their performance. Tracing users' responses to these developments provides insight into the ways in which experiences in and ideas about northern environments were incorporated into the very fabric of these machines.

This understanding of technology picks up on environmental historians' insight that "not all historical processes emanate from humanity."[7] In the context of technological histories, this insight reminds us that even as technologies are social and cultural products, they are part of a material world that guides both their forms and uses. Such an analysis is especially important in the context of northern histories, as the north is so often depicted as marginal to or somehow cut off from the forces and flows of technological modernity—or, if not outside these flows, as the passive recipient of technological incursions. A history of northern Canadian aviation that pays attention to northern environments shows, however, that northern peoples and northern places were actively both entangled in and engaged in forming modern technologies.

The challenge of envirotechnical analysis, however, is to avoid falling into either reductionist materialism or determinism, whether environmental or technical. Rather, we need to remember that, even as neither is fully reducible to culture, neither environments nor technologies are fully autonomous or solely material. As environmental historians and historians of technology have each revealed, both are socio-cultural products. Even as technologies are the outcomes of processes fully imbricated with the social, environments too are produced through the interaction of humans and non-human nature: "even what we call 'wilderness' has been modified by centuries of human contact."[8] In other words, people are also essential components in the processes of producing, shaping, and using both northern environments and technologies.[9]

This social dimension of envirotechnical histories means that the histories of northern aviation are therefore also histories of power. Nowhere is this more obvious than in the silences of the archives. While aviation was an important part of northern peoples' twentieth-century history, Indigenous voices are virtually absent from Canadian Airways' archives and from the personal papers of Mackenzie District bush pilots, attesting to their exclusion from the way in which these aircraft were used and the processes of developing aircraft designs.[10] As we will see, power relationships are also visible in the efforts of more transient northerners—in the persons of bush pilots and flight engineers—to have senior airline managers recognize their operational experiences of the interactions between aircraft and environments as authoritative, and to have these experiences translated into material adaptations. Thus, although an envirotechnical history of northern aviation might begin with a focus on the machine, it is important to remember that envirotechnical histories are fully saturated with the social, and therefore with the political.

Aviation and Ideas of North

When Walter Gilbert went north in 1930, Canadian Airways had been operating along the Mackenzie for only a year, but already there was evidence of the mutually shaping relationship between the environment and the technology. Certainly there is no avoiding the ways in which aircraft deeply influenced perceptions of the Canadian north during the interwar period. These perceptions in turn contributed to the production of particular ideas of north that influenced the airline management's choices regarding the Mackenzie District fleet and its deployment. Indeed, the region had begun to exert its influence even before the airline's aircraft began their work along the Mackenzie.

To begin with, bush flying defined the Shield as a zone of what we would now call natural resources. Aircraft's ability to penetrate the northern interior with seeming ease was central in defining "the north" as a region from which southern Canada could acquire wealth through extractive resource industries. By seeming to annihilate time and distance, aircraft made the resources held in the Shield appear accessible, and therefore realistic to develop. In addition, aircraft were essential tools in identifying which resources were available and where they were located, particularly through aerial surveying techniques such as aerial reconnaissance, photography, and mapping. Airlines, branches of government, and mineral companies all reinforced this image of aviation as a transcendent technology that disengaged the passenger from the arduousness of surface travel while simultaneously compressing northern time and space. Canadian Airways, for example, advertised its services in mining publications with copy such as: "The real exploitation has been made possible by planes, that need no roads and that cover in a few hours a journey that would take days by any other means."[11] Likewise, the company's chief executive, James A. Richardson, presented aircraft as a means of "opening the north"—by which he meant rendering it accessible to industrial development.[12] The fledgling Royal Canadian Air Force also promoted aviation's ability to overcome barriers as a means of transforming the north into an economically productive region.[13] These aerial visions of the north, shared by government bureaucrats and business leaders, informed the northward industrial expansion that so transformed the Subarctic environment in the twentieth century.

The capacity of the aerial gaze to redefine geographies and perceptions of the environment is well studied, exposing the modernist and imperialist dimensions of the synoptic aerial view that creates and reinforces a sense of dominion and possession.[14] In the context of the Canadian north, this aerial vantage point played a pivotal role in constructing the northern environment *as* resource, by transforming "forest" into "timber" and "rocks" into "minerals," and by replacing existing, inhabited geographies with ones defined by the perspective of industrial resource development. At the same time, photography also offered southern administrators a means of rendering the north legible in southern power centres, and administrators were keen to utilize this new technique on the northern land, its inhabitants, and resources.[15] When the technical authority of photography and its apparent ability to capture a putatively objective image of the world was coupled with the aircraft, the combination created a powerful technology for capturing, taming, cataloguing, and possessing the north.

As Tina Adcock discusses in the following chapter, this aerial viewpoint, when applied to the north, depopulated the landscape, erasing or obscuring evidence of existing inhabitants. The aerial photographs produced—particularly the oblique, panoramic images used extensively in northern aerial survey work in order to capture huge swaths of territory—constructed an image of an empty, resource-rich region that could be possessed through the deployment of modern technologies. Even contemporaries of these early bush pilots felt that these aerial images seemed unable to capture the emotional or historical significance of place. The aerial view also disoriented Indigenous northerners, who, transported above the surface, often found they were unable to locate themselves in the landscape, effectively dis-placing them. What appeared through the plane's window or in the aerial photographs were only inanimate rocks, trees, and water that had lost their cultural or emotional meanings. It was a process that rendered the landscape as solely inanimate material, ready to be possessed and exploited.[16]

Selection of the Mackenzie District Fleet

This image of the north as a resource-rich region rendered accessible by aviation in turn affected the choice of aircraft deployed in the Mackenzie. In particular, Canadian Airways' fleet selection reflected a combination of cultural and material factors: managers' appraisals of the regional market and perceived customer requirements, their assessments of regional conditions, and previous experience with bush planes in other districts.

As part of establishing a Mackenzie District service, Canadian Airways sent senior pilot C. H. Dickins to survey the region's potential. Initially Dickins felt that fur traders, whose lightweight, high-value goods made ideal air cargo, would provide the airline's main market, while passengers (by which Dickins meant primarily institutional traffic from government administrators, the RCMP, and church officials) and traffic in perishable goods would provide supplementary markets. Dickins' initial survey also identified a growing potential for mineral traffic, and, even more tantalizing, the promise of a government airmail contract.

These prospective customers fell broadly into two categories. In the one group were the industrial clients: mineral developers (prospectors, developers, and ultimately the new industrial communities) and fur traders. In the other were institutional consumers, particularly the government. General passenger services for local inhabitants formed little part of the company's initial planning.[17] The types of customers identified were evidence of the economic transformations and the extension of government power at work in the north. They were also groups with a particular vision of the north, and their interests in and experiences of the north were central in guiding the selection and adaptation of northern aircraft. Though the individuals within these groups held multiple personal images of the northern environment, as institutions these groups perceived the north primarily as a space of resources.

To meet the perceived needs of this set of customers, airline executives decided they needed a versatile plane, one that could adapt its internal layout to accommodate anything from airmail and passengers to mining equipment and fuel drums. At the same time, their ideal plane needed to carry significant loads; to fly at good speeds and travel significant distances, even when fully loaded; to get into and out of difficult locations with a full load; to operate year round, adapting itself to the changing environment by shifting from floats to skis and back; and to operate reliably under northern conditions.[18]

With these requirements in mind, the company selected the Fokker Super Universal. A high-wing, mid-sized monoplane with flexible cabin fittings that could accommodate up to six passengers or a cargo load of over seven hundred pounds, the Super Universal's flight performance provided good speed, range, and rate of climb, all of which made the plane seem the ideal aircraft with which to open the new Mackenzie route.[19] The company's experience with other Fokker aircraft elsewhere in northern Canada also reinforced this confidence. These experiences had convinced the airline's management of the Fokkers' reliability and freighting capabilities, encouraging them to choose the new Super Universal for their new route.[20] But as we will see, the specific effects on the technology produced by certain characteristics of the Mackenzie District would eventually cause managers to revise that opinion.

Air Routes as Envirotechnical Systems

Once selected, the aircraft's routes emerged out of the complex interactions between geography, environment, technologies, and economics. Canadian Airways' southern base, for example, was located at Waterways, Alberta, near Fort McMurray, because of the natural harbour offered by the Snye (a sheltered side-channel linking the Athabasca and Clearwater rivers), and because it was the northern terminus of the Alberta and Great Waterways Railroad. This location linked the northern air service to the main southern rail network, thus providing easy access to passenger and freight traffic, as well as the parts and personnel necessary for its operations. The nodes along the route north of Waterways exhibited a similar interaction between pre-existing geographies and access to the airline's initial target markets. The traffic in furs, passengers, and small amounts of air freight that the airline sought to tap could all be found at the region's commercial centres: the fur-trading forts. Indeed, the airline's main trunk route included stops at Forts Chipewyan, Fitzgerald, Smith, Resolution, Providence, Simpson, Reliance, and Rae, and, at the northern end, Forts Norman, Good Hope, and McPherson. In addition to traffic, these same forts also provided convenient places to store supplies such as fuel and oil.

Thus, the fur trade's geography, which was formed in part by the need to access a water-based transport network, had important effects on the new aerial transport system.

That said, the fur trade's geography was not the only influence on these routes. Looking into the flight logs of Canadian Airways' first year of Mackenzie District operation, for example, one can see the rising power of the mining industry in the appearance of Dawson's Landing as a regular port of call in early 1929. This jump in activity was tied to a spike of interest in zinc deposits on the south shore of Great Slave Lake. Even as the pre-existing economic geography of the fur trade defined these initial destinations, emerging mineral geographies would quickly exert their own pressures. New nodes appeared and disappeared, and the network of air routes shifted as points of economic activity came and went.

The paths between these nodes also bore evidence of the environment's effects. Despite the air-minded rhetoric about aircraft transcending the obstacles of northern geography, the waterways that had moulded the fur trade's geography continued to be an important component of the new aviation system. Although these aerial routes were not fixed in the same way as a railway or road, they tended to follow the region's waterways, in some ways replicating the region's existing water-based transport systems. In part, this was the result of the location of the airline's main destinations along the shores of the region's major lakes and rivers. More importantly, however, the waterways provided the landing fields for these new "flexible" technologies. Without access to water or smooth, stable stretches of snow and ice, pontoon or ski-equipped aircraft were immobile; if on the ground, they could not take off, and if airborne, they were effectively marooned in the sky, unable to land. Thus, these aerial routes were fundamentally informed by the region's watery environments.

The region's waterways, particularly the major rivers and lakes, also provided an important safety net for northern pilots. Their courses and shorelines offered routes to follow, allowing the pilots to locate themselves in the northern landscape and to find their way from place to place—an important tool in an era when the region was not fully mapped and the pilots navigated visually. In the event of a forced landing, the waterways also offered a place to land, and, especially in the case of the major rivers, provided access to the region's "communication" system. Because these water systems provided the region's main transport routes, even in the winter a pilot forced down on the main rivers could expect to encounter another traveller in a relatively short amount of time, and thus to access help more readily.

The centrality of water also imposed seasonal rhythms on northern bush flying. Planes could land only on open water or solid ice. Thus, the liminal seasons of freeze-up and break-up, when waters were neither fully liquid nor solid, restricted activity, and airlines suspended operations as the seasons changed. As the ice formed soonest and retreated latest at the northern end of the route, when services began or ceased depended on latitude. Likewise, because the surfaces on which a plane could land were dictated by its landing equipment, and because their selection was determined by the water's state at its main southernmost base, the changing seasonal conditions along the route determined the planes' spheres of operation—after all, they could only land on waterways that were in the same state as those at the home base. Consequently, even if the ice had melted at Waterways, planes equipped with floats could not land at Great Bear until the ice on that lake had melted. Thus, the break in service would be longest at the northern end of the route. These seasonal limitations produced spikes of activity, as people sought to get themselves or their cargo into or out of the north before operations ceased, and as business built up over the suspension of operations. Prospectors, for instance, would sometimes try to get into the region just before break-up so that they could begin activities as soon as the snow was off the ground, even though the ice might still be in. These rhythms, along with the other contours of Mackenzie District aviation, exhibit the interaction of technology, environment, and geography, which produced the envirotechnical system that comprised Canadian Airways' Mackenzie District routes.

Material Encounters between Aircraft and Environment

The plane's overall design, in conjunction with the material composition of vital components, responded to its Mackenzie District introduction in ways that would result in reconfigurations of that design, incorporation of new materials, and ultimately a renunciation of the plane's status as the Mackenzie fleet's backbone. As originally designed, the Fokker Super Universals had a somewhat unusual undercarriage design. Where on most aircraft the wheel struts extended from the underside of the plane's body, on the Super Universals the struts dropped straight down from the underside of its wings. To help these struts absorb the force of landing, the plane was equipped with shock absorbers made of rubber cords, almost like bungee cords, which were wound around a set of pins in such a way that when the struts' parts compressed, the cords were stretched, absorbing the impact.

In order to use these aircraft as bush planes, however, their original wheels were replaced with the skis and floats that enabled them to exploit northern waterscapes as landing fields. The consequence was that the planes lost the extra cushioning provided by the rubber wheels, leaving only the rubber cords to absorb the force of the landing. In summer, the water would take up some of the impact, but in winter the planes were landing on hard-packed drifted snow or ice. Because there was no cushioning mechanism built into the planes' skis, the full force of landing was borne only by the rubber shock cords.

The pilots quickly discovered that the region's climate affected the remaining rubber cords in such a way that the struts were unable to withstand these forces, whereupon the aircraft's legs would fail. The first undercarriage collapse, which occurred in January 1929 during the first month of Mackenzie District operations, remained an isolated incident that winter, but the following winter (1929–30) there were three more accidents. C. H. Dickins, now the Mackenzie District superintendent, identified the winter temperatures as the trigger: the cold caused the cords' rubber to become rigid and lose its elasticity.[21] This meant the shocks were no longer able to absorb the landing's impact, and the full force was carried by the thin struts, which were not strong enough to withstand it, causing them to crumple. The struts' collapse evinced the environment's very real interaction with the technology's material composition, as the cold fundamentally altered the rubber's molecular structure. In the context of an environment where the size of the main lakes and rivers, combined with the prevailing winds, produced significant pressure ridges in the ice and drifts of hard, wind-packed snow that translated into hard landings, this loss of elasticity meant that the encounter between the machine and the frozen waterscapes produced forces that caused components of the technology to fail.[22]

In response to Dickins' report, the airline worked with the manufacturer to develop a solution, but the environmental conditions again modified the technology's behaviour, leading to a further round of adaptations. In the autumn of 1930, the planes' rubber cords were replaced with air-filled pistons (or aerol shock absorbers) that could function more effectively at cold temperatures. The adaptation initially seemed to resolve the technical problems, and the pilots were very positive about the change. Walter Gilbert, for one, seemed almost joyous in his praise: "New landing gear is a tremendous improvement—(makes pilot think he is 'learning to land' at last)."[23] The following winter, however, the Super Universals' struts began to fail again. In January 1932 alone, three of the Mackenzie District planes experienced undercarriage collapses, and there were eight accidents in the seven weeks between 15 January and 5 March.

Initially pilots were unsure of the cause, as pre-flight inspections showed no signs of weakness in the struts, but they quickly identified the winter's extreme temperatures as the culprit.[24] As Gilbert pointed out, the shocks had been lubricated with "heavy instead of non-freezing oil, which at the very low temperatures now prevalent here, has a consistency resembling sludgy treacle."[25] In other words, the oil congealed in the cold, and, like the rubber's loss of elasticity, the loss of lubrication could cause the shocks' moving parts to seize, effectively eliminating the plane's ability to withstand the force of landing and causing the undercarriage struts to fail—again. As with the rubber, the Mackenzie's winter conditions altered the very structure of the basic materials that enabled the technology to function. Luckily, as Gilbert's comment indicated, the technical solution was relatively simple: replace the heavy oil with lighter oil that would not congeal in the cold. By the autumn of 1932, the lubricating oil had been adjusted and Dickins had also been successful in advocating to have additional shock absorbers fitted to the tops of the ski pedestals. With these last adjustments, the planes finally seemed adapted to their northern environment.

In the changes to this small but critical part, one can read evidence of the interaction between the technology's materiality and the environment. The oil and rubber's changing properties offer a striking demonstration of the material interpenetration of mechanical bodies and environment. It also highlights the significance of climate and weather in the history of aviation, reminding us that the aircraft's environment is composed of airscapes as much as land and waterscapes.

Translating Northern Experience into Technical Changes

Converting northern operating experience into technical adaptations was not as straightforward as the previous narrative might suggest. District pilots, beginning with Dickins, had highlighted the technology's altered performance in the Mackenzie environment and requested corresponding modifications as early as February 1929, but these requests were not automatically translated into action. Replacing the rubber cords took almost two years from the first accident, and it took another two years before the part was precisely tuned to the northern operating environment.

In part, the delay reflected the airline's scale and structure. Unlike in smaller companies, Canadian Airways' pilots did not deal directly with manufacturers. Instead, their experiences were filtered through two layers of management before being passed on to manufacturers. It was this final management layer, in the person of W. L. Brintnell, the airline's operating manager, that had the power to request design changes. The company's delayed reaction to the pilots' reports about the Fokker Super Universals' performance also illustrates the diminished authority of local experience when transferred outside the Mackenzie District. Tracing the conversion of operating experience into technological change offers insight into the production, circulation, and materialization of ideas about northern environments and, through this, exposes the power relationships that framed these processes.

Given the airline's corporate structure, information about the Fokkers' technical performance made its way out of the Mackenzie and through the company's hierarchy through the instruments of bureaucratic administration: forms, records, and memos—particularly the daily flight reports. Completed by the pilots at the end of a day's flying, these forms recorded the minutiae of daily operations, abstracting their experiences into figures about miles covered, flying times, loads and cargos carried, amounts of fuel added, engine running times, and a host of other details that could be analyzed by central administrators. On these forms, the environment was reduced to measurements of distance and temperature, receding into a field on which the company pursued its profit-making activities. Likewise, the technology was translated into a set of numbers: engine running times, flying times, gas and oil consumed. The very structure of the flight reports expressed the company's view of the north as a commercial space and its focus on profitability, recording details about consumption of resources in relation to distance travelled and paying cargo carried.

Materializing the North as a Space of Economic Opportunity

Granted, it was administratively important to know, for instance, how much of the gas stored at Fort Smith had been consumed in order to ensure an adequate supply. The choice to focus on such quantitative details in these reports, however—and not to record qualitative information, such as passenger experience, ease of maintenance, or qualitative flying characteristics—testified to a managerial focus on operational efficiency: carrying the highest possible payload at the lowest possible cost. In this framework, the north was primarily a site of wealth creation. It was also an empty, depopulated space. While institutional, sojourning passengers dominate the reports, Indigenous inhabitants appear only occasionally as tellingly nameless medical evacuation patients, but are otherwise almost completely absent from official company documents. Thus, the north of the daily flight report was an empty land of economic opportunity that could be capitalized upon by efficient technological tools.

This emphasis on profiting from the north encapsulated in the flight reports, informed the company's response to the planes' performance. Reacting to Dickins' report regarding the first Mackenzie undercarriage failure and to an incident with another aircraft at Regina, for instance, Brintnell focused on the cost of these accidents, which he estimated to be approximately $2,000 each, and emphasized to Dickins that the company "certainly cannot afford to have any more."[26] By the same token, however, he did not authorize the outlay for replacing the shocks.

In addition, Brintnell did not automatically recognize northern experiences as authoritative information about technical performance. Rather than accept the pilots' analyses, for ultimate technological authority Brintnell turned to the aircraft's manufacturer and asked their engineers to review whether or not the struts were in fact strong enough to withstand regular ski flying in the north.[27] In the event, the Fokker engineers confirmed that the plane's struts were not designed to withstand the forces of ski flying, and required adaptation—though they initially suggested thickening the strut walls as the solution, rather than incorporating new types of shocks. Although Brintnell approached the manufacturer regarding potential solutions, and although Fokker supplied a proposal for strengthening the struts and incorporating air-filled shocks, no action was taken.[28] In fact, it was not until over a year later, in March 1930, after the problem had recurred several times, that Brintnell reopened discussions about the modification with the manufacturer.

In order to convince Brintnell of the modification's necessity, the Mackenzie District pilots and Dickins, as the Mackenzie District superintendent, used a variety of strategies to influence his opinion. For instance, when reporting on the second round of undercarriage collapses, Dickins began to highlight the cost implications of poor technological performance over other considerations, reflecting Brintnell's interest in financial matters. Reporting on the events of 1929–30, Dickins pointed out that each problem with the Super Universal's undercarriages cost the company nearly $3,000, not counting loss of time.[29] The need to make these arguments evinced senior management's role as gatekeeper, deciding what information would be passed on to manufacturers, which aircraft would be purchased, and what technical changes would be requested.

In the spring of 1930, however, the costs associated with the undercarriage repairs prompted Brintnell to contact the manufacturer again.[30] These expenses were no different from the costs incurred in 1929, but a changing commercial context had given them added significance. Canadian Airways' regional competitor, Commercial Airways, had begun operations in August 1928, but with only a small, open-cockpit Avro biplane, it had offered little opposition. By late 1929, however, with support from new financial backers, Commercial was able to purchase several larger Bellanca Pacemakers that could compete more directly with Canadian Airways' Fokkers. At the same time, the Canadian government had awarded Commercial the new Mackenzie District airmail contract, and the airline had also been able to secure a good proportion of regional government business. With its new aircraft, airmail contract, and political connections, Commercial began cutting into Canadian Airways' Mackenzie revenues. In fact, bolstered by its airmail income, the smaller company was prepared to undercut Canadian Airways' rates in order to draw off further business.[31] As overall economic conditions deteriorated through the first months of 1930, and as mineral-related traffic in the Mackenzie began to decline, the two airlines were competing for pieces of a smaller pie.[32]

Within this context, the costs incurred by the Fokkers' undercarriage problems loomed much larger for Canadian Airways than they had the previous year. At the same time, Commercial's fleet of Bellancas performed well in the Mackenzie's winter conditions, providing a contrast to Canadian Airways' experience with its own aircraft. Dickins, in particular, was convinced that Canadian Airways had lost work to Commercial because of the Fokkers' problems, and passed this evaluation along to Brintnell.[33] If the north was to be realized as a space of economic opportunity for the airline, Dickins seemed to be suggesting, the technology would need to be adapted to the environment. In this context, when Fokker provided a second modified design for incorporating air-filled struts, Brintnell approved its adoption.

Northern Exceptionalism as a Rhetorical Tool

Interestingly, in addition to emphasizing the cost implications, the pilots also deployed specific images of the northern environment in their efforts to persuade airline management. Dickins, for example, had begun by downplaying the differences between the Mackenzie District and other Canadian norths. When reporting the first Mackenzie District undercarriage failure, he commented that, "In general the landing conditions are not bad [along the Mackenzie], and compare favourably with any in Ontario..."[34] It was a continuation of the language he had used to convince airline managers to open a Mackenzie District service, and, given the route's novelty, suggests that Dickins may have felt an ongoing need to reassure management of the service's viability.

By March of 1929, however, Dickins' rhetoric had changed and he was beginning to pick out the region's challenging operating environment as a reason for the need to alter the technology. Reporting on his first flight north of the Arctic Circle, Dickins described landing conditions on the route north of Simpson as poor, pointing out that between Wrigley and Fort Norman there was nowhere to land on the Mackenzie River's main channel, and that alternate landing sites were restricted by the two mountain ranges squeezing in on either side of the river. According to Dickins, at Fort Norman "the river [was] nothing but a jumble of ice and snow," and there were no safe landing sites on the main river between Norman and Fort Good Hope.[35] Dickins' depictions of the region as an exceptional environment only intensified over the following winter as he continued to point out how difficult the winter landing conditions were.[36] Dickins' descriptions of the Mackenzie's conditions as different from other northern regions was not mere rhetoric—average temperatures, for instance, were colder than on other bush routes—but his language also participated in the exceptionalist discourse that characterizes many descriptions of northern environments penned by sojourners.[37]

Once the new aerol shocks were fitted to the aircraft, the pilots used these same images of extreme environmental conditions to emphasize the level of improved performance. Walter Gilbert, for instance, commented in his flight reports that the new undercarriage made taxiing on the region's hard drifts "remarkably easy," and, after one flight through Great Bear to Coppermine, wrote, "Unbelievably hard rough drifts, following last week's blizzard. Without new type undercarriages operating under these conditions would have been impossible."[38] But when these parts again failed, the same images of exceptionally rough winter landing conditions were redeployed to emphasize the need to properly adapt the technology.[39] Still later, this same image of difficult environmental conditions would be wrapped into the pilots' and engineers' romantic recollections of their time in the region.[40]

Just as the pilots used particular depictions of the environment to support their arguments for technological changes, Brintnell too deployed specific images of the northern environment in specific contexts. When discussing technological performance with the manufacturer, for instance, Brintnell emphasized the north's extreme conditions, questioning whether Fokker's engineering staff properly appreciated the conditions under which their planes operated.[41] However, when discussing the plane's performance with the pilots, both Brintnell and his successor, G. A. Thompson, emphasized the equivalence between the Mackenzie and the airline's other northern districts.[42] This targeted use of specific images highlights the multiple ideas of north circulating within the company and the rhetorical value of these images. In discussions about technological choices, depictions of place were used to support specific aims. In this case, managers used a de-localized image of north to counter the Mackenzie pilots' demands for (expensive) technological changes designed to adapt the technology to their particular experience of the Mackenzie as an exceptional environment more demanding than other northern environments. Simultaneously, however, airline managers used images of the north as an extreme environment in order to convince the manufacturer to design a technology robust enough for the actual operating conditions the planes encountered. Ultimately, when the airline's management briefed the manufacturer to adapt the technology, the resulting changes recorded in the planes' bodies both the interaction between technology and environment, and specific ideas of the north as an extreme environment requiring specialized technologies.

The problems the pilots subsequently experienced owing to the heavy lubricating oil also disclose local variations in ideas about northern environments. Despite Brintnell's consistent presentation of the Canadian north as a cold, hostile, extreme environment, Fokker's engineers still used a lubricating oil that would congeal in cold temperatures. "North," as understood by the aircraft manufacturer, did not match the reality of local conditions. Looking back, Canadian Airways pilot Archie McMullen attributed the design flaw to the fact that the struts had been designed in Detroit and had never been through "real" cold weather testing.[43] McMullen's comment opens up the multiple ideas of north at play in this history. His dismissal of Detroit conditions as not being "real" cold expresses the pilot's sense that the Mackenzie was part of the "true" north, more northern than other norths. This language of "true north," which also appears in Gilbert's memoir, bespeaks a particular image of the region: by "true" north, these men meant a heroic, wild, untamed environment—a place that was still a frontier space of adventure and romance.[44] It was also a view that saw "north" not as an absolute, but as admitting of different degrees of "northern-ness."

The pilots' sense that they needed to highlight the Mackenzie's exceptional environment when advocating for technical changes suggests that, although the pilots experienced it as an environment with specific operating conditions, the airline's management tended to see the northern bush environments across the country as roughly equivalent. Indeed, Brintnell's correspondence with Fokker refers to Canadian or northern winter flying conditions in general, and does not highlight Mackenzie conditions as particularly extreme. The manufacturer's "northern" testing of the parts in Detroit suggests an even more sweeping idea of north, expressing an elision between the American north, of which Michigan was part, and the Canadian north. Thus, the north materialized in the revamped undercarriage strut manifested particular understandings of north. However, as the part's performance made clear, the north instantiated in the physical piece of technology was not the local north of the Mackenzie District. The frustration behind McMullen's comment attests to the difficulties and the amount of work involved in translating the particularities of local experiences of the interaction between technology and environment into shared experiences that could move between locations and across scales. It also highlights how circulation could alter the content of these concepts, and how the ultimate incorporation of those ideas into the technology can reveal these shifting understandings.[45]

Local Adaptations

Although compelled to wait for head-office approval of large-scale changes, pilots and engineers made a series of unofficial, on-location repairs and adaptations designed to enable the technology to continue to function in its environment. In the aftermath of undercarriage failures, for example, pilots reported how flight engineers loosened shock cords in an effort to improve their performance, straightened and reinforced bent struts using wood blocks, fashioned impromptu protective casings out of fur bales, and replaced existing lubricants with a mixture of transformer and coal oils.[46] At the same time, pilots adapted their practice to suit their environment. Walter Gilbert, for instance, remembered learning that "snow is always softest and least drifted in the lee of a northern shore ... because the big winds come from the north and the northwest—and so the general procedure is to land along the northerly shore even in spite of the wind."[47] Pilots also adjusted their practice for summer conditions. Upon learning, for example, that landing in an exposed location on a big lake like Great Slave when the wind was blowing could be dangerous, as the waves could "buck" a plane "all to pieces," pilots would instead attempt to find a sheltered bay in which to land.[48] But these sorts of adaptations were not enough to prevent winter operations having significant consequences for the Supers' undercarriages. In these informal traces, one can read more transient evidence of how local experiences of the interaction between technology and environment were translated into local adaptations, circumventing the formal power structures that circumscribed and sometimes distorted the translation of local experience into "official" technological changes.

Changing Geographies and Changing Experiences of Northern Aviation

Even as the Fokkers were being adapted to suit the environment, aviation was contributing to the reshaping of that environment through its relationship with regional mineral development, specifically by supporting the Great Bear Lake radium boom of the early 1930s. The identification of pitchblende, the mineral that contains radium, at Great Bear Lake sparked a prospecting rush in 1932, which was promptly followed by the expansion of mining activity on the lake's shores. These changes significantly increased aerial traffic in the region and would lead to a re-evaluation of the Super Universals' suitability for Mackenzie District operations.

In the context of the Great Depression, the income from Canadian Airways' bush routes acquired greater and greater significance as revenue on other services dried up. The company's southern airmail routes, for instance, were cancelled as part of the federal government's austerity measures, and traffic dropped more generally. In fact, the Mackenzie, and Great Bear in particular, provided one of the country's few areas of economic growth, and as a result Canadian Airways was desperate to earn as much money as it could from the Great Bear boom. In response, the airline reconfigured its service to offer regular flights to the area. Thus, even as mining reconfigured its terrestrial environment, mineral development altered the Mackenzie District's aerial geography and created new nodes in Canadian Airways' service and new aerial routes.

These new transport patterns in turn altered the physical experience of flying in Canadian Airways' Mackenzie District. For the planes and their passengers, flying to Great Bear meant covering longer distances at a stretch, and therefore increased amounts of time in the air. These longer flight times highlighted some of the physical irritations and annoyances associated with bush flying, such as poorly heated cabins, nearly deafening noise, and the discomfort of being squeezed in (and sometimes on top of) all sorts of cargos. Walter Gilbert, in particular, pointed out that

> Conditions of cold or cramped sitting which may be endured for an hour are much aggravated when met for 8 hours in one day, in stages of 2 1/2 hours each.
>
> ... I do not think that the Company has ever given enough thought to the psychological effect of petty discomforts on the passenger's general attitude toward the Company, especially when the passenger happens to be an important customer for freighting.[49]

Passengers might not have noticed these inconveniences had these flights been their only experiences with flying, but even as the flight times increased, passenger expectations were changing. Southern passenger services were improving, and many Canadian Airways passengers were of a class that was likely to have access to these more luxurious services. Indeed, G. A. Thompson, the airline's new operating manager, reported that the company was receiving more and more complaints about the discomforts of bush planes in comparison to the services available on American airlines.[50]

Even as passenger perceptions shifted, the Mackenzie District's changing competitive and economic context altered the airline's own evaluation of the Super Universal's performance. Although Canadian Airways had purchased its regional competitor, Commercial Airways, in 1931, thus acquiring its fleet of Bellancas, the appearance of the new rival provided a fresh point of unfavourable comparison. W. L. Brintnell, Canadian Airways' former operating manager, had left the airline in late 1931 and had established his own company, Mackenzie Air Services (MAS), serving the same area as Canadian Airways' Mackenzie District service. According to Walter Gilbert, the new airline's aircraft were equipped with "an exceptionally good heating system, a toilet, and a lot of other 'frills,'" and offered passengers "clear, warm, comfortable cabins," all of which Gilbert believed would cause passengers to make comparisons between the two services that would not be favourable to Canadian Airways.[51] Just over a month later, Gilbert's fears seemed confirmed as Dickins reported the circulation of critiques of Canadian Airways' equipment relative to the MAS fleet.[52]

In addition to passenger experience, efficient operation also became a high priority for Canadian Airways, and the Super Universals' performance began to seem increasingly poor in comparison to its other aircraft. The increased distance to Great Bear Lake only served to emphasize the differences in performance. For example, the Fokker was a slower aircraft than the Bellancas acquired from Commercial Airways. Over shorter distances, the effect on flying time was minimal, but on the longer legs to Great Bear, the difference became noticeable. The plane's gas consumption was also less efficient, so in order to cover these greater distances it needed to carry more fuel. This in turn meant it had to carry smaller loads of cargo and could not generate as much profit per flight as other aircraft. The plane's baseline performance had not declined (in fact, it had improved after the undercarriage modification), but as the region's aerial geography changed, perceptions of its performance shifted.

As the company struggled to maintain its market share and capitalize on the region's growing traffic, its managers sought to maximize carrying capacity and provide acceptable passenger accommodation. To do so, in 1932 the airline began introducing new types of aircraft into its Mackenzie fleet to supplement the existing roster of Fokkers and Bellancas: large Junkers for freight cartage, and Fairchilds for passenger transport. At the same time, the airline began to withdraw the Super Universals from the Mackenzie. The combination of changing passenger expectations and altered perceptions of performance meant company managers ultimately deemed the Super Universals inappropriate for the Mackenzie environment. They were thus transferred to districts where distances were shorter and demands lower, to be replaced with aircraft deemed more efficient or more comfortable. Changes in the terrestrial environment, catalyzed in part by aircraft, had reconfigured the region's aerial geography so as to alter both the embodied experience of flying in the Mackenzie and the company's assessment of the planes' performance. Indeed, by the end of 1933, the Fokker Super Universal had all but disappeared from the airline's Mackenzie fleet.

The planes that replaced the Mackenzie Fokkers hinted at significant changes taking place in Canadian aircraft manufacturing. While the bulk of the new Mackenzie District aircraft were American-designed and -built, the basic design of one, a Fairchild 71C produced by Fairchild's Canadian subsidiary, had been adapted for use as a Canadian bush plane. Only a few years later, aircraft designed from scratch specifically for use as Canadian bush planes, including the Noorduyn Norseman, would appear. Like the adapted Fokkers, these designs would incorporate both the material and imagined northern environment, carrying in their material structures evidence of the complex, mutually shaping interactions between technology and environment.

Conclusion

Taking an "envirotech" approach to Mackenzie District aviation history opens up new aspects of the history of northern technologies. To begin with, it demonstrates that, like other technologies, aircraft are simultaneously material and cultural objects, and as such their histories cannot be separated from those of their environments. The decision to adjust the planes' landing gear, for instance, cannot be understood without reference to the behaviour of its constituent elements within the Mackenzie District environment. Likewise, the process of adaptation brings home the various ways in which specific ideas of north were materialized in the bodies of these machines. Similarly, the emergence of the Mackenzie District air routes and their shifting patterns clarifies how northern interwar aviation emerged as an envirotechnical system, shaped by the interactions of technology and environment alongside social factors like cultural images of the north and economic priorities. While the interaction between aircraft and environment is perhaps more easily recognized in the context of a practice such as bush flying, it should also prompt us to consider the place of environment in aviation histories more generally.

This analysis of Mackenzie District bush planes as envirotechnical objects also exemplifies how aircraft are not transcendent machines that, in their aerial lives, exist in a "placeless place." Rather, as the Mackenzie District Fokkers reveal, aircraft histories are deeply connected to their places of operation. Acknowledging the complex interactions between technology and environment, while simultaneously recognizing the contingencies of these interactions and the importance of human actors in these histories, helps us to recognize the active role of the north in technological histories. Complicating our images of a marginal, passive north receiving modern technologies introduced from outside, the history of bush aircraft shows how the north's own, active experience of technological modernity may contribute to our understandings of the wider histories of the north and of technology.

Notes

1   Walter Gilbert, *Arctic Pilot: Life and Work on North Canadian Air Routes* (Toronto: Thomas Nelson and Sons, 1940), 40–42. Jeanne Gilbert herself was a licensed pilot, and so might have been expected to share her husband's excitement about his flying work, but later letters to her aunt, Fanny Downie, though written with tongue-in-cheek humour, indicate life in McMurray as a bush pilot's wife may not have been as exciting an adventure as Walter's experience as a pilot. Quoted in Gilbert, *Arctic Pilot*, 42–47; see also Western Canada Aviation Museum, Mrs. W. E. Gilbert Collection, arc. 697-1, acc. 92-185, Jeanne Gilbert to Fanny Downie, 23 September 1930 and 6 July 1932.

2   Liza Piper, *The Industrial Transformation of Subarctic Canada* (Vancouver; UBC Press, 2009). For a classic overview of the region's economic and political history, see Morris Zaslow, *The Northward Expansion of Canada 1914–1967* (Toronto: McClelland and Stewart, 1988). See also Kenneth S. Coates and William R. Morrison, *The Forgotten North: A History of Canada's Provincial Norths* (Toronto: James Lorimer, 1992); Shelagh D. Grant, *Sovereignty or Security? Government Policy in the Canadian North, 1936–1950* (Vancouver: UBC Press, 1988); H. V. Nelles, *The Politics of Development: Forests, Mines, and Hydro-Electric Power in Ontario, 1849–1941* (Toronto: Macmillan, 1974); Dianne Newell, *Technology on the Frontier: Mining in Old Ontario* (Vancouver: UBC Press, 1986).

3   These concepts draw on Pritchard's notion of envirotechnical systems, applying it to aircraft as envirotechnical systems in miniature. Sara B. Pritchard, *Confluence: The Nature of Technology and the Remaking of the Rhône* (Cambridge, MA: Harvard University Press, 2011).

4   Pritchard, *Confluence*, 5.

5   Martin Reuss and Stephen H. Cutcliffe, "Introduction," in *The Illusory Boundary: Environment and Technology in History*, ed. Martin Reuss and Stephen H. Cutcliffe (Charlottesville: University of Virginia Press, 2010), 1; James C. Williams, "Understanding the Place of Humans in Nature," in Reuss and Cutcliff, *The Illusory Boundary*, 11.

6   Pritchard, *Confluence*, 12.

7   Pritchard, *Confluence*, 11.

8   Williams, "Understanding," 13.

9   Dolly Jørgensen and Sverker Sörlin, "Introduction: Making the Action Visible, Making Environments in Northern Landscapes," in *Northscapes: History, Technology, and the*

Making of Northern Environments, ed. Dolly Jørgensen and Sverker Sörlin (Vancouver: UBC Press, 2013), 1–14.

10   This exclusion in itself has interesting things to say about twentieth century internal state expansion in Canada and the power relationships that lurk behind aviation's functions in supporting this expansion. Further analysis of this history is beyond the scope of this paper, but these absences suggest the very real need for scholars, including myself, to incorporate Indigenous histories of aviation into their accounts of twentieth century northern Canadian histories.

11   Western Canada Airways advertisement, *Canadian Mining Journal* (1929): 280.

12   Archives of Manitoba, Canadian Airways Limited Collection (hereafter cited as AOM, CAL), MG 11 A 34, box 6: Correspondence, April 1934, James A. Richardson, "Canadian Airways Limited, Memorandum," 7 April 1934.

13   Canada, Department of National Defence, *Report of the Department of National Defence for the Fiscal Year Ending March 1923* (Ottawa: 1923), 8, 37; Canada, Department of National Defence, *Report of the Department of National Defence for the Fiscal Year Ending March 1924* (Ottawa: 1924), 45.

14   Stephen Bocking, "A Disciplined Geography: Aviation, Science, and the Cold War in Northern Canada, 1945–1960," *Technology and Culture* 50 (2009): 320–45; Denis Cosgrove, *Apollo's Eye: A Cartographic Genealogy of the Earth in the Western Imagination* (Baltimore: Johns Hopkins University Press, 2001); Denis Cosgrove and William L.

Fox, *Photography and Flight* (London: Reaktion Books, 2010); Mark Monmonier, "Aerial Photography of the Agricultural Adjustment Administration: Acreage Controls, Conservation Benefits, and Overhead Surveillance in the 1930s," *Photogrammetric Engineering and Remote Sensing* 68, no. 12 (December 2002): 1257–61

15   Peter Geller, *Northern Exposures: Photographing and Filming the Canadian North, 1920–1945* (Vancouver: UBC Press, 2004). Stephen Bocking explores this dynamic in the post-war era in "A Disciplined Geography," 267–68.

16   Piper, *The Industrial Transformation of Subarctic Canada*, 37.

17   AOM, CAL, MG 11 A 34, box 1: Correspondence, January–March 1929, C. H. Dickins to W. L. Brintnell, 7 January 1929, 2 February 1929, 13 February 1929, 20 February 1929.

18   AOM, CAL, MG 11 A 34, box 6: Correspondence, April 1934, James A. Richardson, "Canadian Airways Limited, Memorandum," 7 April 1934.

19   Fokker performance figures: speed—approximately 95 mph on floats, 118 mph on skis; range—700+ miles, or approximately four hours flying time; rate of climb—850 feet per minute at sea level. Alberta Aviation Museum, Fokker Super Universal Manual, n.d.; AOM, CAL, MG 11, A 34, box 32: Fokker Aircraft: Individual Fokker Aircraft, CF-AJC, Aircraft Initial Analysis, n.d.; CF-AJC, Certificate of Registration of Aircraft, 25 June 1930; CF-AJF, Aircraft Initial Analysis, n.d.; CF-AJF, Certificate of Registration of Aircraft, 7 January 1930; G-CASN,

Certificate of Registration, 17 December 1928; G-CASQ, Certificate of Registration, 16 January 1928; "Sales Specifications," Atlantic Aircraft Corporation to Western Canada Airways, 16 January 1928.

20 On the Fokkers' performance on daily operations, see AOM, CAL, MG 11 A 34, box 1: Correspondence, November–December 1928, W. L. Brintnell to Fokker Aircraft Corporation of America, 30 November 1928; on the Hudson Bay freight contract: AOM, CAL, MG 11 A 34, box 1: Correspondence post-1918–June 1929, G. A. Thompson, memo, ca. 1930; on the Barren Lands flight, see NWT Archives (hereafter cited as NWTA), N-1992-120-0001, C. H. Dickins, lecture, Ontario Science Centre, 1978; G-1992-041-002, C. H. Dickins, lecture, Prince of Wales Northern Heritage Centre, Yellowknife, NT, September 1979; C. H. Dickins, "Across the Barrens," *CAHS Journal* 8, no. 1 (Spring 1970): 22–24; C. H. Dickins, "The Barren Lands Flight Fifty Years Later," *CAHS Journal* 21, no. 2, (Summer 1983): 56–63; Fred W. Hotson, "Punch," *CAHS Journal* 32, no. 2 (Summer 1994): 40–48.

21 AOM, CAL, MG 11 A 34, box 1: Correspondence: January–March 1929, C. H. Dickins to W. L. Brintnell, 2 February 1929.

22 For a description of ice conditions in this region, see Piper, *Industrial Transformation of Subarctic Canada*, 58–59.

23 AOM, CAL, MG 11 A 34, box 85: G-CASK, Flight Reports, January–December 1930, flight reports, 9 December 1930.

24 AOM, CAL, MG 11 A 34, box 32: Fokker Aircraft: Individual Fokker Aircraft, W. E. Gilbert to G. A. Thompson, 15 January 1932; Box 85: Flight Reports, G-CASK, January–December 1932, flight reports, 27–28 January 1932; Box 32: Fokker Aircraft, Individual Fokker Aircraft, W. E. Gilbert to G. A. Thompson, 4 February 1932; Box 86: Flight Reports, G-CASQ, January–April 1932, flight report, 6 February 1932.

25 AOM, CAL, MG 11 A 34, box 32: Fokker Aircraft, Individual Fokker Aircraft, W. E. Gilbert to G. A. Thompson, 4 February 1932.

26 AOM, CAL, MG 11 A 34, box 1: Correspondence, January–March 1929, W. L. Brintnell to C. H. Dickins, 22 February 1929.

27 AOM, CAL, MG 11 A 34, box 32: Fokker Aircraft, Individual Fokker Aircraft, W. L. Brintnell to A. A. Gassner, 10 January 1929.

28 AOM, CAL, MG 11 A 34, box 32: Fokker Aircraft, Individual Fokker Aircraft, Charles Froesch to W. L. Brintnell, 16 January 1929; W. L. Brintnell to Charles Froesch, 29 January 1929.

29 AOM, CAL, MG 11 A 34, box 42: Edmonton and McMurray Base Correspondence, 2 January–30 June 1930, C. H. Dickins to W. L. Brintnell, 15 April 1930.

30 AOM, CAL, MG 11 A 34, box 32: Fokker Aircraft, Individual Fokker Aircraft, W. L. Brintnell to Charles Froesch.

31 AOM, CAL, MG 11 A 34, box 42: Edmonton and McMurray Base Correspondence #1: 9 February–31 December 1929, L. R. Mattern to W. L. Brintnell, 2 December 1929, C. H. Dickins to W. L. Brintnell, 10 July 1930; box 2: Correspondence 1929–31, C. H. Dickins to

W. L. Brintnell, 29 January 1930, C. H. Dickins to W. L. Brintnell, 27 March 1930; box 42: Edmonton and McMurray Base Correspondence #2, 2 January–3 June 3 1930, C. H. Dickins to W. L. Brintnell, 15 April 1930, C. H. Dickins to W. L. Brintnell, 19 June 1930.

32 AOM, CAL, MG 11 A 34, box 42: Edmonton and McMurray Base Correspondence #2, 2 January–30 June 1930, C. H. Dickins to W. L. Brintnell, 18 April 1930; box 2: Correspondence 1929–1931, C. H. Dickins to W. L. Brintnell, 10 July 1930.

33 AOM, CAL, MG 11 A 34, box 2: Correspondence 1929–31, C. H. Dickins to W. L. Brintnell, 29 January 1930.

34 AOM, CAL, MG 11 A 34, box 1: Correspondence: January–March 1929, C. H. Dickins to W. L. Brintnell, 2 February 1929.

35 AOM, CAL, MG 11 A 34, box 1: Correspondence: January–March 1929, C. H. Dickins to W. L. Brintnell, 10 March 1929.

36 AOM, CAL, MG 11 A 34, box 2: Correspondence, January–March 1930, C. H. Dickins to W. L. Brintnell, 29 January 1930.

37 Environment Canada, Historical Climate Data, http://climate.weatheroffice.gc.ca/climateData/canada-e.html (accessed 1 May 2011); Liza Piper, 'Introduction: The History of Circumpolar Science and Technology,' *Scientia Canadensis* (Special Issue: Comparative Issues in the History of Circumpolar Science and Technology) 33, no. 2 (2010): 1–9; Piper, *Industrial Transformation of Subarctic Canada*, 62–63.

38 AOM, CAL, MG 11 A 34, box 85: G-CASK, Flight Reports, January–December 1931, 12 January 1931, 22 March 1931.

39 AOM, CAL, MG 11 A 34, box 32: Fokker Aircraft: Individual Fokker Aircraft, W. E. Gilbert to G. A. Thompson, 4 February 1932; box 55: Skis, C. H. Dickins to G. A. Thompson, 14 March 1932.

40 Gilbert, *Arctic Pilot*, 53, 63, 66–68; see also NWTA, N-1992-120-0001, C. H. Dickins, lecture, Ontario Science Centre, 1978; G-1992-041-002, C. H. Dickins, lecture, Prince of Wales Northern Heritage Centre, Yellowknife, NT, September 1979; NWTA, G-1988-008-002, Con Farrell, interview, 20 August 1967.

41 AOM, CAL, MG 11 A 34, box 32: Fokker Aircraft: Individual Fokker Aircraft, W. L. Brintnell to Charles Froesch, 21 February 1929, 16 July 1929.

42 AOM, CAL, MG 11 A 34, box 1: Correspondence, January–March 1929, 22 February 1929; box 32: Fokker Aircraft: Individual Fokker Aircraft, W. E. Gilbert to G. A. Thompson, 4 February 1932.

43 NWTA, G-1988-008-0009, Archie McMullen, interview, 16 September 1978.

44 Gilbert, *Arctic Pilot*, 66. See also NWTA, G-1988-008-002, Con Farrell, interview, 20 August 1967; G-1988-008-0009, Archie McMullan, interview, 16 September 1978; and G-1988-008-0003, W. E. Gilbert, interview, ca. 1967.

45 See also Farish's chapter in this volume.

46 AOM, CAL, MG 11 A 34, box 86: Flight Reports, G-CASQ, January–April 1932, Flight Report,

G-CASQ, 6 February 1932; box
32: Fokker Aircraft: Individual
Fokker Aircraft, W. E. Gilbert to
G. A. Thompson, 4 February 1932;
box 75: AJC Logs, 24 June 1930–29
January 1933, CF-AJC log entry, 7
February 1932. Further histories of
these local adaptations are difficult
to trace. Since they were informal
(and not always authorized or
strictly legal), they do not always
appear in official documents. One
might have been able to read them
on objects surviving in museums,
but these aircraft have often been
restored, so many of these traces
have been lost.

47   NWTA, G-1988-008-0003, W. E.
Gilbert, interview, ca. 1967.

48   NWTA, G-1988-008-0002, Con
Farrell, interview, 20 August 1967.

49   AOM, CAL, MG 11 A 34, box 5:
Correspondence October 1932–
August 1933, W. E. Gilbert to G. A.
Thompson, 29 December 1932.

50   AOM, CAL, MG 11 A 34, box 5:
Correspondence October 1932–
August 1933, G. A. Thompson to
W. C. Sigerson and C. H. Dickins, 7
March 1933.

51   AOM, CAL, MG 11 A 34, box 5:
Correspondence October 1932–
August 1933, W. E. Gilbert to G. A.
Thompson, 29 December 1932.

52   AOM, CAL, MG 11 A 34, box 14:
Airmail: Great Bear–Resolution,
16 July 1932–7 June 1934, C. H.
Dickins to G. A. Thompson, 13
April 1933.

5

Many Tiny Traces: Antimodernism and Northern Exploration Between the Wars

Tina Adcock

When, in 1935, the mining engineer George Douglas and the surveyor Guy Blanchet undertook a summer of prospecting together in northern Canada, they should have been ideal partners. Over and above their decade-long friendship, both were experienced northern travellers, and shared a level-headed, professional approach that emphasized careful preparation for fieldwork and its responsible execution. However, their divergent styles of work, rooted in disparate philosophical frameworks, caused friction, as Douglas described:

> Blanchet + I, both competent and experienced in our respective methods, were simply quite hopeless together. Blanchet is like an Indian who knows how to handle an axe and a crooked knife and nothing else, yet with these implements will "get by." While I am like a carpenter who must have many and complicated tools but who knows their purpose, when and how to handle them. And to keep them in good shape![1]

These different manners of being in the north were also evident in their respective attitudes toward maintaining camp. Blanchet again evinced what Douglas termed an "Indian" indifference to comfort, cleanliness, and order, whereas the latter required (though this word is unspoken) a more "civilized" approach to living in the bush. Their contrast crystallized in Douglas' enduring struggle to keep their cooking equipment clean. Many years later, Blanchet wrote amusedly: "We were different in our habits. In washing dishes, George would boil two pails of water if even for one cup, one spoon. I would swish mine in the lake."[2]

Despite these differences, both Blanchet and Douglas travelled, and thought about their northern travels in ways that signal their participation in the historical phenomenon of antimodernism. Emerging first at the *fin de siècle*, antimodernism responded to the political, economic, social, and cultural changes born of industrialization and modernization. Its dominant characteristic, in the classic phrase of Jackson Lears, was "the recoil from an 'overcivilized' modern existence to more intense forms of physical or spiritual experience" thought to exist in Oriental, medieval, or primitive cultures. This desire to retreat to alternative times and places was "ambivalent, often coexisting with enthusiasm for material progress" in Western society.[3]

This chapter casts Blanchet's and Douglas' northern expeditions of the 1920s and 1930s as admittedly vigorous examples of the therapeutic recourse to nature that many middle-class Canadians, particularly their fellow Ontarians, sought from the late nineteenth century onward. After outlining the general characteristics of their early twentieth-century, central Canadian variant of antimodernism, this chapter will delineate the particular antimodern sensibilities these two men displayed during their interwar northern expeditions. Evidence for their beliefs arises from their rich troves of unpublished field notes and letters, and their published articles and books. Blanchet cultivated an active, martial, and playfully "Indian" engagement with the northern environment. His foils to this performance were the Denesuline (Chipewyan Dene), whom he believed had fallen into physical and mental decay through their contact with southern civilization. Douglas made carefully controlled forays into a northern environment that he viewed nostalgically, and which seemed to change little, in contrast to the rest of the world. His private experience of the passage of time in the north increasingly slid away from a public time in which he had less control over his journeys, and which yielded up what was, to Douglas, the lamentable industrialization of the region.

By examining antimodernism within the context of northern modernization and industrialization, I pursue a richer understanding of how historical actors experienced and responded to the momentous social and technological changes to life and work in this region between the wars. Canadians' renewed interest in the economic prospects of the Northwest Territories after 1918 motivated many of these interlocking developments. Excitement flared with Imperial Oil's discovery of a "gusher" forty-five miles north of Fort Norman on the Mackenzie River in August 1920. Private citizens poured down the river the following summer to stake claims, while the Department of the Interior hastened to erect the field office of the newly-minted Northwest Territories Branch at Fort Smith. This was only the opening act in a series of resource-oriented discoveries and rushes throughout the interwar period, in which Douglas (and Blanchet, upon one occasion) participated.

Earlier in this volume, Andrew Stuhl foregrounded the crucial role that travelling scientists played in developing and realizing plans for northern modernization and industrialization between 1918 and 1939. This chapter focuses, in complementary fashion, upon the contributions of travelling technicians, or technical fieldworkers, to the same ends. Much of the fieldwork and travel that occurred in the interwar north was related to the potential or actual development of resources. Although the federal government left such work to the private sector, several of its branches provided maps, surveys, aerial photographs, and other information to prospectors and mining companies to aid and abet industrial activity. The Geological and Topographical Surveys sent teams of fieldworkers north throughout this period to perform geological investigations and control and track surveys, as well as to investigate the incidence of timber, minerals, water powers, and other resources. Such efforts were directed mainly toward the Great Slave Lake and Great Bear Lake areas, where the corporate and industrial gaze rested most keenly. The director of the Northwest Territories and Yukon Branch (as it was known from 1923), O. S. Finnie, also sent special investigators and exploratory engineers on fact-finding missions throughout the region up to 1931. Working in sparsely populated areas on stringent government budgets, these men, including Blanchet, W. H. B. Hoare, L. T. Burwash, and J. Dewey Soper, travelled predominantly by the age-old methods of dog team, canoe, and foot, and occasionally by the newer expedient of the power boat.

Journeys powered by muscle and sinew continued into the 1930s, but they were increasingly overshadowed by newer adventures in what Morris Zaslow terms "the war-induced fashion of directed, organized mass actions attacking difficulties head-on with the latest technology."[4] Between 1928 and 1932, several aviation companies partnered with central Canadian mining interests to initiate large-scale aerial prospecting operations in the Northwest Territories. Companies such as Dominion Explorers Limited (Domex) and Northern Aerial Mineral Exploration (N.A.M.E.) spent unprecedented amounts of money and manpower west of Hudson Bay in 1928, north from the Shield to the southeastern shores of Great Bear Lake and around Bathurst Inlet in 1929, and in the Coppermine Mountains and around Great Bear Lake in 1930, searching for wealth in the northern rock.

Although such levels of expenditure proved unsustainable as the Depression worsened, the increasing value of gold and the already high value of radium set off another wave of rushes after 1932. More prospectors now flew to the shores of Great Bear Lake, Great Slave Lake, and Lake Athabasca to stake claims on increasingly trodden territory. The blossoming interwar symbiosis between the northern mining and aviation industries led to improvements in the north's communication systems through the spread of air mail and wireless services. Perhaps the most significant amelioration to northern life, however, concerned the region's transportation networks. River transport options were expanded through competition between companies and with the airlines, and were modernized, resulting in faster, better fleets and lower freight rates. Regular air service was available in the Mackenzie district from 1929, and airplanes became an effective, reliable means of transportation there in the 1930s.

In her recent monograph, Liza Piper deftly demonstrates how the post-1921 industrialization of the large northern lakes drew that region into global economic networks of capital and production, while simultaneously remaking local environments and societies. This was an integrative endeavour that worked with, not against northern ecological and environmental constraints, and that built upon extant networks of knowledge and movement embedded in the pre-war north.[5] The additive nature of change meant that old ways of being in the north rubbed alongside the new for some time after 1921. The geographer Trevor Lloyd observed in 1943 that "northern travel … is a curious mixture of methods familiar 150 years ago with those of today."[6]

At this point of friction, a critique of northern industrialization and modernization emerged in the interwar era among men like Douglas and Blanchet, southern fieldworkers with many years of northern experience. They were wary of the ways in which the new economic and technological imperatives of these large-scale processes were coming to privilege relationships between northern sojourners and environments predicated upon *distance* rather than *proximity*. One vital factor in this paradigmatic shift was the increasing application of fossil fuels to northern travel. Outboard motors and airplane engines "substituted for and magnified human effort" in a manner that disrupted the traditional framework for experiencing and thereby knowing the north.[7] This was, in Bruce Hevly's excellent phrase, the "authority of adventurous observation," an exploratory notion which construed heroic bodily challenge as a marker of epistemological credibility.[8] Northern travellers who had fought the rapids of sprightly northern rivers, portaged blackfly-plagued trails, and mushed long miles through frigid northern forests earned some authority to speak about the environment and to be believed because they had endured the rigours of the field. By obviating, and thus distancing the human body from such rigours, motor-bound travel challenged, and would eventually make obsolete, older rubrics of authority and experience that had assigned a premium to information gained through proximate, often difficult encounters with northern landscapes.

Airplanes also made panoptical views newly possible in northern Canada. The budding normative relationship between distance and knowledge was fanned by the federal government's program of aerial surveys, which began in the late 1920s and continued into the postwar period. As Stephen Bocking has shown, the airplane made possible a new mode of interaction with the northern landscape, one that asserted an objective kind of authority through its divorce from immediate but sometimes faulty sensory experience.[9] Moreover, as Marionne Cronin describes in her chapter in this volume, the northern bush plane lay at the heart of an emerging modern discourse of colonial knowledge and possession that privileged distant views as the best means toward the codification of the unknown, and ultimately toward the rational, progressive conversion of "empty" northern space into places of settlement and industry. Despite Douglas' and Blanchet's participation in this colonial enterprise and the integration of aerial transport into their fieldwork, they were united in their fundamental ambivalence toward the airplane. Along with many of their sojourning colleagues, they regretted the passing of the active, bodily engagement with the land that older methods of travel had fostered, and mourned the waning authority accorded to close encounters with the northern environment. Unlike the inexperienced southerners bound for the Klondike that Jonathan Peyton describes in his chapter, seasoned northern travellers like Douglas and Blanchet usually found such encounters agreeably stimulating, rather than frustrating or alienating.

The laments of Douglas, Blanchet, and their like-minded colleagues point to the "precarious vulnerability" of the modern colonial and industrial projects pursued at this time in northern Canada. New imperial scholars have taught us to probe with care the disquiets, estrangements, and yearnings embedded within such projects, for such reactions expose the inconsistencies and weaknesses within larger colonial networks of power and knowledge.[10] Antimodernism provided certain structures of feeling through which Douglas and Blanchet could express their doubts about the ultimate benefit of the modernizing project in which they were engaged. Feelings could literally lend structure to places: in the course of their fieldwork, both men created unique personal topographies and time-scapes, which they superimposed upon various northern landscapes. The following analysis unfolds the importance of these spaces partly through the close examination of *traces*, as described in the published and unpublished narratives of Douglas' and Blanchet's northern expeditions. Traces are material inscriptions of past activity on a landscape. As William Turkel notes, they can be read as "indexical signs," or things that signify other things and thus imply physical or causal connections between the two.[11] Douglas and Blanchet used traces upon northern ground to interpret past events and to construct historical narratives that enabled them to make sense of changes there and in the outside world. Traces also bridge the era in which they were created and that in which they are read, "making the past legible and eroding temporal boundaries."[12] They both clarified and blurred experiences of northern time as well as space for this chapter's protagonists, as will be seen below.

The significance of the traces herein lies in their contestation of the emerging hegemony of distance, which the new industrial order was embedding within structures of power, capital, and knowledge in the interwar north. Resources there were severed from local environments, transformed into commodities, and sold in faraway markets for the benefit and enrichment of southerners.[13] Meanwhile, as a new generation of prospectors, geologists, and surveyors flew over northern landscapes in the 1930s, their views from on high literally shrank evidence of a peopled terrain—the tiny settlements, the "clearings, graves, and debris that people leave behind"—to insignificance and irrelevance.[14] Still travelling over ground, Douglas and Blanchet noticed and valued these traces. Their continued attachment to proximate ways of knowing and experiencing the northern environment speaks to a divergent, less well understood set of sensibilities. It offers a point of departure for telling a different story about sojourners' experiences of—and resistance to—change in the modern colonial north.

Full appreciation of these sentiments requires a brief, formal introduction to the persons to whom they belonged. George Douglas (1875–1963), born in Halifax into a family of military officers and medical doctors, had originally wished to become an officer in the Royal Navy. After failing the necessary examinations, he trained instead as a marine engineer in Newcastle-upon-Tyne, and served as fifth engineer with the White Star and Allan Lines until 1900. A stopover in South Africa during the Boer War soured Douglas on the warship and armaments work in which he had apprenticed, and to which he had planned to return. His cousin, the American mining magnate James Douglas, employed him instead as chief power plant engineer at the Moctezuma Copper Company mine in Nacozari, Mexico. Over the next decade, Douglas specialized in the development of gas-powered engines and generators for use in mining operations. He worked as a roving consultant for the Douglas consortium of copper and silver mines in the American southwest, and later for Phelps-Dodge, which bought the Douglas holdings in 1908. In the interwar era, he also worked for the United Verde Company of Clarkdale, Arizona, and the Anaconda Copper Mining Company. Funding from these interests and others enabled Douglas to make a series of exploratory and prospecting trips to Great Slave Lake and Great Bear Lake between 1910 and 1940.

Guy Blanchet (1884–1966), born into a genteel lower-middle-class Ottawa family of thirteen, also longed for an adventurous, mobile career. His degree in mining engineering from McGill led him first to Lille, Alberta, a mining town near the Crowsnest Pass, where he worked as supervisory mining engineer in 1905–06. He then worked on surveying parties based out of Edmonton for several years, and passed the exams to become a Dominion Land Surveyor in 1911. As a newly minted federal civil servant, he spent the next decade as chief of parties surveying baselines and meridians in northern Alberta, Saskatchewan, and Manitoba. He excelled at this difficult work, prompting the Surveyor General to name him, in 1918, "one of the most competent and efficient surveyors in the service."[15] Between 1917 and 1919, while undertaking exploratory work on the headwaters of the Churchill River in northern Saskatchewan, Blanchet designed and implemented modifications to the standard control stadia survey. He did so to facilitate the accurate capture of the complicated shoreline and island topographies characteristic of the northern lakes on which he was travelling. This method was later applied to the Topographical Survey's work north of the sixtieth parallel in the 1920s, in which Blanchet played a leading role. The details of Douglas' and Blanchet's interwar fieldwork in the Territories are embedded in the narrative that follows.

A Southern Antimodernism for the North

Antimodernism can be understood as a complex of related reactions to the rapid structural changes that modernization brought to everyday life in the Western world in the late nineteenth and early twentieth centuries. These included the emergence of secular nation-states and of capitalist, industrial economies, the dramatic adjustment of perceptions of space and time brought about by new systems of communication and transportation, and the emphasis upon new ideals of individualism, science, and technical rationality.[16] While antimodernists were not immune to the widespread enthusiasm for material and moral progress that accompanied these changes, they often proffered dissent, based on their memories of and nostalgia for past modes of existence.[17] Although Lears originally characterized antimodernism as a late nineteenth-century movement prevalent among the middle and upper classes in the northeastern United States, further studies have replicated, confirmed, and built upon his conclusions to the point where antimodernism has been declared a fundamental aspect of the twentieth-century experience.[18] Various North American, Ontarian-Canadian, and northernist strands of antimodernism provided the context for Blanchet's and Douglas' actions.

One pervasive and persistent outcropping of antimodernism, the back-to-nature movement, spanned North America and bridged the late nineteenth and early twentieth centuries.[19] Most commonly found among middle- and upper-class city dwellers, it was rooted in a worry that modern, urban living, often equated with "overcivilization," could be detrimental to one's health. The hectic pace of city life was thought to bring about nervousness and anxiety, while the lack of physical activity in white-collar jobs, in which an increasing amount of the population worked, could cause physical enervation. Either way, the issue was the sapping of one's strength, for which the term neurasthenia was devised, and for which the cure of outdoor recreation was most often recommended. In rural and wild environments, urbanites sought remedial activities that ranged from the gentle, such as cycling, birdwatching, or walking in the countryside, to the strenuous, which included "hiking, camping, canoeing, and alpine climbing."[20] This particular kind of recoil from overcivilization, in its focus on active health and wellness, was an important plank in the therapeutic world view of antimodernism.

The rise of the wilderness holiday was closely connected to trepidation among social critics in the United States, Britain, and Canada that individual cases of neurasthenia heralded the collective physical and mental degeneration of their societies. This reading of events reflected the cultural prevalence at the turn of the century of various theories about race, evolution, and biology based loosely on the work of Charles Darwin. Social Darwinism, as developed through the ideas of Herbert Spencer and others, cast "all sorts of contemporary problems or phenomena as symptoms of racial decline" brought about by overcivilization.[21] Critics fearing for the future concentrated most on the young, white, and male body—that belonging to the paradigmatic citizen and future national leader.[22]

Cultural representations of masculinity attached to Canadian soldiers serving in the First World War stressed the role of the country's wilderness in toughening the minds and bodies of men for European battlefields. The nation's archetypal soldier was conceived to be a "pure and rugged backwoodsman who lived his life far from the stultifying influence of city and university," despite the rather different composition, in reality, of the national corps.[23] In the interwar period, particularly in central Canada, masculine antimodernism was tinged with anti-American and pro-northern sentiments. Degeneration of the Canadian population was commonly associated with the growing evidence of American cultural influence in the country. Personal and national regeneration could be accomplished through recourse to the iconic Canadian therapeutic space, the north, which many thought was the source of the country's racial identity and potency.[24] This was less often the far north of the Subarctic or Arctic than the north of the Canadian Shield, which was relatively accessible to central Canadian cities.

In addition to its characterization as a primeval wilderness playground in which men could regain lost strength and vigour, the north was also regarded as a proto-agricultural and proto-industrial landscape. It became understood as a "second" frontier, following upon the first western Canadian one, that a new wave of pioneers would develop for the benefit of the nation.[25] This view was strengthened by the mineral discoveries made throughout the near and far north in these decades, and the popular and corporate enthusiasm with which these were greeted. Contemporary government policy on the Arctic and Subarctic also reflected this tendency to look to the past and the future simultaneously. In the case of wildlife management, federal bureaucrats combined "the antimodernist desire to preserve wildlife as the most visible remnant of an authentic but fading wilderness and the modern faith in bureaucratic management as a means to cultivate and manage wildlife populations for recreational and commercial purposes."[26]

The following discussion will focus on the backward-looking, nostalgic elements of Blanchet's and Douglas' antimodern sentiments. Yet both also worked toward and believed in the development of the north. After Douglas' first expedition in 1911–12 found traces of copper close to Great Bear Lake, George urged James Douglas, the expedition's sponsor, to fund further exploration in the region, particularly around McTavish Bay. But worldwide copper markets had tumbled in 1912, and James Douglas became more interested in efforts to locate carnotite deposits in Colorado from which uranium could be refined. The great irony, as Douglas recounted regretfully in letters to friends, was that had his cousin sponsored a second pre-war or wartime expedition to Great Bear Lake, the deposits of pitchblende there could have been located fifteen years prior to Gilbert

LaBine's famous encounter with that same mineral. Douglas believed that James Douglas' great desire for radioactive ores, combined with Phelps-Dodge's massive amounts of capital and organizational capacity, would have led to an earlier, more methodical development of industrial mining on Great Bear Lake.[27]

In their publications of the interwar period, both Douglas and Blanchet argue for careful optimism about the productive future of the north.[28] They favoured prudent, long-term development of the country by large mining corporations or by governments, even as the economic boom of the late 1920s fostered an opposing paradigm: a clutch of new, well-funded syndicates with little northern experience and high hopes of striking it rich quickly in so-called "virgin" territory. Domex courted both men as potential fieldworkers, and Blanchet accepted the temporary position offered because the company's owners shared his belief in the north's potential: "I was getting tired of being a prophet in the official wilderness and to meet a group of men who had faith in the country and asked me to help them prove it to be worthwhile—it offered a new outlet."[29]

Personal profit seems never to have figured in their enthusiasm for development. Rather, the chief feature of their northern progressivism was their belief in the region's economic potential, and their corollary desire to prove prevailing myths about the north—its barrenness, emptiness, harshness, and uselessness—wrong. This was particularly so when it came to the northern places they knew best and cherished most. In a letter to his fellow explorer Vilhjalmur Stefansson, Douglas was uncharacteristically boosterish: "With the great plains to the south and east adapted to raising reindeer, with coal, iron, copper, and oil deposits within a few hundred miles of each other, the [Great] Bear Lake country is bound to be one of the greatest mining and industrial districts of the continent. ... It is only a question of time."[30] A passage in one of Blanchet's government reports placed the reader upon an imagined promontory farther to the southeast: "Viewing the so called Barren Lands in August ... enlivened by the colours of its vegetation and animated by the roving bands of caribou, it seems incredible that the country is destined to remain an unproductive waste."[31] Yet even as they constructed future hyperborean empires in their minds, Blanchet and Douglas sought out the interwar north precisely because civilization and industry had seemingly touched the region but lightly. This "pre-modern physical and psychological zone of retreat" permitted them to experience important things that they found lacking in modern life, and to achieve rejuvenation through those experiences.[32]

The Martial North

From 1923 to 1926, Guy Blanchet undertook summertime fieldwork on the western Barrens, heading north, east, and southeast from Great Slave Lake on behalf of the Topographical Survey of Canada. He led parties of four engaged in controlled stadia traverse and exploratory track surveys that aimed to assess, more correctly than previous surveys, the number, sizes, and shapes of the rivers and lakes surrounding the height of land dividing the Arctic Ocean and Hudson Bay watersheds.[33] This work was meant to extend the network of control surveys and magnetic observations in the north, to provide ground control for future aerial surveys, and to enable the immediate publication of new and better maps for public distribution. Blanchet was also tasked with gathering information about the character and resources of the country east and north of Great Slave Lake, about which the federal government still knew very little. His official instructions each year mandated a report upon methods of transport, routes of travel, Indigenous peoples, game and wildlife, mining activity, geological structures, timber, waterpowers, and other notable residents and features of this region.[34] Blanchet also took numerous photographs of the landscapes through which he travelled; other members of his party made studies of flora, bird life, and geology, according to their individual expertise. Blanchet's yearly reports on the Great Slave Lake region were collated and published in pamphlet form by the Northwest Territories and Yukon Branch (NWTYB) in 1926.

In 1928, Blanchet was seconded from his position with the Survey to become coordinator of operations for Domex's new program of aerial exploration. Blanchet worked at their base at Tavane, on the west coast of Hudson Bay, during the year 1928–29. He wrote bimonthly reports to the Topographical Survey, conveying information about that region's geography, topography, natural resources, Indigenous cultures, sea ice and marine wildlife, and ease of aerial operation. This last category included comments about typical wind and weather conditions and visibility during flights made in all seasons, as well as the suitability of various landing and docking places nearby. Blanchet also made aerial exploratory surveys and sketches, and took magnetic observations throughout the winter. Upon his return south, the NWTYB published a report on the resources of the Keewatin and northeastern Mackenzie districts based on Blanchet's missives from the field. Detailed personal journals from all but his 1923 trip are extant, and take the form of separate or continuous letters to his wife, Eileen. Together with the many articles, several pamphlets, and one book that Blanchet published based on his fieldwork in this decade, the diaries reveal several strains of antimodernism that reflect wider cultural discourses about masculinity, savagery, and civilization in early twentieth-century Canada and the United States.

From the late nineteenth century onward, hegemonic ideals of masculinity in North America took on a physically active, vigorous, even martial character.[35] The figurehead of this movement was the American president Theodore "Teddy" Roosevelt. His call for a "strenuous life" in the 1890s became synonymous with any kind of virile, manly endeavour, and antithetical to the overcivilized effeminacy thought to pervade North American culture. Blanchet's journals reflect many of these concerns regarding civilization and chronicle his adopted solution: an active and "natural" masculinity, entwined with antimodern sentiment, which would counteract the banes of modern life. He celebrated certain bodily signifiers of masculinity, a martial engagement with the landscape, the elevation of the "primitive" life over that of "civilized," urban society, and the revitalization of his mental and sensory abilities.

Over these arduous summers of travel, Blanchet gloried in the strengthening of his frame, particularly the conversion of fat into muscle. "I was feeling myself last night to see if I was losing my fat and I think all the soft stuff has gone," he wrote, not long into their travels in 1924.[36] By the end of these seasons, he often felt in top physical condition: "I don't think I ever felt more fit in my life. Lean and my muscles working smoothly…."[37] The sleekness of his middle-aged body in the field prompted positive comparisons with youthful vigour. He enjoyed being "flat + lean as a boy," disparaging by implication the softer contours of his urban body.[38] Blanchet's short, compact figure rendered him peculiarly fit for an iconic display of strength in the north—the ability to pack heavy loads over portages using a "tump line" anchored around one's forehead. While Douglas reckoned conservatively that the average man could stand 70-pound loads,

Fig. 5.1: Guy Blanchet in August 1924, standing beside a cairn his field party had built on the shore of the Coppermine River. Another masculine feature of Blanchet's northern fieldwork was the opportunity to grow a thick beard, such as he sports here. Image J-00323, courtesy of Janet Blanchet and the BC Archives, Royal British Columbia Museum. Cropped from original photograph.

Blanchet regularly took 120-pound loads, and could pack 140 pounds over half a mile.[39]

A connected theme in Blanchet's journals is the sheer force with which he confronted the northern landscape. His task of clarifying the headwaters of the arctic and Hudson Bay watersheds involved frequent, heavy portaging between innumerable small lakes and rivers over boggy ground. The exigencies of the land necessitated recourse to the lighter outfit of older days—the canoe, paddle, and tump line—rather than the large, gasoline-powered boats, towing scows full of supplies, more common to that era. Every year, his journals use explicitly martial language to describe the preparation and execution of his fieldwork. Once both animate bodies and inanimate equipment had been trimmed down to "fighting weight," they journeyed through country that Blanchet characterized as recalcitrant, particularly at the watershed, where rivers could flow in a confusing variety of directions.[40] "We have been fighting this height of land hard and getting so little for it and failure would be so awfully disappointing," he wrote after one particularly frustrating day.[41]

However, he greatly enjoyed strenuous travel, and gained ample satisfaction from being able to navigate through difficult landscapes. Looking back on his 1926 trip, "a hard one through difficult country," he averred that "we fought it without regard to obstacles." His compensation for such struggles was the astonishing beauty of the Barrenlands in high summer, as described in vivid technicolour in his journals and reports. The view was "a pleasing one of gently undulating to moderately rolling country well covered with shrubs and moss on the slopes and grass in the bottoms, colouring it a vivid green. ... After the first frosts have come, a still more striking effect is produced by great splashes of crimson and yellow of the saxifrage, labrador tea and blueberry bush."[42]

In undertaking this work, which doubled as a kind of leisure, Blanchet was fulfilling a deep-seated need for manly testing that the modern world seemed no longer to offer. One strand of antimodernism, in reaction to this civilized ennui, exhorted a martial, activist cult of experience, often identified with outdoor exercise. Such activity served as an antidote to excessive mental work and provided a taste of preindustrial vigour.[43] In Canada, the northern wilderness was the paradigmatic *topos* in which to realize the strenuous life. The more immersed Blanchet became in his travels there, the less appealing did aspects of his normal life appear. Echoing the concerns of critics of overcivilization, he decried the negative effects of office work and urban habitation. Life in Ottawa came to seem confining and constraining: "I can only stand so much of it and then would burst in some way if I didn't get out."[44] By contrast, life in the field was "a life for a man everyday a fight and always facing decisions."[45]

The mention of decisiveness is not incidental. A prominent fear connected to overcivilization was that the thickened social webs characteristic of modern life made it more difficult for a person to take independent steps to shape his or her own fate. In defence of individual autonomy, activist antimodernism celebrated the ability to act decisively.[46] This quality inhered in Blanchet's primary maxim of northern fieldwork: "Make up your mind. Don't stand on one foot: do whatever you think best, but *do* it."[47] This stress on active mental engagement can also be traced back to the core driver of antimodernist sentiment, the longing for regeneration through intense physical and mental experience. The two kinds of challenge were linked: "In spite of all that goes to make up this life it is a good one. You enter soft finicky loving ease + pleasant things and for a while it pounds and bruises you but you emerge with muscles + nerves steeled with the primitive capacity of being able to meet and conquer your difficulties instead of passing them on to some paid agency."[48]

Not only Blanchet's body, but also his mind found a welcome challenge in "stepping off the map" each summer. This phrase, written near the beginning of most of his journals, seems to have symbolized his entry into what he considered "unknown"—and, in an antimodern sense, mentally therapeutic—northern space.[49] Blanchet's journals often contrasted "known" and "unknown" space, invariably favouring the latter. Travelling familiar roads produced lethargy rather than contentment. Mental alertness, by contrast, was linked to the traverse of unknown trails: "It is life after all when you leave the beaten paths and eddys and strike out into any new thing physical mental or moral."[50] While living among urban crowds dulled sensation and emotion, Blanchet found that his fieldwork rendered him newly observant and sensitive. He relished seeing and studying new things, and hoped to discover something completely unknown to his society.[51]

The Primitive North

Entwined with these active, martial, and masculine ideas were those of a related notion: that of the primitive or "Indian." The primitivist discourse of the late nineteenth and early twentieth centuries overflowed with tropes expressing a variety of things about Indigenous peoples real and imaginary, past and present. Blanchet simultaneously sought to acquire and emulate aspects of Indigenous cultures and ways of knowing while maintaining that such peoples were degenerating, and he was not alone in holding such competing views. His doubts regarding the necessarily positive nature of progress, along with his perceptions of the dangers of overcivilization, led him, as it did many others in that era, to play at being an Indian.[52] As Robert Berkhofer notes, non-Indigenous constructions of Indigeneity often arise out of Euro-American impulses to expose and correct the perceived shortcomings of their own societies.[53] By temporarily inhabiting what he conceived as the skin of the Other, Blanchet was able to jettison the aspects of civilization he disliked. Following the classic arc of antimodern feeling, he then drew upon appealing aspects of "primitive" culture in order to regenerate his body and spirit.

However, Blanchet located most of these admirable traits in the lives and cultures of historical northern Indigenous peoples, whom he distinguished from their physically and morally "degenerate" descendants of the 1920s. The notion of the physically or culturally vanishing Indian was the predominant non-Indigenous interpretation of Indigenous peoples on the North American continent in the late nineteenth and early twentieth centuries. Like many of his southern Canadian contemporaries with some experience of northern life, Blanchet considered that Indigenous peoples of that region had fallen prey to the comforts of civilized life at fur trade posts. As trappers dependent on southern goods and prices, they seemed to Euro-Canadian onlookers mere shadows of the independent and virile hunters they had been in former days. Considering Blanchet's avid self-identification with primitive peoples, it was both ironic and rather fitting that he quickly came to be regarded, and to regard himself, as one of another vanishing kind: heroic northern explorers of the pre-industrial age.

Blanchet's Barrenland surveys gave him both the inspiration and the opportunity to incorporate some self-consciously primitive aspects into his conduct of fieldwork. By the mid-1920s, he had developed a "hard" style of northern travel, a significant feature of which was the evanescence of his outfit. He presented this choice as both localized and normative in one article: "There is a general principle when travelling in the North that the less baggage you have, the farther you may go. Moreover, if you know the country and where and how to hunt and fish, there is not much danger of shortage in summer, when living off the country. ... If one wishes to go far, he must go light."[54] Before departing for the summer, Blanchet would cut and recut items from his slim list of supplies: "The final trimming of the cargo, in clearing the paddling places [in the canoe], generally resulted in the sacrifice of a sack of flour or bacon."[55] He would also shed food along the trail if indications of game in an area looked good and if he wished to lessen the number of time-consuming portages necessary to move forward.[56] His luck in finding game held throughout these summers, and he looked back on "blue" or starving periods with fondness: "Home again in the base camp after 11 days of wandering on 4 days grub after all in retrospect I find we were very lucky and lived well. We ... had no hard times and fish came at the critical times."[57]

Blanchet affected a disinterest in food generally while in the field, presumably related to his desire for a slim physique. The only relish he obtained in gustatory matters was the sense of achievement that accompanied the active, martial filling of the meat pot.[58] He went so far as to flirt mentally with the idea of starvation: "I really in the back of my mind wouldn't mind if we had to starve a little to see what the mental reaction would be."[59] In this thought, Blanchet mirrored the intrigue of martial antimodernists with ideas of pain and suffering, which they considered had been largely excised from modern life through analgesic advances in medicine. Encounters with these states of being were yet another way of warding off overcivilization; they proved that the modern body could still handle and overcome physical adversity.[60] Blanchet admitted that, strangely, he loved travel most when it was studded with hardship: "I like better to suffer on the trail for the interest of it."[61]

Blanchet's first foray into the Barren Grounds occurred in the late summer of 1923. He was accompanied by Joseph King "Souci" Beaulieu (1858–1929), a Métis man with ancestral ties to several Dene communities, and "Black" Basile, a Yellowknives Dene (T'satsǫot'ınę) man. Blanchet identified them both as Chipewyan.[62] Blanchet had travelled with Indigenous guides before, but this particular encounter seems to have had an important effect on his subsequent work in the Barrens, which was largely independent of Indigenous assistance. More a reconnaissance trip than an exploratory survey proper, it gave him a model of the primitive past to follow in subsequent journeys, and revealed to him the "decay" that he believed contact with civilization had effected among the Denesuline.

Blanchet told the story of their journey as follows: he had been laying out what he considered minimum supplies for a month when Souci halted him.[63] "What for we carry store food to country where meat abounds?" he reportedly exclaimed. Blanchet came to agree: "At least we should experience life there as did the Indians in the old days. As with them, it might be feasting or there might be times of famine." Their party had both times of plenty, marked by the constant hunting and eating of caribou, and of scarcity, when, in three days' travel, they were only able to catch two fish. Significantly, Souci and Basile were both older men. When Blanchet had inquired after younger guides, the manager of the trading post at Resolution had replied that the young men had no knowledge of the country and preferred to loaf in the settlement during the summer. Blanchet's companions are presented as relics of an earlier age when Denesuline families had travelled more widely over the Barrens to hunt caribou and the rarer, more elusive musk ox. Their journey in 1923 was portrayed as a last, sentimental visit to the land of Souci's and Basile's youthful days, "across the old forgotten hunting grounds of the people where the tent stones of the encampment were almost buried in moss." Similar themes of romanticized, ancient Indigeneity and the presumed degeneracy of then-present Denesuline appear throughout the journals and published narratives that describe the following three years of Blanchet's fieldwork in the Barrens.

Blanchet framed the East Arm of Great Slave Lake, through which his party passed into the Barrens, as a portal to a primitive world of the past.[64] The environment of the Barrens, mirroring the traits of the humans who lived there, appeared to Blanchet to be in the "early stages of natural development."[65] This emphasis on the "long ago and far away" was a key element of primitivist discourse.[66] One of the characteristic ways in which Blanchet deployed this theme was in his constant observation of traces, both ancient and recent, of Indigenous peoples on the land. These relics—old tent poles and stones, arrowheads, spearheads, piles of kindling, scraped caribou bones—signified the permanently anterior nature of the landscape, in which lives had been most vigorously lived in the distant past. From this scant evidence, Blanchet imagined possible actions for these ghosts:

> The trail of the people becomes fainter and fainter ... at one portage their trail was made very human by finding a toy boat whittled out of a stick and one pictured some little brown tot caribou grease to the ears sailing it in the eddy below the falls much as Christopher Robin might be sailing his in the Serpentine and another place—a winter camp of long ago where they made canoes for the spring caribou hunt I saw where the comely squaw of one of the tribe had got a scolding for leaving her awl mounted in a piece of caribou horn.[67]

By conceiving of this environment as a natural and human exemplar of the primitive past, Blanchet indulged his own primitivist wishes to live and travel in such an era and fashion. Not only was he vigilant about recording the remains of old campsites and hunting passes that he encountered, but he also chose sometimes to camp at those places. In motions imbued

FIG. 5.2: Basile (left) and Joseph King "Souci" Beaulieu (right) in the summer of 1923. Image J-00324, courtesy of Janet Blanchet and the BC Archives, Royal British Columbia Museum. Cropped from original photograph.

with practical and romantic sentiment, he retraced and re-enacted past Indigenous journeys over portages and routes, both those still frequented and those that had fallen into disuse. At times, he identified very closely with these bygone actors. One afternoon in the summer of 1925, having shot a moose that he subsequently butchered and hung to dry on wooden racks, Blanchet broached their party's similarity to a "party of Indians," and reflected on the joys of living the "Indian life."[68]

These traces were not evenly distributed across the land of Blanchet's travels, and their patterns suggested social and moral judgments. As they travelled further into the heart of the Barrens, recent signs of Indigenous life gave way to much older traces of people long dead. Blanchet read these latter inscriptions, especially those on the cusp of the old musk ox hunting ranges, as evidence of "the old days when the men were more hardy and adventurous."[69] He interpreted the withdrawal of then-present Denesuline from their former hunting grounds and their loss of knowledge of that region as a mark of decline.[70] This assessment seemed to be confirmed by his interactions with the Dene whom he did meet. At the end of July 1925, his party encountered two large families within one day, both travelling east to meet the caribou migration. When he asked them for information about the country in that direction, the young men of the first group could not reply. An elderly woman named "Soongo," however, became very excited, and indicated that she knew that area well. She could even describe it recognizably, as she had travelled there as a young girl. An elderly man in the second family was similarly well-informed. Blanchet concluded that "the old and blind among the Indians have had more varied experiences, as their lives reach back to the primitive adventurous life of their people, and in thinking over the travel of their early days they keep these routes and landmarks fresh in their minds."[71]

Variations on the same bleak assessment of the Barren Ground Dene run throughout Blanchet's publications of the 1920s. He presents these people not only as barely possessing knowledge about the regions they frequented, but also as fearful of the dangers that lurked there, such as starvation, disappearance, supernatural beings, or conflict with Inuit.[72] Their withdrawal from the Barrens, Blanchet proposed, meant that "life has become for them simpler and more secure, but the change has been accompanied by a certain physical and moral decay. They lack the courage to attempt a long and difficult trip and the stamina to accomplish it."[73]

With such statements, Blanchet used the Denesuline as foils for his own active, martial engagement with their traditional territories. Whereas such people, in Blanchet's view, always selected the easiest route of travel, magnified small obstacles, and could hardly conceive of steady, persistent effort, his own arduous ethic of work and travel, his choice of challenging, sometimes little-travelled routes, and his overall knowledge of the landscape rendered him, it is implied, more "Indian" than northern Indigenous peoples themselves.[74]

The respect accorded to Blanchet's overtly difficult, masculine, and primitive travels waned with the appearance of new technologies of travel in the north. As early as 1929, Blanchet felt himself set apart from the young Euro-Canadian men of the new industrial and airborne age who surrounded him at Tavane. Their easy acceptance of their own dependence on machines would, he believed, lead to the same physical and mental decay that he had always fled to the north to combat: "Some of my trips … demanded all you could give that the youths of today would not attempt without the moral support of gasoline."[75] He continued to believe in the necessity of an activist and martial masculinity to ward off over-civilization, but increasingly perceived that this fight was being lost—or, worse, not even being taken up—by his successors in arctic fieldwork. The weakening effects of modern life that Blanchet had tried to stave off throughout the 1920s seemed to have come to Euro-Canadian sojourners, as well as Indigenous peoples of the north, at last.

The Leisurely North

George Douglas' northern travels, like those of his friend Blanchet, arose from the strand of antimodernism known as the back-to-nature movement. Yet his actions were informed by a different engagement with nature, and were coloured by quite distinct motivations and emotions that require some situation in Douglas' unique life and personality. While Blanchet had always lived in cities or towns and encountered extra-urban environments as a visitor, Douglas had long been a resident of Ontario's Kawartha Lakes district, a focal point for back-to-nature enthusiasts since the late nineteenth century.[76] Growing up on Northcote Farm, situated on Katchewanooka Lake just north of the town of Lakefield, he became an expert canoeist and boatsman who travelled the Kawartha Lakes and the Trent-Severn Waterway often and widely, in all kinds of crafts.

Throughout the interwar era, he returned frequently to Northcote Farm between stints of work in Arizona and Mexico. He owned a smattering of properties on adjoining lakes, which sported shelters ranging from furnished cabins to tents on bare rock, and he moved easily and frequently between these on foot and by canoe, bicycle, and skis—but rarely by car. Douglas' northern expeditions were not vigorous respites from a sedentary urban lifestyle, as were Blanchet's, but rather a natural extension of an unusually active life lived outdoors, often under less than comfortable circumstances, by choice. They were a restorative and therapeutic kind of antimodern leisure even so, given that Douglas' career in arid climes afforded him few opportunities to spend time in a canoe or boat.

Unlike Blanchet, Douglas did not use his time in the north to shore up his sense of masculinity, and he was not captivated by real or imagined Indigenous lifestyles or cultures. But his private reaction to modernity involved a similar sense of alienation from the present predicated upon a longing for types of physical and spiritual experiences found predominantly in the past. His letters, diaries, and *aide-mémoires* throughout the 1920s and 1930s, written in the field and following journeys there, display a persistent nostalgia for and a precise remembrance of his previous northern endeavours. For him, past and present mingled to form a personal northern temporality tied to discrete spaces occupied by the same few beloved sites and traces.[77] His gradual loss of control over his movements and equipment, and thereby his experience of time and space, is mirrored by his representation of the disturbance and destruction of this familiar northern timescape, to which the chronicles of his later trips bear witness.

Because Douglas' antimodern sentiments were yoked so strongly to his personal history in the north, it is important to know something of his first and most celebrated trip there. In 1911, James Douglas sponsored George, his brother Lionel, and the geologist August Sandberg in their search for copper deposits around Great Bear Lake and in the Coppermine region. From Edmonton, the three men travelled to Waterways in northern Alberta, and thence down the Mackenzie River to Fort Norman, where they tracked up the Bear River to Great Bear Lake. They sailed across the lake to its northeasterly corner, where they overwintered. Both that autumn and the next spring, Douglas travelled northeast to the Coppermine

River, which he followed on the latter trip all the way to the Arctic Ocean. Douglas was an excellent photographer, and published some of the earliest photographs of this region in his book-length narrative of this journey, *Lands Forlorn* (1914).

Proper order and forethought were essential to Douglas' comfort and success as a northern traveller. His first expedition was a model trip in that regard. Over the winter of 1910–11, Douglas assiduously researched the environmental and geographical conditions of Great Bear Lake and the Coppermine region. He assembled what he considered to be the ideal outfit, which centered on two canoes that the Peterborough Canoe Company built precisely to his specifications. "I want to give the building of the canoes my supervision + a thoroughly practical trial, afloat, in swift water + on ice," George wrote to James. "Such elaborate attention to detail might be thought hardly necessary but … the more carefully preparations are made and equipment tested out the most sure we are of results."[78] Writing after their trial trip to the Coppermine in the autumn of 1911, Douglas averred that "only the perfection of our equipment pulled us through."[79]

Douglas' methodical approach served him even better as he prepared for his prospecting trips of the 1920s and 1930s. Whereas large mining exploration companies like Domex, N.A.M.E., and their successors of the mid-1930s could afford to fly men out to many different sites in one summer to search for profitable mineralization, this style of fieldwork was quite expensive. When exploring for various American and Canadian mining companies and investors, Douglas preferred to work with smaller budgets. He believed them a fairer investment, given the likelihood that a season's prospecting would yield no valuable result. He also chose to prospect principally by canoe and outboard motor, which restricted him to a relatively small area of operation each summer. In the preceding months, he would pore over geological reports, historical narratives, aerial photographs, and photostat maps to find segments around the great northern lakes with a number of promising outcrops, and that were close enough together to visit in a single intense season of fieldwork.

So much Douglas had control over, but the factors that he could not regulate intruded increasingly upon his leisurely antimodern experiences in the interwar north. To begin, he could not determine the time or manner of his return to the northern lakes. His 1928 trip was intended to be the first in a series of multi-year investigations that never came to pass, for want of money and interest on the part of his investor. The expedition he proposed to lead for Domex in the summer of 1930, for which he had high hopes, was also shelved.[80] Despite his careful selection of sites to investigate, his field seasons, particularly the last two, were unsuccessful. The wasted time, effort, and money put into such endeavours frustrated him greatly. But what disheartened Douglas most were the environmental perturbations brought about by increased interest in the mineral potential of the Northwest Territories. These changes for the worse heightened Douglas' sense of nostalgia for and recourse to his cherished memories of his early expeditions to Great Bear and Great Slave lakes.

The Timeful North

In an insightful essay about the relations between time, modernity, and nostalgia, Kim Sawchuk defines this last state as "a melancholia caused by a protracted absence, a wistful, excessively sentimental and even abnormal hankering for the return of some real or romanticized period or irrecoverable condition or setting in the past." She argues that nostalgia may be regarded as a subconscious reaction to new understandings of space and especially time that emerged around the turn of the century, particularly the Western standardization of time in the 1880s. Noting that nostalgia may be "unintentionally demeaned because of [its] association with the feminine, which is itself associated with irrationality and sentimentality," Sawchuk wonders if this negativity might be replaced with a positive valuation of nostalgia, "a step towards a remembrance of things past, towards a history that includes the senses, a history of home and hearth." By refusing nostalgia, conversely, "we are ... hardened against feeling both the losses in our lives and the changes within our society and culture that are not of our making but that are in the interests of those classes who benefit from the ideology of unrestrained progress."[81]

This commentary provides an excellent context for the unique spatial and temporal mapping of Douglas' nostalgia across the north, as revealed in his sentimental attachments to the items he had used, the places he had camped, and the paths he had travelled while there. They also help to explain the disjuncture between Douglas' occupational identity, which purportedly favoured the rapid development of northern resources, and his private concern over the rate and extent of industrial activity in the north during the 1930s. The twin processes of an evolving northern industrial presence and the devolving of Douglas' control over his northern surroundings may best be understood by perusing his trips of the interwar period in sequence.

Douglas' second trip to the north occurred in 1928, seventeen years after he had first visited the region. Sponsored by the United Verde Copper Company, he and the prospector Carl Lausen made the first geological investigation of Great Slave Lake's southeastern shores. They focused on the territory between Charlton Bay and Stark Lake, and searched particularly for evidence of copper basalt flows. As in 1911–12, Douglas retained full control over his equipment and the rate and means of his travel. He elected to travel north from Waterways by canoe and outboard motor instead of booking passage on a steamship, which was then the choice of most travellers. Douglas wished to introduce his inexperienced companion to travelling by canoe. He also enjoyed the deliberately slow pace of travel, which enabled excellent views of familiar settlements and well-remembered landmarks on the Slave and Mackenzie rivers. On this trip, he established a base of operations on Eagle Island, located about ten miles northeast of the mouth of the Taltson River on Great Slave Lake, to which he would subsequently return.

Recounting the summer's work in an address to the Canadian Institute of Mining and Metallurgy, Douglas exhibited his punctilious attitude toward orderly provisions and preparation by describing and justifying their outfit in minute detail. He also noted the escalating contrast between old and new methods of subarctic fieldwork, as exemplified by other expeditions to the Great Slave Lake area that summer. H. S. Wilson and his party, working for the Nipissing Mines Company, had made a long, traditional overland trip between Great Slave Lake and Hudson Bay, using the Thelon-Hanbury river system to travel across the Barrens. They had "depended on nothing but their own strength, unaided by any later mechanical inventions such as out-board engines and aeroplanes," Douglas reported. Meanwhile, N.A.M.E. had also done extensive work "with its utilization of aeroplanes to an extent and on a scale never attempted before in prospecting."[82] Douglas experienced both these styles of travel in his next two trips, which took place in 1932.

That March, Douglas made a quick, secretive trip by airplane from Edmonton to Great Bear Lake in order to stake some coal deposits that he had noticed on Douglas Bay twenty years earlier, and which Gilbert LaBine's recent finding of pitchblende nearby might render commercially important. Although both Domex and N.A.M.E. had pressured Douglas to reveal the location of these deposits, he staked them for the Sudbury Diamond Drilling Company.[83] Even at this early time of the year, he was one of approximately thirty prospectors heading to Great Bear Lake, and his plane was one of nine either in operation or soon to be so, all in support of prospecting and mining activities.[84] While flying was novel to Douglas, his experiences on the ground were not dissimilar to his past ones there. Writing to Lionel, he affirmed that it felt like he had never been away from the north. Neither certain people, nor certain sights, nor even the taste of the excellent bacon, eggs, and scones in Mason's restaurant at McMurray seemed to have altered.[85]

Douglas' sense of déjà vu grew stronger that summer when he returned to Great Bear Lake once more, this time to investigate his wintertime stakings more carefully, and to examine the south shore of the Keith Arm for float coal. In his unpublished memoir of the summer, Douglas used one of his favourite literary constructions, which juxtaposed with precision his past and present experiences of exact points in the landscape:

> We ... passed through the narrow channel in the lake just as we had done in the "Jupiter" twenty one years before, almost to the day. ... We were now in well-remembered waters. On July 18th, 1912, twenty years before ... we had entered this bay in the "Aldebaran" on our homeward voyage from the Dease River. ... Now on July 17th 1932, in the "Alcyone," I ran along the same shore into the same beautiful bay, and came alongside at the same camping place.[86]

Their campsite from 1912 at Russel Bay, on the west side of Keith Arm, had not been disturbed. The stake to which they had formerly tied the "Aldebaran" still stood near the water, and their cut and stacked firewood from two decades before lit their fire that very night. The place seemed hardly to have changed, partly because the aridity of the lacustrine micro-climate slowed material decay.[87] Neither the extinguished logs of their old campfire nor the boughs that had comprised Douglas' bed of brush looked more than one or two years old: "It was indeed a strange experience to return to this well remembered camp, and find the traces we had left so little changed, while such momentous changes had taken place in the great world without, these frail branches of spruce had outlasted human empires and systems of civilization."[88] Exploring along the shoreline, Douglas discovered many anthropogenic scoriations, the most recent of which was far older—perhaps nearly a century—than their own traces of twenty years before. He conjectured that these had been made by the parties of John Franklin, Peter Warren Dease, Thomas Simpson, and John Richardson, all of whom had travelled along that coast in the first half of the nineteenth century. He delighted not only in his immersion in history, but also in the sensory delights of the lake—"the 'barren ground' conditions with its freshness and exhilaration, its scented atmosphere, its uninterrupted vision, the ease of walking and freedom of movement...."[89]

The summers of 1928 and 1932 had been relatively idyllic, if mostly unproductive in their results. Douglas returned to both great lakes of the north in 1935. With funding from a group of New York investors, he and his partner Bobby Jones sought out silver and copper on the southeastern shore of Great Slave Lake. Meanwhile, the other members of his team, Blanchet and a prospector named René Hansen, searched for gold near Beaverlodge and in the region adjacent to the north shore of Lake Athabasca. In 1934, coarse gold had been found in the latter area, sparking a rush parallel to the one then underway at Yellowknife. Douglas believed that the geological formation in which this discovery had been made might extend a good way north and northeast of Lake Athabasca, into unstaked ground.[90] The team of four also reduced the coal claim Douglas had staked three years prior at Douglas Bay on Great Bear Lake. In order to select the boundaries of the reduced claim, the ground of the original claim had to be traversed again and observed carefully. This smaller claim then had to be surveyed and staked according to federal mining regulations. Douglas, Blanchet, and the others spent nearly a week cutting lines through the bush along the claim's boundaries, running surveying chains down those lines, and digging pits three feet deep in the frozen ground to place iron posts at the corners of the claim.[91]

For the first time, discord and distress tinged Douglas' remembrances of the past and descriptions of the present. The summer's complicated schedule of fieldwork ranging across the north relied heavily upon the timetables of other people, particularly those of the pilots moving them from place to place. Not all of Douglas' coworkers met his exacting standards, or even proved congenial company. As he wrote to Lionel, it was "really a very lonely summer. Bobby was all right but terribly limited + ignorant while Blanchet with all his good qualities was not a pleasant travelling companion."[92] In troubling contrast to previous expeditions, Douglas did not have complete control over all aspects of his fieldwork. He was relieved at summer's close to return to Eagle Island and to find his beloved canoes safely stowed where he had unwillingly been forced to leave them for most of the season.

Eagle Island and its environs proved to be that summer's happiest haven of memories, which Douglas experienced in a manner reminiscent of his re-encounter with Great Bear Lake in 1932. His arrival after six weeks of fieldwork "to this beautiful and now familiar place was almost like coming home." Eagle Island was actually two islands, half a mile long altogether, separated by a narrow, shallow channel. The more westerly island was low, level, and well-covered with spruce, and the easterly was higher, rounded, and rather bare, revealing the purplish-brown quartzite from which the island was fashioned. A little bay made a perfect anchorage for Douglas' canoe, and a thickly wooded glade sheltered the boat he had had to cache. Douglas delighted in setting up his tent in a flat, sandy spot, and constructing a makeshift kitchen atop some smooth rocks. At a nearby camp they revisited, the traces of Douglas and Lausen's presence in 1928 were "unexpectedly fresh and recent." Their campsite's half-burned logs seemed "no more than a few weeks old." A stack of chopped wood, an old packing case, and, most gratifyingly, cans of gasoline and oil lay there undisturbed, ready to serve again as they had done seven years ago. The entire site was pristine, unmarred by any other signs of recent life.[93]

This feeling of stopped or slowed time had been even stronger upon Douglas' arrival, earlier that summer, at Douglas Bay on Great Bear Lake: "It was a curious and overpowering experience to find myself suddenly transported back to this familiar spot. There was nothing to indicate the passage of time: I felt as though I had never been away from the place...." But difference soon became apparent; the happy illusion was dashed.

Fig. 5.3: Eagle Island in the summer of 1935 (detail). The remains of Douglas' camp from 1928, including the packing case and cans of gasoline mentioned, are visible in the background directly above the "Ivaha" and in the foreground to its left. Photo: George Douglas. Library and Archives Canada, MG 30 B95, vol. 5, file 2, reproduction copy no. e011093048.

Douglas' cache of supplies from 1932 had been robbed, a thing that had never before happened to him in the north. It seemed a direct consequence of the changes in that region: "No Indian had done this.... Some of the low-down whites brought into the country by the 'Bear Lake Rush' had been guilty of this robbery." The accompanying destruction had been self-ish and wanton: cans of gasoline were deliberately split with axes to prevent others from using their stores, and the floorboards of the little cabin nearby had been completely riven in search of hidden valuables.[94]

A less shocking but still dispiriting experience was their survey of the reduced area of the coal claim, situated as it was in the densest of landscapes. Douglas described it as "an impenetrable forest of the thickest and toughest willows ... ever seen; a difficult country to walk over, covered with thick moss and tangled dwarf birch."[95] They spent nearly a week struggling with, bending over, standing on, and finally cutting down willows in order to run the boundary line for the claim over hills and through gullies. Their campsite provided neither peace nor respite for Douglas, who was appalled by its disorganization. For the first time, his familiar mental initiation of the comparison between past and present revealed a depressing decline in his circumstances:

> My thoughts turned regretfully to my previous camp in this same place, and the contrast it made with our present camp. Then we had a perfect, fairly light weight outfit. … We had the best of food and plenty of everything.… We were working hard, but not under the same pressure of limited time. There were only two of us, and perfect unanimity of opinion on matters of order and system in camp. … Contrast this with our present situation: confined to the shore, committed to the completion of a heavy job in limited time; dearth of game and faced with acute food shortage; and lastly, instead of a well-ordered camp, what I may describe as the sloppiest and worst run camp I ever had anything to do with in the far north.[96]

This charting of decline was one of two striking similarities between this trip and Douglas' last trip north in 1938, on which occasion he again did not have the luxury of using his own equipment. The other was the forceful appearance of industrialization, in the shape of the mines and mining camps that he saw being constructed on Cameron Bay in 1935 and Yellowknife Bay in 1938. The latter rose, quite literally and to Douglas' amazement, in the four weeks between two of his visits there. A "helter-skelter, hodge-podge city" had replaced the tents.[97] Despite his own participation in mining exploration, he felt distinctly uncomfortable watching the full-fledged apparatus of his profession come to the shores of his beloved northern Canadian lakes. The mining complex at Cameron Bay seemed "out of place, sordid, and discordant."[98] Walking around the Consolidated mine at Yellowknife three years later, he considered that it had "a good layout," but that it was "strange + depressing to see in this country."[99]

Supported by a small syndicate and assisted on the ground by René Hansen and another prospector named Tom Greenfield, Douglas spent his last summer in the north investigating the country at the headwaters

Fig. 5.4: George Douglas at Boulder Point, on Douglas Bay in Great Bear Lake, in the summer of 1935 (detail). He had camped at the same spot during the summer of 1932. Library and Archives Canada, MG 30 B95, vol. 5, file 2, reproduction copy no. e011093047.

of the Thubun River, southeast of Great Slave Lake. There, the three men searched for mineralization along a greenstone belt that hugged the lake's southern shore. A subsequent trip to the lake's North Arm, where more recent gold discoveries had stoked another wave of investment in prospecting, proved "a complete fizzle, a deplorable waste of time, effort, and money."[100] The country around the settlement of Rae, and around Slemon and Russell lakes and Snare River was already saturated with stakings. Douglas sent Hansen and Greenfield by plane to undertake six weeks' exploratory work in the country surrounding Nonacho Lake. Meanwhile, he made a slow, solitary circuit of the eastern and northern shores of Great Slave Lake, stopping to observe certain sites more closely, but otherwise making a farewell tour of his favourite northern people and places.

Upon his last visit to Eagle Island, his own traces, which once had seemed inviolate, were starting to smudge with the increased traffic around Great Slave Lake: "I was surprised to find nothing at our old [1928] camp had been touched. Dr. Lausen's + my old fire looked exactly as it had done when I saw it last (1935). ... My old windbreak still in fair shape also the table + big block of wood. But some Indians had had a winter camp at the head of the Pahie dock + the place looked generally used up."[101] The site's consumption mirrored physiological changes occurring inside of his own body, which had felt increasingly tired, old, and even pained on his last two northern journeys.[102] After returning to Ontario that autumn, aged 63, he retired from northern work altogether.

Throughout the years before and after the First World War, the combination of a favourable climate for preservation, an Indigenous respect for property, and a low population density produced pockets of northern space in which time seemed to move more slowly than in the rest of Douglas' world. In these nostalgic microspaces, things were invested with emotional significance far beyond their quotidian usage through their elevation to "relics," as Douglas termed them.[103] He did so half-humorously, perhaps in order to soften the impression one receives from his unpublished writings—that these were indeed precious, valuable objects to him, redolent of all the sacral connotations of the word, constitutive of a private northern reliquary that held the happiness of his past doings. Relics, like other kinds of ruins, can act as coherent, unified expressions of "all the uncertainties of change in time and the tragedy of loss associated with the past."[104]

While the remnants of campfires and brush beds evoked strong memories in Douglas, the survival of still-useful components of this memorial assemblage allowed the instant bridging of past and present, as though all the time he had spent away from the north had never been. Yet his constant self-insertion into a time that, for everyone else, was quickly ending if not already gone altogether, eventually rendered him as much a relic as the items clustered around him at familiar campsites. Left over from an earlier era of the north, like his friend Blanchet, Douglas felt and mourned its passage, commemorated it as long as he was able to do so, and finally left his traces behind for the last time.

The Royal Road to the North

The experience of time and space in the north underwent a stunning reversal in the late 1920s and throughout the 1930s. The felt, lived gap between the past and future north, into which Douglas and Blanchet found themselves slipping, was the consequence of a sudden rush in modernization brought about particularly by the introduction and quick uptake of the bush plane. Personal encounters with this means of conveyance gave Blanchet and Douglas a chance to reflect on the past, the present, and the future, and the methods of knowing the north that seemed allied to each of these ages. The airplane might be, in Blanchet's phrase, the royal road to the north.[105] But to where did that road ultimately lead? Both men saw the advantages and disadvantages of the methods they had championed, as well as those that new technologies were conferring upon the next generation of fieldworkers.

The airplane solved problems of knowledge, access, and recovery endemic to the north, which had hindered the development, longed for by government and industry alike, of that region's natural resources. It resolved the "paradox presented by a land that was, on the one hand, a source of potential wealth and, simultaneously, an obstacle to exploiting that wealth."[106] As Blanchet observed, the tradition of tough overland surveys stretching back to Samuel Hearne was reaching the limit of its usefulness.[107] At the rate of work then current in the 1920s, it would have taken more than a century to map northern Canada completely using terrestrial methods alone, according to one estimate.[108]

The airplane enabled the comparatively more cost-effective method of aerial surveying, which Marionne Cronin terms "a constellation of techniques clustered around the use of aircraft to acquire information about the country's geography and geology."[109] One such technique was simple aerial reconnaissance of the kind Blanchet had undertaken near Hudson Bay, where a surveyor observed the ground whilst flying and made notes and drawings as he did so. Aerial photography allowed the capture of landscapes from above, using the oblique method that had proved most suited to the northern Canadian environment.[110] Access to northern sites was also eased, as airplanes permitted institutions and companies to deploy teams of fieldworkers with less individual investment of time and money. These craft could move surveyors or prospectors between a series of pre-determined sites at scheduled times, drop supplies and mail to these parties throughout the summer, and then pick them up at the close of the season. Finally, the airplane facilitated the transportation of large quantities of equipment, supplies, and minerals from field sites in the north to southern industrial hubs in Canada or the United States, and thereby overcame the main limiting factor on the development of a northern mining industry.

Given the many aspects of life and work that the airplane ameliorated in the region, it was no wonder that between the wars, "the vision of a vibrant and prosperous North emerged as the centrepiece of air-mindedness in Canada."[111] In the 1930s, the airplane became as central to northern life as it remained peripheral everywhere else in Canada. Northern planes played an essential role in the new networks of transportation that enabled the twentieth-century industrialization of the Canadian Subarctic and Arctic. Workers, materials, and capital regularly moved by air between new northern towns and mining camps, such as Cameron Bay on Great Bear Lake and Yellowknife on Great Slave Lake, and southern centres of production and consumption.

The airplane's compression of time and space would have had a larger relative impact in the north than elsewhere anyway, given the much greater distances and much harsher ground conditions that invariably separated settlements there. What surprised southern visitors to the region most was the ease and speed with which northerners adapted to the collapse of time and distance, the casualness with which they spoke about places five hundred or a thousand miles away as though they were just around the corner.[112] If the north had once been the place where time passed most slowly in Canada, in the space of just a few years it had become the place where hours flew by at their quickest. As if to express this revolution, the juxtaposition of old and new technologies of travel, such as dog teams and airplanes, became the most popular and important visual and literary trope in representations of the north in the late 1920s and 1930s.[113]

Technologies of motion, as Sidonie Smith notes, have the potential to reorder many aspects of travellers' lives, including the narratives they tell, the evidence of their five senses, their relationships pertaining to gender, class, and race, and their perceptions of time and space.[114] Douglas and Blanchet were acutely aware that exploration by airplane had certain drawbacks, particularly in terms of how it positioned travellers differently within time and space. Douglas worried that the airplane's rapid movement over territory traversed much more slowly by travellers on the ground produced brief, superficial encounters that distorted the vagaries of the actual landscape:

It looks so darned easy when you are flying—it makes for erroneous conclusions. The sense of proportion is lost, and even worse than that such important things as common sense seem to lose their significance when travelling at height and speed. … It is a tool requiring discrimination and knowledge. The important things are thought and sight, and there is always more to see than you can take in no matter how slowly and carefully you go. As for sweeping over the country at two or three thousand feet above it and at a rate of 150 miles an hour—So far as learning anything about it you would do better to sit at home and study a map.[115]

Douglas' concerns touched not only upon the compression of the present, but also upon the degradation or severance of relationships between past and present times. Before the airplane's advent, northern travellers had regarded, and many continued to regard, the records and travelogues of their peregrinatory predecessors as invaluable resources. By contrast, Douglas found that some pilots espoused a cavalier neglect of the region's history of exploration; they were ignorant of great funds of information that might have been useful to them.[116]

Douglas' friend and fellow traveller P. G. Downes once commented that air travel gave one's experience of the north a "chattering and particularly inhuman" quality.[117] It was this inhumanity that Douglas and Blanchet deplored in the airplane's distancing of the traveller from the living landscape, human and non-human, below. Travelling at great heights and speeds in a sealed "metallic carapace" bore no resemblance to the "visceral mobility" of travelling by canoe and foot.[118] The aerial camera might allow the complete and accurate mapping of northern waterways, but muted the intimate details of the land, such as its soil, timber, and minor topographical features. This fear of lost intimacy runs throughout Blanchet's writings on the airplane, and extends to the experiences of the inhabitants as well as features of the land: "Much is gained by the wide view of the aerial camera but something is lost, matters important to those who dwell there."[119]

Those matters were also pertinent, if less so, to those passing through the region. Both Blanchet and Douglas agreed that to fly was to sacrifice much of the inherent interest of northern travel: the joy of flexing and pushing one's muscles to their limit, the immersion in vibrant sights, smells, and sounds, and the chance meetings with others in lonely spaces that provided social, mental, and physical nourishment. One of the greatest pleasures they took in flight was the ability to pass over familiar, hard-travelled ground and to see how it appeared from above. At times, the aerial view easily clarified details of topography that had taken much more effort to ascertain on the ground. However, when flying over landscapes they did not know well, they found the scenes unedifying. Regarding his aerial trip from Rae on Great Slave Lake to Cameron Bay, Douglas wrote, "No doubt if I had passed through the succession of lakes by canoe I would have found an interest in recognizing the different places I had camped, or found particular difficulties. But from a plane the effect of these countless lakes is monotonous."[120]

For Blanchet and Douglas, travel by air seemed to diminish the motions and relations that had preceded it in northern space. "From high above," Blanchet wrote, "you lose the pleasant little things that compensate when on the ground. ... It makes the real little things that you know and have been concerned with and valued seem small and unimportant."[121] In prizing such real little things, and in pinning their memories to the ground, these two men displayed an intimate and nostalgic form of imperial sentiment, at once deeply local and particular as well as common to fieldworkers the world over.

Meditating on the exploration of Australia, Paul Carter carefully distinguishes the engagements of imperial administrators and of explorers with unfamiliar landscapes, the latter encountering the land in uncoordinated, unsettling, and deeply personal ways. This essay has demonstrated that viewing northern spaces as "intensely humanized, saturated with local history and meaning" is not a privilege limited only to its inhabitants, as Mary Louise Pratt once thought. But nor should this kind of intimacy be read, as Graham Burnett once read it, as "a point of departure for a history of empire that renounces the imperial point of view."[122] The historical geographies of trace sketched above were neither truer, nor better, nor any less imperial because they were predicated on propinquitous rather than distant encounters with landscapes. Yet such encounters still complicate the predominant historical narrative of sojourners' relations with the interwar northern environment. Such people did not always welcome the lifting of their arduous burden of fieldwork that new networks and technologies of travel enabled. Rather, fieldworkers like Blanchet and Douglas desired the continuation of prolonged, proximal encounters with northern lands and spaces because, within the context of antimodernism, such experiences were coded as both enjoyable and therapeutic.

The activities of Guy Blanchet and George Douglas helped to forward the modern industrial order, in which Indigenous peoples, animals, and minerals of the north were increasingly displaced. As John Sandlos and Arn Keeling assert in their chapter later in this volume, the large-scale industrial exploitation of minerals in the Northwest Territories between the wars both prompted and enabled the Canadian state to begin consolidating its grasp upon this sparsely populated and tenuously held region. On the eve of the Second World War, people, goods, and information circulated by land, water, and air throughout the north, and between the north and the south, more efficiently than ever before, thanks to networks of communication and transportation smoothed and enhanced through liberal infusions of southern capital. Towns with many of the conveniences of modern life had appeared on the shores of the north's great lakes—Yellowknife on Great Slave, and Port Radium on Great Bear. Yet northerners did not always share in the region's increasing prosperity. The Dene, Inuit, and Métis were often indirectly or directly barred from accessing newly available resources and opportunities, a trend that would only accelerate in the 1940s and 1950s.

Douglas' and Blanchet's ambivalence toward the emergence of this new north is evident in their deliberately intimate, romantic, and nostalgic antimodern fantasies about certain northern landscapes. These produced emotional topographies tethered to significant traces that gestured, in synecdochical fashion, toward deeply personal narratives of time and memory that allowed Douglas and Blanchet to work through their anxieties about the modern world. While their topographies drew upon other activities and histories that had been fashioned on the same northern ground, they also displaced them, repeating that larger colonial gesture on a smaller scale. This is why, as Emilie Cameron has argued, historians need to turn away from the opposition of local and universal in favour of "accounting for the multiple, conflicting 'locals' at play in the production of imperial science" and knowledge in, and about, the Canadian north.[123]

The centrality of distance in knowing and experiencing the later interwar north shrank Douglas' and Blanchet's traces to insignificance. Forgotten was Blanchet's imaginative revival, through glimpses of toy boats and awls, of the remnants of lives very different to his own, and Douglas' bed of spruce boughs amidst many other tiny signs of humanity on the shores of Great Bear Lake. These small, soft scenes, smells, and dreams, so pointed and so vital to the bearers of these memories, seemed to have no place in a rapidly modernizing north, the progressive industrialization of which ground over and "devalued the social and economic arrangements it replaced" even as it built upon older material and intellectual networks.[124] Neither man was, of course, aware of all the lived nuances of the places in which they were ultimately sojourners rather than residents, and they certainly translated these landscapes through their own memories, hopes, and fears. But living, however temporarily, and travelling on that northern ground gave them the ability to perceive that, even if they did not understand all of the webs of social and material and natural relations that bent and stretched and sometimes broke with the changes that were happening, such things were still deserving of notice, respect, and, increasingly, mourning.

This essay has taken certain of Douglas' and Blanchet's thoughts and practices as symptomatic of larger currents swirling beneath the surface of early twentieth-century North American society, particularly the disillusionment with modernity and subsequent quest for alternative modes of being that was encapsulated in the movement of antimodernism. This interpretation accounts for the strangely contrasting yet simultaneous views that both held on the north. Their desire to regard it as a pristine outpost of the past coincided with a wish to see it become useful and important through the development of its resources. This double vision paralleled the dual purpose of their presence there: while they worked as surveyors and engineers to advance the region's future, they sought pleasure in returning to a natural world from which they drew different kinds of physical and spiritual sustenance unobtainable in their everyday southern Canadian lives. Blanchet's journeys epitomized a martial, masculine strenuousness nourished by an ideal of historical Indigeneity. Douglas sought encounters precisely ordered in space and time that were refreshed by their own immaculately preserved historical antecedents. Increasingly, however, industrialization and modernization intruded upon their fieldwork and leisure alike. These forces pulled them back from the land and the past, and pushed them to contemplate a future in which their values, methods, and indeed, their sensibilities, seemed to have little presence or influence.

Notes

1 I wish to thank Kathy Hooke, George Douglas' niece, for her warm hospitality and her generous willingness to share her family's historical documents and knowledge, which helped me clarify many aspects of her uncle's life and career. I also wish to thank Janet Blanchet for granting permission to publish photographs from her uncle-in-law's collection, and Gwyneth Hoyle, Guy Blanchet's biographer, for all of her advice and support.

Library and Archives Canada (hereafter cited as LAC), George Mellis Douglas fonds, MG 30 B95, vol. 5, file 7, George Douglas to Jim Douglas, 17 September 1935.

2 Queen's University Archives, George Whalley fonds, Location 1032c, box 2, file 3, Guy Blanchet to George Whalley, 17 January 1963.

3 T. J. Jackson Lears, *No Place of Grace: Antimodernism and the Transformation of American Culture, 1880–1920* (New York: Pantheon Books, 1981), xv.

4 Morris Zaslow, *The Northward Expansion of Canada, 1914–1967* (Toronto: McClelland and Stewart, 1988), 1.

5 Liza Piper, *The Industrial Transformation of Subarctic Canada* (Vancouver: University of British Columbia Press, 2009).

6 Trevor Lloyd, "Activity in Northwest Canada," *Journal of Geography* 42, no. 5 (1943): 162.

7 Piper, *Industrial Transformation*, 75, 287.

8 Bruce Hevly, "The Heroic Science of Glacier Motion," *Osiris* 11 (1996): 66–86.

9 Stephen Bocking, "A Disciplined Geography: Aviation, Science, and the Cold War in Northern Canada, 1945–1960," *Technology and Culture* 50, no. 2 (2009): 265–90.

10 Antoinette Burton, "Introduction: The Unfinished Business of Colonial Modernities," in *Gender, Sexuality, and Colonial Modernities*, ed. Antoinette Burton (London: Routledge, 1999), 1–16; Ann Laura Stoler, *Carnal Knowledge and Imperial Power: Race and the Intimate in Colonial Rule* (Berkeley: University of California Press, 2002). "Precarious vulnerability" is Stoler's phrase by way of Burton ("Unfinished Business," 2); see Stoler, *Race and the Education of Desire: Foucault's History of Sexuality and the Colonial Order of Things* (Durham, NC: Duke University Press, 1995), 97.

11 William J. Turkel, *The Archive of Place: Unearthing the Pasts of the Chilcotin Plateau* (Vancouver: University of British Columbia Press, 2007), 66.

12 Virginia Zimmerman, *Excavating Victorians* (Albany: State University of New York Press, 2008), 8.

13 Piper, *Industrial Transformation*, 7.

14 Piper, *Industrial Transformation*, 37–38, 283.

15 LAC, RG 88, vol. 362, file 16638, E. Deville to W. G. Mitchell, 15 January 1918.

16 Kim Sawchuk, "Modernity, Nostalgia, and the Standardization of Time," in *Antimodernism and Artistic Experience: Policing the Boundaries of Modernity*, ed. Lynda Jessup (Toronto: University of Toronto Press, 2001), 158–59.

17 Lears, *No Place of Grace.*

18 James Smithies, "An Antimodern Manqué: Monte Holcroft and *The Deepening Stream*," *New Zealand Journal of History* 40, no. 2 (2006): 171. Significant contributions to the historical study of antimodernism in Canada include Ian McKay's *The Quest of the Folk: Antimodernism and Cultural Selection in Twentieth-Century Nova Scotia* (Montreal and Kingston: McGill-Queen's University Press, 1994) and Sharon Wall's *The Nurture of Nature: Childhood, Antimodernism, and Ontario Summer Camps, 1920–55* (Vancouver: University of British Columbia Press, 2009).

19 Peter J. Schmitt, *Back to Nature: The Arcadian Myth in Rural America* (Baltimore and London: Johns Hopkins University Press, 1969).

20 George Altmeyer, "Three Ideas of Nature in Canada, 1893–1914," in *Consuming Canada: Readings in Environmental History,* eds. Chad and Pam Gaffield (Toronto: Copp Clark Ltd., 1995), 99–100.

21 Patricia Jasen, *Wild Things: Nature, Culture, and Tourism in Ontario, 1790–1914* (Toronto: University of Toronto Press, 1995), 106.

22 Gail Bederman, *Manliness and Civilization: A Cultural History of Gender and Race in the United States, 1880–1917* (Chicago: University of Chicago Press, 1995), 20–31, 170–215.

23 Jonathan F. Vance, *Death So Noble: Memory, Meaning, and the First World War* (Vancouver: University of British Columbia Press, 1997), 159.

24 The classic analysis of the racial ideology of nordicity in Canada is Carl Berger, "The True North Strong and Free," in *Nationalism in Canada,* ed. Peter Russell (Toronto: McGraw-Hill, 1966), 3–26.

25 Jane Nicholas, "Gendering the Jubilee: Gender and Modernity in the Diamond Jubilee of Confederation Celebrations, 1927," *Canadian Historical Review* 90, no. 2 (2009): 247–74; Janice Cavell, "The Second Frontier: The North in English-Canadian Historical Writing," *Canadian Historical Review* 83, no. 3 (2002): 364–89.

26 John Sandlos, *Hunters at the Margin: Native People and Wildlife Conservation in the Northwest Territories* (Vancouver: University of British Columbia Press, 2007), 11.

27 Dartmouth College, Rauner Special Collections Library (hereafter cited as RSCL), Vilhjalmur Stefansson, Correspondence, 1895–1962, MSS 196, box 80, folder 4, George Douglas to Vilhjalmur Stefansson, 7 April 1954.

28 See George M. Douglas, "Copper Deposits of Arctic Canada," *Engineering and Mining Journal-Press* 118, no. 3 (1924): 85–89; Douglas, "A Summer Journey along the Southeast Shores of Great Slave Lake," *Canadian Mining and Metallurgical Bulletin* 22 (February 1929): 344–60; Blanchet, *Great Slave Lake Area, Northwest Territories* (Ottawa: King's Printer, 1926); Blanchet, *Keewatin and Northeastern Mackenzie. A General Survey of the Life, Activities, and Natural Resources of This Section of the Northwest Territories, Canada* (Ottawa: King's Printer, 1930).

29 British Columbia Archives (hereafter cited as BCA), Guy Houghton Blanchet fonds, MS 0498, box 3, folder 7, Letters re: Dominion

Explorers Ltd. 1928–29, [26] November 1928. Square brackets indicate an insertion or correction of the date as originally omitted or given in the manuscript journals.

30 LAC, MG 30 B95, vol. 2, file 1, George Douglas to Vilhjalmur Stefansson, 1 December 1925.

31 LAC, RG 88, vol. 382, file 18265, Blanchet, "Report on the Country North and East of Great Slave Lake." The Barren Lands (also known as the Barrenlands, Barren Grounds, or simply the Barrens) denote the area of subarctic tundra covering what is today western Nunavut and part of the eastern Northwest Territories. Stretching from Great Slave and Great Bear lakes in the west to Hudson Bay in the east, and from the Arctic coast in the north to the Hudson Bay coastal plain in the south, the region is distinguished by low vegetative cover, many small lakes and rivers, and gently rolling terrain speckled with glacial erratics.

32 Lynda Jessup, "Introduction: Antimodernism and Artistic Experience," in Antimodernism and Artistic Experience, 4.

33 From 1921 to 1924, Blanchet had also supervised the first controlled stadia traverse survey of Great Slave Lake, in which his teams of surveyors corrected many of the errors of previous maps and presented a cartographic result that differed significantly from previous representations.

34 LAC, RG 88, vol. 384, file 18506, Surveyor General to Blanchet, 21 May 1924; LAC, RG 88, vol. 387, file 18707, F. H. Peters to Blanchet, 22 April 1925; LAC, RG 88, vol. 388, file 18822, Peters to Blanchet, 28 April 1926.

35 On masculine culture in America at this time, see E. Anthony Rotundo, American Manhood: Transformations in Masculinity from the Revolution to the Modern Era (New York: Basic Books, 1993); Michael Kimmel, Manhood in America: A Cultural History (New York: Free Press, 1996).

36 BCA, MS 0498, box 5, 1924 Journal—Great Slave Lake survey, [11] July 1924.

37 BCA, MS 0498, box 3, folder 3, Taltson journal, 28 July–19 August 1925, 16 August 1925.

38 BCA, MS 0498, box 3, folder 5, Dubawnt journal, 1–23 July 1926, 18 July 1926.

39 George M. Douglas, Lands Forlorn: A Story of an Expedition to Hearne's Coppermine River (New York: G. P. Putnam's Sons, 1914), 194–95; BCA, MS 0498, box 3, folder 4, Dubawnt journal, 7–30 June 1926, 13 June 1926; Dubawnt journal, 1–23 July 1926, 20 July 1926. The last entry admits difficulty with the top figure: "I sometimes load foolishly. I took about 140 [pounds] over a portage of half a mile and my neck nearly collapsed before I got there."

40 Dubawnt journal, 7–30 June 1926, [16] June 1926.

41 BCA, MS 0498, box 3, folder 2, Taltson journal, 16–26 July 1925, [16] July 1925.

42 Blanchet, "Report on the Country."

43 Lears, No Place of Grace, 98, 108, 118, 138.

44 BCA, MS 0498, box 3, folders 7–8, Letters re: Dominion Explorers Ltd. 1928–29, [27] December 1928, 30 March 1929, May 1929.

45   Dubawnt journal, 1–23 July 1926, 10 July 1926.

46   Lears, *No Place of Grace*, 34, 57.

47   Guy Blanchet, *Search in the North* (Toronto: Macmillan, 1960), 172.

48   Dubawnt journal, 1–23 July 1926, 23 July 1926.

49   As Liza Piper notes, the country around the large northern lakes was, in fact, "very well known to and thoroughly explored by the Dene" (*Industrial Transformation*, 40) by the early twentieth century. Both Blanchet and Douglas enjoyed travelling in areas of the north that were less well known or not at all known to southern Canadians. However, given their extensive research before each field season, they were more than usually aware, as fieldworkers went, of what activity had preceded them in those spaces and what records of that work had been made. See, for example, LAC, MG 30 B95, vol. 4, file 21, Douglas to C. D. LaNauze, 24 September 1933: "…I think with most pleasure of our first voyage [in the summer of 1932], along the south shore of [Great Bear Lake]. This was mostly unmapped country, and so had all the charm of novelty and the unknown. I say 'unmapped,' Franklin's map, the only one ever made, shows that part of the lake in approximate outline…."

50   1924 Journal—Great Slave Lake survey, 29 July 1924; Dubawnt journal, 7–30 June 1926, 26 June 1926.

51   BCA, MS 0498, box 3, folder 8, Letters re: Dominion Explorers Ltd. 1928–29, 6 and 7 June 1929; 1924 Journal—Great Slave Lake survey, [12] and [26] July 1924.

52   On the phenomenon of "playing Indian" or "going Native," see Marianna Torgovnick, *Gone Primitive: Savage Intellects, Modern Lives* (Chicago: University of Chicago Press, 1990); Philip Deloria, *Playing Indian* (New Haven and London: Yale University Press, 1998); Shari M. Huhndorf, *Going Native: Indians in the American Cultural Imagination* (Ithaca: Cornell University Press, 2001); Sharon Wall, "Totem Poles, Teepees, and Token Traditions: 'Playing Indian' at Ontario Summer Camps, 1920–1955," *Canadian Historical Review* 86, no. 3 (2005): 513–44.

53   Robert F. Berkhofer Jr., *The White Man's Indian: Images of the American Indian from Columbus to the Present* (New York: Alfred A. Knopf, 1978), 71.

54   Guy Blanchet, "New Light on Forgotten Trails in the Northwest," *The Canadian Field-Naturalist* 40, no. 4 (1926): 71–72.

55   Blanchet, "New Light on Forgotten Trails," 71.

56   BCA, MS 0498, box 3, folder 1, Taltson journal, 24 June–9 July 1925, 5 July 1925.

57   Taltson journal, 15–26 July 1925, 26 July 1925.

58   Dubawnt journal, 1–23 July 1926, 18 July 1926; BCA, MS 0498, box 9, "The Exploration of the Plateau," in "Beyond the Ranges," unpublished MS.

59   Talston journal, 24 June–9 July 1925, 24 June 1925.

60   Lears, *No Place of Grace*, 44–45, 118, 121–22.

61   BCA, MS 0498, box 3, folder 7, Letters re: Dominion Explorers Ltd. 1928–29, 25 and 30 January 1929.

62 Beaulieu belonged to a prominent northern Métis and Dene family whose history, entwined with that of preceding generations of non-Indigenous northern travellers, offers "an alternative view of northern exploration," according to Enid Mallory. See Mallory, *Coppermine: The Far North of George M. Douglas* (Peterborough, ON: Broadview Press, 1989), 240, and chap. 18 more generally. As a voyageur, Souci's great-grandfather, François Beaulieu I, helped ensure the success of Alexander Mackenzie's late eighteenth-century expeditions to the Arctic and Pacific oceans. His grandfather, François Beaulieu II, had advised John Franklin on travel routes prior to his first overland expedition of 1819–22, and had acted as interpreter and hunter for Franklin at Great Bear Lake during Franklin's second overland expedition of 1825–27. Souci's father, also named Joseph King Beaulieu, had guided the British adventurer Warburton Pike on his travels through the north in the 1890s. Souci had travelled briefly with Pike as a young man, and worked closely with Blanchet during the summers of 1922 and 1923. As pilot of the schooner *Ptarmigan*, and as a fount of local geographical knowledge, he was an essential collaborator in Blanchet's 1922 survey of Great Slave Lake's East Arm. See Gwyneth Hoyle, *The Northern Horizons of Guy Blanchet: Intrepid Surveyor, 1884-1966* (Toronto: Natural Heritage Books, 2007), chap. 3. On the Beaulieu family, see Chris Hanks, "François Beaulieu II: The Origins of the Métis in the Far Northwest," in *Selected Papers of Rupert's Land Colloquium 2000*, comp. David G. Malaher (Winnipeg: Centre

for Rupert's Land Studies, 2000), 111–26.

63 The following account is drawn from Guy Blanchet, "Exploring with Sousi and Black Basile," *The Beaver*, Outfit 295 (Autumn 1964): 34–41. The only contemporary account of this expedition is Blanchet, "Report on the Country." It says almost nothing of his Indigenous guides, terming them simply "two old men."

64 Guy Blanchet, "Narrative of a Journey to the Source of the Coppermine River," *Bulletin of the Geographical Society of Philadelphia* 24 (1926): 168; BCA, MS 0498, box 10, "Mackenzie and Great Slave Lake Explorations," in "These Fifty Years," unpublished MS.

65 Blanchet, *Great Slave Lake Area*, 35.

66 Berkhofer, *The White Man's Indian*, 78.

67 Taltson journal, 24 June–9 July 1925, 4 July 1925.

68 Taltson journal, 28 July–19 August 1925, 28 July 1925.

69 1924 Journal—Great Slave Lake survey, 24 July 1924.

70 That the settlement patterns of the Denesuline had changed since the eighteenth century was, at least, correct. Several factors had combined to draw these people away from their traditional hunting territories on the Barrens to more southerly and forested lands. These included the greater preponderance of both fur-bearers and fur traders in northern Saskatchewan and Manitoba, and the later settlement of treaties with the Canadian government that granted Denesuline families reserves and hunting territories south of the sixtieth

parallel. See James G. E. Smith, "Chipewyan," in *Handbook of North American Indians. Volume 6: Subarctic*, ed. June Helm (Washington: Smithsonian Institution, 1981), 271–84.

71 BCA, MS 0498, box 16, folder 2, Blanchet, "In the Land of the 'Caribou Eaters': An Exploration into the Country Southeast of Great Slave Lake," pt. 2, *Saturday Night*, n.d.

72 Blanchet, "New Light on Forgotten Trails," 71; Blanchet, "In the Land of the 'Caribou Eaters,'" pts. 1–2.

73 Blanchet, "An Exploration into the Northern Plains North and East of Great Slave Lake, Including the Source of the Coppermine River," *Canadian Field-Naturalist* 38, no. 10 (1924): 186; Blanchet, *Great Slave Lake Area*, 39.

74 Blanchet, "In the Land of the 'Caribou Eaters,'" pt. 1; Blanchet, "New Light on Forgotten Trails," 71.

75 BCA, MS 0498, box 3, folder 7, Letters re: Dominion Explorers Ltd. 1928–29, 11 January 1929.

76 Jamie Benidickson, "Paddling for Pleasure: Recreational Canoeing as a Canadian Way of Life," in *Recreational Land Use: Perspectives on its Evolution in Canada*, eds. Geoffrey Wall and John S. Marsh (Ottawa: Carleton University Press, 1982), 329.

77 See Hans Carlson's chapter in this volume for a different and important perspective on the significance of situating personal histories in particular northern places.

78 LAC, MG 30 B95, vol. 3, file 15, George Douglas to James Douglas, 20 May 1910.

79 LAC, MG 30 B95, vol. 4, file 4, George Douglas to James Douglas, 26 December 1911.

80 LAC, MG 31 C6, Richard Sterling Finnie fonds, vol. 7, file 3, George Douglas to Richard Finnie, 5 November 1948; LAC, MG 30 B95, vol. 4, file 1, George Douglas to Lionel Douglas, 16 March 1930.

81 Sawchuk, "Modernity, Nostalgia, and the Standardization of Time," 161–62. See also Fred Davis, *Yearning for Yesterday: A Sociology of Nostalgia* (New York: The Free Press, 1979).

82 Douglas, "A Summer Journey," 358.

83 LAC, MG 30 B95, vol. 4, file 19, George Douglas to Lionel Douglas, 16 March 1932.

84 LAC, MG 30 B95, vol. 4, file 19, George Douglas to Jim Douglas, 19 March 1932.

85 LAC, MG 30 B95, vol. 4, file 19, George Douglas to Lionel Douglas, 10 March 1932.

86 LAC, MG 30 B95, vol. 4, file 22, George Douglas, "Fort Norman, Bear River, Great Bear Lake, Voyage of the ALCYONE, June, July—Memoir with mounted photographs." Although Douglas published neither the technical results of his fieldwork nor personal accounts of his northern trips in the 1930s, he collated several typed and illustrated "memoirs" that describe those summer field seasons.

87 Piper, *Industrial Transformation*, 30.

88 Douglas, "Voyage of the ALCYONE."

89 Douglas, "Voyage of the ALCYONE."

90   LAC, MG 30 B95, vol. 5, file 6, George Douglas, "Report on Explorations, Athabaska Lake and Great Slave Lake, Summer of 1935."

91   LAC, MG 30 B95, vol. 5, file 1, George Douglas, "By Canoe and Plane in the Far North, 1928 Slave Lake, 1935 Great Slave Lake and Great Bear Lake—Memoire with Mounted Photographs" (first draft).

92   Douglas, "Report on Explorations."

93   Douglas, "By Canoe and Plane"; LAC, MG30 B95, vol. 4, file 23, Diary 22 June—14 September 1935, 26 June 1935.

94   Douglas, "By Canoe and Plane."

95   Douglas, "Voyage of the ALCYONE."

96   Douglas, "By Canoe and Plane."

97   Mallory, Coppermine, 226.

98   Douglas, "By Canoe and Plane."

99   LAC, MG 30 B95, vol. 5, file 4, Diary 27 April–6 September 1938, 29 July 1938.

100   LAC, MG 30 B95, vol. 5, file 8, George Douglas, "Our Travels. Season 1938. By Canoe. By Tug and Barge with light canoe. By Plane."

101   Diary 27 April–6 September 1938, 14 June 1938.

102   Diary 22 June–14 September 1935, 24 July 1935; Diary 27 April–6 September 1938, 28 April 1938.

103   See, for example, Diary 22 June–14 September 1935, 16 July 1935.

104   Stephen Kern, The Culture of Time and Space, 1880–1918 (London: Weidenfeld and Nicolson, 1983), 40.

105   Guy Blanchet, "The Caribou of the Barren Grounds," The Beaver, Outfit 267 (1936): 25; Guy Blanchet, "Thelewey-aza-yeth," The Beaver, Outfit 280 (1949): 8.

106   Marionne Helena Cronin, "Flying the Northern Frontier: The Mackenzie River District and the Emergence of the Canadian Bush Plane, 1929-1937" (PhD diss., University of Toronto, 2005), 141.

107   Guy Blanchet, "Conquering the Northern Air," The Beaver, Outfit 269 (1939): 11.

108   J. R. K. Main, Voyageurs of the Air: A History of Civil Aviation in Canada, 1858-1967 (Ottawa: Queen's Printer, 1967), 299.

109   Marionne Cronin, "Northern Visions: Aerial Surveying and the Canadian Mining Industry, 1919–1928," Technology and Culture 48, no. 2 (2007): 306.

110   Cronin, "Northern Visions," 310–11, 316–22.

111   Jonathan F. Vance, High Flight: Aviation and the Canadian Imagination (Toronto: Penguin Canada, 2002), 151.

112   Lawrence J. Burpee, "Where Rail and Airway Meet," Canadian Geographical Journal 10, no. 5 (1935): 239–45.

113   Piper, Industrial Transformation, 73.

114   Sidonie Smith, Moving Lives: Twentieth-Century Women's Travel Writing (Minneapolis: University of Minnesota Press, 2001).

115   George Douglas to Jim Douglas, 17 September 1935 (emphasis original). However, the utility of combining aerial perspectives on northern landscapes with ground-level reconnaissance soon became plain. In 1942, Douglas opined that the best way to select a route for the nascent Alaska Highway was to walk over the country, make flights, and study aerial photographs before and after

the pedestrian trip. See RSCL, MSS 196, box 80, folder 4, George Douglas to Stefansson, 10 April 1942. A decade later, Western Electric technicians took a near-identical approach to collecting geographic information about the landscapes in which the DEW Line would be situated. See Matthew Farish and P. Whitney Lackenbauer's essay in this volume.

116 LAC, MG 31 C6, vol. 7, file 1, George Douglas to Richard Finnie, 30 November 1942.

117 LAC, MG 31 C6, vol. 6, file 5, P. G. Downes to Richard Finnie, 30 March 1956.

118 Smith, *Moving Lives*, 32. But see also Cronin's chapter in this volume, which demonstrates that travel in an interwar bush plane produced similarly intense physical experiences for its passengers.

119 Blanchet, "Across the Southern Plateau," in "Beyond the Ranges"; Blanchet, "Thelewey-aza-yeth," 9.

120 George Douglas, "By Canoe and Plane." As Cronin's chapter shows, the perspective was quite different from the pilot's seat. To those who flew throughout the north regularly, the view from above revealed familiar environments and personal emotional geographies similarly tethered to specific northern places.

121 BCA, MS 0498, box 3, folder 8, Letters re: Dominion Explorers Ltd. 1928–29, [16] May 1929.

122 Paul Carter, *The Road to Botany Bay: An Essay in Spatial History* (London: Faber and Faber, 1987); Mary Louise Pratt, *Imperial Eyes: Travel Writing and Transculturation* (London: Routledge, 1992), 61; D. Graham Burnett, *Masters of All They Surveyed: Exploration, Geography, and a British El Dorado* (Chicago: University of Chicago Press, 2000), 11.

123 Emilie Cameron, "'To Mourn': Emotional Geographies and Natural Histories of the Canadian Arctic," in *Emotion, Place and Culture*, ed. Mick Smith, et al. (Farnham: Ashgate, 2009), 166–67. Cameron's chapter in this volume also engages with multiple "locals," albeit from a contemporary perspective.

124 Piper, *Industrial Transformation*, 42.

PART 2

Transformations and the Modern North

6

From Subsistence to Nutrition: The Canadian State's Involvement in Food and Diet in the North, 1900–1970

Liza Piper

Introduction

The caribou skin, caribou blood, they wouldn't leave that behind. Even caribou guts, they wouldn't leave that behind, either. They took it all to eat, also for dog feed. Around there, they stayed there; ah, it was really nice.

They just ate meat, and there was no grub [store-bought food]. Sometimes somebody had a little tea. Sometimes there was tobacco, too. They boiled meat on the fire. That's all they ate.[1]

—Myra Moses (1884–1984)

Sometimes southern, urban Canadians need to be reminded of the significance of what they eat. As Margaret Lien writes in her introduction to *Politics of Food*, "What appears to be a carrot or a piece of meat is indeed a product with a history and implications more complex and profound than most of us ever want to think about."[2] The apparent simplicity of such foods belies the work done—through the processes of preservation, transportation, distribution, and marketing in southern food systems—to mask the ecosystem of origin and the socio-economic and cultural circumstances of production.[3] In the north, reminders about the context of food are, by contrast, largely superfluous. The high cost and low quality of southern imported foods stand as constant signals of the place of a long-distance, industrial food system in northern diets. Moreover, where subsistence practices persist, as they do across much of the north into the present, there is no reminder necessary of the cultural, social, and economic significance of fish, marine mammals, game, or berries. This different experience of food illuminates more fundamental differences in relations between people and environment in the north compared to southern Canada in any given historical period. These apparent differences go beyond those of rural versus urban experiences (notwithstanding some important parallels between the rural south and the north) because of the profoundly different implications of hunting, gathering, and agriculture for relations with the rest of nature.[4] This chapter draws on an environmental history perspective to consider changes in diet and the rise of nutrition as a new way of thinking about food in the north in the twentieth century. As acknowledged by a 2009 forum in *Environmental History*, food has yet to figure as prominently within environmental historiography as is warranted given how eating intimately connects human bodies to local and global environments. Nevertheless, food history and all that it entails has offered opportunities for the critical study of subsistence and desire, or "needs and tastes," consumption, food commodity chains, and relations of power.[5] Anthropologists have long been interested in studying food for its role in imposing structure and order.[6] This present chapter builds on such an approach, as well as work from across the field of environmental history that considers the role of the state in mediating relations between people and nature, to examine explicitly how the state engaged in "food colonialism"—using what Indigenous northerners drew from the land

and water into their bodies as a means of exerting control over significant social and environmental changes in the twentieth century.[7]

The quotation that opens this chapter was spoken by Myra Moses in 1979. She was a Van Tat Gwich'in woman born in 1884, who lived in the northern Yukon and Alaska. It is but one of countless references to subsistence found in oral histories of northerners conducted in the twentieth century. As Chase Hensel argues in *Telling Our Selves: Ethnicity and Discourse in Southwestern Alaska*, "subsistence is the central focus in the intellectual, material, and spiritual culture of both historic and contemporary Yup'ik society."[8] The same case could be made for communities across the north.[9] Indeed, to hive off food or diet or nutrition as a category of analysis is to scratch only the surface of the significance of northern subsistence, neglecting the fact that this "is not simply [an] activity but [a] socio-economic system."[10] Food itself, in any context, is hardly a simple category. As Lien writes, it is uniquely complex: "food is literally transformed and becomes *part of* the human body. ... The physiological need in humans to eat every day makes access to food a crucial issue." She continues, "It also makes us vulnerable, weak and easy to control. In this way, food is entrenched in structures of subordination, governance and dominance."[11] The categories of "food," "diet," and "nutrition" lie at the heart of this investigation, as their appearance and the concurrent ways in which these categories became distinct from the larger place of subsistence in the north signalled a profound transition in the region in the post-war period. Food, diet, and nutrition became axes along which the southern-based Canadian state could assert control over northern bodies and articulate new standards of healthy and ethical citizenship, especially for Indigenous northerners. In this respect, this chapter builds upon work by Maureen Lux, Mary-Ellen Kelm, and a handful of others, that examines the place of food in changing Indigenous relations with the Canadian state.[12] This chapter, like those by Andrew Stuhl, Tina Loo, and Matthew Farish and Whitney Lackenbauer, also in this volume, considers the different ways the Canadian state sought to manage complicated relations with northern people and environments in the interwar and postwar periods. I broadly outline early nutritional policy (to the extent that such a thing existed), and, more importantly, the changes in subsistence that characterized the first part of the twentieth century before the Canadian government turned, in the latter half of the twentieth century, to use nutrition and nutritional science as the means by which to manage ongoing changes in relationships between northerners and their environments.[13]

By the early twentieth century, tea, bannock or biscuits, molasses, flour, and sugar were well integrated into the diets of most northerners. Nevertheless, and as should be obvious from the paucity of that list, Indigenous and non-Indigenous diets relied upon the resources of the land. These resources ranged from waterfowl, fish, caribou, moose, and sea mammals, to smaller creatures, such as hares, and plants, most notably berries. Fish were of particular importance regardless of whether northerners lived on the coast or inland. As late as 1990, Fikret Berkes estimated that "some 300 000 northern rural people ... may be harvesting on the order of 15 000 t[ons] of fish per year."[14] This was well after the decline of many inland fisheries as a consequence of commercial overharvesting and habitat destruction.[15] All kinds of food, from plants, to animals, to fish, were harvested from the land, but were variously available from year to year, or at different times of the year. A complex food economy thus prevailed across northern Canada.[16] Such harvesting involved an intimate knowledge of the land and animals, but perhaps paradoxically for those who understand close relationships with nature to be focused upon local places, this intimacy extended over a wide area. Such elaborate and extensive food economies served as a strategy for resilience in a highly variable environment often visited by periods of hardship: people could turn to a range of resources to ensure health, particularly during times of scarcity.

Scarcity was neither uncommon nor unanticipated. Across the north, people relied on migratory animals whose migrations shifted, and upon species, such as the hare, with cyclic fluctuations in population. Families and traders froze and dried freshwater fish in the fall to sustain them through the long winter and the spring, the hardest, most vulnerable time of the year. Climate could fluctuate dramatically. Between 1910 and 1920, above-average rain and snow and fluctuating temperatures in the central Arctic drove the caribou herds, to the west of Hudson Bay, away for a decade.[17] In the boreal forest, fires could also push game far from the usual hunting grounds. Historians Gulig, Coates, and Morrison have noted that such fires increased with the arrival of industry, whether as a consequence of prospectors burning off the brush to facilitate mineral exploration, or as a by-product of the presence of more machines.[18]

The periodicity of hardship could range from season to season, year to year, or decade to decade. It meant, at times, going hungry. Occasional and seasonal malnutrition was not uncommon in the Subarctic and Arctic at the turn of the century. Gwich'in elders from Fort Mcpherson and Tsiigehtchic (Arctic Red River) described hungry times from their childhoods in the early twentieth century, and deaths from starvation.[19] At other times, sustained hardship required families or communities to relocate. In northern history, there have long been many instances of places being abandoned, particularly in response to declining environmental conditions and climatic change. With the onset of the Little Ice Age, many Thule turned to new food sources (fish, caribou, and seal) before they and their descendants eventually abandoned sites such as Somerset Island and south Baffin in the 1300s. The south Baffin villages were repopulated by 1500, with residents adopting more mobile harvesting practices than before, such as spending more time in portable skin houses rather than the stone-sod-and-whalebone houses suited to whale-hunting communities.[20] Other sites were famous as ancient gathering places: the village at Kittigaryuit was the site of a natural beluga whale trap, and the length and scale of occupation at that site (estimates of a thousand villagers in the 1820s, for instance), or at the confluence of the Yukon and Klondike rivers, was indicative of the uncommon local wealth of resources.[21]

Dimensions of a Changing Diet: Supply

Between 1870 and 1940, more people exploited northern resources than ever before. Specifically, newcomers to the region from the outside and new activities taking place on northern lands and in northern waters affected subsistence opportunities. From the late nineteenth century on, whalers, scientists, and large research expeditions contributed significantly to the depletion of musk ox, caribou, and walrus populations. David Hanbury, in a journal recorded while he undertook geographical explorations along the western coast of Hudson Bay, wrote, "altho' game may not be so plentiful now as in former times, still there is plenty of it." In April 1902, Hanbury noted more precisely: "Musk ox [my Inuit informant] reports to be scarce both N and S of Backs river. Long ago they were numerous. ... Why have musk ox disappeared?"[22] The answer to Hanbury's query lay not with unsustainable Indigenous hunting (primarily for meat) but rather with the appearance of new sport and subsistence hunters. In 1875, research vessels travelling in the eastern Arctic waters stopped where game was plentiful. Such hunting was in part for meat: men on board Victorian research vessels or those with search parties for the lost Franklin expedition kept fresh meat in the hold by harvesting from northern lands. But hunting was also very much for sport.[23] In the Bellot Bay area, the impact upon large animals by occasional research parties was intensified by the activities of whalers, who began over-wintering in 1864–65 and who relied upon local game harvested by Inuit hunters for their crews' subsistence. Elsewhere in the Canadian north, resource and research expeditions created new demands upon local wildlife for food, trade, and recreation. It is likely, although research remains to be done on this question, that the concentrated scientific efforts associated with the International Polar Years (1882–83, 1932–33) and the International Geophysical Year (1957–58) increased pressures on food resources.[24] To the west, American whalers, using Herschel Island as a base, consumed 12,308 caribou in less than two decades (1890–1908), leaving far fewer caribou for local Inuvialuit.[25] The influx of trappers and traders prior to the First World War intensified demands upon northern furbearers. Trapping, combined with the arrival of men and women working in the surging mining industry, continued the pressure on all kinds of fish, fowl, and game populations well into the twentieth century. Miners at Port Radium on Great Bear Lake, where pitchblende was extracted beginning in 1929, relied on locals who traded fish, moose meat, and other country foodstuffs for variety in a diet that otherwise consisted of canned and preserved goods that had been shipped north.[26]

In the first half of the twentieth century, Dominion government officials facing pressure from hunters and concerned about sustaining the livelihoods of northern Indigenous peoples became attentive to the depletion of game populations.[27] In her chapter in this volume, Tina Adcock further demonstrates the antimodern sensibilities at work in this period, which shaped concerns about both Indigenous peoples and wildlife populations in the 1920s and 1930s. Continental concern about resource conservation influenced the new government interest in the north. Within a decade, the Dominion government introduced the Northwest Game Act (1917) and the Migratory Birds Act (1917), established Wood Buffalo National Park (1922) and the Thelon Game Sanctuary (1927), and expanded the 1924 ban on musk ox hunting in the Northwest Territories to include Indigenous hunters, who had previously been exempt if they were starving. In many respects, the changed relationships between people and the land that had come to the fore in the early twentieth century were to be managed through the regulation of wildlife.

The changes that compelled the Dominion government to introduce new wildlife regulations had direct implications for government policy on and attitudes toward subsistence in the north. O. S. Finnie, then director of the Northwest Territories Branch, articulated this policy in letters addressed "from the Government to the Indian People" (1924) and "from the Government to the Eskimo People" (1926), as well as in correspondence from 1928 between the NWT Branch, the Department of the Interior, the Royal Canadian Mounted Police, and the Hudson's Bay Company. The NWT Branch aimed to "keep the natives strong and healthy, making them self-reliant and independent citizens" and to "keep the native, native."[28] Part of a larger colonial assimilationist project, the specific implications of this attitude for subsistence meant that the NWT Branch wished, insofar as was possible, to keep northern Indigenous people off relief (this was what was meant, in part, by "self-reliant" and "independent"). Finnie emphasized that to ensure independence with regard to food, northern Indigenous peoples had to be taught "to conserve the food supply of the country for [their] own requirements." Indeed, this was the focus of his letters "to the Indian/Eskimo people."[29] Sandlos describes the letter "to the Indian people" as emblematic of the Dominion's paternalism. In it, Finnie detailed what characterized "a good hunter": a hunter who did not kill female caribou with young and who did not kill "more caribou than he needs." Wolves, on the other hand, should be targeted by hunters because they competed with them for caribou. In these ways, Finnie's letter articulated the mantra of early-twentieth-century southern wildlife conservation policy as a new doctrine for northern Indigenous peoples. The letter was interspersed with comments about the value of caribou and the dire consequences of overhunting. The letter to "the Eskimo people" was identical in tone and varied primarily in the details of the advice. Here, Finnie expounded upon the optimal time to hunt seal and the best methods for drying meat and fish, matters in which he clearly presumed himself more expert than the Inuit hunters he addressed. Finnie again emphasized prohibitions against killing pregnant caribou, killing caribou for skins to trade rather than meat, and wasting the meat of animals. In this letter, Finnie went further in terms of the food advice he disseminated. He emphasized, "<u>No matter what you may be hunting always think of what you will leave for your children and their children.</u>"[30] Moreover, he detailed appropriate food to be given to children, appropriate breast-feeding practices, and expectations about cleanliness. In spite of the fact that Dominion officials typically characterized Inuit as more independent than other northern Indigenous peoples, Finnie clearly felt it was necessary to dispense even more detailed food advice to the Inuit than to the "Indians," suggesting that he viewed the Inuit subsistence livelihood as less secure.

Regulation was a response to the depletion of northern food resources as a result of intensified harvesting. However, regulation also reinforced the shortages of country food for northern Indigenous peoples, trappers, and traders, all of whom lived off the land. As John Sandlos and Tina Loo have examined in detail, these new regulations criminalized northern Indigenous subsistence activities. Waterfowl regulations were the most egregious in this regard, as the open and closed seasons were timed in response to the interests of southern, not northern hunters. While most of the new regulations were honoured as much in the breach as in the observance, they nevertheless directly impacted the ability of northerners to continue to obtain subsistence from the land in the fashion to which they had become accustomed as recently as a few years or decades earlier. Harvesting shortfalls and changes in government administration pushed more northerners onto state-supplied relief in the twentieth century. When it came to relief rations, including milk, butter, and bacon, these were characterized as luxuries "from the Eskimo standpoint" and were only to be distributed, if absolutely necessary, to infants and invalids. Otherwise, when relief was necessary, "he should be rationed with his own kind of food and not that of white man's." Yet, if the resources of the land were scarce (which was what pushed many northerners onto relief rolls in the first instance), what—from the perspective of the state—constituted northerners' "own kind of food"? Two examples, suggested by Charles Sale of the Hudson's Bay Company, were seal-meat infused biscuits for Inuit and large quantities of bison meat for distribution across the north.

These, then, can be considered the cumulative pressures on the supply side, when it came to northern subsistence in the early twentieth century.

In addition to environmental variability, which caused fluctuations in the availability of game, fish, and fowl, newcomers increased pressures upon northern food resources, often to the point of depletion, while regulations and surveillance introduced in response to some of these new pressures further affected the ability of Indigenous northerners, in particular, to continue their historical harvesting practices.

Dimensions of a Changing Diet: Demand

From the end of the nineteenth into the early twentieth century, there were also a range of new pressures that affected northerners' harvesting of food resources from the land. Perhaps most notably, from the mid-nineteenth century through until 1960 or so, northern peoples were faced with repeated outbreaks of infectious diseases. While not "virgin soil epidemics" *per se*, these epidemics shared some characteristics with contact-era outbreaks across the Americas. The epidemics tended to affect communities which, due to small population sizes and distance from larger centres, had acquired but limited immunity to crowd diseases such as measles, scarlet fever, and influenza. In turn, the outbreaks often led to significant mortality or had complex social and economic effects upon families and communities.[31] Malnutrition increased mortality from infectious diseases and deepened the social and economic consequences of outbreaks.

Epidemic disease and malnutrition travel hand-in-hand in human history.[32] Malnutrition, by weakening individual immunity, could lead to epidemic outbreaks. Seasonal malnutrition was not uncommon in the North, with the spring being the hardest period: food stores from the winter months were low, supplies from the south had yet to be restocked, and travel for hunting was complicated by the break-up of ice. Spring malnutrition contributed directly to the influenza outbreaks that came with the arrival of the first boats from the south. Epidemics and ill heath also made it much more difficult for people to harvest food in subarctic and arctic environments. The illness itself, whether influenza, typhoid, or another disease, weakened those who were afflicted, and healing demanded considerable energy that otherwise would be put to hunting, fishing, trapping, or harvesting activities. Moreover, by the late nineteenth century, hunting and trapping relied on dog teams that also needed to be fed during an outbreak. Their food requirements were significant. Although dogs happily consumed many of the rough fish discarded from catches, thousands also had to be preserved from fall fisheries for consumption during the winter months.[33] Finally, by the turn of the century, officials increasingly used quarantines to check the spread of diseases across extensive northern territories. Yet quarantines also acted to prevent healthy persons from harvesting by restricting the travel necessary for extensive subsistence practices.[34]

The influenza outbreak of 1928 in the Mackenzie District offers good evidence of the synergistic relationship between infectious disease and hunger in this period. Helge Ingstad, a non-Indigenous trapper living near Lutselk'e (Snowdrift), on the eastern shore of Great Slave Lake, wrote: "[The flu] came at a time when I was living from hand to mouth. Fish was my sole diet, and this I had to procure by hauling in the nets. So far as I was concerned, it might just as well have stayed there till it rotted, for I was unable to swallow a mouthful of food in any event."[35] From his patrol in the Talston River region, RCMP Inspector Gagnon reported that, "these people are practically starving, as they are unable to hunt; there are only three boys attending to the fish nets and the wants of the community."[36] Given the demands of subsistence living, illness could have devastating effects. Later in the century, similar reports were made regarding Inuit who perished in the Garry Lake and Back River districts. There, seven members of one family died from "flu and hunger" in June and July, leaving only one man, Marer, alive. The police report noted that "it is difficult to ascertain whether these people died actually of starvation or sickness."[37] Illness could not only be intensified by malnutrition, but also led to hunger, as it weakened people beyond the point where they could engage in their necessary harvesting activities.

The interrelationship between epidemic illness and nutrition in the early twentieth century encouraged a growing dependence upon rations and foodstuffs (typically preserved) imported from the south at the expense of country-food based diets. Rations were already a part of the treaty process, and thus, with the introduction of regular treaty payments after 1898 in the Treaty 8 area, and after 1921 in the Treaty 11 area, eligible individuals and families could expect to receive ammunition and twine (used for hunting and fishing), as well as rations (in particular tea, sugar, flour, and bacon).[38] These rations and supplies were distributed at the annual treaty gatherings that took place in early summer, and included a visit to the medical doctor assigned to the area. This demonstrates another way in which treaties, as Paul Nadasdy assesses in his chapter in this volume, constituted new relations between people and their home environments in the north. Rations were not exclusively distributed as part of the treaty process, although that was one way in which they became normalized in northern life. They were also distributed by RCMP officers and other agents of the state to people who faced hardship, whether or not they had a formal treaty relationship to the state.

Outbreaks of infectious diseases created increased dependence on rations because of the ways in which they interfered with regular harvesting activities. This disruption could have consequences that extended long after the epidemic had passed. When an epidemic arrived during the summer, it disrupted immediate harvesting activities, as well as the work necessary to ensure subsistence during the fall and winter months (repairs to nets or laying up of winter supplies, for instance). Over the longer term, camps and family groups hit repeatedly by epidemics would find themselves too weak to produce food, and relied upon rations from missionaries and RCMP officers. In the 1928 influenza outbreak, the disease was spread at the treaty gatherings themselves, and, as news of this travelled, some families chose not to travel to the treaty gathering, or fled them before rations and supplies were even distributed. These families were, in some instances, spared the infection—but not always. Some left the treaty gathering only to fall ill afterward, and many died later at their camps elsewhere across the north. But those who had not received necessary rations and supplies faced further difficulties hunting and harvesting the food they needed for the rest of the year. The 1928 influenza thus had a long-term impact upon health and nutrition in the Mackenzie region, and demonstrates how disease accelerated the twentieth-century shift across northern Canada from country-food diets to reliance upon southern, imported foods.

The expansion of the residential school system had similar long-term dietary impacts. The first residential school in the north opened in Providence, on the Mackenzie River just west of Great Slave Lake, in 1867.[39] Following the establishment of this first school, operated by the Roman Catholic Soeurs Grises (Grey Nuns) and the Oblates of Mary Immaculate, further institutions spread across northern Canada. Roman Catholic and Anglican missionaries operated the schools at the outset. The federal government took over both residential and day schools by the early 1960s. Missionaries depended upon local food supplies to feed the resident children, although these local foods were not necessarily indigenous. The children themselves assisted in providing their own food: berry picking in late summer, helping with the fish catch and the potato harvest, and cutting hay to feed the cattle (or the occasional ox) also found at the missions.[40] At Hay River, the Anglican mission hired a Métis father and son, Charlie and Frank Norn, to fish for the mission and the school. Fish dominated the children's diets at the Hay River school, although the missionaries also purchased moose meat for the children from local hunters.

The residential schools acted to create new food relationships with the land by encouraging agriculture and by the very fact that they kept children away from their families, where they would have learned hunting, trapping, and fishing skills. For Dene and Inuit children, education was experiential: learning took place on the land, by doing the things they would need to know how to do in the future. When children were kept in school for part of the year, they missed out on this crucial part of their education. If they only returned to their families in the summer, as was typically the case, they missed out on much of the seasonal harvesting. The residential school system also created a new appreciation for southern food. School menus cultivated new tastes by featuring lettuce, tomatoes, beef, and chicken. In later years, as described below, both residential and day schools also became essential venues for the dissemination of fortified foods and new attitudes toward southern, processed foods. With their focus on children and education, the residential schools directly contributed to the twentieth-century dietary shift across the north.

Finally, there were major changes in northern geography in the first half of the twentieth century: new settlements appeared, their locations guided by new motivations, and northern life became more closely tied to settled communities. On Baffin Island, large, stable polynyas—natural holes in the ice through which seals, walrus, and whales can be hunted in winter—attracted human settlement by ensuring the availability of resources.[41] New environmental rationales underpinned the establishment of newcomer communities. Pond Inlet, for example, offered a good harbour for whaling ships, but was otherwise not important for harvesting—not to mention dark in the long winter months—making it a relatively unattractive site for habitation.[42] Southern demand for resources differed from northern demand, and new communities emerged at rich industrial resource sites: Hay River (commercial fishery), Norman Wells (oil), Yellowknife (gold), and Rankin Inlet (nickel and copper). As more children were sent to residential schools, the communities in which these schools were located (Aklavik, Fort McPherson, Fort Providence, Fort Resolution, Fort Simpson, Hay River, Chesterfield Inlet, Carcross, Dawson, Whitehorse, Shingle Point) became home for the children for at least part of the year. These communities, in turn, became destinations for the parents, who, while prohibited from visiting their children while in school, would come to pick them up for the months that they spent fishing, hunting, and harvesting out on the land. Individuals and families came into communities to trade or to celebrate holidays such as New Year's, and, after 1898 and 1921 in the western Arctic and Subarctic, to receive treaty payments. Increasingly by mid-century, they also came to collect relief or to work at the new industrial operations. Long-distance transport opportunities governed the location of many new communities (Churchill, Simpson, Pond Inlet, and Inuvik, to name a few) and facilitated both the Fort export of northern resources and the import of southern foodstuffs. Whereas many older communities were established close to rich subsistence sites, the twentieth century brought new settlement rationales typically focused on southern demands for resources.

Relocations and Consequences for Subsistence

The other reorientation with significant subsistence implications were the mid-century relocations of Indigenous northerners. These relocations have been addressed in detail by other scholars, most notably Frank Tester and Peter Kulchyski in their 1994 work *Tammarniit (Mistakes): Inuit Relocation in the Eastern Arctic, 1939–1963*. Tester and Kulchyski described the hunger experienced by inlanders, for instance, relocated to a coastal region.[43] Indeed, there are many recollections of hunger and even starvation caused by relocations to areas where subsistence was not assured. Relocations had other impacts upon diet, as well. In their work on northern contaminants, Usher et al. note that, "through keen observation and experience with wildlife, Inuit have their own understanding of the food chain as a vector, and of potential health hazards such as botulism or trichinosis which can be can be associated with country food."[44] Such knowledge was eroded during the relocations, with serious consequences.

In November 1948, long-time northerner L. A. Learmonth, engaged in archaeological work near Fort Ross, sent word to the RCMP detachment at Cambridge Bay that a group of sixteen Inuit had fallen terribly ill at Creswell Bay on Somerset Island during the summer. Nine of the sixteen had died, and, at the time of writing, the other seven remained seriously ill. The news only reached Cambridge Bay in January 1949, at which time the RCMP worked with other government officials to send relief to the survivors and to evacuate them to southern hospitals as necessary. English- and French-language newspapers across Canada reported extensively on these "mercy flights," employing stereotypes that contrasted life in the north to "civilization," characterized Indigenous northerners as "plague-ridden" worriers—who could "die from worrying about a toothache," according to one "pessimistic northerner"—and reinforcing ideas that the Inuit needed to be cared for and "properly fed" by the state.[45] The story held southern media attention in part because there was a lot of uncertainty, even mystery, around the "strange malady" that had struck the group at Creswell Bay.[46] Initial reports suggested influenza or starvation. Then a "plague" of gangrene—this after one of the two survivors was sent south with severe gangrene in both his feet.[47] Typhoid, followed by acute colitis, alternatively described as food poisoning, were the next suggestions. Southern media articulated this last interpretation by noting that "inhabitants of the village had been eating parts of the carcass of a dead whale which had washed ashore."[48] This description was pure speculation on the part of a journalist who aimed to make sense of what was, in fact, a much more complex and challenging story.

The Inuit who fell ill at Creswell Bay had been relocated first to Arctic Bay from Cape Dorset (both Pond Inlet and Pangnirtung) in 1936, and then, the following year, to Fort Ross at the southerly end of the Somerset Peninsula. In 1947, when the post at Fort Ross closed, they were moved again to Spence Bay on the Boothia Peninsula. They made repeated requests to be returned to Baffin Island, but to no avail. In 1948, these Inuit crossed Prince Regent Inlet to Creswell Bay, and en route they evidently consumed some walrus meat. While walrus were common food for the Inuit in their home region and walrus liver regarded by many Inuit as a delicacy, evidently the Netsilingmiut of the arctic coast west of Hudson Bay were "very superstitious about eating the liver of the Bearded Seal or the Walrus, saying that if one eats this the skin will fall off the person's face and arms."[49] This describes, with some accuracy, one presentation of hypervitaminosis-A (or an excess of vitamin A), which can cause excessive skin peeling, particularly on the arms, legs, and face, as well as headache, nausea, and debility.[50] Moreover, this was what was described to have originally happened to one of the surviving Inuit, a teenager, Kayoomyk, who was evacuated with serious gangrene in his feet. Alternatively, the "strange disease" may have been trichinosis, as the group had suffered from serious diarrhea (a contributing cause in some of the deaths). Trichinosis and hypervitaminosis-A are each a consequence of eating the liver of "carnivorous" walrus. Most walrus rely for sustenance on a wide range of benthic organisms, such as shrimp, crabs, molluscs, clams, soft corals, and sea cucumbers. In areas where such food is scarce, walrus have been known to eat warm-blooded mammals such as seals and conceivably even whales; it is these walrus that are described as carnivorous. In eating other marine mammals, and especially in eating their blubber where vitamin A is concentrated, the carnivorous walrus consume much more vitamin A than their benthic-organism-eating counterparts. Their livers become, like polar bear livers, highly toxic to humans. Trichinosis is caused by consumption of the *Trichinella spiralis* parasite, which "is primarily a parasite of carnivores and its transmission is mainly accomplished by one mammal eating the infected flesh of another."[51] It seems, based on the available evidence, that the Cape Dorset Inuit had not worried about walrus livers in their homelands and considered them safe to eat, and that they ate these livers after their relocation to a new environment with tragic consequences. Peter Evans described botulism outbreaks among relocated Inuit in Labrador (caused by food contamination in different environmental circumstances) as evidence of how "traditional ecological knowledge" is place-specific rather than abstracted knowledge.[52] Beyond this, though, the deaths at Creswell Bay and elsewhere across the north speak to the complex legacy of the relocations on Indigenous people's subsistence.

The Federal Government and Post-War Northern Nutrition

In 1944, D. L. McKeand, superintendent of the Eastern Arctic, outlined to his superior, Roy Gibson, the government's position on northern food and health. He emphasized that policy had not changed since the late 1920s, and that the government continued to advocate that Native northerners consume "native foods" to maintain optimum health. He wrote, channelling O. S. Finnie from earlier in the century, "Now is the time to 'keep them native' at the same time introducing articles of food, clothing and shelter which have been especially made or grown for their use." In spite of this assertion of continuity, it is also clear from McKeand's letter that circumstances had changed. McKeand questioned the feasibility of returning to "native foods" on the grounds that "human nature is generally opposed to any return to old customs unless these are carefully disguised." Likewise, he emphasized how particular groups of Inuit were, due to location, "more susceptible to white man's foods (and habits)" and would always be able to secure "white man's" foods and other goods.[53] Thus, while claiming continuity, McKeand nevertheless imparted his understanding that northern Indigenous people had experienced changes in the twentieth century that could not be undone.

These changes to northern subsistence, including those described in the first part of this chapter, were exemplified by a series of food-related health crises at mid-century. The "strange malady" at Creswell Bay arose from the consumption of toxic foods; an outbreak of poliomyelitis at Chesterfield Inlet in 1949 raised the question of whether growing reliance upon imported southern foodstuffs played a direct role in the appearance of this modern, urban disease.[54] Toby Morantz describes the "great famine" among the Cree of James Bay in the 1930s and 1940s.[55] This was followed by the famine in the Keewatin, which Loo discusses in her chapter in this volume, that claimed international attention following the photographs and reporting of Richard Harrington and the publication in 1952 of Farley Mowat's first book, *People of the Deer*.[56] Most important, however, was the prevailing tuberculosis epidemic that came to the fore of northern government policy in the 1940s and 1950s.

It would be false to suggest that the federal government, as a colonial state, stood back and observed the changes underway in the north, merely reacting to events rather than guiding them. On the contrary, through a range of interventions in diet, the federal government played a lead role in moving northerners farther away from subsistence—from one socio-economic system to another that bound the north more intimately to southern markets. The remainder of this chapter examines some of these interventions: in relief, Family Allowances and children's diets, and economic activities. It emphasizes the place of nutritional science in guiding the state's hand, and suggests finally that a profound sense of insecurity about the prospects for a healthy Indigenous diet in the north was at the core of the federal government's policies.

Aleck Ostry has detailed the federal government's role in directing nutrition policy in Canada, and divides this history into five eras from the mid-1870s to the present, with the first three directly relevant to the present discussion.[57] In the earliest period, lasting until 1918, the Dominion government established a system of food safety, inspection, and surveillance within the framework of federal criminal law. In the interwar period, from the creation of the federal Department of Health (1919) to the Canadian Council on Nutrition (1938), the Dominion government engaged with areas formally under provincial jurisdiction by acting on social and health policy matters where they overlapped with nutrition. The government had developed a national dietary standard by the end of the 1930s. This was part of international efforts, led by the League of Nations, which embraced and promoted the new nutritional science, including work on vitamins in the 1920s and the medicalization of nutrition through artificial infant feeding in the 1930s. The Canadian Council on Nutrition (CCN) endured from 1938 to 1972, constituting Ostry's third era, which saw the creation of wartime nutrition policies. This era also saw the CCN, in the post-war period, working together with the Department of Health to fortify the Canadian food supply with various elements and vitamins. This period ended with the CCN coordinating the world's first representative national dietary survey, which reached into the north.[58] Nevertheless, in the context of Canadian nutrition policy history, the north was distinct from the rest of the country as it was the one region where, in the absence of provincial authorities, the federal government had direct responsibility for health and social policy. Yet the region was also characterized by federal government neglect prior to 1940; in the first half of the twentieth century, nutritional interventions fell primarily under the auspices of non-government agencies (the churches and the Hudson's Bay Company).

There was no published or otherwise formalized articulation of the churches' nutrition policy beyond their core mandate of ministering to bodies and souls. Their approach to nutrition was nevertheless manifested in the gardens maintained at the missions, where missionaries cultivated vegetables and flowers (including the ample potato crop at Fort Good Hope on the Mackenzie River), and in the residential schools as discussed earlier. The HBC went further, in 1940 publishing a booklet titled *Your Food and Health in the North*. This publication, penned by Frederick Tisdall, chair of the Canadian Medical Association's Committee on Nutrition, was to be distributed to HBC staff across the north. The booklet was prepared following "a study of post diet [that] convinced us that in many cases diets should be changed, and if they are, improved health will be the result."[59] It included a discussion of minerals and vitamins such as iron, calcium, and iodine, and emphasized the need for Vitamin D supplementation as "sunshine in the North is deficient in Vitamin D." Illustrated with cartoon men and women in parkas, polar bears, and ice floes, the booklet demonstrated a clear belief in the importance of nutrition education as the core of a healthy diet, and asserted, "We can only achieve this [health] improvement if you continue to help and co-operate."[60] The virtues of tinned vegetables and fruits were highlighted, as was the need for gardens at all trade posts (see Fig. 6.1). The booklet also included an extended section on nutrition for children and expectant mothers, emphasizing the importance of breastfeeding and even laying out detailed three- and four-hour schedules for nursing mothers. In these ways, *Your Food and Health in the North* can be seen as an expression of southern Canadian values about nutrition consistent with the dominant policy interests of this period as described by Ostry. This orientation is further evident in the omission of Indigenous people and any extended discussion of country food. The emphasis was on foods that could be imported to the north, and in this fashion the north was elided with southern urban centres, each effectively divorced from their surrounding environments and dependant instead upon extended networks of food supply and distribution—as well as informed science— to ensure adequate vitamins.[61] The Hudson's Bay Company's approach to nutrition, while confined to its employees in the north, nevertheless can be seen as consistent with the federal nutrition policy, which increasingly emphasized fortified southern foods and scientifically informed diets, alongside education, to change individual and family dietary practices.

VEGETABLES

SHOULD BE GROWN AT THE POST

Wherever Possible

Home grown vegetables are the cheapest and have the best flavour. Every post which can raise vegetables should do so. They will pay big dividends in good health.

Salad vegetables, such as lettuce and green onions, give you additional supplies of Vitamin C and add greatly to summer meals, but admittedly are not available at all posts.

Factory canned vegetables, like factory canned fruits, as far as food value is concerned, are equal or superior to cooked fresh vegetables. Remember they are usually processed the day they are picked, which conserves their maximum nutritional value.

Fortunately there is quite a variety of canned vegetables. The following list may be obtained by each post:

Corn	Beets	Sweet Potatoes
Corn on the Cob	Carrots	Asparagus
Baked Beans	Peas	Cauliflower
Lima Beans	Tomatoes	Sauerkraut
String Beans	Macedoine Beans	Celery
Succotash	Spinach	Turnips

14

FIG. 6.1: Page from *Your Food and Health in the North*, prepared by Dr. Frederick F. Tisdall and printed by the Hudson's Bay Company, 1940.

The federal government directed new attention northward during and after the Second World War. This was stimulated in large measure by interest in northern resources, science, and sovereignty in the context of ongoing global conflicts.[62] Yet it also reflected the more activist welfare state of the post-war period in its expanded concern for social and health policy in the region. Federal government interventions in northern nutrition manifested in a variety of ways, including initially in policy and practices regarding relief. At mid-century, the place of relief in northern subsistence was complicated. For decades, the HBC and other traders had supplied grubstakes to trappers, giving them and their families advances on ammunition and provisions for the season on the assumption that the debts would be settled when the hunters returned with fur to trade. As the fur trade declined in the north over the course of the twentieth century, posts were shuttered and the role of the HBC transformed.[63] The federal government instead came to be seen as more responsible for some of the provisioning, which had previously been part of the HBC's trade relations. This shift led Roy Gibson to direct that closer attention be paid to the issue of relief rations in the north. He wrote to the Commissioner of the RCMP that, "what we are trying to avoid is having accounts classed as relief when they should be considered as grubstakes by the traders."[64] Nevertheless, as Tester and Kulchyski argue, by the 1950s the Inuit in the Garry Lake region had shifted from "a condition of total independence and reliance on caribou and fishing, to a reliance on caribou, fishing, and relief to tide them over."[65] The Northwest Territories Branch aimed to avoid such practices. It saw relief as demeaning and wanted to keep northern Indigenous people "independent" and the costs of administering relief low, without being "too niggardly."[66] RCMP officers indicated that they "[made] it a practice of discouraging the issue of relief rations."[67]

Widespread tuberculosis infection among northerners countered efforts to limit relief. Active TB infection could prevent hunters from successfully procuring subsistence for their families; moreover, good nutrition was seen as essential to fighting TB, particularly as antibiotics would not become widely available until the late 1940s. Thus, by the 1950s, there were two categories for relief provision: a relief ration for those who were destitute and required relief to be administered "in part or in full at the discretion of the administering officer," and a supplementary ration for all members of a family for six months "upon discovery and diagnosis of pulmonary tuberculosis in any member of a family," or "upon return from hospital of a T.B. patient."[68] Given the prevalence of tuberculosis in the north, this second category of relief constituted a major intervention by the federal government in the diets of northerners. Likewise, the introduction of regular health checks to screen for TB in the 1940s led to the provision of rations for families who disrupted their regular subsistence practices to come into communities to meet the medical ships.[69]

It is also clear that agents of the federal government viewed relief rations as an opportunity to intervene positively in the diet of northern Indigenous people. In a 1955 meeting on rations for Inuit, Dr. J. S. Willis, of the Indian Health Services branch of the Department of National Health and Welfare, observed that "the percentage of Eskimos suffering from poor nutrition was quite high. … He felt that the relief rations provided an opportunity for us to see that the Eskimos were given foods which provided kinds of nutrition we knew to be lacking."[70] Willis' comments demand two questions: how was Inuit diet perceived to be lacking? And what did the government understand as appropriate rations under the circumstances? There is much evidence that southern administrators and others viewed country-food diets, and especially "transitional," diets as inherently inadequate. The "post diet" was a particular concern, as L. B. Pett, chief of nutrition services with the federal Department of Health and Welfare, noted in a letter to Gibson. Pett characterized Indigenous northerners as moving from "a true native diet (that seems to be pretty adequate) [to] a 'trading post' diet of flour, lard, salt, baking powder and tea."[71] In spite of Pett's description of the "pretty adequate" Native diet, it was apparent from many of the government discussions about rations that "food" meant southern food, not country food. At the same meeting on Inuit rations where Willis spoke, Mr. J. Cantley, of the Arctic Division of the Department of Northern Affairs, pointed out that there were "many more foods … available in the North now than there were five years ago," by which he meant southern foods.[72] Ben Sivertz, Acting Director of the Department of Northern Affairs and National Resources, likewise noted, in his memo to L. A. C. O. Hunt, the Mackenzie District administrator, that greater discretion could be exercised in issuing rations to "white and halfbreed persons" relative to Indigenous people. "For those persons who are accustomed to broader selection of foodstuff than laid out in Schedule 1," he noted, "a cash equivalent in the form of a food voucher may be issued."[73] Northern Indigenous people would most certainly have been likewise accustomed to a "broader selection of foodstuff" than specified on ration lists, but in his writing here, Sivertz reveals the unspoken assumption, common in state correspondence on nutrition, that country food was less adequate in a variety of ways—here, specifically, less diverse—than southern food.

Across Canada, administration of food relief was guided by the principle that the standard of living provided in relief should not exceed that which could be obtained through "economic effort"—in other words, through subsistence harvesting or wage labour. Yet there was a perception that northern Indigenous peoples in general maintained only a very low standard of living. This reflected attitudes toward country-food diets, discussed above, and functioned to lower the bar for relief in the north.[74] Relief rations were universally applied in the north: the lists of rations did not vary according to whether the recipient was Inuit, Métis, non-Indigenous, and so on. Most lists resembled this one:

Flour

Rolled Oats

Rice

Sugar, Jam or Molasses (one or the other)

Lard, fortified margarine or beef fat

Beans, dried, or extra rolled oats

Tea

Baking Powder

Salt

Cheese

Milk

Tomatoes, canned (where available)[75]

Meat, specifically beef and typically tinned rather than fresh, was also commonly found on relief lists. Nevertheless, ration lists emphasized grains, dairy, and vegetables to a much greater extent than country-food diets. What is not immediately apparent from this particular list was that the identification of appropriate rations was directly informed by scientific research, particularly with regard to fortified foods and vitamins. Multivitamin capsules and cod liver oil appeared on lists and in discussions about rations, as did "special flour" or "bannock mix"; the latter (manufactured by the Canadian Doughnut Company of Trenton, Ontario) was flour to which skim milk powder, vegetable shortening, baking powder, and salt had been added. Much like the seal-meat-infused biscuits of the 1920s, bannock mix aimed to fortify northern diets, in this instance with milk added into an otherwise familiar food.[76] There were repeated concerns that a taste for fortified foods, such as enriched flour or the bannock mix, needed to be inculcated amongst northern Indigenous peoples. In addition to fostering a taste for such foods among children, nutritional researchers also conducted experiments with northerners on the best way to prepare such mixes to ensure their optimal palatability.[77]

Northerners did not passively accept these decisions about relief and rations as constructed by agents of the federal government. In 1956, Inuvialuit chiefs and the Citizens' Committee in Aklavik sought an increase in the caloric content of rations from 2,800 calories to 8,400 calories per day. They called for more food and more varied foods. The chiefs and the Committee included fresh meat and fresh fish on their lists, as well as matches, dried fruit, ham or bacon, and vegetable soup. The chiefs and the Committee, moreover, drew on scientific research to make their argument. The state representatives, in turn, used military research that looked at soldiers, air crews, and lumberjacks—people working hard, outdoors, and in northern climates, but who did not require such a high calorie intake—to justify keeping the caloric content of rations lower.[78] State officials opposed race-based differences in rations on the grounds that this impeded their fundamental assimilationist project. The argument was resolved by keeping the caloric content of rations higher than it was in southern Canada, and approving additional rations for those with active cases of TB and their immediate family.[79] This demonstrated the emphasis upon environment over "race" within the rationale for improved nutrition. It also exposed the new emphasis on scientific authority. Northerners had long argued for improved rations; now they deployed the language of nutritional science to do so, although ultimately they remained unsuccessful in having their needs fully met.

The foodstuffs on ration lists were only one prong in a strategy of nutritional interventions that also included education and surveillance, with a specific focus on Indigenous people. This was most clearly articulated by Sivertz to district administrator Hunt in a letter dated 22 August 1956, in which Sivertz directed that, "it will be the responsibility of the administrator in the area, upon discovery of T.B. in a member of a household, to investigate the family circumstances. The administrator will be required to insure that there is not only sufficient food in the home, but that it is the type which is nutritionally sound. ... If there is a shortage of adequate foodstuffs ... immediate steps must be taken to insure that there is a balanced diet in the home."[80] In a letter dated the following February, F. J. G. Cunningham in Ottawa wrote to the district administrator at Fort Smith emphasizing the importance of sitting down with ration recipients individually to "point out the necessity of a balanced nutritional diet."[81] The language of "adequate foodstuffs" and "a balanced diet" masked assumptions about what constituted good and healthy food. Cunningham further acknowledged that educating the Inuit population in this fashion would take a lot of time and effort, but that the ultimate goal was that they receive the same rations as their non-Indigenous counterparts. He thus simultaneously articulated the nutritional dimension of the assimilationist project and the educational objectives of the ration and relief system.

In contrast to relief, Family Allowances were a universal welfare program intended to raise the standard of living for Canadian children across the country.[82] Family Allowance credits could be used to purchase foods specified on a list "recommended by the Department of National Health and Welfare as beneficial to the health of children."[83] Nursing and pregnant women also fell under the purview of the Allowances. Recommended foods included milk, Pablum, flour (fortified with vitamin B), rolled oats and oatmeal, sugar, eggs, canned or fresh meat ("issued only when game is scarce"), peanut butter, cheese, fruit, canned tomatoes and tomato juice, green or dehydrated vegetables, rice and beans, and more. Milk and Pablum were considered particularly important, and Gibson supplied the general manager of Hudson's Bay Company's fur trade department with a list detailing the number of children at each post across the north registered for Family Allowances, and who therefore should have been receiving milk and Pablum regularly.[84] Family Allowances dramatically increased milk consumption in the north. In one HBC report, where no milk had

Fig. 6.2: "Mr. and Mrs. Sigvaldasson issuing Family Allowance to Inuit woman in the form of powdered milk, Cape Smith, N.W.T., 1948." This photograph was taken and captioned by S. J. Bailey, then Regional Director, Family Allowances for the Yukon and Northwest Territories. As is apparent in the positive framing of this photograph, Bailey was convinced of the value of Family Allowances for helping Indigenous northerners (see Tester and Kulchyski, *Tammarniit (Mistakes)*, 87). Source: S. J. Bailey/Library and Archives Canada, PA-167637.

been stocked for Indigenous people in the eastern Arctic prior to 1946, 655 pounds had been distributed among Inuit by 1948.[85] Gibson expressed concern that Family Allowances, like relief, could cultivate dependency in Indigenous northerners, and he stressed that they were to be for the sole benefit of children. He simultaneously embraced the potential of Family

Allowances to help cultivate new dietary practices. He noted, for instance, that many northerners held an "unfounded bias" against vitamin B-enriched flour, "but it is believed that, as its use becomes more general by issue through Family Allowances credit, its popularity will grow."[86] Targeting children with the fortified foods and nutritional advice available as part of the Family Allowances program reflected a larger effort focused on early diet, which included encouraging parents to administer cod liver oil and a synthetic vitamin D preparation called viosterol.[87]

Economic advantage was part of federal government food interventions in the north, even if it was not spelled out as a nutritional policy measure. The bannock mix, for instance, was seen not only to improve diet by increasing the amount of milk consumed, it was also thought possible to link this new diet to new commerce. One proposal suggested shipping the ingredients in bulk to Rankin Inlet, and repackaging the mix into five-pound bags for distribution across the north.[88] This proposal foundered given that the mix could be supplied much more cost-effectively through the existing HBC transportation and post network. By the 1950s, the state also encouraged greater employment of northern Indigenous people in industrial operations. This included work in mines, on hydroelectric developments, and in the construction of new northern infrastructure, such as highways and Cold War–era defensive lines, including the DEW, Mid-Canada, and Pinetree lines. Contractors on the Mid-Canada Line hired Inuit to work at Great Whale River in the mid-1950s. These men initially took time off for hunting, but the contractors found this arrangement unsatisfactory as it interfered with "getting the work done on time."[89] One alternative was to encourage the Inuit to buy seal meat from other Inuit, but this was not well received. As Sivertz wrote to a James Bay district HBC manager, "of course this is something that might be overcome in time as the Eskimos become accustomed to wage employment and buying their food requirements."[90] Where northern Indigenous people did participate in the larger market for southern goods, they were typically seen as making poor consumer decisions. One RCMP officer suggested that rather than controlling the sale of food, the goods that arrived in the north should instead be more closely controlled to ensure that there were fewer non-essentials and that more be "invested in foods of high nutritional value."[91] Even the use of the word "consumers" to refer both to those who purchased or ate new southern products demonstrated how the new foods were bringing with them new socio-economic arrangements that sought to replace subsistence with markets for food.[92]

The earliest government nutrition survey conducted in northern Canada was completed in 1943. The Bureau of NWT and Yukon Affairs distributed a questionnaire to post managers and RCMP officers asking about subsistence practices and the availability of food. Seventeen questionnaires were returned from the central and western Arctic, but none from the east (although at least one post manager from the east mailed his comments in a letter to the Bureau).[93] This nutritional investigation coincided with the earliest northern TB survey, conducted along the Mackenzie River. The simultaneity of these two surveys spoke both to the federal government's increased interest in the north and the interconnections between TB—already understood to be the foremost northern health issue—and nutrition. As mentioned above, good nutrition was central to tuberculosis treatment, especially prior to the development of effective antibiotics. In the north, malnutrition was seen as having played a direct role in the tuberculosis epidemic.[94] As part of the 1943 TB survey, men, women, and children were systematically X-rayed to identify active tuberculosis infections. The results, published in 1945, revealed the staggering scale of the problem: for the Inuvialuit in the Mackenzie Delta, the physician responsible for the survey found that TB had "a death-rate of 314 per 100,000 population compared with 53 per 100,000 for the rest of Canada."[95] Eastern Arctic, western Arctic, and Yukon surveys followed shortly thereafter. The aim of these surveys was to X-ray the entire population, Indigenous and non-Indigenous, and then to isolate those who had active tuberculosis and send them south for treatment. Given the size of the region, the fact that most northerners lived off the land rather than being concentrated in communities, and the relative inexperience of southern doctors with subarctic and arctic environments, these comprehensive surveys were massive undertakings. New surveys followed every few years, as the federal government—including bureaucrats with the Department of National Health and Welfare (DNHW) and the Department of Northern Affairs and National Development (DNAND), and its successors—sought to assess whether or not the problem was under control.

The large-scale, comprehensive surveillance techniques and research methods developed in the TB surveys soon came to be applied to the issue of nutrition in the north. The 1943 investigation had relied on a small number of questionnaires from non-Indigenous police officers and post managers, supplemented with descriptions from officials, and advice from doctors and dentists.[96] By 1961, DNAND and DNHW officials agreed to survey nutrition at large. The new surveys entailed:

(a) A full clinical examination of selected sections of the native population;

(b) The collection of blood and urine samples for detailed biochemical examinations for vitamin and mineral contents;

(c) The collection of samples of native foodstuffs for analysis of protein, fat, carbohydrate and mineral content to see whether they are similar to the values obtained in Alaska studies; and

(d) A dietary survey.[97]

Some of the health information, such as blood and urine samples, was collected during the annual tuberculosis surveys. The dietary surveys interrogated aspects of home and family life that pushed beyond the simple question of "what do you eat?" The 1961 surveys were much more comprehensive than those of 1943, with questionnaires distributed throughout the school system reaching children in residential schools and the families with children attending day schools. Detailed interviews were conducted with schoolchildren, family groups, and communities. Although the quotation above only mentions "native foodstuffs," elsewhere researchers measured the nutritional content of a range of non-Indigenous and country foods. The new research methods closely probed northern bodies, both human and those of the non-human animals consumed, as well as social and cultural dimensions of diet, with particular attention to Indigenous people.

To survey a community, researchers would question local merchants, area administrators, and RCMP officers. Families were given money to purchase food (which also served as an incentive to participate in the survey), and DNHW workers monitored their food consumption for one week each month over a six-month period. In the day schools, teachers distributed survey booklets to all children who could write. The children took the booklets home to complete for one week out of every month over a one-year period. Teachers returned the completed booklets to the Department of National Health and Welfare for analysis.[98] In general, the children cooperated with the survey process. Some children, teachers complained, lost their forms or forgot to fill them out. While it is not surprising that young children might lose their booklets or be less than assiduous record keepers, it is also likely that this carelessness occasionally reflected opposition to the survey. Active opposition was clearly articulated by some parents. Mrs. Cockney, an Inuvialuit mother in Inuvik, wrote the following note to a teacher, "Sister C.:"

> I just want to know if Margaret has to write what she eats all the time cause I don't think its not anybodys business to know what our children eats as far as I know I always give my children what's good for them.
>
> So please let the Principal know.
>
> Regards, Mrs. L. Cockney[99]

In her short note, Mrs. Cockney captured the larger significance of the dietary surveys: that an examination of what children ate went beyond the caloric and protein constituents of particular foodstuffs to an evaluation of "what's good for them." Inuvialuit mothers knew what was good for their children, and therefore did not see the need for teachers and Health and Welfare workers to interrogate their diets. Other northerners evidently expressed "opposition and hostility" to the survey, as reported by government workers in 1965 and 1966.[100] One school principal in Fort Resolution pointed out that the requirement for students to include their names and treaty numbers led "pupils and parents ... [to] look upon [the survey] as some type of government 'spying.'"[101] The same principal went on to observe that the children often recorded what they thought teachers would like to see them eat, and not what they were actually eating.[102] By the latter half of the twentieth century, the state had become increasingly intrusive in the lives of northerners—and northern Indigenous people, in particular—and Mrs. Cockney's letter, alongside evidence of wider opposition, reveals the range and spirit of resistance to these intrusions. In their correspondence, survey administrators expressed concern that such opposition would lead to inaccurate survey results and apathetic or antagonistic attitudes toward proper nutrition. Officials with the Department of National Health and Welfare tried to assuage community concerns by holding meetings, but unsurprisingly, such consultations neither led to changes in the overall program nor alleviated individual and community resistance to escalating state and scientific surveillance.

The 1965–66 survey revealed a preference among northerners for store-bought foods (hard biscuits, lard, jams, tea, and dried milk), even as country foods (moose meat, caribou, seal, fish, and berries) continued to make up a significant portion of their diets. In most northern communities, those attending residential schools (also called hostels) ate very differently from those who continued to live with their families. The latter typically ate much more country food, but also more store-bought "junk food," including "candy, pop, chocolate bars, bubble gum, suckers."[103] The children at the hostels were characterized as "eat[ing] better" and having "more nutritionally adequate" diets.[104] In general, the residential school system was praised for the ways in which school administrators drew upon nutritional science in preparing weekly menus and daily meals. All the children in the hostels were, moreover, given vitamin A and D supplements in fortified bannock, although it was duly noted that the children did not like these biscuits. Among children outside the residential schools, and some inside, researchers found a range of vitamin deficiencies, most commonly anaemia.[105]

In designing these nutrition surveys and responding to their findings, researchers and federal government officials articulated repeatedly what can only be described as insecurity about country-food diets. Even as they praised the richness of country foods and the good health of those who continued to live in "traditional hunter societ[ies]," such comments were always qualified with concerns about the move away from such traditional diets, from the land into settlements and from "native" to "white man's" foods.[106] As Stephen Bocking notes in his chapter in this volume, anxieties about country foods would continue to evolve later in the twentieth century, with mounting evidence of the contamination of such foods by a range of toxins.

A material foundation for mid-twentieth-century concerns about Indigenous diets and traditional foods lay in the environmental changes that had come to the north in the twentieth century. Game animals, fur-bearers, marine mammals, fish, and birds had all suffered population declines. Access to these and plant-based foodstuffs met with new restrictions, as subsistence practices competed with industrial, scientific, conservationist, and military activities on northern lands and waters. Likewise, new ties to settlements (to participate in health surveys, obtain relief, pick up children from school, trade, meet with friends, and so on) interfered with seasonal rounds. Yet the perceived insecurity was not simply a response to the wholesale cultural, economic, and environmental changes that had reached across the Subarctic and Arctic. It also reflected deeply held southern Canadian values about what were reliable foodstuffs and what constituted a safe and secure diet. Foremost, this was a diet predicated upon agriculture. That trade posts and missions invariably established garden plots (and that the HBC's nutritional booklet strongly encouraged such practices) was itself evidence of the hold that agriculturally derived foodstuffs had upon southerners relocated to the north. Likewise, descriptions of northern diets often made reference to the lack of agricultural opportunities. In correspondence with Roy Gibson, J. G. Wright noted that, "In the case of the Indians, there is some suggestion that they might raise gardens and depend more upon vegetables. This solution, of course, could not apply to the Eskimos in the Arctic. It seems to me that the Eskimo problem is an even greater one than that facing the Indian Affairs [sic]."[107] By the 1960s, a safe and secure diet was also increasingly a diet that was informed by nutritional science. Hence the perceived need to study blood and urine samples from northerners, and to supplement their diets with vitamins and minerals. Finally, attitudes about appropriate diet have to be seen within their colonial context. While southerners regularly praised the ability of northern Indigenous people to obtain subsistence from the land, they did so within a profoundly racist culture that understood Indigenous people as primitive and their cultures as rooted in the past, and not appropriate to the present or the future. To rely wholly upon subsistence from the land was not a practice that came easy to relocated southerners; therefore it is not surprising that it was not seen by them as a reliable foundation upon which to build a healthy northern future.

Conclusions

The authors of the 1965–66 nutritional survey report, including distinguished physician Otto Schaefer, wrote:

> Experience in other parts of the world has indicated that cultural change is almost invariably accompanied by a nutritional inadequacy of diet and the consequent appearance of clinical disorders in the native people resulting from malnutrition and metabolic change.[108]

Schaefer was very sympathetic to the experiences of Indigenous northerners and was well respected by northerners and southerners alike.[109] Nevertheless, he and his fellow survey authors articulated views held widely across the federal administration that saw northerners as caught up in a unidirectional and inevitable process of modernization: a series of changes that led away from the land and into the communities, and which were already apparent across the north beginning in the late nineteenth century. Perceptions of the process of modernization among northern Indigenous peoples were rooted in culturally dominant ideas about race and culture, namely the forward progress of northern Indigenous peoples from primitive, traditional lives to civilized, modern ones. Yet the process that was underway was neither inevitable nor solely cultural; it was, instead, historical, and therefore could potentially have moved along any of a number of paths, as directed in the twentieth century through the actions of the state, missionaries, trading companies, resource operators, and northerners themselves.

In the early part of the century, the Dominion government's influence in the north was relatively circumscribed. Its nutrition policy dovetailed with wildlife regulation efforts and a basic goal to "keep the native, native." State intervention before 1940 was primarily in response to epidemics, where its role in providing rations ultimately grew into a significant intervention in northern diets. Government efforts were directly complemented by the role of the churches in establishing and operating the residential schools. Countless other newcomers influenced northern subsistence practices and diets in the early part of the twentieth century through increased and changed demand upon the range of northern resources.

New pressures were placed on game animals and fish for subsistence, recreational, and commercial harvesting. Demand for other northern resources, whether fur-bearers, minerals, or oil, intensified pressures on subsistence resources while simultaneously introducing new foodstuffs and influencing changes in settlement and transportation geographies.

The relocation of northern Indigenous peoples—such as the Inuit from Baffin Island, who found themselves at Creswell Bay in the summer of 1948—signalled a turning point in the federal government's role in the north. The relocations exemplified an interventionist approach by Ottawa, one that sought to manage directly the changing relations between northerners and their environments. The health consequences of these relocations, including hunger, starvation, and the illness suffered by those at Creswell Bay, highlight two key aspects in the history of northern diet:first, that subsistence represented an intimate relationship with the land and animals, and one not easily transferred from place to place; and second, that the federal government would meet with greater success in achieving its objectives when it did not intervene so aggressively, but rather pursued a more bureaucratic approach.[110] Such an approach would rely upon regular interactions between government officials and northerners, such as relief payments and Family Allowances; would focus upon children and educational efforts among northerners; and would see the formulation of policy that was directly informed by comprehensive scientific research.

These latter measures came to the forefront after the Second World War, when the federal government turned to diet and nutrition as tools to manage the changing relations between northerners and nature. State administrators used relief, rations, and Family Allowances to guide northerners away from subsistence and country foods, and toward diets heavy in dairy products and grains that depended upon agriculture and new relationships with southern producers and food networks. Northerners resisted this shift in different ways, ways that were particularly evident in their responses to the large-scale nutrition survey undertaken in the early 1960s. Children, the main target of the survey and of dietary interventions in general, recognized that the survey questionnaires were not objective inquiries into what they ate, but rather loaded with expectations about what they should eat. At times, they completed the questionnaires accordingly. Mothers articulated their anger at state surveillance methods, while community leaders—both Indigenous and non-Indigenous—relied on the same science that informed state policy to develop alternative representations of dietary health. Ultimately, resistance was also apparent whenever northerners continued to practice subsistence and to eat country foods.

A profound sense of insecurity about the possibilities for healthy living in the north permeated federal government nutrition policy. This insecurity, although rooted in broad cultural perceptions about what constituted food, and good food in particular, was markedly reinforced by the health crises at mid-century, including the famine in the Keewatin and the belated recognition of the northern TB epidemic. This insecurity was also inseparable from understandings of northern environments as imperfectly controlled by the Canadian government. Federal policy was not solely predicated upon negative perceptions of Indigenous diets, however, but also upon aspirations that Indigenous northerners could be assimilated into the mainstream of Canadian social and economic life. This would be realized through participation in wage economies that enabled northerners to procure foods through a shared marketplace overseen by federal regulators and informed by nutritional science. Government officials focused on children who could be educated to comply with what were understood to be the best dietary practices, and among whom new tastes could be most effectively cultivated. Diet was thus a dimension of ethical citizenship, a means to bring northerners into the mainstream of Canadian society by eating in common with their southern counterparts, and doing so under the state's watchful eye.

Notes

1   The author gratefully acknowledges the Inuvialuit Cultural Resource Centre and the Gwich'in Tribal Council Department of Cultural Heritage for granting access to unpublished interview transcripts with Inuvialuit and Gwich'in elders as part of the research for the larger project of which this chapter is a part. In particular I wish to thank Cathy Cockney and Ingrid Kritsch for their support and assistance. The Social Sciences and Humanities Research Council of Canada and the Killam Trusts funded this research. I also thank Stephen Bocking, Brad Martin, and the workshop participants (in both the June 2009 and October 2011 workshops) for their suggestions and insights into northern environmental history.

Myra Moses as quoted in Vuntut Gwichin First Nation and Shirleen Smith, *People of the Lakes: Stories of our Van Tat Gwich'in elders = googwandak nakhwach'ànjòo Van Tat Gwich'in* (Edmonton: University of Alberta Press, 2009), 74.

2   Marianne Lien, "Introduction," in *The Politics of Food*, ed. Marianne Elisabeth Lien and Brigitte Nerlich (Oxford and New York: Berg, 2004), 5.

3   Environmental historians have explored the consequences of such masking to the environments of origin for foodstuffs and natural resource products. See, for example, L. Piper, *The Industrial Transformation of Subarctic Canada* (Vancouver: UBC Press, 2009), 102.

4   The salience of these divides was highlighted during the writing of this chapter with the visit of the United Nations Special Rapporteur on the Right to Food, who grouped some of the issues facing Aboriginal and northern Canadians. See "Aboriginal access to food under scrutiny," *Edmonton Journal*, 13 May 2012.

5   See Robert N. Chester III and Nicolaas Mink, "Having Our Cake and Eating It Too: Food's Place in Environmental History, A Forum," *Environmental History* 14, no. 2 (2009): 309–11; Nancy Shoemaker, "Food and the Intimate Environment," *Environmental History* 14, no. 2 (2009), 341.

6   Sidney Mintz, *Sweetness and Power: The Place of Sugar in Modern History* (New York: Viking, 1985); Mary Douglas, *Purity and Danger: An Analysis of Concepts of Pollution and Taboo* (New York: Praeger, 1966).

7   "Food colonialism" is a variation upon Nancy Shoemaker's use of "food imperialism." Shoemaker, "Food and the Intimate Environment," 342.

8   Chase Hensel, *Telling Our Selves: Ethnicity and Discourse in Southwestern Alaska* (New York and Oxford: Oxford University Press, 1996), 3.

9   And is in the anthropological literature: See, for instance, George Wenzel, *Animal Rights, Human Rights: Ecology, Economy and Ideology in the Canadian Arctic* (Toronto: University of Toronto Press, 1991), especially ch. 3, "The Culture of Subsistence," 56–63; Carol Zane Jolles, with the assistance of Elinor Mikaghaq Oozeva, *Faith, Food, and Family in a Yupik*

Whaling Community (Seattle: University of Washington Press, 2002); June Helm, *The People of Denendeh: Ethnohistory of the Indians of Canada's Northwest Territories* (Montreal: McGill-Queen's University Press, 2000), especially "Part I: Community and Livelihood at Midcentury."

10   Peter J. Usher, "Socio-economic effects of elevated mercury levels in fish on sub-arctic native communities," in *Les Contaminants dans l'environnement marin du Nunavik : actes du colloque, Montréal, 12 au 14 septembre 1990 = Contaminants in the marine environment of Nunavik : proceedings of the conference, Montreal, September 12-14, 1990* (Sainte-Foy, QC: Université Laval, Centre d'études nordiques, 1992), 47.

11   Lien, "Introduction," 6. Emphasis in the original.

12   Mary-Ellen Kelm, *Colonizing Bodies: Aboriginal Health and Healing in British Columbia* (Vancouver: UBC Press, 1998); and Maureen Lux, *Medicine that Walks: Disease, Medicine, and Canadian Plains Native People, 1880-1940* (Toronto: University of Toronto Press, 2001). See also Margery Fee, "Stories of Traditional Aboriginal Food, Territory, and Health," in *What's to Eat? Entrées in Canadian Food History*, ed. Nathalie Cooke (Montreal: McGill-Queen's University Press, 2009), 55-78; Margery Fee, "Racializing Narratives: Obesity, Diabetes, and the 'Aboriginal' Thrifty Genotype," *Social Science & Medicine* 62 (2006): 2988-97; Ian Mosby, "Administering Colonial Science: Nutrition Research and Human Biomedical Experimentation in Aboriginal Communities and

Residential Schools, 1942-1952," *Histoire sociale / Social history* 46, no. 91 (2013): 145-72; James Daschuk, *Clearing the Plains: Disease, Politics of Starvation, and the Loss of Aboriginal Life* (Regina: University of Regina Press, 2014); as well as ongoing work by Paul Hackett on diabetes and First Nations populations.

13   The issue of toxic contamination of country foods in the north falls outside the scope of this study. It came to the fore after 1970, when mercury first appeared as a significant contaminant in freshwater fisheries. For discussions of food contamination, see Usher. "Socio-economic effects of elevated mercury levels," and Peter J. Usher, Maureen Baikie, Marianne Demmer, Douglas Nakashima, Marc G. Stevenson, and Mark Stiles, *Communicating about contaminants in country food: The experience in Aboriginal communities* (Ottawa: Inuit Tapirisat of Canada, 1995).

14   Fikret Berkes, "Native Subsistence Fisheries: A Synthesis of Harvest Studies in Canada," *Arctic* 43, no. 1 (1990), 40.

15   See Piper, *Industrial Transformation*, ch. 7.

16   Frank Tough, *"As Their Natural Resources Fail": Native Peoples and the Economic History of Northern Manitoba 1870-1930* (Vancouver: UBC Press, 1996), 24-25.

17   Renée Fossett, *In Order to Live Untroubled: Inuit of the Central Arctic, 1550-1940* (Winnipeg: University of Manitoba Press, 2001), 190.

18   Anthony G. Gulig, "'Determined to Burn off the Entire Country': Prospectors, Caribou, and the Denesuliné in Northern Saskatchewan, 1900-1940," *American Indian*

Quarterly 26, no. 3 (2002): 335–59; Kenneth S. Coates and William R. Morrison, *The Alaska Highway in World War II: The U.S. Army of Occupation in Canada's Northwest* (Toronto: University of Toronto Press, 1992), 86–88. See also Stephen J. Pyne, *Awful Splendour: A Fire History of Canada* (Vancouver: UBC Press, 2007).

19   Michael K. Heine and the Elders of Tsiigehtshik, et al., *Gwichya Gwich'in Googwandak. The History and Stories of the Gwichya Gwich'in, As Told by the Elders of Tsiigehtshik* (Tsiigehtshik, NT: Gwich'in Social and Cultural Institute, 2007).

20   Fossett, *In Order to Live Untroubled*, 33.

21   For details on the rich fishery and whaling in the Mackenzie Delta, see David Morrison, "Inuvialuit Fishing and the Gutchiak Site," *Arctic Anthropology* 37, no. 1 (2000): 1–42.

22   Scott Polar Research Institute (hereafter cited as SPRI), BJ, MS 699/2, David Theophilus Hanbury, Journal Kept During his Explorations of the Keewatin District of Canada, 18 April 1902; vol. 2: 2 December 1901–12 May 1902.

23   See Greg Gillespie, *Hunting for Empire: Narratives of Sport in Rupert's Land* (Vancouver: UBC Press, 2007). For a description of a hunt for musk ox by members of the British Arctic Expedition, see SPRI, BJ, MS 41, George Gifford Journal—British Arctic Expedition, Bellot Bay, Lady Franklin Straits, 25 August 1875.

24   See, for example, meteorological research conducted by the British Polar Year Expedition to Fort Rae under J. M. Stagg, with its

relationship to land and communities detailed in journals kept by expedition member Alfred Stephenson (SPRI, BJ, MS 432/2).

25   John R. Bockstoce, "The Consumption of Caribou by Whalemen at Herschel Island, Yukon Territory, 1890–1908," *Arctic and Alpine Research* 12 (1980): 383; T. Max Friesen, *When Worlds Collide: Hunter-Gatherer World-System Change in the Nineteenth-Century Canadian Arctic* (Tucson: University of Arizona Press, 2013), 188.

26   I use the term "country food" anachronistically in this essay (as it was not employed until the late twentieth century) to refer to foods harvested from the land in contrast to food purchased at the HBC store or elsewhere. For a description of the decline in animal and bird life, see Library and Archives Canada (hereafter cited as LAC), George Mellis Douglas Fonds, George Douglas to P. G. Downes, 24 April 1955.

27   This concern was for both northern and southern interests. For example, the Migratory Birds Act was a response to pressure from southern hunters. To see these subjects addressed in detail, see John Sandlos, *Hunters at the Margin: Native People and Wildlife Conservation in the Northwest Territories* (Vancouver: UBC Press, 2007); Tina Loo, *States of Nature: Conserving Canada's Wildlife in the Twentieth Century* (Vancouver: UBC Press, 2006); Kurkpatrick Dorsey, *The Dawn of Conservation Diplomacy: U.S.-Canadian Wildlife Protection Treaties in the Progressive Era* (Seattle: University of Washington Press, 1998).

28   LAC, RG 85, vol. 1416, file 252-1-2, Charles V. Sale, HBC, London, to O. S. Finnie, 30 July 1928; O. S. Finnie to W. W. Cory, Deputy Minister, Department of the Interior, 25 June 1928. Finnie was apparently quoting comments made by Frederick Banting where he advocated for "keeping the native, native."

29   LAC, RG 85, vol. 768, file 5208, O. S. Finnie, "A Letter from the Government to the Indian People," 1 April 1924; and Yukon Archives (hereafter cited as YA), COR 261, file 4, O. S. Finnie, "A Letter from the Government to the Eskimo People," 19 November 1926.

       LAC, RG 85, vol. 1416, file 252-1-2, O. S. Finnie to C. H. French, Fur Trade Commissioner, HBC, Winnipeg, 22 June 1928.

       Finnie to Cory, 25 June 1928.

30   Emphasis in the original.

31   A fuller examination of these processes is the main focus of the larger project from which this essay draws.

32   For the most devastating overview of this relationship, see Mike Davis, *Late Victorian Holocausts: El Niño Famines and the Making of the Third World* (London, New York: Verso, 2001).

33   For an estimate of winter fish requirements for dog teams at Lac La Martre, see Helm, *The People of Denendeh*, 57. On the need to account for dog food on relief expeditions, see LAC, RG 85, vol. 1118, file 1000/128–1, H. H. Cronkhite, Insp., O.C., "G" Division telegram to Officer Commanding, RCAF, Northwest Air Command, Edmonton, 25 January 1949.

34   CBC Radio 1 North, *Trailbreaker*, Radio Interview with Marc Winkler, Tuktoyaktuk elder Tom Thrasher, and Liza Piper on tuberculosis and influenza epidemics in northern history, 6 October 2009.

35   Helge Ingstad, *Land of Feast and Famine*, trans. Eugene Gay-Tifft (New York: A. A. Knopf, 1933), 149–55.

36   LAC, RG 85, C-1-a, vol. 789, file 6099, Extracts from the report of Inspr. M. Royal Gagnon dated at Fort Smith N.W.T., 11 Aug 1928, on his summer inspection patrol to Reliance.

37   LAC, RG 85, vol. 640, file 552-1-1-3, Const. C. J. Dent, Baker Lake Detachment, "Destitution and Deaths Amongst Eskimos Garry Lake and Back River Districts, N.W.T.," 14 September 1954. The Garry Lake famine is also described in Frank J. Tester and Peter Kulchyski, *Tammarniit (Mistakes): Inuit Relocation in the Eastern Arctic, 1939–63* (Vancouver: UBC Press, 1994), ch. 6.

38   See, for example, René Fumoleau, *As Long as this Land Shall Last: A History of Treaty 8 and Treaty 11, 1870–1939*, rev. ed. (Calgary: University of Calgary Press, 2004), 225, 240.

39   The earliest residential schools in Canada opened in the 1840s. For a comprehensive history of the residential school system, see James R. Miller, *Shingwauk's Vision: A History of Native Residential Schools* (Toronto: University of Toronto Press, 1996). For more specifics on food in residential schools, see John Milloy, *A National Crime: The Canadian Government and the Residential School System* (Winnipeg: University of Manitoba Press, 1999), 121; Fee, "Stories

of Traditional Aboriginal Food, Territory, and Health," 66–68.

40   L. Piper and J. Sandlos, "A Broken Frontier: Ecological Imperialism in the Canadian North," *Environmental History* 12, no. 4 (October 2007): 778–79.

41   James E. Woollett, et al., "Palaeoecological Implications of Archaeological Seal Bone Assemblages: Case Studies from Labrador and Baffin Island," *Arctic* 53, no. 4 (2000): 409.

42   NWT Archives (hereafter cited as NWTA), N-1992-012, file 1–4, Doug Wilkinson Daily Journal, 6 December 1953.

43   Tester and Kulchyski, *Tammarniit (Mistakes)*, 359; this example is also cited in a brief discussion in Fee, "Stories of Traditional Aboriginal Food, Territory, and Health," 68–69.

44   Usher et al., *Communicating about Contaminants*, iii.

45   "Order Mercy Hop After Dog Team Brings Out Message," *Ottawa Evening Journal*, 22 January 1949, 1, 11; "Groom Two Planes To Fly To Remote Arctic Village," *Edmonton Journal*, 31 January 1949, 1–2 (quotation on p. 2); "Two Planes Fly to Aid Plague-Ridden Eskimos," *Edmonton Journal*, 1 February 1949, 1–2 (toothache quotation on p. 1); "North Mercy Flight Enters 2nd Phase," *Edmonton Bulletin*, 17 March 1949, 13 ("properly fed" quotation).

46   "Two Planes Fly to Aid Plague-Ridden Eskimos," *Edmonton Journal*, 1 February 1949, 1–2 (quotation on p. 1).

47   "Eskimo 'Plague' Gangrene, Two of Stricken Flown Out," *Winnipeg Citizen*, 15 February 1949, [n.p.].

48   "Whale Meat Blamed for Mystery Ills," *Edmonton Bulletin*, 22 February 1949.

49   LAC, RG 85, vol. 1511, file 1000/128-1, pt. 2, Extract from report of Mr. A. Stevenson, Northern Administration, Western Arctic, March 1949.

50   Francis H. Fay, "Carnivorous Walrus and some Arctic Zoonoses," *Arctic* 13, no. 2 (1960): 114.

51   Fay, "Carnivorous Walrus," 115.

52   Peter Evans, "Aunt Kate's Map, or, How the Moravians Made the Labrador Inuit Legible to the Liberal Welfare State," ASEH Conference, Portland, OR, March 2010 (audio recording available at: http://niche-canada.org/resources/conference-workshop-archive/american-society-for-environmental-history-annual-meeting-2010/). One of these instances is described in John C. Brocklehurst, "Fatal Outbreak of Botulism Among Labrador Eskimos," *British Medical Journal* 2, no. 5050 (19 October 1957): 924.

53   LAC, RG 85, vol. 1416, file 252-1-2, pts. 1–2, D. L. McKeand to [R.] Gibson, 3 March 1944.

54   See materials in LAC, RG 29, vol. 203, file 311-P11-22, pt. 2, "Epidemiology Diseases Poliomyelitis—Poliomyelitis Epidemic, Chesterfield Inlet, NWT."

55   Toby Morantz, *The White Man's Gonna Getcha: The Colonial Challenge to the Crees in Quebec* (Montreal: McGill-Queen's University Press, 2002), 5.

56   One of the photographs was included in a 1955 exhibition, *Family of Man*, at New York's Museum of Modern Art. See Harrington's obituary by John Goddard, "Richard

Harrington, 94: A photographer to the end," *Toronto Star*, 20 December 2005. Tester and Kulchyski describe the impact of Mowat's book in *Tammarniit (Mistakes)*, 56–57.

57   Aleck Samuel Ostry, *Nutrition Policy in Canada, 1870–1939* (Vancouver: UBC Press, 2006), 3–6.

58   The last two eras cover the decades from 1973 to the present, which saw the publication of modern dietary guidelines in the 1990s, and the paradigm shift from concerns about malnutrition to concerns about overeating and its health consequences. The period from the 1990s to the present has been dominated by international pressures on Canadian food supply and policy, with the internationalization of markets and efforts to similarly internationalize food standards and dietary guidelines.

59   Frederick F. Tisdall and Associates, *Your Food and Health in the North* (Winnipeg: Hudson's Bay Company, 1940), Introduction.

60   Tisdall, Introduction.

61   Tisdall, Introduction.

62   For discussions of the dramatic changes that attended and followed the Second World War, see Stephen Bocking, "Science and Spaces in the Northern Environment," *Environmental History* 12, no. 4 (2007): 867–94; P. Whitney Lackenbauer and Matthew Farish, "The Cold War on Canadian Soil: Militarizing a Northern Environment," *Environmental History* 12, no. 4 (2007): 920–50; Edward Jones-Imhotep, "Nature, Technology, and Nation," *Journal of Canadian Studies* 38 (2004): 5–36; Matthew Farish, "Frontier Engineering: From the Globe to the Body in the Cold War

Arctic," *Canadian Geographer* 50, no. 2 (2006): 177–96; Shelagh D. Grant, *Sovereignty or Security? Government Policy in the Canadian North, 1936–1950* (Vancouver: UBC Press, 1988).

63   Arthur Ray, *The Canadian Fur Trade in the Industrial Age* (Toronto: University of Toronto Press, 1990), *passim*.

64   LAC, RG 85, vol. 103, file 253-2/145, pt. 1B, R. A. Gibson to The Commissioner, RCMP, Ottawa, 28 July 1945.

65   Tester and Kulchyski, *Tammarniit (Mistakes)*, 238.

66   LAC, RG 85, vol. 463, file 1003-1-8, pt. 1, "Relief to Destitute Eskimos [circular]," 22 April 1953.

67   LAC, RG 85, vol. 103, file 253-2/145, pt. 1B, E. L. Hadley, Coppermine Detachment, to O.C. Fort Smith, 24 April 1945.

68   LAC, RG 85, vol. 463, file 1003-1-8, pt. 1, B. G. Sivertz, Memo for L.A.C.O. Hunt, District Administrator, Fort Smith, 22 August 1956. The specifics of this policy were revised over time, extending the period of rations from six months to one year and changing who was eligible.

69   LAC, RG 85, vol. 463, file 1003-1-8, pt. 1, "Minutes of a meeting on rations for Eskimos held at 3:00 PM Tuesday, October 25, 1955, Room 302, Vimy Building, Ottawa."

70   Minutes of a meeting on rations.

71   LAC, RG 85, vol. 1416, file 252-1-2, pts. 1–2, L. B. Pett, Chief, Nutrition Division, to R. A. Gibson, 24 April 1950.

72   Minutes of a meeting on rations.

73   Minutes of a meeting on rations.

74   See, for example, LAC, RG 85, vol. 463, file 1003-1-8, pt. 1, G. W. Rowley, Study on the Issue of Relief, 11 May 1956.

75   "Relief Schedule of Monthly Rations," printed in "Relief to Destitute Eskimos [circular]."

76   Pett to Gibson, 24 April 1950.

77   LAC, RG 85, vol. 1416, file 252-1-2, pt. 4, J. Rutherford to J. M. Saulnier, 29 March 1962.

78   LAC, RG 85, vol. 463, file 1003-1-8, pt. 1, "Council of the Northwest Territories Report, Relief and Rehabilitation Ration for Eskimos and Indians," n.d.

79   See LAC, RG 85, vol. 463, file 1003-1-8, pt. 1, letter re: Relief Rations for Eskimos, 27 January 1956.

80   Sivertz to Hunt, 22 August 1956.

81   LAC, RG 85, vol. 463, file 1003-1-8, pt. 1, F. J. G. Cunningham, Memorandum for C. L. Merrill, District Administrator, Fort Smith, 20 February 1957.

82   Dennis Guest, *The Emergence of Social Security in Canada*, 3rd ed. (Vancouver: UBC Press, 1997).

83   LAC, RG 29, vol. 203, file 311-P11-22, R. A. Gibson, Circular Letter to Traders, R.C.M. Police, Missionaries and Doctors in the Eastern Arctic, 21 July 1947.

84   R. A. Gibson, Circular Letter to Traders, R.C.M. Police, Missionaries and Doctors in the Eastern Arctic.

85   Extract from LAC, RG 85, vol. 1416, file 252-1-2, pts. 1–2, S. J. Bailey's Report, Eastern Arctic Patrol, dated at Chesterfield Inlet, 27 July 1948.

86   LAC, RG 29, vol. 203, file 311-P11-22, R. A. Gibson, Instructions to District and Sub-District Registrars for Family Allowances and Vital Statistics, 20 March 1948.

87   LAC, RG 85, vol. 1416, file 252-1-2, pts. 1–2, B. H. Harper to P. E. Moore, Indian Health Services, [re:] Cod Liver Oil versus Viosterol, Powdered Milk versus Lactogen, 18 February 1948; LAC, RG 85, vol. 1416, file 252-1-2, pts. 1–2, S. J. Bailey to Mr. Wright, 8 March 1948. See also LAC, RG 85, vol. 1416, file 252-1-2, pt. 4, C. M. Bolger, Memorandum Rehabilitation, Keewatin Region, Improved Bannock Mix, 18 September 1962.

88   LAC, RG 85, vol. 1416, file 252-1-2, pt. 4, Memorandum for the Administrator of the Arctic, 18 June 1962.

89   LAC, RG 85, vol. 98, file 252-3-1, pt. 5, B. G. Sivertz to D. A. Wilderspin, 22 February 1956.

90   Sivertz to Wilderspin, 22 February 1956.

91   LAC, RG 85, vol. 1416, file 252-1-2, pts. 1–2, L. F. William, Coppermine Detachment to the Officer Commanding, Fort Smith, 27 December 1943.

92   Rutherford to Saulnier, 29 March 1962.

93   LAC, RG 85, vol. 1416, file 252-1-2, pts. 1–2, Winifred Hinton, Nutrition Services, Dept. of Pensions and National Health, "A study of the food habits and supplies in the Northwest Territories," February 1944; L. F. Willan, I/C Coppermine Detachment, to The Officer Commanding, Ft. Smith, 27 December 1943; L. Budgell, Manager, HBC, Wolstenholme Post, to Dept. Mines and Resources, Ottawa, 31 May 1945.

94   Pat Sandiford Grygier, *A Long Way from Home: The Tuberculosis*

Epidemic among the Inuit (Montreal: McGill-Queen's University Press, 1994), 55.

95 As cited in Grygier, *A Long Way from Home*, 64.

96 Hinton, "A study of the food habits."

97 LAC, RG 85, vol. 1416, file 252-1-2, pt. 4, H. A. Procter, Director, Medical Services, to F. A. G. Carter, Director, Northern Administration Branch, 5 July 1965.

98 LAC, RG 85, vol. 1416, file 252-1-2, pt. 4, B. Thorsteinsson, Chief, Education Division, to B.C. Gillie, District Superintendent of Schools, Mackenzie District, 23 July 1965.

99 LAC, RG 85, vol. 1416, file 252-1-2, pt. 4, L. Cockney to Sister C., n.d. [1965].

100 LAC, RG 85, vol. 1416, file 252-1-2, pt. 4, J. Maher to Mrs. E. Ellis, 15 February 1966.

101 LAC, RG 85, vol. 1416, file 252-1-2, pt. 4, Paul A. Dufort, Principal, Federal Day School [Fort Resolution, NT], to B. G. Thorsteinson, Chief, Education Division, 30 October 1965.

102 Dufort to Thorsteinson, 30 October 1965. The idea of good nutrition as a form of ethical citizenship is taken from Lien, "Introduction," 10.

103 LAC, RG 85, vol. 1956, file A 1003-20, pt. 3, Report on Northwest Territories Nutrition Survey 1965–66, p. 6.

104 LAC, RG 85, vol. 1416, file 252-1-2, pt. 4, R. J. Orange, Regional Administrator, Memo for the A of the A, Frobisher Bay, NWT, 2 January 1963.

105 LAC, RG 85, vol. 1416, file 252-1-2, pt. 4, H. A. Procter, Director General, Medical Services, Re: Nutrition Survey, 5 July 1965.

106 See, for example, LAC, RG 85, vol. 1956, file A 1003-20, pt. 2, Otto Schaefer to Alec Stevenson, Director Arctic Division, 23 November 1964.

107 LAC, RG 85, vol. 1416, file 252-1-2, pts. 1–2, J. G. Wright to [R.] Gibson, 18 March 1947.

108 LAC, RG 85, vol. 1956, file A 1003-20, pt. 3, Report on Northwest Territories Nutrition Survey 1965–66, p. 1.

109 Gerald W. Hankins, *Sunrise over Pangnirtung: The Story of Otto Schaefer, M.D.* (Calgary: Arctic Institute for North America, 2000).

110 The examples discussed in this essay were only some of the unintended consequences of the relocations. For a fuller discussion, see Tester and Kulchyski, *Tammarniit (Mistakes)*.

7

Hope in the Barrenlands: Northern Development and Sustainability's Canadian History

Tina Loo

In the early months of 1950, documentary photographer Richard Harrington set off on his third trip to the Canadian Arctic in as many years. This time his destination was Churchill, the starting point for a journey along the west coast of Hudson Bay and into the Keewatin (now Kivalliq) region, or the "Barrenlands" (see Fig. 7.1). By the time he returned, he could no longer take pictures—his fingers had been frozen one too many times.[1] But Harrington was lucky: he would recover. Long before he got home, many of the people he had met were dead. His photographs had become a collective obituary.

More than records of the passing of a way of life, Harrington's images were meant to provoke a complacent public. Confronted with starvation among the Padleimiut, a group of Caribou, or inland Inuit, Harrington could do no more than bear witness to a tragedy. When death came, tea and tobacco were all he had to greet it.

> February 11 [1950]: For the first time I realize how serious it is when caribou don't come. ... Dogs are dying everywhere.

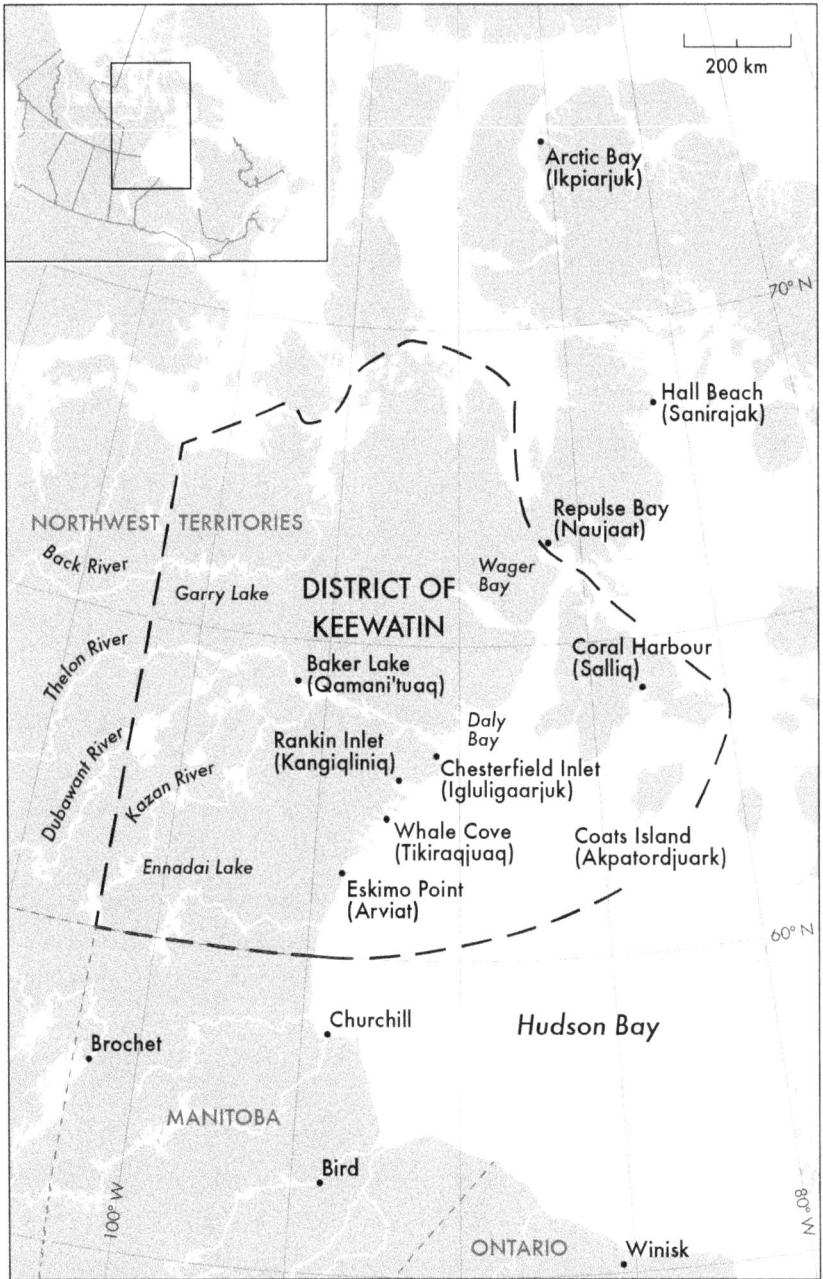

FIG. 7.1: Map of the Keewatin region by Eric Leinberger based on Doug Schweitzer, *Keewatin Regional Dynamics: A Research Report* (1971), 11.

Remaining dogs: skin & bones, shivering, listless. Since Eskimos must travel to obtain food & furs, it means they cannot move around anymore. ... By & by, no more tea, coal-oil, matches. Real hardships begin. ...

February 25: Left alone in a cold, iced-over igloo, Arnalukjuak sat hunched over, in threadbare clothing, her hair frosted over, saying nothing. I put some tobacco in her soapstone pipe, but she was too weak to suck on it. By next morning, she had died of cold & hunger. Her relatives sealed off the igloo.[2]

Two years later, when his photographs were published as *The Face of the Arctic*, Harrington gave public voice to what he had seen. Arnalukjuak was not named in his "portrait of famine," but is identified only as a "starving Padleimiut." Another of his subjects did not merit even this designation. The cold and lack of food had reduced her identity to a state of being. Beyond anonymity, she was neither Alaq, the name he recorded in his diary, nor a Padleimiut. Instead, she was "Near Death." The next sentence turned that description into an indictment, aesthetics into politics: "Near Death. Note Government Identification Tag."[3]

As other scholars have argued, knowledge is the key to statecraft, especially in the modern period.[4] People have to be visible, or "legible," to the state in order to be subject to its power. In Canada, the census and system of Social Insurance Numbers are two commonplace techniques of tracking and control, and so too were the government tags Harrington referred to. With no standardized spellings of Inuit names and in the absence of surnames it was difficult for authorities to keep accurate trade accounts and police records. Implemented in 1941, the E-number (E for "Eskimo") identification system was meant to distinguish Inuit by issuing each a unique number worn on a tag around the neck. It became especially important for the delivery of Family Allowances, the first of Canada's universal social programs, established in 1944.[5] But as Richard Harrington made clear, legibility was no guarantee of social security.

Whereas Harrington chose to be a witness, his contemporary, Farley Mowat, assumed the job of prosecutor in the court of public opinion. The same year the public was shown *The Face of the Arctic*, Mowat published his first book. Also set in the Keewatin, *People of the Deer* (1952) was an explicit condemnation of the Canadian government's deadly neglect of the north and its peoples.[6]

The work of both Harrington and Mowat came at a time when Canadians and their government were taking a more active interest in the north. As Matthew Farish and P. Whitney Lackenbauer's chapter in this volume discusses, that interest was sparked in part by the Cold War. The Atlantic Charter and Canada's commitment to social security also led to greater state intervention in the lives of all Canadians, including northerners. In addition to housing, hospitalization, unemployment, and care for the elderly in the form of universal old age pensions, the government also interpolated itself into areas as fundamental as people's diets, as Liza Piper shows in her chapter.[7] Despite the growth of the welfare state, its safety net did not adequately protect the Indigenous peoples of the region.[8] When thirty-three Inuit in the Keewatin died as a result of starvation in the winter of 1957–58, things changed—rapidly.[9] Famines did not happen in Canada, and people were certainly not supposed to die from a lack of food.

Death by starvation was not the kind of publicity the newly established Department of Northern Affairs and National Resources had in mind for the region it had charge of, especially since its objective was, in the words of Minister Jean Lesage, "to give the Eskimos the same rights, privileges, opportunities, and responsibilities as all other Canadians, in short to enable them to share fully in the national life of Canada."[10] Scandalized by what happened in the Keewatin, Prime Minister John Diefenbaker echoed the mandate of Northern Affairs in directing the civil service to act so that "no more Canadians will starve!"[11]

Once a predictable, if tragic, end for "primitive" peoples, in the late 1950s death from hunger was no longer tolerable for any of the country's citizens. The rhetoric of citizenship was prominent in the post-war period: Parliament passed legislation in 1947 making Canadian citizenship a legal reality, but what did that mean in the face of increased levels of immigration and a growing Indigenous population, which brought new challenges to the "unitary model" of belonging created by the Citizenship Act?[12] The universality implied by citizenship was also challenged by existing inequalities, particularly among regions. The starvation deaths in the central Arctic brought the issues of social and economic inequality into sharp focus, but north was also politically unequal. As territories under the tutelage of Ottawa, it was a colony of Canada, a situation that would become increasingly untenable in the post-war period.[13]

The shift in the semantics of starvation and citizenship was matched by a shift in policy. For Ottawa, the question was not just about how to feed the inland Inuit or conserve the barren-ground caribou on which they depended. Instead, any new policy would have to grapple with the more fundamental and challenging issue of what we would now call "sustainable development": how could the region be made a viable place that could support northerners into the future, allowing them to live modern lives with the same degree of social security enjoyed by other Canadians?

While the federal government tried to answer that question all over the north, the Keewatin region was a notable and early proving ground. The decade after the starvations saw a variety of initiatives undertaken in the Barrenlands to address the immiseration that Harrington and Mowat had brought to the public's attention. These initiatives and the people who implemented them are my focus. Important themselves in understanding the history of Canada's north, they also speak to a new and broadly shared desire among the old and new colonial powers of the world in the postwar period to elevate the condition of those deemed unfortunate and downtrodden. Motivated by humanitarianism and Cold War geopolitics, an encompassing "will to improve" led countries around the world to invest in "development."[14]

As its critics argue, the word naturalized a process that was anything but natural. Development—"growth with change," as the United Nations put it—was a normative concept; to governments in the West, improvement meant progressing through stages of economic, social, and cultural growth marked by the acquisition of liberal democratic values and an embrace of individualism and the market.[15] The "humane internationalism" that led Canada to intervene in development after the Second World War emerged from its own history, as well as its wealth and political culture of liberalism. As David R. Morrison notes, like the countries of the global south, Canada suffered from the problems of foreign investment and control and an economy oriented toward the export of natural resources. This experience situated it somewhat differently in taking up development, and perhaps in how its international development initiatives were received.[16]

During the United Nations' "Development Decade" (1960–70), the Canadian state's energies of improvement were focused inward as well as outward, toward bettering the condition of those who lived in its "under-developed" regions, including the north. For Ottawa, improving the lot of northerners meant making sure the postwar resource boom Canada experienced was not limited to its southern reaches. Although previous governments had hardly ignored it, the Diefenbaker regime made the region a priority, claiming it would end the "absence of mind" that characterized northern administration prior to the Second World War.[17]

Announced in 1957, his "Northern Vision" was an aggressive plan to open Canada's neglected frontier to development by building "Roads to Resources." Highways and railways, as well as improved navigation on waterways, would lure venture capitalists anxious to exploit the oil, gas, and mineral resources north of sixty. In addition to infrastructure, the federal government provided loans to help finance the greater costs of building in the north, granted tax holidays to mining companies, and relaxed permitting conditions.[18] Both prior to Diefenbaker's election and after his defeat, Ottawa also directed the attention of its public servants toward facilitating the operations of private enterprises like the North Rankin Nickel Mining Company, which established the northernmost base metal mine in the world in the Keewatin in 1953.[19] The investment corporations like it brought to bear on the north would, the government believed, allow residents to prevail over the limits of their environment: the cold, the distance, and the unpredictability of the animals on which they depended would all be overcome by capital and the global market for commodities.

But Ottawa also came to realize that developing the north required putting it on a more stable and enduring footing than mining allowed—and that necessitated being attentive to the exploitation of renewable resources and taking a more spatially differentiated approach to development. The "north" was a big place with big problems that could only be solved by taking its diversity seriously. With the closure of the North Rankin mine in the early 1960s, the Department of Northern Affairs and National Resources did just that, undertaking a number of new regional and community development initiatives in the Keewatin.

A recognized part of Canada's political and economic history, regional and community development is also an important, if unexplored, part of the history of sustainability, pre-dating both the World Commission on Environment and Development (1987) and the United Nations Conference on Environment and Development, also known as the Rio or Earth Summit (1992), widely acknowledged to mark the origins of sustainable development.[20] Sustainable development emerged from concerns about the global environmental impacts of industrialization. Although they worried about issues like water and air pollution, many people in the global south bridled at being told by the countries that had benefitted from industrial development that they would have to check their own—and forego the growth that came along with it. First popularized by *Our Common Future* (1987), a report of the World Commission on Environment and Development, the concept of "sustainable development" was meant to address just such concerns about the uneven distribution of the environmental, economic, and social costs and benefits of industrialization. It was "development that meets the needs of the present without compromising the ability of future generations to meet their own needs. … Sustainable development requires meeting the basic needs of all and extending to all the opportunity to satisfy their aspirations for a better life."[21]

The articulation of sustainable development was an important moment in the history of environmentalism. From the late nineteenth century to the Second World War, environmentalism was largely concerned with the impacts of economic growth, manifesting itself in both efforts to preserve wilderness areas from the ravages of industrial development and to conserve resources through expert, scientific management. According to Samuel P. Hays, as North Americans became more affluent, urban, and well educated in the second half of the twentieth century, public values about the environment changed. Post-war environmentalism was less about experts managing the conduct and effects of material production and more about "beauty, health, and permanence." The environment became an amenity, consumed for aesthetic and health reasons, and a sensibility, a way of seeing the world more holistically, in ecological terms.[22]

By raising questions of fairness and justice, advocates of sustainable development further differentiated environmentalism by bringing people and questions of poverty and power to the forefront. Rather than balancing environment and economy, sustainability is measured by a "triple bottom line," one that is not just attentive to the planet and profits—as had been the case for much of the twentieth century—but to people, as well, and specifically to questions of equity.[23]

Well before sustainability became part of the landscape of global environmental politics in the 1990s, however, bureaucrats working for Canada's federal government grappled with meeting that triple bottom line on the frozen ground of the north. Their efforts at regional and community development in the 1960s were aimed at putting northerners on the same footing as other Canadians by finding ways for them to remain in place, to live in an environment transformed by capitalism and colonialism.

Like those implemented internationally, the development schemes the Canadian state tried in the north were techniques of governance, imposing a disciplinary power on the Inuit in the name of overcoming colonialism and facilitating self-determination. They effected change by problematizing the region in particular ways—namely in terms of what its residents lacked. For the bureaucrats and fieldworkers at Northern Affairs, development was a matter of addressing questions of "capacity"—of the Inuit and their land—as well as raising northerners' consciousness. Doing so brought new expertise to bear on the north, ushering in what Timothy Mitchell calls "the rule of experts."[24]

If the development initiatives undertaken in the Barrenlands after the starvations are part of the history of sustainability, they are also chapters in a broader history of hope. It may seem jarring to discuss hope in conjunction with the state and its agents, and it is meant to. Hope is something historians tend to reserve for those who were the objects of power, not a force motivating those who exercised it. The dissonance helps historicize and complicate my subject in a way the "will to improve" does not, measuring how far we have travelled from a time when people believed that an activist state could and should improve the human condition.

But framing development as a history of hope is also meant to nationalize and globalize the region's past. Like much regional history in Canada, the history of the north is not integrated especially well into other narratives. Situating the Barrenlands in a history of hope is meant to help overcome the exceptionalism that can characterize studies of place. What happened in the north was part of the history of postwar Canada and the emergence of welfare liberalism. Born of the belief in the power of the redistributive state to make the conditions for a good life, the development initiatives I discuss were efforts to put northerners on a more equal footing with other Canadians when it became apparent that the universal social programs successive governments had implemented had not done so. More broadly, what happened in the Keewatin was part of a moment in international history when a range of peoples living in different places came to be perceived similarly: they were deemed by the state as in need of improvement. Hope effected change around the world. Hope was power.

***

In the aftermath of the killing winter of 1957–58, the starvation survivors were relocated to settlements on the west coast of Hudson Bay where a new future awaited them. It was not one located on the land, but was to be found instead in permanent settlements. There, new expertise could be brought to bear on improving the condition of Inuit, like that possessed by Walter Rudnicki.

Charged by the Department of Northern Affairs and National Development with interviewing the starvation survivors, Rudnicki arrived in Eskimo Point (Arviat) in March 1958. It was a telling choice: Rudnicki was not the usual northern hand, a man with connections to the Hudson's Bay Company, the Royal Canadian Mounted Police, or the Christian churches. Instead, he was a professional social worker. Specializing in psychiatric social work, Rudnicki was especially interested in mental illness among recent immigrants and had worked with them in Vancouver before joining the civil service.[25] This was someone whose chosen career involved working across cultures, and with vulnerable people—both things he would continue to do as chief of the Welfare Section of the Department of Northern Affairs and National Resources.

At Eskimo Point, Rudnicki put his professional training to work, administering the Thematic Apperception Test in a modified form to the starvation survivors. One of the most commonly used psychological tests, it involves showing participants a series of provocative yet ambiguous drawings about which they are asked to tell a story. Psychologists believe that the stories participants tell reflect their state of mind—their sense of self and the world.

Instead of using the usual set of standard pictures, Rudnicki, a talented sketch artist, created his own depicting some of the events leading to the starvation. He reported that there was no evidence of "mental pathology," and, insofar as he had reason for concern, it lay in the attitudes the survivors had toward non-Inuit. In pictures where there were only

Inuit present, the starvation survivors perceived the people to be happy and helping each other. But when white men were depicted, the response was different: "...the general reaction seemed to be that the Eskimos were unhappy, sad, or frightened. ... White men were not differentiated, that is, police, northern service officer, etc. All were regarded as 'big bosses' and seemed to be equally viewed with fear and suspicion."[26]

What Rudnicki identified was a psychology of colonialism created by regular interactions with Hudson's Bay Company traders, missionaries, and members of the Royal Canadian Mounted Police at small and scattered settlements across the north. After the war, those interactions had increased for certain groups and individuals who were drawn into construction work along the DEW Line. For Rudnicki, overcoming the fears borne of their encounters with colonial power would be the key to rehabilitating the survivors.

More broadly, grappling with this legacy of colonialism by rebuilding capacity would have to be a central part of any northern development policy. While trained social workers were important, doing so in the north also required other kinds of experts, people who knew something about the Inuit. As the Deputy Minister of Northern Affairs and National Development observed, "one of the greatest difficulties facing those responsible for the health, welfare, and education of the natives of northern Canada is a lack of basic information on their social and cultural patterns."[27] Thus, in Canada, as in other parts of the world at the time, anthropology as well as social work came to be implicated in development, influencing the design of initiatives in the north and shaping the state's understanding of what development was.[28] While anthropology wielded its influence through the Northern Coordination and Research Centre (NCRC), established by Northern Affairs in 1954 to support and coordinate scholarly research about the north, in Keewatin it also made an impact through the more informal interventions of Ottawa's agent in the field.[29]

In the Barrenlands, Northern Affairs' man on the ground was an amateur anthropologist hired to do social work. Almost immediately after Rudnicki visited Eskimo Point in 1958, the Department appointed a new welfare officer to the region whose responsibilities included overseeing the rehabilitation of the starvation survivors and attending to the welfare needs of the growing Inuit population of Rankin Inlet, where he was based. Staffordshire-born Robert G. Williamson arrived in the Keewatin community just as the ink had finished drying on his undergraduate anthropology degree.[30]

Although his diploma was new, Williamson was not—to fieldwork or the Canadian north. He immigrated to Canada as a young man and made his way to the western Arctic, working on the Mackenzie River barges. While wintering at Fort Simpson he recorded Dene folklore, and later published his findings in the scholarly journal *Anthropologica*. After a year-and-a-half in the western Arctic, he moved to take a job with the Eastern Arctic Patrol. He became fluent in Inuktitut and continued his ethnological investigations around Pangnirtung and Cumberland Sound. In 1954 he moved to Ottawa, and while working there enrolled at Carleton University, where he earned an undergraduate degree in 1957. He later went on to earn a doctorate in anthropology at Uppsala, Sweden, in 1974, and to have an academic career at the University of Saskatchewan—after having done a great many other things, including a stint in the civil service with Northern Affairs.[31]

Looking back, Williamson considered himself both an exemplar and a proponent of applied anthropology, a sub-discipline committed to applying the methods of anthropology to the solution of practical problems. His undergraduate education coincided with the field's emergence in the post-war years, when hope reigned supreme about the prospects for a new world order in which all nations would enjoy the benefits of democracy and modernity. Anthropologists, no less than economists, became implicated in this global project of transformative change. As James Ferguson points out, "as experts on 'backwards' peoples, anthropologists were well placed to play a role in any project for the advancement of such peoples."[32] Funding and positions opened up in government and non-governmental organizations for anthropologists willing to use their skills in the service of what became known simply as "development."

Williamson's early career with Northern Affairs provided him with the opportunity not just to see social change, but also to intervene in it as a welfare officer, using some of the tools of anthropology. What he saw and did would form the basis of his doctoral dissertation, a study of socio-cultural change in the Keewatin. As a field of study and a form of practice, applied anthropology was controversial, particularly by the time Williamson undertook his graduate studies. Its detractors considered it "second rate, both intellectually and morally"—a form of neo-colonialism. Its practitioners shot back that their critics were "irrelevant, both theoretically and politically."[33] If Williamson were aware of these debates, he likely would have dismissed them: for him there was no contradiction between intellectual rigour and political engagement. After his undergraduate degree, he leapt at the chance to leave "the quiet contemplative corridors of the National Museum" for a post with the federal government in the north. "There was so much to be done that one could not sit, eyes cast to the ceiling, finger-tips together, thinking only abstractly," he recalled. "One had to respond to one's responsibility to make use of one's knowledge."[34]

Once on the ground, the amateur anthropologist got to work. Although part of his time was spent with the starvation survivors, he devoted a good portion of his energies to helping manage the Inuit working at the recently established North Rankin Nickel Mine, and in that sense assisted with the economic development of the region. Like the starvation survivors, the Inuit miners and their families were also relocatees, drawn from the Keewatin's coastal communities by the company and the federal government as a labour force.

Having recruited and trained Inuit, the mine's operators were flummoxed when some employees withdrew after three or four days, or dropped their pickaxes, picked up their rifles, and headed off to sea or to the floe edge when a whale or a group of seals was sighted.[35] Without enough shift workers, operations ground to an expensive and annoying halt. Despite "their natural quickness to learn" the technical aspects of the work, the Inuit persisted in this seemingly undisciplined behaviour.[36] It was the company that changed its practices, and according to Williamson, it did so on his advice.[37]

For Williamson, the key to solving North Rankin's labour problem lay in re-creating the pattern of "cultural commuting" he had first seen in the western Arctic. The Indigenous people of the Mackenzie River and Delta seemed to experience the least social disruption when they were able to shuttle back and forth between life on the land and life in trading posts—between their old lives and their new ones.[38]

Although the physical distance between social worlds had collapsed in Rankin, he still hoped to create a space that would act in the same way, as a buffer. He did so not by physically removing the Inuit from the settlement, but by suggesting how the work regime at the mine might be reconfigured. Specifically, Williamson convinced management to train more men than they would employ at any one time and to redefine what a "shift" was. Indeed, doing the first allowed for the second. Instead of particular individuals, a shift came to consist simply of certain number of people. An Inuit "straw boss" would have the responsibility of ensuring there was the necessary number of men to fill each one.

In essence, Williamson convinced the mine to work with cultural difference and to treat Inuit labour as a collective endeavour. While his innovation did not win him many friends in North Rankin's accounting department, his proposed restructuring freed individual Inuit to hunt or mine as they wished and could negotiate with the group. Modified for an industrial setting, Williamson's cultural commuting was a structure of practice that combined the new time-work discipline governed by the clock with an older pattern of work, one that was sensitive to and shaped by environment and opportunity.

Despite the apparently successful adaptation on the part of both the Inuit and North Rankin's management to mining in the Canadian north, operations wound down in 1962, the victim of falling commodity prices. The news, when it came, was not a surprise; the threat of closure seemed to hang over its operations almost as soon as they started.[39] With 520 people, or about thirty per cent of Keewatin's population, dependent on the mine, its shutdown provoked Northern Affairs to declare a "state of emergency."[40]

Again the anthropological ambulance responded to the call, with Robert Williamson at the wheel. Just as he had helped facilitate the transition to industrial employment in Rankin, he intervened again to ease the disruption associated with its end. The year before the mine closed, Williamson conducted a survey of fifty-nine men who worked in the mine to ascertain what they wanted to do. Some told Williamson they would happily go back to hunting. But most, he reported, wanted to pursue wage work, concerned about whether the hunt could sustain their families reliably. That said, they were quite specific about the conditions under which they would labour: for at least one man, "work in white man's land [was] not a happy thought." Another told Williamson he "would go elsewhere in Esk[imo] country to work, but not to the white man's land." A third made a distinction between the kinds of mining work he wanted to do, noting "outside work happiest. Underground work worst, especially as no extra pay. V. frightening."[41] In sum, most wanted to continue to work for wages, in Rankin if possible, or at other mines in the north. Only a few wished to go back to their home communities, or to other ones where they could hunt and fish.[42]

To meet the wishes of those who wanted to work for wages, as well as to capitalize on a trained labour force and diminish a potentially crippling welfare bill, Williamson made the case for relocating Rankin's Inuit with his superiors—but he did so with an anthropologist's sensitivity to context. Only those most likely to succeed in their new jobs would be moved, and then only to other mines in the north. To identify the most promising candidates, Williamson assessed each family's "adaptation potential": he constructed genealogies, believing that relocation would only be successful if a miner's family went with him to a new job, and he noted whether husbands and wives were competent in English. While facility in the language contributed greatly to an Inuit family's potential to adapt, Williamson also recorded whether they possessed things like stoves, fridges, washing machines, radios, and record players. For the anthropologist, these consumer goods were another indicator of a family's acculturation.[43]

With Williamson's recommendations in mind, Northern Affairs worked with various companies through the 1960s to send Rankin miners and their families to Tungsten and Yellowknife in the Northwest Territories, to Lynn Lake in Manitoba, and to Asbestos Hill, in Arctic Quebec. While the numbers of Inuit relocated were never large—three to twelve families—the amount of attention directed at them was great, speaking perhaps to the importance the Department attached to their success or failure and to the reach of the state.[44] Williamson's insights about the need to take culture seriously in crafting employment policy and practice circulated well beyond the central Arctic. The Department repeated them in advising companies thinking of operating in the north, particularly as resource extraction sped up in the 1970s.[45]

The results of the government's selective, anthropologically informed relocation of Inuit for industrial employment were mixed. Although there was rarely a problem with the quality of the work the miners performed, there were signs of "mal-adjustment," including absenteeism, drunkenness, and a high turnover of labourers, all familiar problems in resource extractive communities. More worrying to Northern Affairs officials was the situation outside the workplace. Charged by the Northern Coordination and Research Centre with assessing the relocation initiative, anthropologist David Stevenson reported that alcohol abuse among Inuit women, neglect of children, and indifference to maintaining functional households were, to different degrees, common among the relocatees. Moreover, insufficient housing, a lack of familiarity with managing daily expenses, and an inability to comprehend the informal social rules governing white society made adjustment all the more difficult. For many Inuit, relocation was a fundamentally alienating experience.

When people talked about their alienation, they made no distinction between the social and the environmental. Yellowknife felt more "remote" than Rankin when distance was measured in terms of exotic presences and gaping absences—when spiders and heat replaced family and kin. For one Rankin man, the trees around Yellowknife signalled his separation, preventing him from seeing very far. "It's just like looking at the floor under you."[46] Asked why they repeatedly went on alcoholic binges, two Inuit women living in Hay River told David Stevenson it was "'because I have no place in this land.'"[47]

Encompassing the social and environmental, "place" was not something that could be factored easily into the calculus of adaptive capacity and incorporated into development planning. While Williamson was right to think relocation was a family matter and those who were more acculturated would have fewer problems, successful adaptation hinged on a variety of factors, some of which, like attachment to place, could not be measured.

Equally importantly, success depended on considerations that lay outside the boundaries of his analysis. For Northern Affairs, development was a matter of addressing a deficiency in the Inuit. The focus was squarely on improving them, rather than the communities they were joining. For all their sensitivity to culture, neither Stevenson nor Williamson turned their attention to ascertaining the kinds of social settings that would facilitate improvement best. Their concern was almost exclusively on the capacity of the Inuit to adapt, not on the ability of the receiving community to incorporate new members. Yes, better housing would help, and yes, the negative attitudes of the business people and landlords in Yellowknife were an obstacle to successful relocation and the development of a mobile labour force. But beyond acknowledging the existence of racism and its corrosive effects on the project of improvement, neither the experts engaged by Northern Affairs nor its own officers chose to tackle it, perhaps recognizing it couldn't easily be addressed, while Inuit capacity could—or so they thought.

While the possibility of industrial employment animated development policy through the 1960s and beyond, Northern Affairs also recognized that it could not be the sole basis for improving the condition of the region and its peoples. Its own research arm was telling it as much. Even as Robert Williamson worked with North Rankin's management, some professional anthropologists questioned whether industrial employment could ever be the basis of a stable and healthy society. In their report for the Northern Coordination and Research Centre, Robert and Lois Dailey argued that just because the Inuit at Rankin had been integrated into the wage economy did not mean they had achieved equality. Far from it: they were paid less than white men for similar work, and they were subjected to blatant discrimination. In 1958, Rankin was a segregated community, with Inuit and Qallunaat (Inuktitut for "non-Inuit") sleeping and eating in different facilities and using the one rec room on different days. To the Daileys, each Inuk on the mine's payroll was being "trained to be a labourer—not a citizen."[48] What had been achieved at Rankin was the creation of a workforce, not a community. It was no model for the future.

Four years later, things had not improved. Like the Daileys, Jean Malaurie's work was supported by the NCRC, and, like the Daileys, he too worried about the corrosive effects of industrialization. The adventurer's views were shaped by his relationship with an Igloolik man who worked as a labourer in the North Rankin Nickel Mine and—unusually—kept a diary. While for Malaurie the mere act of keeping a diary was a sign of the Inuk's distress, he found its contents even more disturbing. As the hunter became an alienated wage labourer, there was a slow but inevitable closing of his "diaphragm of expression" until "the man of before is replaced not by a new man but by a void pure and simple." Suffocation and annihilation—that was the effect of "deculturation."[49]

Perhaps with these warnings in mind, and with little prospect of enough new mining activity to support the region through industrial wage labour in any case, the bureaucrats at Northern Affairs struggled with how to make the Barrenlands viable. While there were discussions about relocating the entire population to southern Canada, Ottawa chose instead to try to make the region and its communities "self-generating."[50]

While "sustainable development" was not a term the federal government used to describe what its officials were doing, the initiatives they undertook in the north bore many of its hallmarks. First and foremost, they were informed by a respect for environmental limits, albeit one born in part of a concern for the economic bottom line. In the aftermath of the starvations, Northern Affairs and the Canadian Wildlife Service implemented a number of conservation measures designed to curb and regulate the use of barren-ground caribou. These ranged from providing emergency food caches, conservation education, and community freezers to prevent meat from being wasted, to supervising hunts to forestall the "wanton slaughter" of animals.[51] These measures represented a change in conservation policy. Whereas wildlife management in the first half of the twentieth century was about maintaining a food supply for northerners— like the one Andrew Stuhl talks about in his chapter in this volume—in the wake of the starvations, the emphasis shifted to a "more rigid preservationist philosophy."[52] Such an approach was not effective in stopping the decline of the caribou. Not surprisingly, Ottawa concluded what Inuit had known for generations, that northerners could only sustain themselves if there were enough renewable resources to do so.

In the absence of caribou, Ottawa encouraged Indigenous people to get their protein elsewhere, particularly from fish and marine mammals. Helping them to do so effectively required in-depth and systematic censuses of a region's natural and human resources. To that end, Donald Snowden, chief of Northern Affairs' Industrial Division, directed his department's planners to conduct area economic surveys across the north, enumerating each region's marine and terrestrial resources, as well as its human population. The *Keewatin Mainland Economic Survey* (1963) identified areas where there was a "mal-distribution" of population and resources, and made the case for relocation where the number of people exceeded the capacity of the land to support them.[53] As the Brundtland Report put it more than twenty years later, "sustainable development can only be pursued if demographic developments are in harmony with the changing productive potential of the ecosystem."[54]

Identified as one of the communities with more people than resources, Rankin Inlet became the site of what one civil servant jokingly called the Department's "Back to the Land with Joy" program. In 1964, Northern Affairs engineered the removal of fifty-four Inuit, approximately ten percent of Rankin's population, relocating them to Daly Bay, more than 150 kilometres north.[55] Making their case for relocation, the survey's authors argued that unless Inuit moved to other "areas of opportunity" in Keewatin many of them would have "no hope at all for social and economic advancement."[56] The approach Northern Affairs took to putting the region on a sustainable footing was not dissimilar to the management techniques their counterparts in the Canadian Wildlife Service used: Inuit could be herded like bison or caribou to better ranges.

In fact, the logic of sustainable development in the Keewatin rested on the same powerful and flawed idea that animated scientific resource management, namely "carrying capacity." Originally developed in the mid-nineteenth century by engineers to designate the payload a vessel was designed to transport, carrying capacity emerged in the twentieth century as one of the central concepts of population biology, used to understand the relationship between population and habitat. Taken up by range and game managers in the first half of the twentieth century, and later by ecologists, it was used to describe, and, more importantly, to prescribe the number of organisms that could be supported in a given environment without degrading it or themselves. The power of carrying capacity lay in its calculability and its promise of certainty.[57]

What made the initiatives undertaken in the Keewatin instances of sustainable development rather than conservation was the explicit agenda of equity, social change, and empowerment that came with them. While advocates of both conservation and sustainability argued that people needed to live within limits, proponents of sustainable development parted company with conservationists in their insistence that environmental problems could not be addressed separately from the social context in which they occurred: addressing human poverty and injustice was crucial to addressing environmental degradation.[58] In addition, unlike progressivist conservation, sustainable development required the active participation of ordinary people, as well as the intervention of "experts." In other words, the process through which sustainable development was to be achieved also distinguished it from conservation.

While these aspects of sustainable development gained formal expression in the late 1980s and early 1990s with the Brundtland Commission and the Earth Summit, they were also in evidence in the development initiatives undertaken by the Department of Northern Affairs and National

Development. Its pioneering efforts in the Keewatin in the 1960s were consistent with the federal government's growing emphasis on "regional economic development," even if they were at odds with the conservation policies that agencies more directly involved in resource management, like the Canadian Wildlife Service, continued to pursue.[59]

The focus on region emerged from a recognition in the discipline of economics that development required a spatially differentiated approach. Informed by neo-classical approaches, the thinking about development had not fully acknowledged the friction of distance, geography, and climate, and how it shaped growth. The work of French economist François Perroux did just that: his "growth pole" theory influenced the postwar development policies of governments around the world, including Canada's. Although it was one of the most regionalized of the world's industrialized nations, it was not until the middle of the twentieth century that Canadian economic policy began to be conceived of in a framework other than a national one. In the 1960s, Ottawa initiated a number of regionally differentiated development initiatives that targeted particular sectors and geographic areas, and that were eventually extended across the country under the auspices of the new federal Department of Regional Economic Expansion, established in 1969.[60]

At the same time, the federal government recognized that regional inequality meant social inequality. Disparities among regions affected the delivery of public services, not just economic development. To make good on the promise of universality, the government introduced federal equalization payments in 1957. A recommendation of the Royal Commission on Canada's Economic Prospects, equalization payments were meant to reduce disparities between regions by ensuring that all provinces had the ability, if they chose, to provide comparable services at comparable tax rates. In doing so, transfer payments entrenched the idea of social security as spatial justice. As Prime Minister Pierre Trudeau put it a decade later, "every Canadian has the right to a good life whatever the province or community he lives in."[61]

Informed by these shifts in economic thinking, Northern Affairs' approach to development was also premised on the belief that northerners had a central role to play in determining their own futures and that of their region. Its job was to position them to do so. As important as attending to the carrying capacity of the land was, putting the region on an ongoing, viable footing was also a matter of building human capacity: people needed to be taught organized resource harvesting, to think regionally, and to govern themselves. In the Keewatin, sustainable regional development involved facilitating social change and empowerment: it was an exercise in social engineering.

For instance, Northern Affairs did not move Inuit from Rankin Inlet to Daly Bay so they could subsist from the land as they always had. Instead, they were moved to participate in an "organized" char fishery and work in the cannery it had built. What made the resource harvesting conducted there and at other Keewatin communities "organized" was the scale at which it occurred. Under the direction of the Department's field officers, Inuit were taught to hunt and fish more intensively, producing enough not just to feed themselves, but other communities in the region, as well as supply commercial markets.

The federal government believed that canned char, whale, and seal might, if properly processed, packaged, and promoted, find a lucrative market outside the region. With that in mind, when the Inuit working with Northern Affairs officer Max Budgell caught a hundred whales off Eskimo Point in 1961, the Department sent its specialty foods officer to investigate what might be done with them. A German who spent some time in a Canadian internment camp during the Second World War, Erich Hofmann had "a positively wild interest in preserving food" to the extent that his colleagues believed that "no living thing is safe from him."[62] With his experience processing traditional or "country" foods (the term Inuit and Indigenous peoples give to foods harvested from the land) in Wood Buffalo National Park and the Mackenzie Delta, Hofmann went to the Keewatin in 1962, working with Inuit women at Whale Cove to make muktuk sausage.

By the next year, Hofmann was convinced that his program "could not only help the Eskimo achieve a degree of self-sufficiency through preservation of foods for local use, but in areas of surplus could also generate income by means of export to southern markets." He set about proving it at Daly Bay and Rankin Inlet. By 1965, flash-frozen, smoked, and canned char and herring were making their way south, some of them ending up as samples in Donald Snowden's briefcase (Fig. 7.2). As Edith Iglauer recalls, the head of the Industrial Division was so committed to the project that he went from restaurant to restaurant in Montreal flogging "his" canned

FIG. 7.2: The results of "organized resource harvesting." Library and Archives Canada, Donald Snowden Fonds, MG 31 D163, vol. 14, file 34.

goods.[63] When the decade closed, more than one hundred thousand tins of seven different kinds of country food were being produced yearly. Inuit consumed half of this, and Canadian Arctic Producers, a non-profit marketing company, distributed the other half.[64]

From the perspective of Northern Affairs, organized resource harvesting addressed one of the central challenges of sustainable development in Keewatin and the Arctic generally: the problem of capital.[65] The outward flow of capital from the north stifled its development. The challenge was to find ways of keeping it circulating within the region, among its communities. If Whale Cove could supply Eskimo Point or Baker Lake with fish, and those communities could supply other products, capital would remain in the region and be available for local investment—in more Peterhead boats or nets or traps, which would allow for greater or more efficient returns and potentially increased exchanges among communities. A regional economy would be born.

As important as forging economic connections among Keewatin's settlements was, Northern Affairs also recognized that regional development depended on a "regional consciousness"—a collective sense of belonging among Inuit that extended beyond the bounds of any one of the communities in the Barrenlands. "The people of Keewatin today can be described aptly as 'different-place-miut,'" observed planner D. M. Brack in 1962. "Adjacent houses for example in Rankin are occupied by families who may not know each other's names, and in other parts of Keewatin there are many instances of group cohesion inhibiting full community action and consciousness. ... If regional planning is to be really effective as a medium of social development then the people themselves must become region conscious."[66]

Through its use of communications technology, Northern Affairs tried to facilitate a sense of cohesion among the Inuit of Keewatin, one that would overcome the fragmentation caused by starvation, relocation, distance, and linguistic and cultural difference. Radio could connect people and communities, especially when broadcasts were in Inuktitut. Again, Robert Williamson played a role: having pushed Northern Affairs to publish a magazine wholly in Inuktitut in 1959, he remained committed to the idea that language played a central role in creating community cohesion. As a private citizen living and working in Rankin Inlet, he wrote and produced two Inuktitut-language programs that ran weekly on CBC North from 1962 to 1964.[67]

Equally importantly, Williamson supported Rankin Inlet as the site of the Department of Communications Northern Pilot Project, which would test the "comminterphone" (Fig. 7.3). Developed by Bell Northern Research Laboratories, the "community interaction telephone" was their answer to "a growing concern for the communications needs of the social, cultural, and political groupings which characterize Northern Canada." It combined the features of a party line with those of radio: by dialing in, up to four callers could participate in a conversation with a radio host that was broadcast over a low-power AM transmitter in a five-mile radius.[68]

Comminterphone service was initiated in 1971, and within a year had become a popular source of local information. Although it had not become a vehicle for discussing community issues or for consensus building, both communications researcher Gordon Wensley and Robert Williamson argued that the comminterphone had created "a sense of involvement in what is going on around the settlement, an association—even if passive—with the events and feelings of the day." Over time, they believed "living in an atmosphere of overheard activity would appear to add to the ambiance of community feeling amongst a collection of migrants heretofore somewhat fragmented and semi-isolated."[69]

In part, the mixed results of the comminterphone experiment spoke to the internal dynamics of the settlements in Keewatin that D. M. Brack

Rankin Inlet resident Willie Adams makes adjustments to the relays of the Comminterphone conference unit housed in the telephone exchange building of the small northern community. The interface unit permits up to four telephone calls from the exchange (background) to be aired simultaneously through a low power radio transmitter.

Bell-Northern research

January 1972

FIG. 7.3: Willie Adams at work on the "comminterphone" in Rankin Inlet, January 1972. University of Saskatchewan, University Archives and Special Collections, Institute for Northern Studies Fonds, INS-808.

had hinted at. Building social connections among settlements might have been crucial to sustaining the region, but there also was work to be done building relationships within them, among the "different-place-miut." For that reason, the federal government's development initiatives also focused on the community, as well as the region.

As a strategy, "community development" relocated change. Rather than being animated by outside investment or connections to markets, it proceeded on the assumption that the transformation of the north would begin from the inside out, grounded in the human resources in each of its settlements. As community development workers, the job of Northern Affairs' officers in the field was to help people identify their collective wants and the means to achieve them. Stated simply, community development was "the process of helping people to help themselves."[70]

In outlining its approach, Northern Affairs drew on models being used internationally. According to Welfare Division chief F. J. Neville, the Department was inspired by the 1964 "War on Poverty" waged by the government of the United States in its inner cities and rural areas, and by the development work being undertaken in the global south by a number of industrialized nations of the world.[71] Indeed, the language of its policy directive on community development was taken from a 1957 United Nations report discussing its work in under-developed countries.[72] But the books and reports in the Northern Affairs library suggest the intellectual genealogy of its community development initiatives stretched back further and was entangled with empire: in addition to a variety of studies by the United Nations, there were works dealing with community development in Fiji, Ghana, Jamaica, and South and Southeast Asia carried out by governmental and non-governmental organizations, including the British Colonial Office and Christian missions.[73]

The influence of Christianity was particularly visible in the main instrument of community development in the north and elsewhere: the cooperative. As institutions organized for the mutual benefit of their members, cooperatives have roots going back to medieval Europe—but the cooperative movement is of more recent vintage. Originating in mid-nineteenth-century England and Europe, it was a reaction to industrialization and the economic hardships it visited on urban workers and small farmers. The cooperative movement came to Canada in the early twentieth century as part of the broad culture of reform initiated in part by the middle-class members of its Christian churches.[74] As locally owned businesses, cooperatives kept capital in communities through profit-sharing with their members, took direction from their membership in how business was conducted, and ensured fair prices. In establishing them in the north, Industrial Division chief Donald Snowden drew from international examples, but also on the work of Fathers Moses Coady and Jimmy Tompkins and the Antigonish Movement in Nova Scotia to alleviate rural poverty in the 1930s and '40s.[75] Taken up globally, their program of social reform, centred on adult education oriented toward cooperative action, drew from the ideas of liberal Catholicism, as well as papal encyclicals dating to the late nineteenth century, and enjoyed broad support from clergy of all denominations.[76] For one development officer in the Northwest Territories, the philosophy of cooperation was "simple and easy to understand. It is found in the Sermon on the Mount and in the Golden Rule."[77]

To their promoters, co-ops were instruments of social, as well as economic, development. They built capacity by schooling Inuit in the processes of formal democracy and self-government, as well as the workings of the market. As Northern Affairs' Supervisor of Cooperatives put it in 1960, "the practice of democracy in the economic sphere ... set an example for democracy in the larger sphere."[78] Gordon Robertson agreed. The Deputy Minister considered co-ops to be an "incubator from which political leaders emerged"; during his time in office, he observed how Inuit men moved from being directors of their community's cooperative to the most prominent members of its "Eskimo Council."[79]

In essence, using cooperatives to develop northern communities was a political project. It involved nothing less than creating civil society from the ground up, helping Inuit to govern their own lives, convincing them "that their world does not begin and end with Government action or its lack."[80] To Northern Affairs, "community development and self-determination are inseparable."[81]

From their start in Arctic Quebec in 1959, co-ops sprang up quickly. Just five years later, nearly twenty per cent of Canada's Inuit were members of one of nineteen such institutions across the north.[82] By 1970, northern co-ops handled over $2.5 million dollars in sales of goods and services yearly, returning close to $1.25 million dollars to members in the form of salaries, purchases from members, and patronage dividends.[83] Ten years later, in 1980, sales amounted to an astonishing $27 million dollars, and payouts $9.1 million dollars.[84]

While it might be easy to see co-ops simply as instruments of assimilation, the intentions of Northern Affairs in promoting them as part of community development in the north were more complex. Co-ops were vehicles for capital accumulation *and* redistribution; they were meant to discipline the Inuit to Western forms of democracy *and* teach them how to subvert power. For Alexander Laidlaw, Coady's colleague at St. Francis Xavier University and head of the Co-operative Union of Canada, the latter aspect of cooperatives was especially significant, given how large the state had come to loom in people's lives in the mid-twentieth century. "Welfare measures are being pushed farther and faster all the time in all parts of the world—and rightly so," he wrote in *North* magazine in 1963. "But in order to prevent domination by government bodies and the official mind, citizens must be strongly organized to do things for themselves. … Cooperatives are proving to be one of the most effective agencies in this role."[85]

Those working in Northern Affairs shared Laidlaw's concerns about the need to counter domination by government bodies. Not only were the Inuit unfamiliar with formal participatory democracy, but cross-cultural differences regarding communication and the impact of colonialism also left them unlikely to challenge authority. There was no better example of the disastrous consequences of such miscommunication than what happened in the winter of 1957–58. Misunderstanding over whether they had agreed to be resettled to Henik Lake and their knowledge of the resources of the area had contributed to the deaths of eight of the thirty-three Inuit who died of starvation that winter. Privately, senior bureaucrats in Northern Affairs felt that "their decision to move … was probably because they regarded it as a command of the white man."[86]

Although officials could exercise more care in interpreting Inuit responses, the long-term solution was to end Inuit diffidence. That was why co-ops were so valuable: as Donald Snowden told Edith Iglauer of the *New Yorker*, "I don't believe that the government is infallible, and the co-ops make it possible for the Eskimos to give us hell."[87] And they did. The Inuit interviewed about their participation argued co-ops gave them "a way to regain some of the control [over our lives] we previously had." Others went further, pointing out their long-term political consequences; according to former Inuit politician Thomas Suluk, co-ops were "underground governments" that provided the foundation for a "pan Inuit solidarity that had no historic precedent."[88]

Although the emphasis was principally on Inuit to change, there was some sense that development also required changes in Qallunaat. Specifically, it required the active engagement of civil society and not just government. In praising the establishment and growth of cooperatives in the north, Alexander Laidlaw criticized the cooperative movement he led for not extending assistance to the Inuit. Not only did its disinterest run counter to the ethic of cooperation, but it also invited the kind of excessive government intrusion into the civil sphere that he worried about.[89]

As well, development called on those in government charged with overseeing and facilitating it to change—to begin to decolonize themselves.

Walter Rudnicki spoke to this realization when he clashed with his colleagues in the Industrial Division, taking issue with their paternalistic reluctance to give Inuit control over government funds. In his view, their hesitancy stemmed from "the well known, well worn and outdated thesis that 'them folks ain't ready yet for responsibility.'"[90] His arguments resonated. In setting up the first co-ops, Donald Snowden took pains to check his own tendencies to tell Inuit what to do. He sometimes seemed to work as hard at convincing himself that he and his colleagues had not become the great white fathers of yore as he did persuading the Inuit they were their own bosses. "It is important that we should all understand what happens in a co-op," he told a gathering of Inuit at Frobisher (Iqaluit) in 1963. "I'm not sure whether the Eskimos do things because the white man thinks they should or because they want them themselves?"[91]

The sentiments and self-consciousness expressed by Laidlaw, Rudnicki, and Snowden speak to the nature of the hope that animated the development project. As the environmentalist Bill McKibben points out, hope is a word whose meaning has been debased. It now seems to mean "wishing"—wanting something that might not happen. But "real hope implies a real willingness to change," and to be changed.[92] That more robust meaning of hope was visible in the efforts of Northern Affairs in the 1960s to build the capacity for Inuit to talk back, govern themselves, and in so doing put limits on the very state that cultivated that capacity in them. The political project of sustainable development in the north complicates our notion of the "will to improve" that underpinned the "development decade."

For Snowden and his colleagues at Northern Affairs, development was very much "the management of a promise"—the promise in all humans to be who they were and could be.[93] It was a promise contained in Jean Lesage's assertion that the Inuit would share fully in the national life of Canada, and in the Daileys' use of the word "citizen." Given the state's reluctance to undertake a wholesale relocation of Inuit to southern Canada, it was also a promise that had to be fulfilled in place.

But was it? In the early 1970s, Doug Schweitzer, a researcher at the University of Saskatchewan, assessed the results of a decade of development initiatives in the region, measuring the distribution of income by source. The figures suggested that the Keewatin crisis had been averted: the destitution caused by the closure of the North Rankin Nickel Mine and the decline of caribou had not created a population entirely on the dole. In fact, although welfare transfers tripled over the decade, they contributed only a small portion of individual incomes, falling from a high of about thirty-five percent after the mine closed to seven percent in 1969. Instead of welfare, organized resource harvesting and handicrafts contributed a significant, if fluctuating, amount of revenue. Perhaps most surprisingly, wages made up the largest proportion of per-capita incomes in the Keewatin: by 1969, three-quarters of the people in Keewatin could claim a wage income from their labours, one that averaged $600 yearly.[94]

While Schweitzer's figures suggest the Keewatin was a development success story, the narrative of change is somewhat more complex. Although the wages that supported three-quarters of the Keewatin's residents came in part from the employment northern cooperatives offered, most were drawn from government coffers—principally from the budgets of the Royal Canadian Mounted Police, the Department of National Health, and the Department of Northern Affairs and National Resources.

Ironically, the state's regional and community development initiatives had initiated a new kind of dependency, even as they strove to build capacity. For better or worse, government had become the main source of revenue and the motor of growth for the region. Insofar as the Keewatin was on a more sustainable footing at the end of the 1960s than it was at the beginning, this was so because of the transfusion of government money in the form of wages. Sustainability was a matter of public subsidy, and as such it was a political choice, reflecting the triumph of Keynesian economics and a belief that the state should take an active role in development. The growth of the Canadian state was reflected in increasing public expenditures and the size of the public sector in the post-war years: public expenditures more than doubled in the 1960s, and the public sector grew from just below twenty percent of GDP in 1960 to approximately thirty percent in 1970.[95]

In many ways, this is not surprising; the state was the only institution capable of meeting the challenges posed by distance and the market. But even its interventions were no guarantee of success. Two years after Inuit from Rankin Inlet were moved to Daly Bay to work in the char fishery and cannery, they had to be moved back to Rankin. While operations there continued through the 1960s, they too were eventually wound down. The failure was not due primarily to problems with the workforce or even the resource. Instead, organized resource harvesting failed because of the geography of capitalism; whatever its nutritional value, arctic char wasn't worth enough on the market to collapse the distance between an arctic cannery and Canadian dinner tables.[96] Albeit unequally, these structural forces entrapped both the Inuit and the civil servants whose job it was to develop the north. In the 1960s, the agents of the liberal welfare state could often only wish for the kind of transformative power history sometimes ascribes to them.

***

Writing in 1970, Liverpool-born Jim Lotz, who worked for Northern Affairs during its early days and later became a specialist in community development, reflected on what he had learned—not just about bureaucracy, but about the country he had chosen to make home. "The further north we go in Canada," he mused, "the more national we become."[97] The north was where people would encounter those things that were truly national in scale, and which defined the country: boreal forest, shield, and Indigenous peoples.

But Lotz did more than reiterate a truism—that "the north is Canada." He wanted to make a point about the history of the region and country, underscoring how entangled they were, and not just with each other, but with the world. "The further north we go in Canada, the more national we become, and yet, strangely, the more international the problems tend to be," he wrote. "The north awakens our own humanity and makes us consider the humanity of others." For Lotz, working in Canada's north called on people to act with an understanding of how the region and its peoples were shaped by larger forces and to think across geographic scales and cultures—in short, to be citizens of the world.

The Barrenlands in the 1960s are a case in point. In dealing with the Keewatin, the federal government brought new people, new ideas, and a new optimism to bear on achieving social security. Even as they lived through it, some in Northern Affairs had a sense that theirs were unusual times—and fortunate ones. "Whenever I get angry with this country, disturbed by its hibernation, worried by its gentle ways, angry at its vacillations, frustrated by its indecision, and ALL READY TO LEAVE one thing always brings me back to my senses," Donald Snowden confessed. "That in this nation, at this time, there is a genuine interest in our north, and most of all in its people. And I say thanks to those I will never know who foot the bills, not so much because they keep me alive, but because they are willing to spend their money to make it possible for others of this country to learn to come through a time of confusion and change."[98]

These were the Department's "freewheeling, 'elastic-band-off-the-bundle' days," when everything seemed possible, including bureaucrats changing the world.[99] Informed by international debates, alive to the challenges of working cross-culturally, and armed with the insights of social work and anthropology, civil servants approached the problem of the north as one of development, of exploiting its mineral resources and its renewable ones. Aimed at the scale of region and community, the latter efforts at sustainable development were meant to help Inuit live within environmental limits and govern their own lives in a world transformed by colonialism and the market. A few in Northern Affairs, like Snowden and Walter Rudnicki, were aware of some of the contradictions inherent in what they were doing, of the fine line between facilitating change and imposing it. Yet they still felt compelled to act—however imperfectly—hopeful, if not always entirely convinced, that they were doing the right thing.

For Northern Affairs, putting the Keewatin on a sustainable footing was an issue of "capacity"—that of the land and its peoples. Sustainability was largely a technical matter, one requiring expertise to unleash what was already there, to realize the potential of the land and its peoples, lest it be wasted. As such, sustainable development was a moral project, but one oddly beyond politics. Its apolitical character was what ultimately limited its effectiveness, preventing a recognition of the larger forces that posed a fundamental challenge to the ongoing viability of the north and the efforts to develop it. If sustainability's Canadian history has any lessons for us, it is that its achievement is a matter of structure as well as agency, of engaging capitalism as well as capacity, and confronting the liberal assumptions embedded in it.

Notes

1   "Richard Harrington's Photos," *Globe and Mail*, 5 December 2009, at http://www.theglobeandmail.com/news/arts/richard-harringtons-photos/article1390245/ (accessed 17 September 2011).

2   Richard Harrington, *Padlei Diary, 1950: An Account of the Padleimiut Eskimo in the Keewatin District West of Hudson Bay during the Early Months of 1950*, ed. Edmund Carpenter (New York: Rock Foundation, 2000), 28, 66.

3   Richard Harrington, *The Face of the Arctic: A Cameraman's Story in Words and Pictures of Five Journeys into the Far North* (New York: Abelard Schuman, 1952), 261, 237.

4   James C. Scott, *Seeing Like a State: How Certain Schemes to Improve the Human Condition Have Failed* (New Haven: Yale University Press, 1998), 2.

5   Sarah Bonesteel, *Canada's Relationship with Inuit: A History of Policy and Program Development* (Ottawa: Public History, Inc., for Indian and Northern Affairs Canada, June 2006), 38–39.

6   Farley Mowat, *People of the Deer* (Boston: Little Brown, 1952).

7   Alvin Finkel, *Social Policy and Practice in Canada: A History* (Waterloo: Wilfrid Laurier University Press, 2006), ch. 7–11.

8   For an overview of the welfare measures that were undertaken for the Inuit, see David Damas, *Arctic Migrants, Arctic Villagers: The Transformation of Inuit Settlement in the Central Arctic* (Montreal: McGill-Queen's University Press, 2002), ch. 5.

9   For a discussion of the starvations, see Frank James Tester and Peter

Kulchyski, *Tammarniit (Mistakes): Inuit Relocation in the Eastern Arctic, 1939–63* (Vancouver: UBC Press, 1994), ch. 5–7. They are also covered more briefly and in a different interpretive frame in Damas, *Arctic Migrants, Arctic Villagers*, 87, 91–92.

10 Quoted in R. A. J. Phillips, *Canada's North* (Toronto: Macmillan, 1967), 169.

11 Cited in C. S. Mackinnon, "The 1958 Government Policy Reversal in Keewatin," in *For Purposes of Dominion: Essays in Honour of Morris Zaslow*, ed. Kenneth S. Coates and William R. Morrison (North York: Captus University Publications, 1989), 166.

12 For an introduction to the rhetoric of citizenship in the postwar period and the ways in which the Canadian state implemented it with respect to both immigrants and Aboriginal peoples, see Heidi Bohaker and Franca Iacovetta, "Making Aboriginal People 'Immigrants Too': A Comparison of Citizenship Programs for Newcomers and Aboriginal Peoples in Postwar Canada, 1940s–1960s," *Canadian Historical Review* 90, no. 3 (2009): 427–61. The phrase quoted is from Peter Russell, "The Constitution, Citizenship and Ethnicity," cited in Bohaker and Iacovetta, "Making Aboriginal People 'Immigrants Too,'" 433.

13 See, for instance, Ken Coates, *Canada's Colonies: A History of the Yukon and Northwest Territories* (Toronto: Lorimer, 1985).

14 Tanya Murray Li, *The Will to Improve: Governmentality, Development, and the Practice of Politics*

(Durham, NC: Duke University Press, 2007).

15 Gustavo Esteva, "Development," in *The Development Dictionary: A Guide to Knowledge as Power*, ed. Wolfgang Sachs (Johannesburg and London: Witwatersrand University Press and Zed Books, 1992), 13. Also see Gilbert Rist, *The History of Development from Western Origins to Global Faith* (London: Zed Books, 1997), ch. 1 and 4.

16 David R. Morrison, *Aid and Ebb Tide: A History of CIDA and Canadian Development Assistance* (Waterloo: Wilfrid Laurier University Press, 1998), 2.

17 The quote is from Prime Minister Louis St. Laurent in the House of Commons, introducing the bill that would create the Department of Northern Affairs and National Development in December 1953. Cited in R. Gordon Robertson, *Memoirs of a Very Civil Servant: Mackenzie King to Pierre Trudeau* (Toronto: University of Toronto Press, 2000), 114. Also see Morris Zaslow, *The Northward Expansion of Canada, 1914–1967* (Toronto: McClelland and Stewart, 1988), 333ff.

18 Zaslow, *The Northward Expansion of Canada*, ch. 12.

19 Damas notes that the federal government also granted the North Rankin Nickel Mining Company $20,000 to assist it in its early operations. See *Arctic Migrants, Arctic Villagers*, 94. For details on the mine, see The Mines Staff, "North Rankin Nickel Mines," *Canadian Mining Journal* 70, no. 8 (August 1957).

20 See, for instance, Jennifer A. Elliott, *An Introduction to Sustainable Development*, 3rd ed. (New York and London: Routledge, 2006), ch. 1; Desta Mebratu, "Sustainability and Sustainable Development: Historical and Conceptual Review," *Environmental Impact Assessment Review* 18, no. 6 (1998): 493–520; and Michael Redclift, "Sustainable Development (1987–2005): An Oxymoron Comes of Age," *Sustainable Development* 13, no. 4 (2005): 212–27.

21 Gro Harlem Brundtland and the World Commission on Environment and Development, *Our Common Future: Report of the World Commission on Environment and Development* (1987), ch. 2, para. 1 and 4 (http://www.un-documents.net/ocf-02.htm#I).

22 Samuel P. Hays and Barbara D. Hays, *Beauty, Health, and Permanence: Environmental Politics in the United States, 1955–1985* (Cambridge: Cambridge University Press, 1987).

23 The term "triple bottom line" was coined by John Elkington. See his "Towards the Sustainable Corporation: Win-Win-Win Business Strategies for Sustainable Development," *California Management Review* 36, no. 2 (1994): 90–100. Also see his *Cannibals with Forks: The Triple Bottom Line of 21st Century Business* (Gabriola Island, BC: New Society Publishers, 1998).

24 Timothy Mitchell, *Rule of Experts: Egypt, Techno-Politics, and Modernity* (Berkeley and Los Angeles: University of California Press, 2002).

25 Walter Rudnicki, "Mental illness among recent immigrants: a social work study of a sample group of hospitalized patients in British Columbia" (master's thesis, University of British Columbia, 1952).

26   University of Manitoba, Walter
      Rudnicki Fonds, box 81, folder 1,
      Walter Rudnicki, "Report—Field
      Trip to Eskimo Point, 1958," Ap-
      pendix C: "Thematic Apperception
      Test," 2.

27   Robertson to E. L. Harvie, 17 June
      1957, cited in Richard J. Diubaldo,
      "You Can't Keep the Native Na-
      tive," in Coates and Morrison, eds.,
      For Purposes of Dominion, 180.

28   For an overview, see Katy Gardner
      and David Lewis, Anthropology,
      Development and the Post-Modern
      Challenge (London: Pluto Press,
      1996).

29   V. F. Valentine and J. R. Lotz,
      "Northern Co-ordination and
      Research Centre of the Canadian
      Department of Northern Affairs
      and National Resources," Polar
      Record 11 (1963): 419–22.

30   University of Saskatchewan
      Archives, Robert G. William-
      son Fonds (hereafter cited as
      USA-RGW), MG 216, box 3, file:
      Transfer of Williamson to Rankin
      Inlet, Memorandum for Mr. A. C.
      Wimberley, from F.A.G. Carter,
      for the Director, Re: Transfer of R.
      G. H. Williamson to Rankin Inlet,
      n.d. [March 1958].

31   The information in this paragraph
      is from R. G. Williamson, "A Per-
      sonal Retrospective on Anthropol-
      ogy Applied in the Arctic," 1988,
      and the biography that is part of
      the finding aid for the Robert G.
      Williamson Fonds at the Univer-
      sity of Saskatchewan Archives. See
      USA-RGW, MG 216, box 4, and
      http://scaa.usask.ca/gallery/north-
      ern/en_finding_aid_display.php?-
      filename=williamson&title=Rob-
      ert%20Williamson%20fonds
      (accessed 5 May 2011). Also see
      "Robert Williamson (1931–2012),"

      Etudes Inuits/Inuit Studies 36, no. 1
      (2012): 231–33.

32   James Ferguson, "Anthropology
      and its Evil Twin: 'Development' in
      the Constitution of the Discipline,"
      in International Development and
      the Social Sciences: Essays on the
      History and Politics of Knowledge,
      ed. Frederick Cooper and Randall
      M. Packard (Berkeley: University of
      California Press, 1997), 159.

33   David D. Gow, "Anthropology and
      Development: Evil Twin or Moral
      Narrative?" Human Organization
      61, no. 4 (2002): 299.

34   Williamson, "A Personal Retro-
      spective," 46.

35   USA-RGW, MG 216, Box 24, file
      I7-1, International Nickel Co., R.
      G. Williamson, "Eskimos in the
      Modern World,", 1970, 9.

36   The Mines Staff, "North Rankin
      Nickel Mines," 97.

37   Williamson, "A Personal Retro-
      spective," 46–47. Williamson went
      on to make the Rankin miners the
      subject of his doctoral research. For
      his comments on the work rotation
      system at the mine, see William-
      son, Eskimo Underground, 116–19.

38   USA-RGW, MG 216, box 12, Robert
      G. Williamson, "The Notion of
      Cultural Commuting: Evaluation
      of Short-Term Feasibility," n.d.

39   Northern Service Officer D. W.
      Grant reported that the mine
      could close by the end of the 1959
      shipping season. Library and
      Archives Canada (hereafter cited
      as LAC), RG 85, vol. 1962, file
      A-1009-10/184, Memorandum for
      Mr. C. M. Bolger, Administrator of
      the Arctic, from D. W. Grant, Re:
      North Rankin Nickel Mines, Ltd.,
      16 March 1959, 1.

40   D. M. Brack and D. McIntosh, *Keewatin Mainland Economic Survey and Regional Appraisal* (Ottawa: Projects Section, Industrial Division, Department of Northern Affairs and National Resources, March 1963), 135; and LAC, RG 85, vol. 1148, file 1000/184, "Confidential [no title; recommendations of the Keewatin Conference, 1962]," vol. 6, p. 1.

41   USA-RGW, MG 216, Box 32, "Rankin Inlet Questionnaires— Post Close-Down Interviews, Genealogies."

42   A numerical breakdown of the survey results is given in Williamson, *Eskimo Underground*, 131.

43   USA-RGW, MG 216, Box 32, "Rankin Inlet Questionnaires— Post Close-Down Interviews, Genealogies."

44   For overviews of these relocations, see D. S. Stevenson, *Problems of Eskimo Relocation for Industrial Employment: A Preliminary Study* (Ottawa: Northern Science Research Group, Department of Indian Affairs and Northern Development, 1968); and R. G. Williamson, *Eskimo Relocation in Canada* (Saskatoon: University of Saskatchewan, Institute of Northern Studies, 1974).

45   Prince of Wales Northern Heritage Centre, Alexander Stevenson Fonds (hereafter cited as PWNHC-ASF), N-1992-023, box 25, file 9, Speech by the Honorable Jean Chrétien, P.C., M.P., Minister of Indian Affairs and Northern Development, to the Yellowknife Board of Trade, Monday, 10 November 1969, 3; and PWNHC-ASF, N-1993-023, box 23, file 7, A. Stevenson, "Sociological Aspects—Northern Development," presented at the Arctic Petroleum

Operators Association Arctic Environmental Workshop, Gulf Theatre, Gulf Oil Building, Calgary, AB, 27–28 October 1971.

46   Williamson, *Eskimo Relocation*, 54.

47   Stevenson, *Problems of Eskimo Relocation*, 8.

48   Robert C. Dailey and Lois A. Dailey, *The Eskimo of Rankin Inlet: A Preliminary Report* (Ottawa: Northern Coordination and Research Centre, Department of Northern Affairs and National Resources, 1961), 94–95.

49   Jean Malaurie, *Hummocks: Journeys and Inquires among the Canadian Inuit*, trans. Peter Feldstein (Montreal: McGill-Queen's University Press, 2007), 321, 325.

50   Relocation as a way to effect the redistribution of population was a strategy for northern development in general, not just for the Keewatin. In the context of northern development, the Keewatin was targetted for "depopulation" and "cold storage." How serious an option it was is unclear: a 1969 policy paper said it was being articulated as a way to generate discussion that might lead to concrete strategies. See PWNHC-ASF, N-1992-023, box 28, file 8, Economic Staff Group, Northern Economic Development Branch, Department of Indian Affairs and Northern Development, "A Strategy for Northern Development—Discussion Draft No. 2—CONFIDENTIAL," 15 September 1969, 62–64.

51   John Sandlos, Peter Kulchyski, and Frank J. Tester, and, to a lesser extent, I have written about this. See Sandlos, *Hunters at the Margin: Native People and Wildlife Conservation in the Northwest Territories* (Vancouver: UBC Press,

2007), 205–30; Peter Kulchyski and Frank J. Tester, *Kiumajut (Talking Back): Game Management and Inuit Rights, 1900–1970* (Vancouver: UBC Press, 2007), ch. 2 and 4; and Tina Loo, *States of Nature: Conserving Canada's Wildlife in the Twentieth Century* (Vancouver: UBC Press, 2006), 132–33 and 159–61. Also see Damas, *Arctic Migrants, Arctic Villagers*, 166–67.

52   Sandlos, *Hunters at the Margin*, 201.

53   Brack and McIntosh, *Keewatin Mainland Economic Survey*, 117.

54   Brundtland, *Our Common Future*, ch. 2, para. 7.

55   LAC, RG85, 1997–98/076 20, file 251-9/1059: Specialty Foods, Daly Bay Cannery (Project File), Memo from R. Mulligan re: Move to Daly Bay, January 1964.

56   Brack and McIntosh, *Keewatin Mainland Economic Survey*, 137.

57   Nathan F. Sayre, "The Genesis, History, and Limits of Carrying Capacity," *Annals of the Association of American Geographers* 98, no. 1 (2008): 120–34.

58   As Brundtland noted, "we recognize that poverty, environmental degradation, and population growth are inextricably related and none of these fundamental problems can be successfully addressed in isolation. We will succeed or fail together." From "Making Common Cause," US-based Development, Environment, Population NGOS, WCED Public Hearings, Ottawa, 26–27 May 1986, cited in Brundtland, *Our Common Future*, ch. 2, para. 10.

59   For instance, in the 1960s and 1970s, the Canadian Wildlife Service continued to design and implement policies informed by older conservation ideas that had the effect of displacing Indigenous peoples. It was not until the early 1980s that some species started to be managed more jointly in ways that were consistent with the principles of sustainable development. See Peter Kulchyski and Frank J. Tester, *Kiumajut (Talking Back): Game Management and Inuit Rights, 1900–1970* (Vancouver: UBC Press, 2007); Tina Loo, *States of Nature: Conserving Canada's Wildlife in the Twentieth Century* (Vancouver and Seattle: UBC Press and University of Washington Press, 2006), and John Sandlos, *Hunters at the Margin*. On the first wildlife co-management initiative, which involved the caribou of the Central Arctic, see Peter J. Usher, "The Beverley-Kaminuriak Caribou Management Board: An Experience in Co-Management," in *Traditional Ecological Knowledge: Concepts and Cases*, ed. Julian T. Inglis (Ottawa: International Program on Traditional Ecological Knowledge, International Development Research Centre, Canadian Museum of Nature, 1993), 111–20.

60   On regional economic development and the influence of Perroux in Canada, see Bernard Higgins and Donald Savoie, *Regional Development Theories and Their Application* (New Brunswick, NJ: Transaction Publishers, 1995), 271–88; and Donald J. Savoie, *Regional Economic Development: Canada's Search for Solutions*, 2nd ed. (Toronto: University of Toronto Press, 1992).

61   Pierre Elliott Trudeau, *Federalism and the French Canadians*, ed. John T. Saywell (New York: St. Martin's Press, 1968), 147.

62 Quoted in Edith Iglauer, "A Change of Taste—A Reporter at Large," *The New Yorker*, 24 April 1965, 122.

63 Edith Iglauer, "Donald Snowden," obituary, *Inuktitut Magazine* (Summer 1984), nos. 56–57.

64 PWNHC-ASF, N-1992-023, box 30, file 6, "Arctic Circle Specialty Foods," n.d. [1969]. Also see PWNHC-ASF, N-1992-023, Box 30, file 5, A. W. Lantz, "Fish Cannery in Northwest Territories," *Trade News* (Department of Fisheries of Canada), October 1965, 12–13. For details on the char fishery operating from Rankin Inlet, see LAC, RG 85, vol. 1448, file 1000/184, vol. 7, H. M. Budgell, Report on the Keewatin Project, Ottawa, 4 June 1962.

65 Brack and McIntosh, *Keewatin Mainland Economic Survey*, 133.

66 LAC, RG 85, vol. 1448, file 1000/184, vol. 8, D. M. Brack, "The Keewatin Region—Preliminary Report," 3 October 1962, 14. Brack is making a play on words; in Inuktitut, the suffix "-miut" means "people of" (e.g., "Padleimiut" refers to the people from Padlei).

67 See, for instance, USA-RGW, MG 216, box 8, file 1, "CBC—Rankin Inlet News and Commentary." Also see Williamson, "A Personal Retrospective," 48–49.

68 Gordon R. Wensley, *Comminterphone—Rankin Inlet: A Report of Research for the Department of Communications, Government of Canada, Ottawa* (Saskatoon: University of Saskatchewan Institute for Northern Studies, June 1973), 1, 4.

69 Wensley, *Comminterphone*, 2, 82.

70 LAC, RG 85, vol. 1948, file A560-1-5, pt. 4, Northern Administration Branch. Branch Policy Directive No. 32. Subject: Community Development in the North, n.d. [1964], 1.

71 LAC, RG 85, vol. 1948, file A-560-1-5, pt. 4, Community Development Committee, Second Meeting, 7 June 1966, 3.

72 Northern Administration Branch. Branch Policy Directive No. 32. Subject: Community Development in the North, n.d. [1964], 3.

73 LAC, RG 85, vol. 1946, file A-501-1, pt. 1, Bibliography of Community Development, Area and Community Planning Section, Industrial Division, 23 May 1961. For a brief overview of the history of community development, see Alice K. Johnson Butterfield and Benson Chisanga, "Community Development," in *Encyclopedia of Social Work*, vol. 1 (New York: Oxford University Press, 2008), 376–77.

74 Ian MacPherson, *Each for All: A History of the Co-operative Movement in English Canada, 1900–1945* (Toronto: Macmillan, 1979), Introduction. The companion book is Ronald Rudin, *In Whose Interest? Quebec's Caisses Populaires, 1900–1945* (Montreal: McGill-Queen's University Press, 1990).

75 PWNHC-ASF, N-1993-023, box 30, file 10, Paul Godt, "The Role of Co-operatives," 25 August 1960, 5.

76 MacPherson, *Each for All*, 130–32.

77 Address to the Yellowknife Education Conference, 3 February 1966, cited in Glenn Fields and Glenn Sigurdson, *Northern Co-operatives as a Strategy for Community Change: the Case of Fort Resolution* (Winnipeg: Centre for Settlement Studies, University of Manitoba, May 1972), 5.

78 Godt, "The Role of Co-operatives," 5.

79 Robertson, *Memoirs of a Very Civil Servant*, 179, 190.

80 LAC, RG 85, vol. 1962, file A-1012-9, vol. 2, Director, Northern Administration Branch. Branch Policy Directive No. 13, Community Development and Local Organization, 25 October 1961, 3.

81 Northern Administration Branch. Branch Policy Directive No. 32. Subject: Community Development in the North, n.d. [1964], 3.

82 PWNHC-ASF, N-1992-023, box 30, file 10, "Northern Cooperatives," Department of Northern Affairs and National Resources Information Services Division, December 1964, 1.

83 PWNHC-ASF, N-1992-023, box 30, file 10, R. J. Wickware, "Northern Co-operative Development," Paper Presented to the NAACL Annual Meeting, 5–8 July 1971, St. John's, NL, 1–2.

84 M. Stopp, *The Northern Co-operative Movement in Canada*, Submission Reports, vol. 2, Report Number 2009-22, Spring 2009 Meeting (Ottawa: Historic Sites and Monuments Board of Canada, 2009), 670.

85 Alexander F. Laidlaw, "Cooperatives in the Canadian Northland," *North/Nord*, November–December 1963, 12.

86 Frédéric Laugrand, Jarich Oosten, and David Serkoak, "'The saddest time of my life': Relocating the Ahiarmiut from Ennadai Lake (1950–1958)," *Polar Record* 46, no. 2 (2010): 121–22; the quote is from R. A. J. Phillips, Memorandum for the Director Re: Movement of Henik Lake Eskimos, 15 January 1958, cited on p. 125.

87 Edith Iglauer, "Conclave at Frobisher—A Reporter at Large," *The New Yorker*, 23 November 1963, 192.

88 Stopp, *The Northern Co-operative Movement in Canada*, 671, 684.

89 A. F. Laidlaw, *Co-operatives in Canada's Northland: A report based on the first conference of Arctic co-operatives, held at Frobisher Bay, N.W.T, March 12–18, 1963* (Ottawa: Co-operative Union of Canada, April 1963), 6, 23.

90 LAC, RG 85, vol. 1447, file 1000/184, pt. 5, Walter Rudnicki, Chief, Welfare Division, Memorandum for the Director, re: Rankin Inlet, et al., 11 October 1961, 2.

91 Edith Iglauer, "Conclave at Frobisher," 202. Later Snowden confided to Iglauer that, "if the Eskimos had felt they were not making their own decisions, I would have considered our whole technique wrong, and would have changed it" (203).

92 Bill McKibben, *Hope, Human and Wild: True Stories of Living Lightly on the Earth* (Minneapolis: Milkweed Editions, 2007), 5.

93 Jan Nederveen Pieterse, "After Post-Development," *Third World Quarterly* 21, no. 2 (2000): 176.

94 Doug Schweitzer, *Keewatin Regional Dynamics: A Research Report* (Saskatoon: University of Saskatchewan, 1971), 65.

95 J. Stephen Ferris and Stanley L. Winer, "Just How Much Bigger Is Government in Canada? A Comparative Analysis of the Size and Structure of the Public Sectors in Canada and the United States, 1929–2004," *Canadian Public Policy* 33, no. 2 (2007): especially 15–16.

96 Government of the Northwest
Territories, *Keewatin Arctic Char
Plants: An Investment Concept*
(Yellowknife: Government of the
Northwest Territories, Economic
Development and Tourism, Natural
Resources Section, 1991), 2–4.

97 Jim Lotz, *Northern Realities: the
Future of Northern Development
in Canada* (Toronto: New Press,
1970), 31.

98 LAC, Donald Snowden Fonds, MG
31 D 163, vol. 14, file 34: Northern
Affairs, Miscellaneous, 1961–64,
Snowden to Edith Hamburger,
n.d. [however, it appears to be in
response to a letter from her dated
1 November 1963], 1.

99 Williamson, "A Personal Retrospective," 48.

8

Western Electric Turns North: Technicians and the Transformation of the Cold War Arctic

Matthew Farish and P. Whitney Lackenbauer

The October 1959 issue of the *Illinois Technograph*, a periodical produced out of the Civil Engineering Building at the University of Illinois Urbana-Champaign, was rife with advertisements for work in the American military-industrial complex: atomic technology at the Sandia National Laboratories in Albuquerque, New Mexico; fighter planes at Convair in Fort Worth, Texas; and missile systems at the Garrett Corporation and the Jet Propulsion Laboratory in Los Angeles and Pasadena, respectively. The presence of these and other recruiters was testament to the growing prominence of a "gunbelt" landscape spreading across the south and west of the United States.[1] The magazine also contained an ad for a corporation with prominent Illinois roots (Fig. 8.1). Opposite an image of that month's "Technocutie," Judy Stephenson, was the headline "ENGINEERS explore exciting frontiers at Western Electric." Below a photo of two glasses-and-tie-wearing "defense projects engineers" hard at work, the text began:

W.E. DEFENSE PROJECTS ENGINEERS are often faced with challenging assignments such as systems testing for the SAGE continental air defense network.

ENGINEERS explore exciting frontiers at Western Electric

If guided missiles, electronic switching systems and telephones of the future sound like exciting fields to you, a career at Western Electric may be just what you're after.

Western Electric handles *both* telephone work and defense assignments . . . and engineers are right in the thick of it. Defense projects include the Nike and Terrier guided missile systems . . . advanced air, sea and land radar . . . the SAGE continental air defense system . . . DEW Line and White Alice in the Arctic. These and other defense jobs offer wide-ranging opportunities for all kinds of engineers.

In our main job as manufacturing and supply unit of the Bell System, Western Electric engineers discover an even wider range of opportunity. Here they flourish in such new and growing fields as electronic switching, microwave radio relay, miniaturization. They engineer the installation of telephone central offices, plan the distribution of equipment and supplies . . . and enjoy, with their defense teammates, the rewards that spring from an engineering career with Western Electric.

Western Electric technical fields include mechanical, electrical, chemical, civil and industrial engineering, plus the physical sciences. For more detailed information pick up a copy of "Consider a Career at Western Electric" from your Placement Officer. Or write College Relations, Room 200D,

Western Electric Company, 195 Broadway, New York 7, N. Y. And sign up for a Western Electric interview when the Bell System Interviewing Team visits your campus.

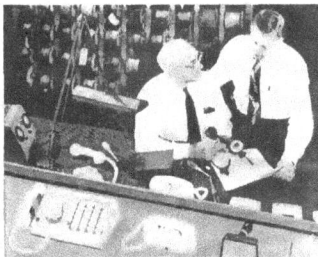

TELEPHONES OF THE FUTURE—Making telephone products for the Bell System calls for first-rate technical know-how. Tomorrow's telephone system will demand even more imaginative engineering.

Western Electric

MANUFACTURING AND SUPPLY ⬡ UNIT OF THE BELL SYSTEM

Principal manufacturing locations at Chicago, Ill.; Kearny, N. J.; Baltimore, Md.; Indianapolis, Ind.; Allentown and Laureldale, Pa.; Burlington, Greensboro and Winston-Salem, N. C.; Buffalo, N. Y.; North Andover, Mass.; Lincoln and Omaha, Neb.; Kansas City, Mo.; Columbus, Ohio; Oklahoma City, Okla.; Teletype Corporation, Chicago, Ill. and Little Rock, Ark. Also Western Electric Distribution Centers in 32 cities and installation headquarters in 16 cities. General headquarters: 195 Broadway, New York 7, N. Y.

FIG. 8.1: Western Electric recruitment ad. *Illinois Technograph* 75, no. 1 (October 1959): 37.

If guided missiles, electronic switching systems and tele-phones of the future sound like exciting fields to you, a career at Western Electric may be just what you're after. ... Western Electric handles *both* telephone work and defense assign-ments ... and engineers are right in the thick of it. Defense projects include the Nike and Terrier guided missile systems ... advanced air, sea and land radar ... the SAGE continental air defense system ... DEW Line and White Alice in the Arc-tic. These and other defense jobs offer wide-ranging opportu-nities for all kinds of engineers.[2]

Western Electric, "the manufacturing and supply unit of the Bell Tele-phone System," recruited regularly in the *Technograph*.[3] In the December 1959 issue, its ad focused on the wonders of computer technology and promised "8,000 supervisory jobs" for engineers "in the next ten years," along with "corresponding opportunities for career building within re-search engineering."[4] In February 1960, beneath an extraterrestrial vista, the text suggested that Western Electric employees "may engineer instal-lations, plan distribution of equipment and supplies," or join a growing group of field engineers "whose world-wide assignments call for working with equipment we make for the Government."[5]

This chapter is concerned with one set of installations that had recent-ly been engineered by Western Electric employees sent out on those world-wide assignments, namely the radar stations established across Canada's arctic reaches in the early years of the Cold War. As the *Technograph* ad-vertisement indicated, Western Electric had "handled" the Distant Early Warning (DEW) Line, which the company—with the help of Pentagon partners, almost three thousand sub-contractors, and a host of geologists, oceanographers, meteorologists, and other scientists—built from Alaska to Iceland in the 1950s and early 1960s.[6] The prospect of a high-arctic radar network that could detect hostile bombers, a project US Air Force officials had dismissed as excessively expensive and technologically dubi-ous when it was first contemplated in 1946, received presidential approval in the aftermath of the Soviet detonation of a hydrogen bomb in 1953. The United States paid for and established the radar chain across the seventi-eth parallel, three-quarters of which was in Canada; a formal agreement on the matter was settled by the two states in November 1954. The DEW Line was completed by 1957, its creation an extraordinary feat of "geographical engineering" that altered the military, logistical, environmental, and social characteristics of the North American Arctic.[7]

The consequences of these endeavours have recently achieved notoriety in Canada. Over the last two decades, the remnants of the DEW Line have been targeted by a massive, $500 million effort, undertaken by contractors and northern residents, to remove debris and decontaminate sites rife with toxic waste.[8] Our focus here is on the conception, siting, and initial construction of the Line, and expressly on the roles played by Western Electric technicians—to borrow a useful term from the period—in the Canadian and Alaskan north during an era of dramatic arctic militarization.[9] One announcement in *Life* listed the corporation's duties as "development, design, engineering, procurement, transportation, construction, installation, testing and training of operating personnel."[10] It also participated in the broader northern extension of the state discussed more fully in Tina Loo's chapter in this volume—or in this case, *two* states, whose arctic ambitions were markedly similar when it came to defence.

Crafting a set of enduring military structures resulted in, and was also premised on, a fundamental reconfiguration of arctic environments. This reconfiguration was focused on the dozens of DEW Line sites, but it ultimately represented a systematic alteration of the "southern" human presence on northern sea and soil. The consequences for Indigenous northerners, who received minimal attention from the DEW Line's designers, were profound. While particular substances and stories undoubtedly remain elusive or under-discussed, evidence of the Line's impact on land and northern lives is now reasonably well documented across an eclectic group of remediation reports, media accounts, and community-based oral histories produced over the last two decades.[11]

We are interested in and indebted to these sources, and along with our own archival and field-based research, they are central components of a broader study that is in progress. In focusing on a largely unknown and understudied group of workers, however, this chapter situates their physical labour alongside an equally significant imaginative revaluation of arctic geography.[12] While variants of the "idea of north" are common tropes in Canadian scholarship on the Arctic, these studies have remained largely separate from regional or community histories—or, for that matter, histories drawing connections between such sites and places ostensibly beyond "the north." The case of Western Electric technicians is instructive precisely because it renders these spatial separations impossible. How the "handlers" of the DEW Line *approached* the north—how they were trained to understand it, how they saw and described it, in both distant and more grounded ways—was inextricable from their numerous northern activities during a brief period when extraordinary power was bestowed upon and exercised by their employer. As Edward Jones-Imhotep puts it in a rare discussion of northern technicians after the Second World War—specifically, employees of Canada's Radio Physics Laboratory who travelled to Churchill, Manitoba, to study auroral displays—"visions of a remote and isolated" region "were in fact underwritten by the assertions of science."[13]

Western Electric's promotional films and publications frequently treated the terrain of the DEW Line as wilderness or wasteland replete with "harsh and unrelenting disciplines," and this narrative would certainly have influenced the company's employees, who faced an "imposing array of problems" on their northern assignments.[14] Such unsurprising characterizations, hardly limited to the DEW Line case, have been both durable and consequential. According to a leader of the recent Canadian remediation effort, "It was remote, hardly anyone lived there, these sites were in the middle of nowhere far from anyone's imagination and therefore the environment wasn't at the fore."[15] But this sort of historical and geographical distancing should be treated with scepticism. After all, the "mysterious north" was a realm of tremendous cultural and political significance in North America during the 1950s.[16] Equally, as Andrew Stuhl and others in this volume document, the role of the Canadian Arctic as a target for combined state and scientific experimentation was certainly not new after the Second World War. While the dimensions of the DEW Line were unprecedented, a consistent historiographical point remains: even as the Arctic was and still is frequently treated as a distinct social or environmental region, it is imperative that scholars challenge such exercises in strict boundary-making.[17]

It is precisely for these reasons that the histories of concern, here, are not exclusively "northern"; nor is the north itself particularly easy to fix in the genealogy of something as simultaneously monumental and dispersed as the DEW Line. This story must extend beyond radar stations and adjacent communities to a sprawling set of bases, boardrooms, field sites, and laboratories—and to the pages of periodicals such as the *Technograph*. In these locations, powerful forms of environmental knowledge, along with elements of the Line itself, were being generated, debated, and tested. To speak of the Cold War transformation of the north, then, means blurring the materiality of places with their discursive counterparts, and requires movement across a range of scales and relationships.

In the next section of the chapter, we introduce two terms, *high modernism* and *technopolitics*, which help to locate the DEW Line both conceptually and historically. A project of tremendous physical and symbolic scope, the Line was consistently treated by its champions as an unprecedented and masterful achievement in the long human struggle with a forbidding northern nature. Building on related scholarship in environmental history, historical geography, and the history of science, we suggest that such epic assertions were made possible by the more focused efforts of engineers, planners, and related workers, who established and maintained the conduits between site-specific and more indiscriminate knowledge of arctic geography. This role, and the authority granted to it, allowed the Line, and "the north" as a whole, to be treated in technical terms—as a space for the receipt of devices and practices perfected in a laboratory setting. How this came to be, and who these northern technicians were, are interesting questions, and we address them in the middle of the chapter. But they must be placed alongside two additional queries: what was missing from this technical vision of the north, and what were its limits? Given that the DEW Line was a military initiative, and given that its consequences for the north continue to be discussed and experienced today, these latter questions are particularly acute, and we turn to them in the final pages.

High Modernism, Technopolitics, and "Industry's Defense Mission" in the North

Frequently characterized as one of the largest and most challenging engineering initiatives in human history, the DEW Line—"so spectacular, so awe-inspiring, so nature-defying in concept"—was a paradigmatic example of a high modernist megaproject.[18] If its military value was debatable, the Canadian observer C. J. Marshall wrote in 1957, "there is no doubt" that the Line was "a dramatic engineering achievement; and because of it, life in the Canadian Arctic will never be the same again."[19] In the wake of James C. Scott's influential book *Seeing Like a State* (1998), scholars from a variety of disciplines have documented and debated the global history of high modernism, or what Scott defines as "a sweeping, rational engineering of all aspects of social life in order to improve the human condition." This concern for social improvement was paired with a "belief that it was man's destiny to tame nature to suit his interests and preserve his safety."[20] Neither of these imperatives was new in the first decade of the Cold War, but the planetary extent of that conflict, the gathering power of multiple military-industrial complexes, and the particular conflation of technoscience and geopolitics spawned by the Manhattan Project all contributed to the belief that a megaproject such as the DEW Line was both attainable and necessary.

DEW Line planners were far more concerned with the "safety" of a threatened North American industrial heartland than they were interested in the sites that would be dramatically altered by the arrival of the Line. Similarly, the taming of *northern* nature, however challenging, was only a late stage in a series of governmental moves that fused science and security during the 1940s and 1950s. Dreamed and designed in laboratories described by their patrons as contributors to the "field of safety engineering," the Line was a mere component—if a significant one—of a massive new defence network treated as a manageable system precisely because this terminology countered the profound uncertainty of the Cold War.[21]

In James Scott's wake, environmental historians and geographers have provided nuanced accounts of widespread landscape transformations proposed and conducted in the spirit of high modernism, from nuclear testing to the building of large dams.[22] In one exemplary study of the latter, Tina Loo and Meg Stanley note that high modernism was characterized by a particular "way of seeing," a "synoptic" view that treated territory schematically, without the messy details of "biophysical and social contexts."[23] This way of seeing is also a symptom of a depoliticized, administrative treatment of geography.[24] The "success" of a megaproject depended on the combination of such abstracted hubris with a more practical form of knowledge about places and people. Yet it would be a mistake to oppose these two modes of understanding. Following historians of science, who have argued persuasively that all science is marked, in Steven Shapin's words, by the "spatial circumstances of its making," Loo and Stanley argue instead for what they call "high modernist local knowledge," found most prominently in the agents who designed and built large dams. These workers were distinct from Scott's detached protagonists. Engineering studies generated quite intimate information about dam sites, in the process making those environments "legible," ready for reconfiguration.[25] Such reports might be considered "the engines of change" for high modernism—alongside the relatively obscure figures who prepared them and set them in motion.[26]

This framework is well suited to understanding Western Electric's DEW Line role. But the admirable search for a more contextual history of high modernism must also consider the traffic between multiple sites of knowledge production—and, as Marionne Cronin reminds us in her chapter in this volume, the irrevocable entanglements of technologies and environments. In addition, there are, in our account, many intersecting variants of "the local," and Emilie Cameron's chapter encourages careful consideration of that term's uses and limits. Moreover, the DEW Line was a distinctive type of high modernist endeavour: it was treated as a technical solution to a military dilemma, and as such represented the dual mid-century "optimism in both technology and state authority," a faith channelled into the Cold War competition for the dominance of terrestrial and extra-terrestrial space.[27] Unlike many of the high modernist examples introduced by Scott, the DEW Line was not an explicitly social project. Its proponents and producers devoted a tiny fraction of their energy to the consequences of radar construction for northern residents. Even so, as a new communications network, a demonstration of apparent environmental mastery, a source of wage income for certain northerners, a repository of southern expertise and cultural practices in the Arctic, and a "shield" built to secure North American "social life" (in Scott's words), the Line was irrefutably an exercise in *modernization*.[28]

If Western Electric engineers were building "installations" and planning "distributions" of equipment, by the 1950s it made particular sense that these were military installations and items. The practice of calling on "the resources and know-how of the large communication companies to act as prime contractors in planning and supervising" defence projects, according to one AT&T executive in a 1958 speech, was to be celebrated as "The American Method."[29] In the midst of Western Electric's DEW Line responsibilities, an in-house magazine article on the topic referred straightforwardly to "Industry's Defense Mission."[30] As Shapin notes drily, "Big Science had remarkably few apologists, just because it had so little *need* of apologetic defense."[31] By the end of the 1950s, there were some eighteen thousand Western Electric employees engaged exclusively in defence work.[32] But *how* they participated in their "world-wide assignments" is a more complicated matter.[33]

During the Second World War, "Army ground and air forces, Navy ships, submarines, and planes, and Marine landing forces" were all using Western Electric-built radars.[34] Similar collaborative "Big Science" was underway at MIT's Radiation Laboratory (Rad Lab), and of course in the various facilities of the Manhattan Project. Elements of both, including some individuals, soon migrated into MIT's Lincoln Laboratory, the crucial interdisciplinary site for research on Cold War continental defence. One historian of science describes all of this effort as defined not just by accomplishments, or magnitude, but by the "efficient integration of emerging technologies into goal-directed systems, along with the organizational arrangements" that accompanied these systems.[35] Fabricating durable radar networks, as S. P. Schwartz of Western Electric put it, was ultimately a question of "the capability of man and equipment to function" together.[36]

To speak in the language of goals, functions, and human-machine hybrids is to deploy the vocabulary of *technopolitics*. The creation of the North American continental defence network, with the DEW Line along one edge, involved what Gabrielle Hecht describes as "the displacement of power onto technical things."[37] Technopolitics, for Timothy Mitchell, is "a particular form of manufacturing ... so that the human, the intellectual, the realm of intentions and ideas seems to come first and to control and organize the non-human."[38] The influence of this arrangement was certainly on display in the North American Arctic after the Second World War, an era of far-reaching, state-driven social experiments, some of which blended human development aims with infrastructural initiatives.[39] Nevertheless, as Mitchell notes, "the intentional or the human is always somewhat overrun by the unintended," and this maxim is a reminder of technology's powerful instability and a useful check on the triumphant narratives of military modernization found in period newsreels, print journalism, industry reports, government documents, and even academic sources.[40]

As such, we should carefully consider how continental defence technology was designed or used "to enact political goals," but also how the material of the DEW Line—as it was conceived and put in place—led to unpredictable results.[41] The crucial hinge is the relationship between the expert "perfection" of radar devices at locations like the Lincoln Laboratory and their subsequent conversion into "field-ready combat tools, capable of being operated by military personnel with only limited scientific training."[42] More than any other agency, it was Western Electric that managed this conversion from laboratory to field.

To embark on this task did not mean completely leaving the laboratory milieu behind. As Scott Kirsch notes, laboratories "have travelled ... as methods; that is, as experimental technologies and epistemologies adapted to the field, but also, it should not be overlooked, as communities."[43] Much of the northern work carried out by Western Electric employees was premised on attempts to apply consistent, learned practices in a distinctly different environment. The north, as many visitors expressed both informally and directly, was itself a laboratory for science, engineering, and military exercises—statements that imply geographic distinction, but also a certain confidence of universality, backed by a clear political mission.[44]

In the newsreel language of the period, a "grim and forbidding Arctic waste" or "barren frozen wilderness," "where man must learn to live with nature before he can defend himself against a living enemy," might challenge the capabilities of technopolitics, but for Western Electric and its boosters, the "frigid latitudes" certainly did not overturn them. The "test of men and machines" was just that—an opportunity to further the reach of a technopolitical worldview.[45] The result, another film boasted, was that a "3000-mile long strip ... began to lose its timeless identity and take on its new identity: DEW Line."[46]

Techne, Technicians, and Training

As radar stations and other military installations were constructed across the north in the 1950s, they emerged as part of a particular *techne*, or what Michel Foucault called "a practical rationality governed by a conscious goal"; they were produced by a combination of art and artifice, creativity and labour.[47] In the familiar tones of one archival account, Western

Electric's many DEW Line responsibilities are followed by results: "Special buildings and construction techniques were devised to withstand severe arctic weather. Stations were exhaustively evaluated, designs were modified, equipments were changed to combat polar magnetism, effects of constant wind and cold were measured—*both on men and the complex devices they would have to keep operable.*"[48] The production of the Line—its bringing-forth—was thus done with and by such men (and the masculine, or more precisely a particular variant of masculinity, is significant here). They had been trained in these special techniques of representation, calculation, development, and maintenance. Such consistent and yet flexible practices were more generally being put to use by teams of workers across the facilities of the growing military-industrial complex.[49]

"Technician" had been on the lips of many North American educators and scientific advisors since the Second World War. The word's meaning in English as a "person skilled in the mechanical arts" apparently dates from 1939. In an influential 1945 paper, the American geographer Edward Ackerman grappled with the extraordinary "data for wartime research" generated in agencies like the team-driven Office of Strategic Services, where he had worked. This data covered such a "wide range of subjects" that it required the skills of new "systematic specialists." Ackerman's words were prophetic: these specialists turned out to be the high-profile "experts" who led post-war area studies institutes, or who headed interdisciplinary teams devoted to the problems of continental defense. Meanwhile, Ackerman suggested, those "technicians" whose skill with detail was offset by poor interpretative abilities should, like laboratory employees in physics or biology, be put to use on "mechanical work of the mind and eye."[50] While Ackerman had geographers and cartographers in mind, similar changes were unfolding in other fields—a sharper, more vertical version of an older and very gendered division of labour, perhaps, but also the recognition that both "specialists" and "technicians" were necessary to understand, to diagnose, a complex, interrelated post-war globe.[51] Ackerman's conundrum was epistemological, but it was also one (to borrow from his article's title) of "Geographic Training" and "Professional Objectives."

In his essays on American physics and physicists after the Second World War, David Kaiser notes the importance granted to the production of new scientific workers during a period of dramatic increases in funding for the sciences. A disproportionate amount of this spending arrived from the Pentagon and related branches of the federal government.[52] These dollars paid for instruments and equipment, but also for education itself.[53] The equation of national security with science was backed by a "Cold War logic of wartime requisitions" and repeated references to scientific *manpower*.[54] This discourse bubbled along steadily through the late 1940s and 1950s, and grew more urgent after the Soviet launch of Sputnik I in October 1957. The context for the *Technograph* advertisement and similar notices was thus a furious drive to produce engineers for Cold War work.

Only some of these physicists and engineers were destined for academic careers; most found their way to well compensated positions at both national and corporate laboratories.[55] Most of the Western Electric advertisements in the *Technograph* concluded with a pledge to fund a reader's graduate education in engineering. As these readers probably recognized, and as historians of science have shown, any prestige and authority granted Western Electric employees was not the same as that ascribed to the caricatured individual academic scientist; instead, it was firmly rooted in the rewards of a job with a prominent industrial firm that served the state, the nation, and corporate capitalism all at once.[56] The Bell System's "most valuable resource," a Western Electric executive wrote in a 1956 discussion of the DEW Line, was "trained manpower." When it came to northern work, however, Western Electric technicians required an extra set of lessons pertaining to life in an environment understood as markedly different from their "peaceful state-side towns."[57]

The Laboratory in the (Corn) Field

The extension of the laboratory into the field, in a world "of testing and simulation, experiments and proving grounds," was a critical concern for students of arctic engineering and arctic warfare during the early Cold War.[58] For Western Electric, this concern was addressed in the middle of an actual field: a swath of corn five miles from the town of Streator, Illinois, about 160 kilometres southwest of Chicago.

In 1952, fresh with instructions from the military, Western Electric employees arrived in the cornfield to erect a prototype radar facility, an "inflated 55-foot hemisphere," which was operational by the following year.[59] The company "provided overall management of the planning, design, and construction" of the Illinois "pilot plant," which was built to "ensure smooth and efficient operation when the actual DEW Line construction was undertaken."[60] In Illinois, and at a three-station "test line" also set up in 1953 by Western Electric along the Beaufort Sea between Barter Island, Alaska, and a Canadian location just east of the Alaska border, attacks were simulated using a variety of military and civilian aircraft, along with the insertion of "artificial data" into a rudimentary tracking system.[61] The arctic sites were established to test "DEW Line-like operations … under arctic conditions," while the Streator location "served as both a proving ground for testing experimental prototype equipment, and as a training area for personnel."[62] But it was not clear—in Illinois, on the Beaufort coast, or on the finished Line—where the laboratory ended and the field began. Were the lines drawn back at MIT's Lincoln Laboratory, at the New York headquarters of Western Electric, inside a radar station, at the southern borders of the amorphous region known as the "Arctic," or at the edge of a cybernetic continent? After all, as one *Western Electric Engineer* article put it, the company's technicians shared a "common objective," identifiable in an "esprit de corps … which touched alike the surveyor in the field who was trying to keep his eyelid from freezing to his transit and the engineer working late into the night in a New York office." Was this objective the defence of North America from Soviet attack, or was it the pursuit of "inspired engineering" in the face of "an uncooperative Nature"?[63] The two were not necessarily identical.

Hired by Western Electric in 1953, the Canadian radar engineer William Barrie recalled meeting with a team of psychologists from Northwestern University, who subjected him and other recruits to aptitude tests, questions about solitude and survival, and political conversation (this was the McCarthy era, after all). One Bell System publication described this "processing" by stating that "any of the men would confirm that the armed forces never gave a physical any stiffer than the one they had to pass—in addition to a psychological test to insure their successful coping with rigorous living and isolation."[64] Having passed, Barrie was sent north to Barter Island to "calibrate and evaluate" the equipment of the test line. Instructions and questions arrived daily via teletype from New York City, from Bell Labs, and from other Western Electric plants. Security was so severe that he was "prevented access to my own test results." But he "had all the amenities of an isolated US military base," including a model rail-road and a darkroom.[65]

The Cold War itself was being staged as a series of experiments. In the Illinois cornfield, these were managed by the Western Electric (and later Federal Electric) technicians who were "transported to and from the site each day in a blue military bus with US Air Force markings."[66] Meanwhile, at a Bell Labs site in Whippany, New Jersey, a full "DEW Line Radome" was built to "eliminate the need for many trips to the Arctic by engineers."[67] Even so, this experimental use of new radar technology for testing and training was also changing and circumscribing the arctic environment in important ways.[68] The nuances of northern nature were undoubtedly important for the DEW Line's builders, but only in a secondary sense—as hindrances to be overcome or held at a remove from the technical devices placed in the north for reasons of military exigency.

Siting the Stations

The Streator testing and training centre and the western arctic test sites were vital parts of Western Electric's Project 572, the code name for the preliminary and secretive surveying, planning, and experimental work conducted from late 1952 to 1955 that led directly to the construction of the DEW Line.[69] According to one account, it was "a problem of logistics to the nth degree," requiring the creation of an entirely new "organiza-tion."[70] By early 1953, a "siting engineers" group had been created to over-see the experimental locations and to recommend a suitable route for the future full DEW Line. The qualifications for membership on the small crews dispatched by this group to the Arctic were precise: unmarried male Bell System engineers under thirty-five years old who were war veterans, outdoor enthusiasts, familiar with radio and radar technology, and of the appropriate "disposition" for long hours of labour in a challenging mi-lieu.[71] While these teams "lived and worked under the most primitive con-ditions," they nonetheless approached their duties, and the north itself, with "scientific means."[72]

Meanwhile, the core cluster of siting engineers began to collect geographic knowledge on the Arctic—knowledge that combined high modernist visualization with grounded nuance. While prior surveys and military construction projects had produced a wealth of usable information for the north coast of Alaska, a "similar review of the available data of Arctic Canada did not meet with as much success." Western Electric proceeded to bombard branches of the Canadian government for additional detail, but much of it was deemed inadequate. The exception was the trove of aerial photographs—many of them trimetrogon prints suitable for cartographic use—held at the Royal Canadian Air Force's 22 Photo Wing Headquarters in Rockville, Quebec.[73] These were new images, produced since the end of the Second World War, when northern Canada had become "the major part of the photographic survey program."[74] The resulting synoptic views of arctic space—turned into similarly influential topographic maps—were emblematic of the rich, entangled relationship between aviation, science, governance, and militarization in the post-war north.[75] Although the RCAF's photographs were considered "excellent," and were used (along with soil reports and other environmental studies) to prepare "paper layouts" of stations before actual construction began, at this early stage the images were *too distant* for Western Electric and contained inadequate environmental detail. They were therefore combined with "low-level aerial reconnaissance, and on-the-ground inspections" in the summer of 1953.[76] All of these approaches to arctic geography prioritized "strategic location and topography," fitting individual locations into a logistical network of supply and operations.[77]

In addition to tests of communications and radar devices, Western Electric also experimented with forms of shelter along its western arctic test line in Alaska. The best result, according to an official history, "was the 'module' unit," a flat-roofed building made from panels of prefabricated plywood. Placed together and set on pilings that were driven into the permafrost, these modules formed "trains" that were suitably enclosed and separated from an external environment.[78] Such structures were modest but influential additions to northern architecture in the period after the Second World War, sitting alongside the more dramatic DEW Line radome, also designed precisely to contain its mechanical organs from a hostile exterior. In the 1950s, few structures were more redolent of "industry's power to transform nature" than the radome, but the accompanying "train" was also a hallmark of military modernization in the north (Fig. 8.2).[79]

FIG. 8.2: DEW Line site POW-M, Barrow, Alaska, December 1962. David E. Chesmore Photo Collections, Archives, University of Alaska-Fairbanks, Alaska and Polar Regions Collection, Item 2004-0171-00163). Reprinted with permission.

All of this effort—not to mention many conversations with Arctic "experts" and an extensive literature review—was funnelled into the centre of calculation that was Western Electric's New York City corporate headquarters.[80] By October 1953, it had prepared a "consolidated report" for the US Air Force, asserting that "a feasible route exists for construction of a distant early warning line using … methods and techniques as now proposed by Western Electric Company."[81] Receiving this and other documents, the recently formed bi-national Military Study Group (MSG) of strategists "could not escape the conclusion," according to a June 1954 memorandum, "that there was a need for the establishment of the Canadian Arctic segment of the distant early warning line … a start should be made at once."[82] Similar technical studies were discussed at the October 1954 meeting of the Permanent Joint Board of Defence, where American participants used Western Electric material to confirm that "the necessary data to start work on the sites during the 1955 construction season [is] available."[83] The following month, Western Electric collaborated with a spin-off of the MSG, a Locations Study Group, to settle on a route for the Line, with its eastern end at Baffin Island's Cape Dyer rather than further south at Resolution Island. By the end of 1954, Western Electric had secured the full contract for the DEW Line. "Consummated" in July 1955, it was a "package plan" based on a "cost, plus fixed fee" arrangement, which ranged from design to testing, with a completion date set for mid-1957.[84]

Western Electric had already created a "siting manual" for additional, more place-specific operations across the Arctic in 1955. With the help of Canadian and American aerial photography, surveying, and engineering firms, it quickly prepared detailed location reports for each future station. These texts relied on a variety of confident forms of geographic description and visualization. Each one included a section titled "Acquisition of Lands," and this was tellingly a process rendered in strictly technical and contemporary terms.[85] While Western Electric established a DEW Line Project Office in New York City, contractors quickly finished all of the siting work in Canada, and moved on, in the second half of 1955, to the Alaskan locations.[86]

"Conquered by Degrees"

When Western Electric received the contract to build the DEW Line, one promotional film portrayed the project as a "weapon system."[87] Months later, in one of the first media accounts of the initiative, journalist Leslie Roberts described it as "the first mass assault on the Arctic."[88] Despite the clear defence imperatives and the accompanying propensity for martial terminology, however, the DEW Line's military purpose was simultaneously complemented and superseded by another discourse in the popular press, corporate journals, and government literature: the Line was merely a matter of "technical feasibility," and the targets of "assault" were northern landscapes.[89] According to a typical account in *North*, the bi-monthly magazine produced by Canada's Department of Northern Affairs and National Resources, "trials and related work" revealed that "the practical DEW Line could be built across the Arctic, despite rigorous climatic conditions and difficult supply routes."[90] With construction nearly completed in the spring of 1957, "over 1,000,000 tests were performed to prove out the system to Western Electric's satisfaction."[91] The DEW Line was consistently presented as a logistical matter of overcoming environmental challenges. If political questions entered into this description, they had either been settled beforehand or were simply negated by the realization of technical success.

If success was defined in technical terms, it was nonetheless *made* in improvisational, intimate, and even awkward ways. One Western Electric history suggested that "the biggest threat to the project did not come from the Soviets, but from the forbidding Arctic weather," and described workers wearing "30 pounds of clothing" and awkwardly carrying bulky sleeping bags "whenever going out for a stroll."[92] C. J. Marshall recalled "watching the weary and disgruntled members of a pioneer construction crew laboriously moving their tents, supplies and equipment across the five miles which separated their camp from the site eventually chosen for the station." In such accounts of bringing-forth, stories of "miscalculations made under the pressure of the moment"—and eventually corrected—are legion.[93]

Unsurprisingly, many Bell System publications featured a more triumphant rendering, employing intriguing phrases and figures.[94] Western Electric's slim volume *The DEW Line Story*, distributed in 1958 to members of "the Bell System team responsible for the planning and construction of this history-making project," arrived with an accompanying letter from William Burke, the company's Vice-President for Defense Projects. His gracious note of thanks also nodded to a clandestine, affective affinity: "Only those who have been intimately associated with the project," Burke wrote, "can know the full extent of the difficulties our people overcame, the hardships they endured and the intense effort they applied."[95] *The DEW Line Story* bolstered this sense of environmental adversity, but it also tabulated unprecedented arrivals from the south in the north. "*460,000 tons* of materials were moved from the U.S. and Canada to the Arctic by air, land, and water," the commemorative book exclaimed, as if the terminus was a distinct continent (Fig. 8.3).[96]

Even as it characterized the "north" as a destination *for* labour and material, *The DEW Line Story* stitched arctic places into a continental network whose extraordinary complexity was statistically signalled: 113,000 purchase orders, 818,000 oil drums, 45,000 commercial flights, 22,000 tons of food, and so on.[97] Such tabulation, drawn directly from Western

THE
DEW
LINE
STORY for Mr. G. O. Ekstedt

Western Electric Company

FIG. 8.3: Cover of *The Dew Line Story* (New York: Western Electric Company, 1958).

Electric's reports to the military, suggested smooth progress, an operation of enormous scale, and a northern realm bereft of modern artifacts.[98] All of these elements were amalgamated with the aid of a new "military technology": the science of logistics. It allowed the Arctic to be "conquered by degrees," ruled ultimately by a generic, dispassionate DEW Line station, with its "one eye" and radio "ears" and "voice." This cybernetic entity, the publication implied, was barely northern: it was made and maintained by southern workers with southern tools and intelligence. As if to illustrate this principle, an image in *The DEW Line Story* shows "a pair of Eskimos" ice-fishing "in the shadow" of a generic, forbidding radome.[99]

Writing for a corporate audience in *Bell Telephone Magazine*, William Burke captured the technopolitical version of the Line. "Only through the skill, the scientific knowledge, the superlative mechanical equipment which the latter half of the twentieth century have made available, and the teamwork of many military and civilian agencies of both the United States and Canada, could so stupendous a project be accomplished at all," he boasted.[100] Burke's prose melded human and machine, military and civilian, and American and Canadian to *complete* a "project" whose geographic location was less important than what history had passively "made available." Yet it was ultimately "up on the DEW Line" that all of these factors came together to form a "single, closely integrated team, welded together by the sense of the urgency of the job."[101] For all of the hyperbole and the language of accomplishment, the Cold War Arctic and its identity as a proving ground for military experimentation were never far from view.

The Limits of the Technical Arctic

Western Electric's approach to the DEW Line combined dominance over non-human nature with a displacement of military and corporate power onto radar devices and the equally technical work of erecting and maintain them.[102] Northern communities and the Canadian government are still reckoning with land around DEW sites that was treated "like a vast garbage dump." Photographs of vast numbers of bags filled with contaminated soil, waiting to be removed from Nunavut's Cape Dyer (the location of the DYE-MAIN station) in the summer of 2012, serve as a dramatic reminder of high modernism's deleterious legacies.[103] It is too simplistic to ascribe this recklessness to the "remoteness" of the north without first understanding how such distancing was manufactured, not just from a southern location such as Streator, but in northern locations, as well. Alongside the detailed and yet disconnected site reports, or the consistent language of conquest, popular documentation on Western Electric's DEW Line efforts placed Indigenous northerners precariously at the edges or completely outside of the technical geography of the Arctic. They are strikingly absent from many of the studies that prepared the ground for the implementation of the Line's infrastructure. At best, Indigenous people are treated as a "small, but valuable reservoir of labor which is acclimated to the Arctic Zone," "baffled at first by modern machines and construction methods ... [but] quick to catch on."[104]

The social and cultural context into which Western Electric employees entered was deliberately conditioned by southern expectations and experiences. To limit isolation and improve comfort, new arrivals to DEW Line stations received copies of military manuals on "polar operations"— handbooks that drew extensively from the conclusions of recent military exercises, and, in tellingly circumscribed and anecdotal ways, from a certain form of local knowledge.[105] A 1954 *Western Electric Magazine* article cheerfully titled "Next Door to S. Claus" noted that "W.E. Arctic dwellers ... benefit from the experiences of the Armed Forces ... to say nothing of the Eskimos who know from practical, if not scientific, experience, how to dress properly in the cold."[106] Reflecting a common (if crude) view shared by some Canadian northern administrators, C. J. Marshall contrasted the "precarious existence of the hunter and trapper" with the "life of routine and financial security" on the Line.[107]

While many of the Arctic's Indigenous communities were undoubtedly altered by the "routines" of Western Electric's technopolitical exercise, this process was not one of unidirectional assimilation. Project 572 arrived on Barter Island, in the northeast corner of Alaska, in the winter of 1952–53, and quickly moved east across the Canadian border. This was a startling development for the Inuvialuit people of the area. As David Neufeld writes, with a nod to oral histories conducted on the Yukon's North Slope, "The lack of local consultation and the consequent failure to attend to issues of local importance affected people deeply." Still, the proximity to DEW Line construction meant that complicated "social and economic connections developed." Many families eventually congregated close to DEW Line sites, where one member (the adult male) might gain work, and where other informal relationships were forged. In sum, according to Neufeld, "The project initiated a host of changes in Inuvialuit lifestyle and activities," and other government and corporate initiatives across the North American Arctic paralleled Western Electric's arrival. The Canadian government recruited and sent Northern Service Officers northward to manage relationships between Indigenous residents and new arrivals. For all of their individual efforts, the NSO position came with a paradoxical combination of patronization and powerlessness in the face of larger environmental and social changes.[108]

At Cambridge Bay in the central Arctic, where a large DEW Line station (CAM-MAIN) was under construction in 1956, the young social scientist J. D. Ferguson, on assignment for the Canadian government's Northern Research Coordination Centre, perceived that the Inuit socio-economic hierarchy had been completely upended.[109] Residents were looking to the contractors building the station—who were distributing water, delivering mail and other items free of charge, offering casual employment, and offering their equipment to help with other construction—as the proximate "seat of authority," even as the general superintendent claimed that he did not want to get "mixed up" in community politics. (This person's professed interest, presumably, was only technical.) Although Ferguson invoked a familiar frontier motif to characterize what he witnessed, like a number of scholars travelling across the Arctic in the same period he could not check his concern with the damaging effects of "economic laissez-faire" and an "emphasis on individual enterprise at the expense of community."[110] His hope that the DEW Line would bring stability, once northerners had adjusted to its presence, was a rather lofty one. The result, instead, was a subset of Indigenous people who had received some training, but were not granted full-time, lasting work.[111]

Conclusion: "Electronic Outposts" on a Cold War Frontier

In an account of Hilton International Hotels and the building of the Cold War, Annabel Wharton shows how these monumental structures were explicitly positioned as symbols and repositories of American culture, detached from but clearly intended to modernize foreign societies through the elites, national and foreign, who accessed and invested in such projects. As Conrad Hilton himself put it in his 1958 autobiography, "we, as a nation, must exercise our great strength and power for good against evil. If we really believe this, it is up to each of us, our organizations and our industries, to contribute to this objective with all the resources at our command."[112]

The comforts of the DEW Line facilities were not quite equivalent to those of Hilton's hotels. But radar sites were similarly representative of a normalizing culture—in domestic and community spaces, but also because it was "a thrill to fly from one side of the continent to the other and hardly be out of sight of friendly lights."[113] Alongside Conrad Hilton, we might consider H. G. Ross, the DEW Line project manager for Western Electric. The megaproject, Ross said on the occasion of the 1957 handover to the US Air Force, was a "full-scale attack on the Arctic unparalleled in military history," which "broke all the rules of the book and made a frontier which, for many years to come, will play a major role in keeping alive a way of life which makes the DEW Line worth every effort put into it."[114] This convoluted sentence captures the northward extension of the state, both geographically and historically. And the technicians who "built these electronic outposts in the unexplored Arctic" carried out much of this extension.[115] But when the universality of this technopolitical language, and the "way of life" that it conveniently secured, is placed in proper perspective, we are left with another set of uncertain, complicated, *lived* histories that are all the more consequential for their modesty.[116]

When planning the DEW Line project, military and corporate employees reimagined the Arctic as a functional realm, placing their unquestioned faith in the transformative powers of modern technology to reconfigure or simply overcome a "hostile" nature. To become part of the continental defence web, the Arctic was in one crucial sense rendered analogous to other locations—from laboratories in New Jersey to cornfields in Illinois. The result was that the north became a technical space upon which certain designs and assumptions were inscribed. Realizing these aspirations also required the production and transmission southward of details derived from arctic site visits, yielding new forms of usable geographical knowledge. And yet, as they transcended and blurred national, regional, and local scales, planners, engineers, logisticians, and technicians also implicitly muddied the boundaries between north and south, as they devised and implemented practices that would overcome the particularities of climate and terrain. The DEW Line embodied the hubris of technopolitics and the assimilation of the remotest reaches of North America into a truly *continental* environment: one of military science and state control. This environment was not equivalent to "the Arctic"—however we define that term. Nor did it entirely supersede it. Our suggestion, instead, is that the two are indivisible.

Notes

1   Ann R. Markusen, et al., *The Rise of the Gunbelt: The Military Remapping of Industrial America* (New York: Oxford University Press, 1991).

2   *Illinois Technograph* 75, no. 1 (October 1959): 37 (original emphasis). See the similar advertisement in the *Wisconsin Alumnus* 60, no. 10 (February 1959): 36. In 1949, Western Electric was also asked to manage the Sandia Laboratory complex in New Mexico—the principal site for the development of American nuclear weapons after Second World War. Stephen B. Adams and Orville R. Butler, *Manufacturing the Future: A History of Western Electric* (Cambridge, UK: Cambridge University Press, 1999), 149.

3   "Contractor Chosen for Radar in Arctic," *New York Times*, 22 February 1955, 19.

4   *Illinois Technograph* 75, no. 3 (December 1959): 27.

5   *Illinois Technograph* 75, no. 5 (February 1960): 35.

6   The number is from "Western Electric: A Brief History," www.porticus.org/bell/doc/western_electric.doc (accessed 1 April 2011), 12. The bureaucratic history of the DEW Line is dizzying. Western Electric drew on multiple components from within the Bell System, including its Canadian affiliate, the Northern Electric Company, which at the time of construction was both separating from and still closely tied to AT&T. The major Canadian construction subcontractors were Northern Construction Ltd. and James W. Stewart Ltd., both in the west, and the Foundation Company of Canada, responsible for the eastern section of the Line. Once completed in 1957, the Line was staffed by yet another American military contractor, the Federal Electric Corporation. See *The DEW System* (Paramus, NJ: Federal Electric Corporation, n.d.).

7   On the Cold War origins of "geographical engineering," see Scott Kirsch, *Proving Grounds: Project Plowshare and the Unrealized Dream of Nuclear Earthmoving* (New Brunswick, NJ: Rutgers University Press, 2005).

8   For an example of news coverage, see "Northern DEW Line cleanup bill hits $500M," *CBC News*, 8 August 2011 (http://www.cbc.ca/news/technology/story/2011/08/08/north-dew-line-bases.html). The

clean-up project has been led by Defence Construction Canada, a Crown corporation whose sole client is the Department of National Defence. Ironically, the construction of the Line was often estimated to cost a similar amount—in 1950s American dollars.

9   For more on the physical transformations wrought by northern militarization, see P. Whitney Lackenbauer and Matthew Farish, "The Cold War on Canadian Soil: Militarizing a Northern Environment," *Environmental History* 12, no. 4 (2007): 921–50.

10 The advertisement, headlined "New Radar Sky-Watch to Guard Arctic Frontier," is in the 22 August 1955 issue of *Life* (p. 11).

11 In addition to the sources cited below, see collective histories such as *Inuit Recollections on the Military Presence in Iqaluit*, http://www.tradition-orale.ca/english/pdf/Inuit-Recollections-On-The-Military-Presence-In-Iqaluit-E.pdf (accessed 25 August 2012); *Paulatuuq Oral History Project: Inuvialuit Elders Share Their Stories* (Inuvik, NT: Parks Canada Western Field Unit, 2004); and Murielle I. Nagy, *Yukon North Slope Inuvialuit Oral History* (Whitehorse, Heritage Branch, Government of the Yukon, 1994). Also see individual oral histories, such as Abraham Okpik, *We Call it Survival: The Life Story of Abraham Okpik*, ed. Louis Mc-Comber (Iqaluit: Nunavut Arctic College, 2005); the work of Parks Canada's David Neufeld, such as "Commemorating the Cold War in Canada: Considering the DEW Line," *Public Historian* 20, no. 1 (1998): 9–19; and media accounts, such as Sandro Contenta, "DEW

Line: Canada is cleaning up pollution caused by Cold War radar stations in the Arctic," *Toronto Star*, 4 August 2012, http://www.thestar.com/news/insight/article/1236806--dew-line-canada-is-cleaning-up-pollution-caused-by-cold-war-radar-stations-in-the-arctic.

12 See Derek Gregory, "Imaginative Geographies," *Progress in Human Geography* 19 (1995): 447–85.

13 Edward Jones-Imhotep, "Nature, Technology, and Nation," *Journal of Canadian Studies* 38, no. 3 (2004): 7.

14 The first quote is from M. S. Cheever and J. D. Brannian, "Building the DEW Line," *Engineering and Contract Record* 706 (1957): 80; the second is from C. J. Marshall, "North America's Distant Early Warning Line," *Geographical Magazine* 29, no. 12 (1957): 616–17. A different version of the wilderness motif was staged in Lois Crisler's *Arctic Wild* (New York: Harper and Brothers, 1958): "The DEW Line, obsolete while on paper, would keep rolling, wiping out wild habitat and animals in the biggest, best-armed invasion of this fragile life zone, the Arctic, ever performed" (64).

15 Dave Eagles, from Defence Construction Canada, quoted in the short film *Undoing the DEW* (Environment Canada, 2009).

16 Pierre Berton, *The Mysterious North* (New York: Knopf, 1956); Matthew Farish, *The Contours of America's Cold War* (Minneapolis: University of Minnesota Press, 2010), 173–89.

17 This point is made well in Robert McGhee, *The Last Imaginary Place: A Human History of the Arctic*

World (Chicago: University of Chicago Press, 2007).

18   Herbert O. Johansen, "World's Toughest Building Project," *Popular Science*, August 1956, 86.

19   Marshall, "North America's Distant Early Warning Line," 616–17. Marshall was listed as an officer in Canada's Department of Northern Affairs and Natural Resources who "represented Canada on the DEW Line Site Survey Team in 1955 and has since visited the line several times" (616).

20   James C. Scott, *Seeing Like a State: How Certain Schemes to Improve the Human Condition Have Failed* (New Haven: Yale University Press, 1998), 88, 94–95. For a more detailed discussion of high modernism in a northern context, including some of the limitations of Scott's formulation, see Matthew Farish and P. Whitney Lackenbauer, "High Modernism in the North: Planning Frobisher Bay and Inuvik," *Journal of Historical Geography* 35, no. 3 (2009): 517–44.

21   MIT President James Killian, quoted in "National Safety and the Universities," *Technology Review* 56, no. 7 (1954): 357; Farish, *The Contours of America's Cold War*, 153–54.

22   Important examples include Benjamin Forest and Patrick Forest, "Engineering the North American Waterscape: The High Modernist Mapping of Continental Water Transfer Projects," *Political Geography* 31 (2012): 167–83; Kirsch, *Proving Grounds*; Allen Isaacman and Chris Sneddon, "Toward a Social and Environmental History of the Building of Cahora Bassa Dam," *Journal of Southern African Studies* 26, no. 4 (2000): 597–632; Daniel

MacFarlane, *Negotiating a River: Canada, the US, and the Creation of the St. Lawrence Seaway* (Vancouver: UBC Press, 2014). For a useful reminder that not all megaprojects were built, and that the resulting "unbuilt environments" have also been consequential, see Jonathan Peyton, "Corporate Ecology: BC Hydro's Stikine-Iskut Project and the Unbuilt Environment," *Journal of Historical Geography* 37 (2011): 358–69.

23   Tina Loo with Meg Stanley, "An Environmental History of Progress: Damming the Peace and Columbia Rivers," *Canadian Historical Review* 92, no. 3 (2011): 402.

24   See Patrick Joyce, "What Is the Social in Social History?", *Past and Present* 206 (2010): 213–48.

25   Loo, with Stanley, "An Environmental History of Progress," 406–08 (including Shapin). While the distinction between planners and engineers is a useful one, it does not seem necessary, historically speaking, to make it too sharply.

26   Joyce, "What is the Social in Social History?," 243.

27   Forest and Forest, "Engineering the North American Waterscape," 169.

28   Forest and Forest, "Engineering the North American Waterscape," 169; Farish and Lackenbauer, "High Modernism in the North."

29   C. C. Duncan, "Communications and Defense," *Bell Telephone Magazine*, Spring 1958, 15.

30   "Continental Defense," *WE Magazine*, July/August 1955, 27.

31   Steven Shapin, *The Scientific Life: A Moral History of a Late Modern Vocation* (Chicago: University of Chicago Press, 2008), 80 (original emphasis).

32 Adams and Butler, *Manufacturing the Future*, 150.

33 A more thorough answer would begin in an earlier age of American militarization. Western Electric's history is twinned to that of the telegraph—the technology, along with the railroad, that was crucial to rapacious national expansion and what the geographer Cole Harris calls the "struggle with distance" in nineteenth century North America. According to David Noble, by the early twentieth century, "the large corporations of the electrical industry sought to dominate not only markets for their products but the manufacture of those products as well," through strategies of "industrial warfare." This language is appropriate: during the First World War, Western Electric employees and other representatives of the National Association of Corporate Schools supervised and advised the Army on training and rating of soldiers "according to their industrial abilities." After the war, the same group devised plans "for the reinfiltration of the soldiers into industry." Harris, "The Struggle with Distance," in his *The Resettlement of British Columbia: Essays on Colonialism and Geographical Change* (Vancouver: UBC Press, 1996), 161–93; Noble, *America by Design: Science, Technology, and the Rise of Corporate Capitalism* (Oxford University Press, 1977), 95, 208.

34 "Western Electric: A Brief History," 10.

35 John L. Rudolph, *Scientists in the Classroom: The Cold War Reconstruction of American Science Education* (New York: Palgrave, 2002), 85–86.

36 Quoted in Kent C. Redmond and Thomas M. Smith, *From Whirlwind to MITRE: The R&D Story of the SAGE Air Defense Computer* (Cambridge, MA: MIT Press, 2000), 298.

37 Gabrielle Hecht, "Introduction," in *Entangled Geographies: Empire and Technopolitics in the Global Cold War*, ed. Gabrielle Hecht (Cambridge, MA: MIT Press, 2011), 3.

38 Timothy Mitchell, *Rule of Experts: Egypt, Techno-politics, Modernity* (Berkeley: University of California Press, 2002), 42–43.

39 Farish and Lackenbauer, "High Modernism in the North"; William J. Rankin, "Infrastructure and the International Governance of Economic Development," in *Internationalization of Infrastructures: Proceedings of the 12th Annual International Conference on the Economics of Infrastructure*, ed. Jean-François Auger, et al. (Delft, Netherlands: Delft University of Technology, 2009), 61–75.

40 Mitchell, *Rule of Experts*, 43; see also Farish and Lackenbauer, "High Modernism in the North."

41 Hecht, "Introduction," 3.

42 Rudolph, *Scientists in the Classroom*, 86.

43 Scott Kirsch, "Laboratory/Observatory," in *SAGE Handbook of Geographical Knowledge*, ed. John A. Agnew and David N. Livingstone (Thousand Oaks, CA: SAGE, 2011): 80.

44 For one example, see Jim Lotz, "The North as a Laboratory," in *People of Light and Dark*, ed. Maja Van Steensel (Ottawa: Department of Indian Affairs and Northern Development, 1966), 68–71.

45 The quotes are from "Arctic Sentinels: Building Rushed on Radar

Defense," *Universal International News*, n.d., http://www.youtube.com/watch?v=cg4phLghz2U& (accessed 17 August 2011); and Western Electric, *Arctic Mission: A Report on Dew Line Activities* (1955), http://www.youtube.com/watch?v=nMZl6Xlm6ak (accessed 17 August 2011).

46 *Building the DEW Line* (New York: Audio Productions Inc., n.d.). A copy is in the Yukon Archives, Whitehorse (V-357).

47 Michel Foucault, "Space, Knowledge and Power," in *Power: Essential Works of Foucault, 1954–1984, Volume 3*, ed. James D. Faubion (New York: New Press, 2000): 364; see also Martin Heidegger, "The Question Concerning Technology," in Heidegger, *Basic Writings*, ed. David Farrell (New York: Harper and Row, 1977), 287–317.

48 AT&T Archives, Warren, NJ (hereafter cited as AT&T), Corporate Collection, Box 139-11-03, "DEW Line: Engineering in the Arctic," 1 (our emphasis).

49 On the multitudinous geographies of the American military-industrial complex, see the special issue of *Antipode* 43, no. 3 (2011).

50 Edward A. Ackerman, "Geographic Training, Wartime Research, and Professional Objectives," *Annals of the Association of American Geographers* 35, no. 4 (1945), 121–43.

51 On "vertically structured research organizations," particularly in astronomy, and the industrialization of science before Second World War, see Kirsch, "Laboratory/Observatory," 80.

52 See Stuart W. Leslie, *The Cold War and American Science: The Military-Industrial-Academic Complex at MIT and Stanford* (New York: Columbia University Press, 1993).

53 Bruce E. Seely, "The Other Re-engineering of Engineering Education, 1900–1965," *Journal of Engineering Education* 88, no. 3 (1999): 289.

54 David Kaiser, "Cold War Requisitions, Scientific Manpower, and the Production of American Physicists after World War II," *Historical Studies in the Physical and Biological Sciences* 33, no. 1 (2002), 131–59.

55 David Kaiser, "The Postwar Suburbanization of American Physics," *American Quarterly* 56, no. 4 (2044): 858. This was additionally a period when the demand for "applied science" was drawing engineering and physics much closer together in the United States. According to Seely, the University of Illinois's physics department was in the College of Engineering, and the university offered a degree in engineering physics during the 1940s. Seely, "The Other Re-engineering of Engineering Education," 290.

56 Shapin, *The Scientific Life*; see also Rudolph, *Scientists in the Classroom*, 85.

57 W. E. Burke, "The DEW Line: Sentinel of the Northland," *Bell Telephone Magazine*, Summer 1956: 73.

58 Kirsch, "Laboratory/Observatory," 77.

59 "Our People in the Far North," *The Reporter*, October 1955, 4. "Cooperation of the local farmers was not of the best, since security regulations precluded advising them of the true nature of the project and its military importance." J. D. Brannian, "Siting the DEW Line

Radar Stations," *Engineering and Contract Record* 70, no. 7 (1957): 55.

60 See the "Findings of Fact" document for the "USAF DEW Line Training Station," 5 June 2007, at www.radomes.org/museum/documents/Seward%20-%20E05IL334300_01.10_0001_a.pdf (accessed 5 April 2011).

61 AT&T, Radar Collection, box 172 06 02, folder 2, "Projects 572 and 540: Dew Line Engineering Report," 1 October 1957, 58. See also AT&T, Corporate Collection, box 139 11 03, "Dew Line: Engineering in the Arctic," 1. The Parks Canada historian, David Neufeld, who has carefully documented the history of the DEW Line in the Yukon, notes that from Barter Island, "the components for one station," originally pieced together and then disassembled outside of Streator, "were loaded on cat trains and hauled to the Canadian site through the spring of 1953." The experimental stations "were in operation by early 1954." Neufeld, *The Distant Early Warning (DEW) Line: A Preliminary Assessment of its Role and Effects upon Northern Canada* (revised for the Arctic Institute of North America, 2002), www.stankievech.net/projects/DEW/BAR-1/bin/Neufeld_DEW-Linehistory.pdf (accessed 17 September 2011).

62 Thomas Ray, *A History of the DEW Line*, Air Defense Command Historical Study 31 (Maxwell Air Force Base, AL: Air Force Historical Research Agency, 1965), 13.

63 James D. Brannian, et al., "W. E. Engineering for the DEW Line—I: Siting and Construction," *Western Electric Engineer* 1, no. 3 (1957): 3. These comments are from the editorial note accompanying this article.

64 "The DEW Line," *195 Bulletin*, November 1953, 7–8.

65 William Barrie, "Project 572 and WE8GY," 19 February 2008, http://qcwa70.org/PROJECT%20572%20&%20VE8GY.ppt.ppt (accessed 30 March 2011).

66 Brian (Simon) Jeffrey, "Adventures from the Coldest Part of the Cold War: A DEW Liner's Memoirs, 1960–63," http://www.dewlineadventures.com/wp-content/uploads/2013/02/Cold-War-Adventures-Rev2.5.pdf (accessed 13 June 2016), 11. For a fictional treatment of the Federal Electric years at Streator and on the DEW Line, see Charles Flynn, *Green Grass Fever* (London: Minerva Press, 2001). Contracted to train at Streator by the end of April 1956, Federal Electric personnel began arriving on the Line in October of that year, assisting "Western Electric installation crews in finalizing and check out of the communications and electronics equipment." Lynden T. (Bucky) Harris, "The DEWLINE Chronicles: A History," http://lswilson.dewlineadventures.com/dewhist-a.htm (accessed 13 June 2016); Ray, *A History of the DEW Line*, 33.

67 "Dew Line 'Radome' Constructed at Whippany Laboratory," *Bell Laboratories Record* 34, no. 8 (August 1956): 308.

68 See Nick Cullather, "Miracles of Modernization: The Green Revolution and the Apotheosis of Technology," *Diplomatic History* 28, no. 2 (2004): 227–54.

69 While Western Electric staff used "Project 572," the US Air Force referred to the Project as

COUNTERCHANGE, and later
CORRODE.

70   "Dew Line: Engineering in the
      Arctic," 2.

71   Brannian, "Siting the DEW Line
      Radar Stations," 54–55.

72   *The DEW Line Story*, 6.

73   Brannian, "Siting the DEW Line
      Radar Stations," 173–75.

74   R. I. Thomas, "Photographic
      Operations of the Royal Canadian
      Air Force," *Arctic* 3, no. 3 (1950),
      150–65. See also Moira Dunbar
      and Keith R. Greenaway, *Arctic
      Canada from the Air* (Ottawa:
      Defence Research Board, 1956).

75   See Stephen Bocking, "A Dis-
      ciplined Geography: Aviation,
      Science, and the Cold War in
      Northern Canada, 1945–1960,"
      Technology and Culture 50, no. 2
      (2009), 26590.

76   Brannian, "Siting the DEW Line
      Radar Stations," 175, 197.

77   Ray, *A History of the DEW Line*, 16.

78   Ray, *A History of the DEW Line*,
      14–15.

79   Alex Soojung-Kim Pang, "Dome
      Days: Buckminster Fuller in the
      Cold War," in *Cultural Babbage:
      Technology, Time and Invention*,
      ed. Francis Spufford and Jenny
      Uglow (London: Faber and Faber,
      1996): 168.

80   The oft-used phrase "centre of
      calculation" is Bruno Latour's; see
      his *Science in Action: How to Follow
      Scientists and Engineers through
      Society* (Cambridge, MA: Harvard
      University Press, 1987).

81   Latour, *Science in Action*, 175–76;
      Directorate of History and Her-
      itage, Ottawa, file 79/87, *Project
      Corrode: Report on Summer
      Survey of Arctic Routes* (New York:

Radio Division, Western Electric
Company, 16 October 1953), 3. See
also Neufeld, "The Distant Early
Warning (DEW) Line."

82   *Documents on Canadian Exter-
      nal Relations*, vol. 20-462 (4 June
      1954), http://epe.lac-bac.gc.
      ca/100/206/301/faitc-aecic/histo-
      ry/2013-05-03/www.international.
      gc.ca/department/history-histoire/
      dcer/details-en.asp@intRefid=635
      (accessed 13 June 2016).

83   *Documents on Canadian External
      Relations*, vol. 20-482 (12 Novem-
      ber 1954), http://epe.lac-bac.gc.
      ca/100/206/301/faitc-aecic/histo-
      ry/2013-05-03/www.international.
      gc.ca/department/history-histoire/
      dcer/details-en.asp@intRefid=656
      (accessed 13 June 2016). This
      document is an extract from the
      minutes of Canada's Cabinet
      Defence Committee, and includes
      a report from the recent Permanent
      Joint Board on Defence meeting.

84   Ray, *A History of the DEW Line*,
      20–21. "Cost, plus fixed fee" is from
      "Contractor Chosen for Radar
      in Arctic." Resolution Island had
      been proposed as the terminus by
      Western Electric.

85   For an example, see Air Force
      Historical Research Agency,
      Maxwell Air Force Base, AL, file
      K243.0482-1, Western Electric
      Company, *Siting Report, Distant
      Early Warning Line Station Bar-B*
      (September 1955).

86   Ray, *A History of the DEW Line*, 21.

87   *Building the DEW Line*.

88   Leslie Roberts, "The Great Assault
      on the Arctic," *Harper's*, August
      1955, 37.

89   W. E. Burke, "The Arctic Distant
      Warning System," paper presented
      at the 26th Annual Meeting of

the Institute of the Aeronautical Sciences, 27–30 January 1958, 3. A copy of this document is in the library of Air University, Maxwell Air Force Base, AL.

90 R. B. Wybou, "The Dew Line," *North* 7.4-5 (1960), 11.

91 Ray, *A History of the DEW Line*, 34.

92 "Western Electric: A Brief History," 12.

93 Marshall, "North America's Distant Early Warning Line," 623, 628.

94 While helpfully detailed, many of the publications authored by Western Electric employees are also strikingly similar and anodyne, tales of inexorable triumph over an environmental foe. See, for instance, V. B. Bagnall, "Operation DEW Line," *Journal of the Franklin Institute* 259, no. 6 (1955): 481–90; Bagnall, "Building the Distant Early Warning Line," *Military Engineer* 47, no. 320 (1955): 429–32; Brannian et al., "W.E. Engineering for the DEW Line"; and the three part-series in the *Engineering and Contract Record* (June, July, and August 1957 issues).

95 W. E. Burke to G. O. Ekstedt, 1 May 1958, letter attached to a copy of *The DEW Line Story* (New York: Western Electric Company, 1958), at http://www.beatriceco.com/bti/porticus/bell/pdf/dewline.pdf (accessed 21 January 2016). Burke's comments were contradicted in the attached book, but in an equally intriguing manner: "The area along the DEW Line may be desolate, but it is steeped in the history of Arctic exploration" (8).

96 *The DEW Line Story*, 6 (original emphasis).

97 *The DEW Line Story*, 24–25.

98 See Ray, *A History of the DEW Line*, 31–32. As this formerly secret document notes, there were also twenty-five fatalities from aircraft accidents during DEW Line construction in 1955 and 1956 (32).

99 *The DEW Line Story*, 19, 6.

100 Burke, "The DEW Line: Sentinel of the Northland," 69.

101 Burke, "The DEW Line: Sentinel of the Northland," 83.

102 Joyce, "What is the Social in Social History?", 245.

103 Contenta, "Dew Line."

104 *Project Corrode: Report on Summer Survey*, 8–9; *The DEW Line Story*, 12. This treatment of Indigenous northerners is certainly evident in the Western Electric-affiliated publications cited in note 94.

105 For example, Harold A. Edgerton, *Personnel Factors in Polar Operations* (Washington, DC: Office of Naval Research, 1953).

106 "Next Door to S. Claus," *WE Magazine*, November/December 1954, 47.

107 Marshall, "North America's Distant Early Warning Line," 628.

108 Neufeld, "The Distant Early Warning (DEW) Line." See also Burke, "The Arctic Distant Warning System," 19–20; and the 1955 report of Graham W. Rowley, representing the Canadian government on the US Navy's sea supply of the DEW Line's western sector, quoted in Yukon Archives, box GOV2858, file 4, David Neufeld, "History and Operation of Station BAR-1" (draft manuscript). Writers such as C. J. Marshall ("North America's Distant Early Warning Line," 627) hinted at a triangular relationship between DEW Line employment, increased Indigenous populations,

and declining "game resources." This is a claim worth pursuing (and challenging) further, given the documented concerns around increased hunting that accompanied the arrival of DEW Line workers, and the relationship between modernization, relocation, and conservation during the 1950s. See Peter Kulchyski and Frank James Tester, *Kiumajut (Talking Back): Game Management and Inuit Rights, 1900–70* (Vancouver: UBC Press, 2007), 105–9; John Sandlos, *Hunters at the Margin: Native People and Wildlife Conservation in the Northwest Territories* (Vancouver: UBC Press, 2007), 239.

109 The six MAIN stations were "where the most equipment and largest contingent of personnel were positioned for round-the-clock manning." They were "focal points for the operation, administration, maintenance and communication of the entire DEW Line," receiving data from smaller sites. Ray, *A History of the DEW Line*, 25.

110 J. D. Ferguson, *A Study of the Effects of the Distant Early Warning Line Upon the Eskimo of the Western Arctic of Canada* (Ottawa: Northern Research Coordination Centre, Department of Northern Affairs and National Resources, 1957), 26–27, 43; on social science and modernization, see Farish and Lackenbauer, "High Modernism in the North."

111 For more details and alternate perspectives on Inuit work at DEW Line sites, see Maxime Steve Begin, *Des Radars et Des Hommes: Memoires Inuit de la Station Fox Main de la DEW Line (Hall Beach, Nunavut)* (master's thesis, Université Laval, 2004). Sizing up the

predicament in his bluntly titled 1964 survey *Eskimo Administration*, anthropologist Diamond Jenness recommended that the military should play a more interventionist social role. Diamond Jenness, *Eskimo Administration II: Canada*, Technical Paper 14 (Montreal: Arctic Institute of North America, 1964), 175, 183. Given the documented environmental and social legacies of the DEW Line, this suggestion seems troubling in retrospect. On military perceptions of social intervention, see P. Whitney Lackenbauer and Ryan Shackleton, "Inuit-Air Force Relations in the Qikiqtani Region during the Early Cold War," in *De-Icing Required: The Canadian Air Force's Experience in the Arctic*, ed. P. W. Lackenbauer and W. A. March (Trenton, ON: Canadian Forces Air Warfare Centre, 2012), 73–94.

112 Quoted in Annabel Jane Wharton, *Building the Cold War: Hilton International Hotels and Modern Architecture* (Chicago: University of Chicago Press, 2001), 8.

113 Burke, "The Arctic Distant Warning System, 19.

114 "First DEW Line Sites Turned over to Air Force," *Bell Laboratories Record* 35, no. 5 (May 1957): 194.

115 W. E. Burke to G. O. Ekstedt, 1 May 1958.

116 See Emilie Cameron, "New Geographies of Story and Storytelling," *Progress in Human Geography* 36, no. 5 (2012): 573–92; Mona Domosh, "American Capitalist Experiments in Revolutionary-era Russia," *Journal of Historical Geography* 39 (2013): 43–53.

PART 3

*Environmental History and
the Contemporary North*

.

9

"That's the Place Where I Was Born": History, Narrative Ecology, and Politics in Canada's North

Hans M. Carlson

History, Meaning, and Place

I'm sitting in camp this evening, here on the southernmost lands of the James Bay Cree, watching the dragonflies diving and feeding on the mosquitoes that are getting thicker as the light fades. A Canada jay is calling out into the dusk—*wiishkachaanish* in his forest—and the light is fading behind the black spruce across this wide section of the Brock River, here in Eeyou Istchee, the People's Land. There are few more evocative images of the Canadian north than this dark, boreal skyline of thin black spruce spires and witches-brooms. For so many, they signify isolation and wilderness, yet, even with the deep, settling quiet tonight, my thoughts are at variance with any perception of the "lonely land" here in the north.[1]

Our campsite is less than twenty kilometres west of the Quebec mining town of Chibougamau and northeast of the Cree village of

Ouje Bougoumou—"the place where the people gather." The name Chibougamau is likely a mispronunciation of Ouje Bougoumou, whose people once inhabited the land around Lac Doré where the French town is now located. Their displacement happened in the 1940s and 1950s with the development of copper mines around the lake, and, though the mines are closed now, Chibougamau is still a good-sized town. Lumber mills became the economic base in the 1980s, though old mines may find new life, along with newer operations, thanks to the recent boom in mineral development here on the Canadian Shield—a subject that Arn Keeling and John Sandlos examine in their chapter in this volume. There's a yard, at the south end of Chibougamau, full of thousands of rock core samples. Helicopters fly in two or three times a day to add to the collection, and I've been shown dozens of places where smaller prospectors have used the network of logging roads to bring in equipment and take their samples. This is the next big push up here.[2]

So, as peaceful and remote as this scene seems tonight, this place is anything but isolated, for the weight of events is heavy on this land. The early mines around Chibougamau had local impact, but Quebec has now been exploiting resources region-wide for nearly four decades. Since Hydro-Québec and then-premier Robert Bourassa began developing hydroelectric power here, there has been an uncomfortable dissonance between the solitude and peacefulness of places like this campsite and the momentum of political and economic forces working just beyond that line of trees. The Cree people that I am travelling with, and this forest, have been deeply affected by those forces, so this is an important part of narrating environmental history and issues of justice in this place. Yet the "history" happening beyond the trees is really only part of the story. This was brought home to me, earlier this afternoon, at the other end of this portage trail where we are camped.

There is another old campsite up there, and when we landed, Solomon, one of the three Cree men I'm travelling with, gestured toward a long-used tent site and rather casually announced, "that's the place where I was born." It was an almost offhand remark, but of course it struck me—how could it not? Solomon knows enough about my culture to know that this fact was likely to be evocative and unusual in my experience—thus the wry smile and twinkle in his eye, I think—so while it was done with humour, he did mean it to move me and make me think. These men all have a clear sense of who ought to have rights on this land. They are not shy about saying what they think about the wealth that has been taken, but theirs is more than just a claim to resources or real estate—or even a claim to political sovereignty. I took Solomon's statement as an intellectual challenge to think more holistically about this place, its history, and the current situation here, and to try to understand how it reaches out beyond this forest to engage that story of development and change. That is what I am focused on tonight, as the dark settles over this northern land.

The Cree's is not a claim to land only, but to history and meaning in place. This is clearest when considering things on the ground, because history is so much closer than you think in the boreal north. History and "place" are not really separable at ground level, in fact. On the one hand, I mean that metaphorically, in that disruption here has its analogies in the Native histories of other places and other times, but I mean it more literally, too. I was trained to think chronologically—diachronically—to see the past as receding from the present, but as I've become more and more focused on the meaning of place in my thinking—Eeyou Istchee as a place, that tent site as a place—it has been getting harder not to see time cycling, as well as advancing. It's all still right here, in some sense, cycling in the stories on this land.

We miss this present-ness of events because of a problematic cultural blind spot, for we would say that Solomon's birth "took place" at that site sixty years ago. This is accurate in a way, yet we have to be careful of those easy turns of phrase. As Kiowa author N. Scott Momaday tells us, stories and events like this one quite literally "take place"; they take possession as they attain personal and historical meaning, and this does not pass with time.[3] When Solomon made a point of showing me his birthplace, it was a historical narration, because in those few words, delivered on that spot, there was a storyline. It began in a forest bush camp, in a region controlled almost exclusively by the traditional stewards of Cree lands—the Cree word is *Kaanoowapmaakin* (pronounced *Gah*-new-whap-*mah*-gan). It ended this afternoon, in the same place, but within a landscape massively altered by flooding and cutting. It was a personal history of a past event, but one connected intimately with present ones, and so I would say that it was a political statement as well as a history.

Solomon's intellectual challenge to me was not purely academic, then, for his story is still "taking place" there, as he returns and tells it within this changed landscape. In his words, I can't help but hear an echo of novelist Thomas King's comment concerning that kind of narration: "it's yours [now]. Do with it what you will. Tell it to friends. Turn it into a television movie. Forget it. But don't say in the years to come that you would have lived your life differently [or written history differently] if only you had heard this story. You've heard it now." Stories like Solomon's, in other words, come with responsibility. This is fundamental to the history of this land and its people, particularly if you try to understand the power of these places and the forces that have shaped them. It's a matter of doing justice—or some semblance of it—by acknowledging that, as powerful as outside forces may be, this land and its history are still defined from within.[4]

History is still "taking place." Solomon's connection is one small aspect of that, but while it may seem locally meaningful, when we start thinking about the Cree reaching out to the larger world, as they have often done in defense of this place, then his story is more than it seems at first glance. Cree men and women are still the primary connection between culture, land, and history, and all three of these are politically central in the north. In speaking of the Cree, Grand Chief Matthew Coon Come once wrote, "we have discovered that our way of life, our economy, our relationship to the land, our system of knowledge, and our manner of governance are an inter-linked whole. Remove us from the land, and you destroy it all." This is the lesson of Cree history, as well.[5]

In one sense, this is the kind of scaling issue that Tina Loo highlights in her history of justice and damming on the Peace River in British Columbia. There she assesses our understanding of history and environmental social justice by considering the different scales on which people understood what projects like that would do to the land. At the scale of provincial politics and economy—where dams brought growth and increased provincial power for British Columbia—damming the river looked very different as a justice issue than it did when scaled down to the local view, where land and lives were changed forever, mostly for the worse. She also makes the important point that, when considering local First Nations peoples, there needs to be temporal scaling in historical understanding. A dam seen as a modern environmental issue, in the 1960s, has one historical meaning; seen on a longer, temporal scale, one that highlights centuries of colonization for Indigenous people on this continent, a dam takes on a much deeper and more sinister meaning. And one might add that, on a temporal scale that comprises human cultural occupation in North America, those dams look different yet again. This for me leads to the larger historical lesson of stories like Solomon's.[6]

Part of the intellectual challenge here in this forest is to think about land, people, and history in a way that does not neglect those other scales, but also acknowledges that Solomon returned to that tent site, not only within the context of recent resource development and environmental change, or the history of colonization, but also within the much larger context of Cree cultural understanding of place and story. Travelling this land with the Cree is to traverse both the landscapes of personal memory and the geography of cultural connection. It's not just the fact that people continue to travel the lands of their birth, but again that they are connected to a culture that has been placing itself in this forest for the better part of four millennia. As Matthew Coon Come also once said, "our land is our memory, that is why it is so important to us." Memory, the stories of births or the struggles for rights, is held in the land. I am here researching the environmental history of the last few decades, and, while much of what I will write about resides in various libraries and archives—that story of massive change—the sources are more diverse.[7]

Whether we are talking about storied places like that tent site, or the place that Dave, Solomon's brother-in-law, showed me this morning—the scene of an old legend he has told me several times in winter camp—the stories attached to them are important to the way people act in relation to the land. They are important to the ways they talk, listen, and think about the land, as well. As storyteller Jeannette Armstrong says of her own lands and stories far to the west of here, in them "I understand I am being spoken to, I'm not the one speaking. The words are coming from many tongues and mouths of Okanagan people and the land around them. I am a listener to the language's stories, and when my words form I am merely retelling the same stories in different patterns." These stories, and this perspective on stories, shape the land everywhere on this continent.[8]

Living and narratives are connected, so this is not simply a statement about stories or language, but a powerful ontological statement about the nature of the world we live in—another echo of Momaday and King—that the truth about these stories, "is that that's all we are."[9] Here is a kind of philosophic scaling, in that these stories represent another intellectual tradition in this forest and give a very different view of what "environmental history" or environmental justice might look like. It's not a matter of telling history from within those traditions—something outsiders are incapable of doing—but of giving them the kind of weight or agency that they deserve. We ought to keep our eye out for the way they have shaped, and continue to shape, the history that unfolds in the places we study. And this is not only for the sake of doing justice to Native people living on their lands, but because we might learn something about our own connection to these places.

Narrative Ecologies and the Polyphony of Meanings

Ethnographers, translators, and folklorists have done a great deal of work with Native people interpreting these kinds of meanings on the land. There are lessons here for historians. What the best of this work shows, in the words of translator and poet Robert Bringhurst, is that "a coherent system of storytelling is like a system of science or mathematics. And like a forest, it is more than the sum of its parts. So long as it remains alive, [it is] not just a collection of stories or myths. It is a system that can be used to regulate and to record transactions with reality,"[10] and thus shape history. These systems of stories shape human action over time, most importantly our relationships with the land, and here in Eeyou Istchee—all across the north—Indigenous systems of stories are still very much alive.[11]

Again, while we may never fully explore these stories in our histories of people and land, we should not forget their presence. They have helped people live here successfully for thousands of years, after all, and are at the heart of all their histories of place. Nor should we forget the relationships between our whole system of stories and those already placed here on the ground; histories, too, are transactions with reality, after all. They can claim no transcendent understanding, for they are part of something more complex. As an environmental historian, I believe all these narratives have shaped land and human culture here, just as they themselves have been shaped in that dialog. All of this makes up the history of this land.

Like Bringhurst, I have come to believe that this is best thought of as an ecology of story in this forest—a fabric of personal, cultural, and political narratives of place and time—related to, and every bit as complex as, the ecology of plants and animals that make up the boreal north. Ecology, as a science, is the study of the interactions between abiotic and biotic parts of the Earth's systems (and the distribution of those parts across the globe), and human stories are no part of its business. And ecology is also a term that is currently very trendy and probably thrown around in a lot of ways that it should not be. My point is not to be trendy, but to complicate our understanding, both of the term "ecology" and of the ways that Western paradigms of thought, like science, shape our thinking about the narrative interactions between people and land.

Human stories may not be part of the science of ecology, but that science is founded on a narrative understanding of the world that sees a clear distinction between what it studies (nature) and the stories I've been thinking about (culture), and this has shaped the way we tell our stories of this land. There is a heated and unresolved debate in environmental circles over the relationship between culture and nature, and over history's relationship with the science of ecology. What's important for me, however, is that the culture/nature dualism—no matter how you position yourself in that ongoing debate—is not indigenous to or normative in this place. It is really an unhelpful abstraction for people who occupy this land at the scale that many Cree people still do. Their understanding of networks and interactions, both local and global, are focused on another set of meanings, and all this makes up environmental history here in this forest.[12]

This narrative ecology is a metaphor, of course, but so too is the science of ecology, at some level. And given that the stories here began as soon as there was a forest to be storied, which ecology are we going to call natural and which one cultural? It's not that the distinction between culture and nature is unhelpful at some level—it is fundamental to the intellectual transactions that are reshaping this forest today—but it is a distinction newly brought. So, while physical ecology may be abstracted temporarily for the sake of study, it is a tricky business for the historian. Part of that cultural blind spot that I was thinking about earlier is that we too easily fall into what Alfred North Whitehead called "the fallacy of misplaced concreteness," forgetting that our abstractions are only creations of convenience and not real in any universal sense. For my purposes, abstracting local historical meaning out of this forest—calling it simply "tradition," or some such thing—allows us to ignore the ecological inseparability of people and land held in Cree thinking. We thus lose sight of much of the real meaning of Solomon's or any other story that people use. Here is where injustice lies.[13]

Without sliding down the rabbit hole of postmodern linguistic relativism, we have to acknowledge that stories are part of this forest's makeup. They nest inside one another, like the various ecosystems, presenting a conceptual diversity as vital as any biodiversity in these woods. Bringhurst would tell us that these stories are musical, that their diversity is polyphonic, "a subtle, flexible, trim, and self-policing form with room for many voices." In "polyphonic music, several voices sing or play at once. They sometimes say very similar things in several different ways; they sometimes contradict each other. Each voice has its own melodic line, its own simultaneous path through musical space. Dissonance can occur; it may even be sought, though it is rarely expected to last. Some voices may say more than others, but no one voice is allowed to dominate the whole."[14]

The ecology of this forest is not harmonious; that is the romanticized and abstracted ideal, born of that nature/culture dualism. This forest is polyphonous, and to understand how history plays out here demands, like Jeannette Armstrong's claims about language and story, a different understanding of who is doing the singing or the storytelling. Seeing the forest like this points to the full meaning of that tent site back there, and to the way that the Cree still think about place. Historically the Cree have not thought of this land as an abstract set of ecological components on which they lived. Nor have they thought of their stories as discreet bunches of disconnected words with no putative meaning outside of their narrative context.

The Cree have thought of the land as a collection of communities of other beings *in* which they live. In many ways, they still see these communities of plants and animals as being made up of individual other-than-human beings, all of whom communicate with one another and with humans, just as people do within human communities. These communities of other beings feed the Cree in the gift exchange of the hunt, giving of themselves materially as the hunter gives back in respectful acknowledgment of the gift. Stories and food together, then, create networks of exchange on which the Cree depend for their survival, and abstracting one away from the other is a dangerous thing. The successful hunter not only skillfully uses the material resources at hand, but more importantly navigates these communities of very talkative and animated other-than-human beings who also live in the bush.[15]

So here is another metaphor, a Cree understanding of narrative ecology or Bringhurst's polyphony. In a sense, it is Geertzian cultural understanding writ large, one that takes the "thick description" of human cultures—the Weberian web—and spreads it out into the forest, encompassing all of what we would call the "environment" here, but what the Cree simply call "the bush." In translation it is easy to forget that "the bush" is not some quaint, colloquial term, but a precise ecological description that is imminently practical. Living in the bush for the Cree is not living in culture or in nature separately, but operating within an ecological relationship with the food and materials one needs, while communicating with the other residents of this forest. Within this way of thinking, I would say that "the bush" is also a historical description, because this polyphonic, animistic understanding shaped Cree reaction to the arrival of Europeans in this forest, and shaped the experience of this forest for those Europeans, as well. It is still one of the narrative voices here.[16]

Much of what I have written about Cree history previously underscores how Europeans carried out their fur trading and other activities within this Cree understanding of the bush—that both Cree narratives and Cree practices defined history in this forest into the twentieth century.[17] Contact was not the sudden advent of large numbers of non-Natives streaming into these lands, but hundreds of years of Europeans negotiating a narrative world that they struggled to understand. The Cree did not completely avoid the imported diseases and ecological change brought by Europeans, but Francis Jennings' description of a "widowed" rather than a virgin land—while true in many places—is not accurate in James Bay.[18]

For centuries here, there was a mixing of cultural economies, a tenuous, often contentious balance of negotiated power by people who perceived that they needed one another. In actuality, the same thing played out down the St. Lawrence Valley and the Atlantic coast in the sixteenth and seventeenth centuries. There, however, the sway of Native thinking lost much of its agency with the imperial and revolutionary conflicts of the eighteenth century.[19] It all lasted a great deal longer in Eeyou Istchee, and finding the agency of these ideas continues to be my goal in this place, though it has become more difficult as my focus has moved toward the more recent past. All of the politics and sudden change, all of the new people living here in the north, have reshaped the land, but also reshaped the ways we can speak and think about the land and its history.

Northern Forests and Western Frontiers

When writing about events before the 1970s, ethnohistorical accounts of those earlier encounters between Natives and newcomers were important to my thinking. Most ethnohistory is primarily concerned with early religious, trade, and military engagement between Natives and Europeans, yet the records of fur traders, missionaries, and bureaucrats are full of environmental history, too.[20] Ethnohistory's boundary-jumping methodology was highly useful in finding this out, and crossing the lines "of time and space, of discipline and department, of perspective, whether ethnic, cultural, social, or gender-based," helped me look at and think about the environmental history of that prolonged story of the bush.[21] This was not something historiographically novel; Calvin Martin had done something similar back in the 1970s, but while he and his detractors raised important environmental questions concerning Native involvement with the fur trade, the larger and long-term meaning of the bush strangely did not penetrate into environmental historical thinking. Environmental history still largely misses the importance of Native ideas in its interpretations.[22]

Environmental historians over the decades have too often portrayed Native people as being swept along into tragic decline, from "natural" freedom and "traditional" subsistence into poverty and political irrelevancy, without considering how Natives acted according to their own understandings and affected the larger story by doing so. Historians have used First Nations as the canary in their narrative coalmines—judging environmental decline by Native decline and vice versa—without acknowledging the deep sense of interrelationship that made Natives stay in place, suffering along with the land; they have not often acknowledged the political fact that they are still in place, living their histories there, either. This was particularly true when environmental histories were driven by the idealization of past pristine wildernesses, where historians once sought this ideal. And, while recent works have radically qualified our notions of wilderness, Native peoples still too often seem to get essentialized when they are living "traditionally," or ignored when they are not.[23] This goes back to that debate over culture and nature. In leaning on the science of ecology, or getting too enthralled with the narrative aspects of culture, environmental historians have missed a lively and living intellectual perspective. They have missed the intellectual articulation of living in the bush, which takes their arguments concerning narratives and nature and turns them on their head.

Missing the importance of the community of the bush has also been a matter of getting swept up in that outside history, which I was thinking about earlier. Environmental writing, in the US particularly, has been dominated by the post-revolutionary rush west, so it has been easy for generations of environmental historians to read back into the colonial period, or onto places like Eeyou Istchee, the same nationalist/capitalist juggernaut that rolled out on the western frontier. Arguably, this has kept environmental history from exploring the nuances of places like the north, where people like the Cree did not face the expansion that the Cherokee, or Lakota, or Yurok faced in the United States. Places like this forest, until recently, were not the focus of the transformative economic and political activity that marks the speed of US expansion.

Here in Canada's north, a lingering colonial heritage, slower economic expansion over the nineteenth and twentieth centuries due to staples dependency, and technological limitations in moving into this territory, made for a much different history for both land and people. The fact that Canadian growth, historically anyway, lacked most of the ideological imperative that marks so much of US history also made a great deal of difference.[24] For much of what I have wanted to say about Cree history before the 1970s, environmental history's focus on rapid transformation and dispossession has been a poor fit, and I have found myself reaching out not only to ethnohistorians, but to historical geographers and colonial historians as well. These disciplines in Canada have had a much better eye for subtle changes in the land.

The question that arises now is whether recent events here have changed the way we should look at history. In the last four decades, roads and outsiders have proliferated, hunting lands have been flooded and forests clear-cut, and non-Native hunting cabins have been built by the hundreds—all in a way that looks a lot like that older frontier onslaught. The Cree now are certainly also facing an intensity of change that threatens communities, families, and individuals in a new way, and that historical view of long, slow, negotiated change within a Cree understanding of the bush seems to lose much of its interpretive power.

Across the whole of the North, beginning in the 1960s and 1970s, advances in technology, as well as changes in the market value of resources, have made resource exploitation possible and profitable in new ways. Northern development has shaped, and is still shaping, federal and provincial economies in ways that look very similar to what happened in the American west. This has once again made "Staples Thesis" history [25] relevant economically in Canada. It has also added some of the ideological motivation so strong in US expansion, so maybe we need to start thinking in ideological terms about the lands and people here, in addition to the economic imperative.

Certainly the James Bay Projects were an outgrowth of the Quiet Revolution, and so this land has been part of the history of cultural emancipation and sovereign expansion of the Québécois technocratic state, as well as its economic growth. Here is some of that ideology lacking in earlier Canadian expansion. When Bourassa announced "the project of the century," in 1970, he was riding the tiger of a growing separatist movement in Quebec and was fighting for his political life. He celebrated the expansion of French ethnic and political identity in this northern environment and began a story that has been carried forward by boosters from all sides since that time.[26] One has only to look at former premier Jean Charest's April 2011 announcement of "Plan Nord"—a twenty-five-year, $80 billion international investment scheme to develop any and all of the resources in Quebec's north—to see the continuing power of northern expansion in shaping the economic and political landscape here in the province.

Robert Bourassa only made the argument that the James Bay Projects would create a stronger and more self-reliant Quebec within a changed Canadian confederation; but he might well have been speaking about all of the provinces, their resources, and their power relationships with both Ottawa and the north since the 1960s. Across the north now, one can define a nationalist, as well as a resource frontier, and this is reshaping the Canadian political landscape. Now, those environmental histories critiquing expansionism and its costs are, in fact, more applicable to this place and its story than they ever were before – and for much the same reason when looked at from the other side of development.

In the US, those kinds of environmental histories were greatly needed as a counter-argument to celebratory nationalist stories of manifest destiny, "improvement," and "progress" on the frontier. And histories of sweeping and destructive environmental and cultural changes during nineteenth-century expansion and development in the trans-Appalachian and trans-Mississippian west certainly speak directly to many things happening in Eeyou Istchee today. One has only to look at the scale of change, here, to make the connections obvious. Frontier environmental histories also inverted Frederick Jackson Turner's thesis—that open spaces create national character—and the triumphalism of conquering wilderness and the Natives that inhabited it. Since the 1970s, Québécois rhetoric concerning the north has often been the same blithe Turnerian vision of progress, and is arguably in need of the same kind of inversion—the same is true all across Canada since the early 2000s. Turner's stages of frontier growth and his individualistic settler, imbued with nature's democracy, are not a perfect fit in Quebec or Canada, but environmental history's meta-narrative of exploiting untouched landscapes, thus foreshadowing the bad end of modern experiments in massively altered landscapes and national ideologies, is a very tantalizing paradigm in which to think these days.[27]

Looking to James Bay, in the nineteenth century Curé Antoine Labelle called on the "founders of this future North American empire" to travel north and "conquer this land of America against the English philistines," and thus give new life to a French nation.[28] A hundred years later, Robert Bourassa styled himself the "conqueror of the north" for much the same purpose, and, though Charest was more staid with his rhetoric—focusing more on economics—"Go North Young Man" was clearly the sentiment and political message at the heart of Plan Nord. So frontier conquest is useful to think with in Eeyou Istchee, because in a great many respects Quebec's north is a frontier. This forest, like the American west in previous centuries, can be seen as a place of massive environmental disruption and conscious cultural destruction by both government and business—a colonial possession in the worst sense of the word, with all the ugliness that this implies. As a friend of mine over in Waswanipi told me in reference to clear-cutting, "my impression is that the Quebec government is out to occupy Cree territory as quickly as it can."[29]

The initial temptation, then, is to take up uncritically those same interpretive tools of frontier, both in writing history and thinking about justice. Yet, as useful as these ideas are, focusing too heavily on historical frontiers at some point ceases to educate and begins to obfuscate. Quebec's north is not the American west—neither is any of the Canadian north, even with Stephen Harper's legislative agenda—and none of what has happened in the last few decades happened in the context of the militaristic expansion of the eighteenth and nineteenth centuries. It has all happened in the legalistic and technocratic twentieth and twenty-first centuries, and so, while the process of negotiated meaning changed in the 1970s, the north has not been conquered.[30]

Here in Quebec, for decades now—and this is not to say that the province has not tried to take as much as it could without negotiating—everyone has been engaged in bargaining over this territory. The Cree, who have been anything but passive in this, have exploited both the tensions in Confederation, over Quebec sovereignty, and also the tension between provincial control of resources and federal responsibility for First Nations. They have also exploited ideas of justice and multiculturalism within modern Canadian and world politics. Within this process and amidst all the dramatic changes brought by development, the Cree have been kept busy trying to define a form of Native sovereignty, and this is central to understanding the continued negotiation over the meaning of the land.

In this struggle they have hard-won victories to their credit, which have mitigated some of the bad results of development, and, importantly, put them in a somewhat better position than other First Nations in Canada's north. The Cree have not always been obstructionist—this has been part of their success—but when they have been, it's been dramatic and history-making. When they have conceded, it has been history-making too, and so in this sense Eeyou Istchee has become a very political place over the last four decades. I mean this in the sense that provincial and national politics have a place here that they never did before, but also that the politics of development have become a part of Cree culture now. The older ideas of the bush have not been abandoned, but they have been nuanced and challenged by the need to negotiate politically and to codify Cree rights and legal standing. The meaning of Eeyou sovereignty now sits side-by-side with stories like Solomon's and the two together illustrate something more than just a replaying of frontier history.

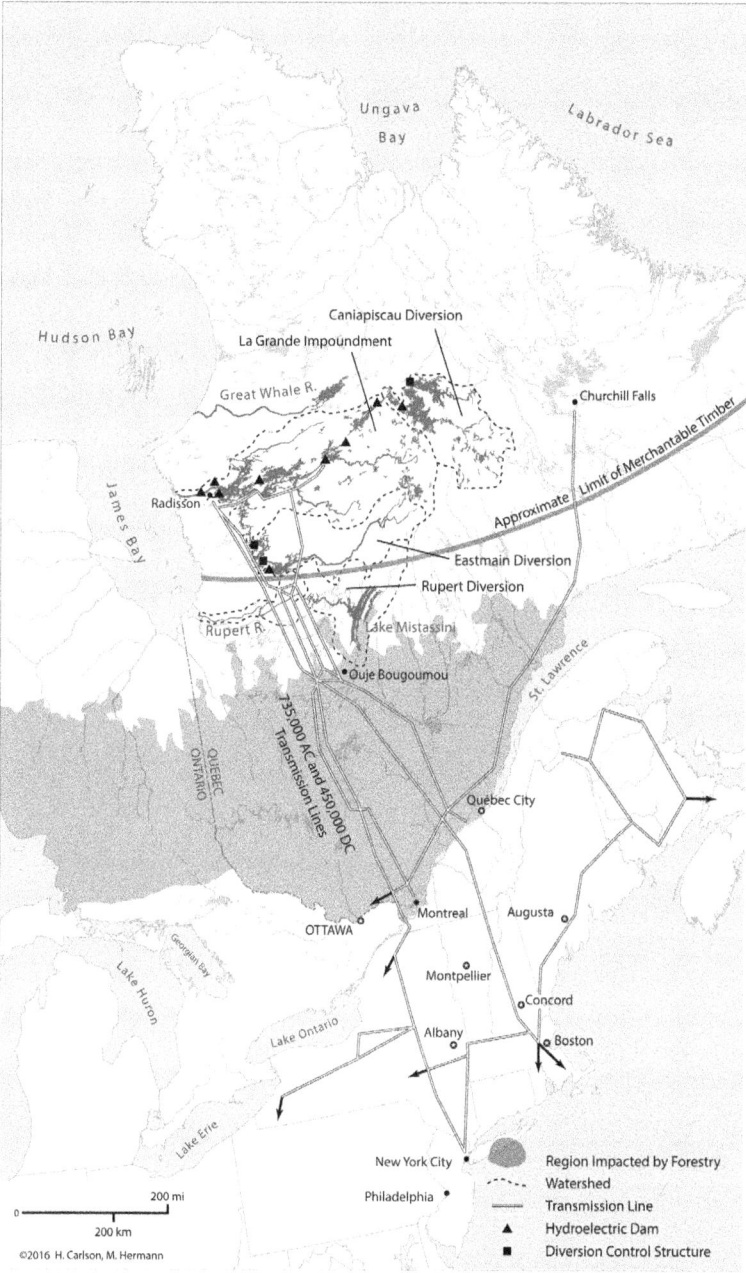

FIG. 9.1: Map of James Bay region by Hans M. Carlson showing extent of hydroelectric development and logging on Cree lands.

The Politics of Development and the Development of Politics

The Cree began this new political history under the 1975 James Bay and Northern Quebec Agreement (JBNQA), and this is still the foundational legal framework for them as a people and the land on which they live. The JBNQA compensated them, in some measure, for the damming of the La Grande River by creating a good deal of self-determination and control for each Cree village—now usually identified as individual nations. Most importantly, it created the Cree Regional Authority, which is the corporate vehicle through which the Grand Council exercises its authority and negotiates with other governments. All of the communities of James Bay were incorporated under it, and all share equal representation. Much of the document does not deal with the land, but lays out the structure of local and regional government, giving the Regional Authority direct control of some services that, until 1975, were controlled by the federal government: health services in the region, under the Cree Board of Health; education, under a unified school board; and police forces, both in the individual communities and special Cree units of the Sûreté du Québec.[31]

The key provisions concerning the land were those that subdivided the territory into three categories, for purposes of development and legal control, and those that protected Cree hunting. Category I lands, which surround the nine Cree communities, were designated to be held most exclusively within Cree control. Here Cree authority is strongest, and, while these lands can be taken for development, there are strict guidelines and compensation regimes for doing it. Category II lands, shared by all the communities, are less controlled by the Cree. Here they have exclusive rights to hunting and fishing, but it is easier for Quebec to develop resources without compensation. On these lands, hunting is controlled by the traditional Cree stewards of the land—the Kaanoowapmaakin—and the Cree have fought over the years to keep these carriers of traditional knowledge within the legal framework of negotiation with outsiders. Category III lands are open to use by all parties and are controlled by the province. The Cree are not excluded from using these lands—here, too, the Kaanoowapmaakin maintain their stewardship within Cree culture—but non-Natives have access.

In addition to this categorization, the Income Security Program (ISP) administered by the region-wide Cree Trapper's Association was set up, by which hunters who continue to live a significant portion of their lives on the land are guaranteed an income from the province. This more than anything has helped maintain not only a Cree presence on the land, but also has enabled the continuation of a Cree understanding of their lands. Without the ISP, the continued stewardship of the Kaanoowapmaakin would have been much more difficult to maintain. Something akin to the "territoriality" described by Paul Nadasdy in the following chapter might have developed had the communities or the Cree regional government had to take control of hunting, but this kind of regulation has not happened. The Kaanoowapmaakin still look to social and cultural sanction, rather than government regulation, to exercise their stewardship.

All that said, there is no doubt that the JBNQA established a new language of environment through which the Cree have had to speak of their land to non-Natives. This has changed the way the Cree can conceptualize their lands in certain contexts. Legally, land, water, air, and people—and how they should be interrelated ecologically and economically—are categorized in a system of understanding that thinks in terms of resources. This now has as much meaning as traditional thinking, so that hunters, under the agreement, have the "right to harvest game resources" under Western conservation principles. Hunting is now defined as "the pursuit of the optimum natural productivity of all living resources and the protection of the ecological systems of the territory." It is here that Western divisions between culture and nature come face-to-face with the Cree understanding of the bush—the difference between resources and relations.[32]

Hunters on the ground are thus challenged by resource exploitation, but also by new ways of thinking, and Cree politicians stand between two worlds. This makes for a complicated situation. The money in the ISP helps insulate hunters by keeping them on the land where they still have some measure of autonomy. They are not as independent as their grandfathers, but they still have room to maneuver intellectually, as well as spatially, and this shapes events in the region. Their presence has given Cree leaders a foundation to stand on, as they have asserted a Cree presence in Canada and Quebec, and traditional use creates moral and political high ground from which politicians can negotiate with governments. This can be seen as a way of mitigating change from outside, but it should be seen as a way of controlling change from inside, as well. The culture of traditional hunting has been used to bring pressure to bear on Cree leaders, as they navigate the political landscape that has been evolving since the JBNQA was signed, and this is still happening.

Provincial and federal politics fully entered Cree territory with the initial treaty, and continued with the ratification of the JBNQA, which was not passed by the Quebec legislature until 1978. Provincial politics—the fight between Bourassa's Liberals and the increasingly powerful Parti Québécois—became a Cree problem when Bourassa was defeated in 1976. They had to lobby with the new government to get the JBNQA ratified, and to get Ottawa and Quebec to live up to their ends of the agreement. Neither government made it a priority, in the strained atmosphere of the time, though the Cree were helped by an official inquiry, the Tait Report, which severely criticized the government's inaction in seeing the JBNQA through, and by the UN Conference on Indigenous Peoples in Geneva.[33] In the end, the Cree managed to make the whole issue simply too uncomfortable, and, with ratification and implementation, they began to inhabit a new and powerful political structure. The Cree speak of their rights under the JBNQA as other Canadians speak of their Charter rights. It makes sense for them to do this in relation to the Canadian constitution, but the implications of this are still developing, in large measure because of Quebec's continued demand for the resources on Cree land.

Robert Bourassa and his federalist Liberals came back to power in 1985, and with them came a plan for further expansion in James Bay. The next phase of hydroelectric development was to be the Great Whale River, to the north of La Grande, and when the announcement came that construction was slated to begin in 1991, the Cree organized to stop it. Here is where their story reaches out beyond Canada, as a large part of the justification for the new project was to sell power to the United States. The Cree actively engaged people in New York and New England, trying to bring understanding to people whose desire for power was going to change their lives yet again. The launch of the campaign was the arrival of *Odeyak*—a Cree/Inuit hybrid boat—at Earth Day celebrations in Manhattan in April 1990, where the Cree spoke about what they faced. After that they travelled extensively and repeatedly put the case in no uncertain terms to Americans: "a project of this kind involves the destruction and rearrangement of a vast landscape, literally reshaping the geography of the land. This is what [we] want you to understand: it is not a dam. It is a terrible and vast reduction of our entire world. It is the assignment of vast territories to a permanent and final flood. The burial of trees, valleys, animals, and even the graves beneath tons of contaminated soil."[34] Many in both Canada and the United States worked with the Cree, and this was really the moment when Native issues in James Bay entered most clearly into outsiders' consciousness. The expansion was stopped, in part because US states cancelled power contracts, but it's important to see what a double-edged victory this was in many ways.

First, the Parti Québécois really called a halt to Great Whale because, after defeating Bourassa again in 1992, they wanted to spend their political capital on another referendum on separation. This was a threat to the Cree, too, as an independent Quebec would not be bound by the legal relationships negotiated in treaties like the JBNQA. This began another fight for rights, but the cancellation of Great Whale also opened up a rift between most people's understanding of James Bay and the ongoing process of change that was happening on the ground. While there was a great deal of well researched and well considered reporting done about the land, people, and issues in Eeyou Istchee during the Great Whale fight, the dominant theme of this work was the destruction of untouched "wilderness" and "traditional Cree culture."[35] This wasn't entirely wrong, but it missed the many ways that land and people had worked to adapt to the situation in the 1990s. Here are those issues of scaling in clear terms, because in the minds of most people who fought the dams, and who lived far away, we had won a great environmental battle, and they presumed that the Cree could go back to the way they had "always been." In Eeyou Istchee, this was not the case.

The Great Whale was only one river, and halting its diversion did not slow the province's use of other resources. They added hydroelectric capacity along the Eastmain River, dams to which the Cree government agreed, despite a good deal of opposition in certain Cree communities. That river was already dammed, and the additional environmental impacts were balanced against the money that the Cree needed to deal with other pressing issues. One of those issues was Quebec increasing its logging on Cree lands in dramatic and devastating ways. This was something that most of us here in the south missed, just as we missed the fact that the building boom in the United States—the bubble that has only recently burst—was fed largely on Canadian lumber.

During the 1990s, the forestry operations that had been slowly expanding, here in the southern sections of Cree land, grew exponentially in the matter of a few years. Massive clear-cutting became the mirror image of the vast landscape changes that the river diversions brought to the north, and roads advanced on a yearly basis, devastating much of the bush. The Cree had Quebec in court over this issue for much of the late 1990s, and it was pressure from cutting that moved them toward negotiating yet another treaty. The 2002 Agreement Respecting a New Relationship Between the Cree Nation and the Government of Quebec, or "La Paix des Braves," introduced a new system of cutting meant to alleviate problems, though the Cree only got Quebec to the negotiating table by offering the damming of the Rupert River as a precondition.[36] The Paix des Braves has largely been misunderstood outside of Eeyou Istchee, both in terms of the outside forces at play on the land and the fact that it was an initiative of the Cree government.[37]

When the negotiations came to light, people in the south asked about fighting the Rupert diversion, as they had the Great Whale, but the Cree government wanted to negotiate, even if environmentalists and outdoor enthusiasts were bothered by the loss of the Rupert. Many were deeply disappointed in the Cree, not understanding that life in Eeyou Istchee was no less changed by the decade of the 1990s than life in the United States and southern Canada had been. In fact, going back to that theme of imagined wilderness, most did not understand that change was even a part of the James Bay story. The problem for the Cree was, and is, that most people were unaware—unaware of the history here, of the current threats to this land, and especially of their own connections in creating the threats and driving them forward.

Many thought the Cree sold out. In 2010, Vermont signed new long-term contracts for Hydro-Québec power, and this was done with almost no public consultation. This was largely because people did not understand why the Cree government did what it did. I am from Vermont, and this conspicuous lack of public interest was troubling for me personally because, whether we know it or not, we are connected to this land in terms of energy and other natural resources. You will find few Cree who will celebrate the damming of the Rupert, and many of them who are critical of their government's actions—but many of them also voted to ratify the treaty because of the reality of life in the north today.

Vermonters did not understand recent historical events on the lands of Eeyou Istchee, or the processes by which change happens in many places, so we missed the real meaning of actions like those new contracts. Today, metaphorically anyway, every fourth time I flip the light switch in my house, I am drawing power off the Cree lands. Hydro-Québec and other Quebec companies are buying up substantial sectors of the electric grid south of the border in order to make the connections stronger still. The Rupert diversion is a *fait accompli*, but long-term contracts will be used to capitalize further damming all over Quebec's north—and the same is true of other resources. When I go to the lumberyard back home and buy a two-by-four, there is a good chance that it came out of this forest, and an increasing amount of minerals used in electronics are coming from here, too. Quebec would like all these connections to be stronger and more permanent, and so, in its relationship to the northeastern economy particularly, northern Quebec is the closest illustration we have for all the disruptive effects of our resource demands all around the world.

When I said there were analogies to other frontier histories to be made, I could have said other geographies and ecologies, too. As I travel the roads of James Bay and visit the communities, I can easily make analogies with more far-flung places, like the Niger River Delta, the boreal forests of Russia, or the Australian Outback; in all these places, the ever-increasing demand for energy resources and raw materials—oil, lumber, uranium, and other minerals—is reshaping the land and the lives of the people who live there. Many of these people, like the Cree, are the Indigenous occupants of traditional territory, and are trying to save their culture as well as land in the face of this expansion. They, too, deal with the tremendous lack of knowledge that most of us have of their places and their histories. There, too, the forces of distant power and the lack of public understanding are forcing a singular history upon them. This is the ongoing legacy of colonization.

Since the Rupert diversion, the Cree have negotiated several additional agreements-in-principle with the province and the federal government. In an attempt to clear up lingering issues concerning how the JBNQA has been implemented, Ottawa has agreed to give the Cree money and a great deal more control over their internal political lives—infrastructure, justice, and more. This makes Eeyou Istchee look increasingly sovereign in many ways. The Grand Council has also become much more open to the idea of development, even tentatively signing off on Charest's Plan Nord in exchange for Cree economic participation. Charest was eager for the opportunity to advertise his cooperation with First Nations, and the administration of Matthew Coon Come was a willing partner. The 2012 change in government put Plan Nord in question, and it is unclear at the moment if the 2014 return of the Quebec Liberal party will give it new life. In any case, Cree demands for participation in development have remained constant.

One can feel a number of different ways about this, and Cree people are not universally happy with the actions of their government or with Cree involvement in many of these projects. There is a near-unanimous consensus against uranium mining in Eeyou Istchee—there is a huge deposit about one hundred kilometres north of here—but some Cree are fully in favour of more development. Others see development as problematic and troubling, though inevitable; they favor the Cree getting what they can, but they worry about the consequences. Still others see participation as an abandonment of Cree values and the relationship with the land that has always supported them. This has become a tremendously complicated political situation—a far cry from forty years ago, when opposing the first dams was a simple matter, if a nearly impossible task.

We cannot ignore these changes. We should not romanticize or essentialize the Cree as we did during the Great Whale fight, for they are not the same people they were when the development began. At the same time, we have to be wary of any simple-minded denial of that traditional culture, that history, which people like Solomon, David, and Lawrence still very much embody. There are those, often thinking of themselves as political "realists," who dismiss what they see as "idealized" notions of Indigenous culture and relations to the land. They point to increasing Aboriginal engagement with resource development in areas around the north as support for this position, but there is essentialism here, as well. At best, these they-are-just-like-us-now arguments miss a good deal of what is happening on the lands of the north. They miss the many reasons individuals might have for participating in the whirl of activity brought by development, yet also working to stay connected with older ways of thinking and living. They miss the historical and current agency of people living on the land, the ways that they are conceptualizing events. At worst, one might see these arguments as a continuation of the colonial process. So, how do we think and write about this increasingly complex situation in a way that does not essentialize people on the ground, yet does do some justice to all the current situation's nuances?

History and Political Ecology

In her chapter in this volume, Tina Loo frames the history of development in the post-war Keewatin District in terms of hopeful bureaucratic action on behalf of First Nations people. One can point to similar processes in Eeyou Istchee during that same period, and this is an important part of the history of this place. But the bureaucrats here were not alone in their hopefulness, for traditional Cree culture is founded upon a belief in the efficacy of hope. Hunting success relies on the hopeful attitude of the hunter, in his or her relationship with the hunted animals, and hope, as a cultural force, is something that I feel defined how the Cree have encountered Western culture over the last three hundred years. It is still a powerful presence in Eeyou Istchee, and should probably influence the way that we see Cree involvement in current events, as well.[38]

I had dinner not too long ago with an older couple whose land was the focus of the early mining activity here in this southern region of Cree territory. Matthew, who was Kaanoowapmaakin until he turned the responsibility over to his son, had helped with the prospecting and other development. He had done so for a number of reasons, he told me. First, he believed the land had been given him to share, and at the time he did not make a distinction between uses. Like all mining, the extraction of copper in the Chibougamau area has left a legacy of polluted water and land, but this was not something that Matthew could have foreseen.

Matthew now feels a great deal of responsibility for not understanding, but at the time he believed that mining would offer him and others the opportunity to feed their families, make good lives for themselves, and continue to hunt. There is little doubt that he entered into the process with a great deal of hope for what the land could provide. It was not economic hope that the land would make him rich, but the active and effective hope that is the basis of Cree hunting. This is the point, as we try to conceptualize the many ways that Cree people take part in what is going on around them presently. The hope with which Cree hunters engage an animate world of other beings is still important for understanding recent events.

Matthew's is another story to set alongside Solomon's, and neither can be understood in isolation from all of the larger changes brought by development, or the treaties and politics that now define parts of Cree life. Because all this has happened. Eeyou Istchee now defines political boundaries and some facets of sovereignty for the Cree; the decades that the Grand Council has spent working to wrest more Cree control from this situation have not been wasted, despite the high price that has been paid and the compromises made. This Eeyou Istchee is now where all Cree live, and their stories must be viewed within this nested set of political contexts, which begin at places like that tent site up the trail, but in the end reach out far beyond the Canadian north. Not to connect these things together is to risk intellectual irrelevancy.

The changes I'm trying to capture—those since 1970—are aspects of the global economy and the part that Eeyou Istchee now plays in that economy. They are also part of the history of Canada, because all northern First Nations have and still do face everything that the Cree have faced: challenges to traditional life, a loss of control over land base, and all the social and cultural stress that this brings. Like the Cree, they are trying to negotiate ways forward within the context of the Canadian state, as well as the global economy. Like the Cree, too, they know that the history of this land is theirs in large measure, as much as the land itself, and in both these ways environmental history encroaches on current politics and needs to be attuned to this fact. The whole north is politically charged; one cannot avoid this, but one must also work to show the continued historical importance of the Cree understanding of the bush in an era of industrial development. Narrating this complexity requires some methodological adaptation.

I mentioned earlier that the mix of ethnography and history was valuable in understanding many aspects of history here, but that this approach fails to get at everything that has happened in recent decades. Frontier history offers something, too, but understanding the full meaning of historical change means focusing instead on the interplay between larger forces and the power of local stories like Solomon's, on the context of politics and economic development, as well as the meaning of traditional culture and local events. It means focusing on ecological change, but also on the ecology of stories on the land, and so I find the current situation moving my environmental history toward the discipline of political ecology, in search of new ways to frame historical questions.

Developed originally to challenge apolitical ecological explanations of environmental change—behaviorist and functional understandings of human-environmental relations, which limited their explanations to the local and the cultural—political ecology looks for the global context of local environmental impacts. In this, it offers insight into issues of development here at James Bay, looking at those connections between environment and the larger forces of political economy. It does this with an eye to the meaning of local action, and in ways that are useful for understanding the current complexity in Eeyou Istchee. As "a field of critical research predicated on the assumption that any tug on the strands of the global web of human-environment linkages reverberate throughout the system as a whole," affecting distant and seemingly isolated places, it applies well to what I've been thinking about here in Canada's north.[39]

Sitting here in this campsite, this seems a natural fit for this region—but interestingly, political ecology as a discipline has focused largely on the developing world outside of North America and Europe. It has looked at what happens to the traditional systems of South American peasants and their forests when the bottom drops out of the coffee market, for example. Or at what happens when the World Bank funds massive afforestation projects that enclose land and restrict traditional use.[40] The geography of the north has not been its focus, because Canada is a "developed" country. Yet this approach lends itself readily to asking what happens to land and people when the United States wants to buy power or lumber from Quebec's northern forests—or what happens when governments and industry try to provide them.

Interestingly, too, political ecology is a cousin to the kind of environmental history I have written about this region, delving into current uses and adaptations of local ecology and traditional environmental knowledge of places like this forest. It has done this, however, with little historical analysis as part of its method. The past, for most political ecologists, has been secondary to present events and their causes. Oddly, much political ecology positions itself theoretically in ways that dichotomize present subaltern practice and past globalizations from its interpretation of current events.[41] Ethnohistory and environmental history, which might illuminate current cultural responses to global pressure, seem only lightly understood. This dichotomy, as political geographer Karl Offen writes, "assumes that local, peasant/indigenous perspectives, ambitions, memories, ideas, consciousnesses, and resource uses are somehow separated from global-local continuums in the past." To use Eeyou Istchee as an example, many aspects of Cree cultural traditions grew from three hundred years of fur trading and other past "globalizations." Many will fail, however, to connect this with the fact that Cree traditions continue to grow within the current context I have described.[42]

To achieve a more historical political ecology, Offen suggests a "field-informed" perspective on the past, by which he means "lengthy field immersion that includes ethnography, surveys, participant observation, mappings, and often biophysical research." This means getting to know the land and people involved, the various stories of place, as well as digging in archives. In addition, he wants to see "an explicit linkage between social justice and the management of natural resources, a broadly conceived 'nature conservation' that takes into account the health and viability of the non-human world." This call for justice touches back upon that dichotomy between traditional culture and political aspirations, and highlights the need to really understand the relationship between people like Solomon, here on the ground, and larger issues outside, in their historical context. Both at ground level and in the larger political arena, Offen emphasizes that, "any notion of 'social justice' is historically contingent and culturally specific, it should include a respect for cultural difference, customary rights and ways of knowing the world, as well as an equitable mode of resource distribution, economic opportunity, and political representation." He means that both the cultural traditions of local people and their political hopes are important.[43]

By incorporating historical perspectives and methodology, political ecology's perspectives can help contextualize the last forty years on the ground in Eeyou Istchee. Together they can focus attention on local ecological facts and stories—that narrative ecology with which I began—and the political/economic context of development. Together they can also help show how all of this has developed over time in relation to the varied ways this land has been and is now being used. Industrial use, including current Cree participation, is obvious wherever one looks up here, but there are other equally important uses still happening. In the same way that the marriage of history and ethnology into ethnohistory illuminated the meaning of Native actions in the more distant past, so, too, can this union help us understand more recent change. All this together gives us a way of contextualizing the various stories here on the land.

Getting History back to the Land

The men I'm travelling with are all Kaanoowapmaakin. They are stewards over hundreds of square miles of territory, and the knowledge that they have about their lands is detailed: they know where the good wood is, where fish can be caught in abundance, where medicine grows, and, most important, where the animals can be found. They also know where the stories are placed, and all of this knowledge must be tended. Because of this, the Cree often speak of the bush as being "like their garden."[44] Here is another metaphor, which the Cree use to communicate that the land is more than a set of material resources for daily living; that Cree tradition is of the land, just as the plants and animals upon which the Cree tradition-ally rely. The garden is a conscious simile, which they use to describe the bush to outsiders, knowing full well that the word skates the line between the biblical and the horticultural.[45] They use it because it's a relationship with their food and a story about their land that resonates with non-Cree, and gives them some entrée into the communicative reciprocity of stew-ardship I described above. "Place," in all its many aspects, is tended "like a garden" by people who are in turn cared for by the land.[46]

Dave, Solomon, and Lawrence have a responsibility within this un-derstanding. They have to make sure that these lands are as productive, as useful, and as whole when they pass them on as they were when they took charge of them. Others may hunt, fish, and trap these lands, but under their guidance—if the rules are being followed—and the land is thus kept in order. This cultural system of stewardship, not dissimilar to the old ru-ral rights of the shared commons, is known as "Weeshou Wehwun," and it has shaped the lands of Eeyou Istchee for many generations. This is what we would label Cree tradition, and this is what is threatened by everything industrial development brings. Yet here is where historical research is so important—in addition to the tools of political ecology—because if we focus too much on the traditional ecological knowledge of this system in the present, then we risk missing the fact that the *indoh-hoh istchee* system is also a historical creation.

There is a lively anthropological discussion as to whether Native hunting grounds like these predate European contact, for the Cree are not the only Algonquian people to have this kind of land-use system. All their provenances are an open question.[47] This debate is interesting in its way, but in Eeyou Istchee defined hunting lands have been around for at least a couple of centuries, as fur-trade records show that there were certain hunters in "possession" of certain lands from the beginning. Whites did not understand this system of possession (as was the case with most Indigenous divisions of land), so they wrote little about it. A system of some kind existed, however, from an early time, and the *indoh-hoh istchee* are certainly part of Cree culture and tradition today in all the ways I have described; nobody here would say that they or the Kaanoowapmaakin are not fundamentally part of Cree traditional use. All of this is an important foundation of understanding, but it is not the most interesting historical aspect of these territories, nor does it connect Cree tradition historically to issues of development as it should. To make that connection one has to understand that, from the 1930s to the 1950s, the *indoh-hoh* system was reinvented by both Cree hunters and non-Native outsiders.[48]

In the early twentieth century, regional beaver populations were decimated. This was caused first by white trappers overhunting fur resources, but then by the abandonment of traditional boundaries as individual Cree tried to get what they could before it was gone. The resulting hunting free-for-all affected other animals, and this became a cultural as well as an ecological crisis, as people began to starve in the region. The hunger and deprivation in James Bay did not receive the national attention that drove governmental response in the Keewatin District, but it did inspire action in the creation of beaver reserves on Cree lands. In writing previously about the longer history of the region, I described this response as a manifestation of the hope within Cree hunting culture, but I think now that it also speaks to the historical and political issues I've been discussing here.[49]

In the decades surrounding the Second World War, the Kaanoowapmaakin responded to environmental crisis by adapting the conservation ideas of fur traders and government officials to their traditional ideas of stewardship through political action. They did this with the aid of a few cooperative non-Cree actors, through petition and negotiated agreement, and thus got their traditional system sanctioned under federal and provincial laws.[50] In essence, they allowed their stewardship to be bureaucratized, reporting to the Ministry of Natural Resources on populations and receiving quotas for beaver. For this they regained exclusive Cree use of the land, and—importantly—a great deal of control over its regulation. The Kaanoowapmaakin knew the government formula used to come up with quotas for the next year, and they were the only ones in control of the data, so in essence they negotiated a system where they set their own quotas. In all other respects, officials were happy to let them regulate their land as long as they could put the government stamp of approval on the numbers.

The process was useful for individual hunters, but more importantly it made the *indoh-hoh istchee* system a legal manifestation of political hopes in that era, as well as a cultural artifact. Today it's the Grand Council that negotiates and signs treaties and agreements, but back then it was hunters on the land who began the modern political engagement with both governments. It's important to remember that these agreements were not formal treaties, but the codification of a system of land use, which helped the land and the Cree rebound from a period of hardship caused by the breakdown of Kaanoowapmaakin control. Everyone at the time agreed that the system was part of the *longue durée*, here in Eeyou Istchee, but this new use of tradition created part of the current political context.

If we step back, then, from the larger story driven by resource extraction, if we jump scales, then history and political ecology together allow us to see that Cree political existence begins with the creation of beaver reserves, rather than with the JBNQA as I claimed earlier. The Kaanoowapmaakin role is changed now, but it's important to see that these men are both the archive of traditional knowledge and land use—this aspect of political ecology—and a key to the political history of Eeyou Istchee. Kaanoowapmaakin were and often still are leaders in Cree communities, and their actions from the 1930s to the 1950s speak to the adaptability of Cree land tenure. They also speak of the Cree fight for their land since the damming and logging.

The Kaanoowapmaakin continue to interact with the Ministry of Natural Resources and other government bureaus, though things have changed. The recognition of the Kaanoowapmaakin was one of the concessions that Quebec made in the Paix des Braves agreement to get the Rupert diversion. These men are now "consulted" over what will be done by forestry companies on their lands, though consultation has turned out to be a deeply one-sided affair—despite the fact that the province likes to talk about Cree participation. The Kaanoowapmaakin can at best only blunt the effects of development, but they continue to try in different ways, and this is the point.[51]

In continuing to hunt their lands, these men are carrying on a long cultural tradition, but in continuing to negotiate, the Kaanoowapmaakin are also continuing a political tradition that is part of the more-obvious struggle for political control and sovereignty. Here is where Offen's "field-informed" historical research is vital, because the history of this process is here on the land, and with these people, and in stories like Solomon's. Here is the history and ecology that is found in the archive of the land—the history at which other archives only hint.

Concluding Thoughts

As I think about this place where I'm sitting tonight, as I think about that tent site at the other end of the portage trail and the story connected to it, I want to be clear about the ways that land and story are being acted upon by the larger forces of political economy. I also want to make clear how they continue to exist as places that are defined within Cree culture and Cree history. These two processes are connected, and I think they ought to be connected in both our historical and our present understanding of this place. Adapting academic methodology is part of this, but I want to be clear that in many ways the Cree themselves are pushing me to grapple with the full meaning of the narrative ecology with which I began. They push me to do this in the ways they themselves think about both their land and their stories; and they do it through the actions they take to protect both land and culture together.

Since 2011, the community of Waswanipi has been trying to protect an area north of the Broadback River, which the province and Eacom (the parent company of Domtar) aim to clearcut. This is the Assinica Valley, one of the last unbroken valleys in Quebec's boreal forest and one of the last intact habitats for the threatened woodland caribou. Many environmental groups want to see it protected for these reasons, and thus seek to work cooperatively with the Cree. There is a great deal of room for cooperation, but for the Cree the area is known as *Mishigamish*, the Big Sea, and it takes in four or five Waswanipi *indoh-hoh istchee*. These are hunting lands that have not been touched by logging, so the Waswanipi Cree want to protect this area for its environmental value, certainly—but they also want to protect its cultural and political value in maintaining their traditional presence on their land.

Traditional practice on these *indoh-hoh istchee* has not been affected by development; hunters there have not had to make compromises because of cutting or the changes brought by roads. Nor have they been tempted to change their hunting practices because of the convenience of roads. The Cree know that people change for reasons other than being forced. They have a nuanced understanding of how and why culture and land change, and they also understand that cultural adaptation is the heart of their relationship with the land. This is one of the lessons in the adaptation of the *indoh-hoh istchee* in the twentieth century: adaptation, participation, even politicization, are not loss of tradition. In fact, not only should they be seen as signs of a healthy tradition, they actually explain much about what is going on in Eeyou Istchee today.

There are issues, however, in protecting the land this way. Provincial environmental agencies and independent scientists are increasingly moving toward a model of ecosystem management in dealing with issues of preservation and conservation. There is a good deal to recommend this approach in purely ecological terms, but the ecosystems that science defines are not the same as the *indoh-hoh istchee* defined by Cree culture. Ecosystems are not the bush either, and part of the Cree desire to protect these places where older hunting methods still exist is also to protect those older ideas on the land, as well. Like the caribou, these ideas need an intact forest to thrive.

Ecosystems do not define the cultural or political space that is central to Cree culture and to their desire for protection. There will have to be another negotiation over land, how it will be defined, and this is not simply a matter traditional culture versus Western thinking. It's not that simple anymore—if ever was that simple—and this is true of many issues in Eeyou Istchee today. This is true across the whole of the Canadian north, and it's the challenge for those of us writing about this land, its history, and its people. In all these places, stories like Solomon's continue to cycle, continuing the present-ness of First Nations' culture and history.

The north comprises a wide variety of communities of place and all of them have deep traditions of seeing the land as central to the idea of community. They see the land as community. The Kaanoowapmaakin and all the people living on the lands of Eeyou Istchee are the continuation of traditional Cree practice, but also embody a larger intellectual tradition concerning how humans live in these places. They represent a political argument, too, and now, nearly everywhere, the press of development and its concomitant politics is part of the local ecology of story. The challenge is telling our histories in a way that does justice to this reality.

Notes

1   I am referring specifically to Sigurd Olsen, *The Lonely Land* (Minneapolis: University of Minnesota Press, 1997), but generally to all romantic portrayals of the Canadian North.

2   The Ouje Bougoumou Cree were not recognized and given land to build a new village until the 1990s. They did not achieve full status under the James Bay and Northern Quebec Agreement—and subsequent treaties with Quebec—until November 2011.

3   N. Scott Momaday, *The Man Made of Words: Essay, Stories, and Passages* (New York: St. Martin's Press, 1997), 187.

4   Thomas King, *The Truth About Stories: A Native Narrative* (Minneapolis: University of Minnesota Press, 2003), 29. See also Donald L. Fixico, "Ethics and Responsibilities in Writing American Indian History," 84–99, and Vine Deloria Jr., "Comfortable Fictions and the Struggle for Turf: An Essay Review of the *Invented Indian: Cultural Fictions and Government Policies*,"

65–84, both in *Natives and Academics: Researching And Writing about American Indians*, ed. Devon A. Mihesuah (Lincoln: University of Nebraska Press, 1998).

5   Matthew Coon Come, "Survival in the Context of Mega-Resource Development: Experiences of the James Bay Cree and the First Nations of Canada," in *In the Way of Development: Indigenous Peoples, Life Projects and Globalization*, ed. Mario Blaser, Harvey A. Fiet, and Glenn McRae (New York: Zed Books, 2004), 158.

6   Tina Loo, "Disturbing the Peace: Environmental Change and the Scales of Justice on a Northern River," *Environmental History* 12 (October 2007): 895–919; George Towers, "Applying the Political Geography of Scale: Grassroots Strategies and Environmental Justice," *Professional Geographer* 52 (2000): 25; Andrew Herod and Melissa W. Wright, "Placing Scale: An Introduction," in *Geographies of Power: Placing Scale*, ed. Andrew

Herod and Melissa W. Wright (Oxford: Blackwell Publishing, 2002).

7   Matthew Coon Come, "Shafted By Hydro-Québec: The Cree, the Environment, and Quebec," *Canada Speeches* 5, no. 5 (September–October 1991).

8   Janet Armstrong, "Voices of the Land," in *Speaking for the Generations: Native Writers on Writing*, ed. Simon Ortiz (Tucson: University of Arizona Press, 1998), 181.

9   King, *Truth About Stories*, 92.

10   Robert Bringhurst, "The Polyhistorical Mind," in *The Tree of Meaning: Language, Mind, and Ecology* (Berkeley, CA: Counterpoint, 2006), 28.

11   The best of these are: Keith H. Basso, *Wisdom Sits In Places: Landscape and Language Among the Western Apache* (Albuquerque: University of New Mexico Press, 1996); Julie Cruikshank, *Do Glaciers Listen? Local Knowledge, Colonial Encounters, and Social Imagination* (Vancouver: UBC Press, 2005); Julie Cruikshank, *The Social Life of Stories: Narrative and Knowledge in the Yukon Territory* (Lincoln: University of Nebraska Press, 2000). For a more philosophical discussion, see Tim Ingold, *The Perceptions of the Environment: Essays on Livelihood, Dwelling and Skill* (New York: Routledge, 2000).

12   For a primer on this debate, see Donald Worster, "Doing Environmental History," in *The Ends of the Earth: Perspectives on Modern Environmental History* (New York: Cambridge University Press, 1988), 289–307; Donald Worster, "History as Natural History: An Essay on Theory and Method," *Pacific Historical Review* (1984):

1–19; William Cronon, "A Place for Stories: Nature, History, and Narrative," *Journal of American History* 78 (March 1992): 1347–76; "In Search of Nature," in *Uncommon Ground: Rethinking the Human Place in Nature*, ed. William Cronon (New York: W. W. Norton, 1995), 23–68.

13   Alfred North Whitehead, *Science and the Modern World* (Cambridge: Cambridge University Press, 1926), 64.

14   Bringhurst, "The Polyhistorical Mind," 33–34.

15   Harvey Feit, "Gifts of the Land: Hunting Territories, Guaranteed Incomes and the Construction of Social Relations in James Bay Cree Society," *Senri Ethnologica Studies* 30 (1991): 223–68; Harvey Feit, "James Bay Crees' Life Projects and Politics: Histories of Place, Animal Partners, and Enduring Relationships," in *In the Way of Development: Indigenous Peoples, Life Projects and Globalization*, ed. Mario Blaser, Harvey A. Fiet, and Glenn McRae (New York: Zed Books, 2004); Adrian Tanner, *Bringing Home Animals: Religious Ideology and Mode of Production of the Mistassini Cree Hunters* (New York: St. Martin's Press, 1979); Richard Preston, *Cree Narrative* (Montreal: McGill-Queen's University Press, 2002).

16   Clifford Geertz, *The Interpretation of Cultures: Selected Essays* (New York: Basic Books, 1973); for his ideas on "thick description," see pp. 1–29; also see Gilbert Ryle, "The Thinking of Thoughts: What is *le penseur* Doing?", *University Lectures* 18 (1968).

17   Hans M. Carlson, *Home Is The Hunter: The James Bay Cree and Their Land* (Vancouver: UBC Press, 2008).

18   Alfred W. Crosby, *The Columbian Exchange: The Biological and Cultural Consequences of 1492* (Westport, CT: Greenwood, 1972); Francis Jennings, *The Invasion of America: Indians, Colonialism, and the Cant of Conquest* (Chapel Hill: University of North Carolina Press, 1975), 15–31.

19   See, for example, Bruce Trigger, *Natives and Newcomers: Canada's "Heroic Age" Reconsidered* (Montreal: McGill-Queen's University Press, 1985); Karen Kupperman, *Indians and English: Facing Off in Early America* (Ithaca: Cornell University Press, 2000); Kenneth M. Morrison, *The Embattled Northeast: The Elusive Ideal of Alliance in Abenaki-Euroamerican Relations* (Berkeley: University of California Press, 1884); James D. Drake, *King Philip's War: Civil War in New England, 1675–1676* (Amherst: University of Massachusetts Press, 1999); Evan Haifeli and Kevin Sweeney, *Captors and Captive: The 1704 French and Indian Raid on Deerfield* (Amherst: University of Massachusetts Press, 2003); John G. Reid, ed., *The "Conquest" of Acadia, 1710: Imperial, Colonial, and Aboriginal Constructions* (Toronto: University of Toronto Press, 2004).

20   For an understanding of the roots of ethnohistory as a sub-discipline, see Alfred G. Bailey, "Retrospective Thoughts of an Ethnohistorian," *Historical Papers* (1977); James Axtell, "Ethnohistory: An Historian's Viewpoint," *Ethnohistory* 26 (Winter 1979); Bruce Trigger, "Ethnohistory: Problems and Prospects," *Ethnohistory* 29 (Fall 1982); and Francis Jennings, "A Growing Partnership: Historians, Anthropologists and American Indian History," *Ethnohistory* 29 (Fall 1982).

21   Jennifer S. H. Brown, "Ethnohistorians: Strange Bedfellows, Kindred Spirits," *Ethnohistory* 38, no. 2 (Spring 1991): 117; see also James Taylor Carson, "Ethnogeography and the Native American Past," *Ethnohistory* 49, no. 4 (Fall 2002): 770–88; and R. Cole Harris, *Making Native Spaces: Colonialism, Resistance, and Reserves in British Columbia* (Vancouver: UBC Press, 2002).

22   Calvin Martin, *Keepers of the Game: Indian-Animal Relations in the Fur Trade* (Berkeley: University of California Press, 1978); Shepard Krech, *The Ecological Indian: Myth and History* (New York: Norton, 1999); Ken S. Coates and Robin Fisher, eds., *Out of the Background: Readings on Canadian Native History* (Toronto: Copp Clark, 1996); Daniel Francis and Toby Morantz, *Partner in Fur: A History of the Fur Trade in Eastern James Bay, 1600–1870* (Montreal: McGill-Queen's University Press, 1983); Jennifer S. H. Brown, *Strangers in Blood: Fur Trade Families in Indian Country* (Vancouver: UBC Press, 1980); Sylvia Vankirk, *Many Tender Ties: Women in Fur-trade Society in Western Canada, 1670–1870* (Norman: University of Oklahoma Press, 1980); Deny Délage, *Le pays renverse: amerindiens et europeens en Amerique du nord-est, 1600-1664* (Montreal: Boreal Express, 1985).

23   In works like William Cronon, *Changes in the Land: Indians, Colonists, and the Ecology of New England* (New York: Hill and Wang, 1983); Richard White, *The Roots of Dependency: Subsistence, Environment, and Social Change among the Choctaws, Pawnees, and Navajos* (Lincoln: University of Nebraska Press, 1983); Carolyn Merchant, *Ecological Revolutions: Nature, Gender, and Science in New England* (Chapel Hill: University of North Carolina Press, 1989); Timothy Silver, *A New Face on the Countryside: Indians, Colonists and Slaves in South Atlantic Forests, 1500–1800* (Cambridge University Press, 1990); and David Rich Lewis, *Neither Wolf Nor Dog: American Indians, Environment, and Agrarian Change* (New York: Oxford University Press, 1994), we are given thoughtful regional studies that show the agency of Native people and their cultural understandings of the land they lived on. At the same time, these works perpetuate a narrative fall from grace, as Native people's worlds disappear in the face of European colonization. Native people themselves often disappear, if they are not the direct focus of a work, leaving the impression that there are no Native people left in places like New England. This declentionist narrative trope is something that William Cronon wrestled with in "A Place for Stories: Nature, History, and Narrative," *Journal of American History* 78 (March 1992), but it's still a powerful force in the writing of history. This is not just a problem of environmental history, and has deep roots; see Jean M. O'Brien, *Firsting and Lasting: Writing Indians Out of Existence in New*

England (Minneapolis: University of Minnesota Press, 2010).

24   Paul Sutter, "Reflections: What Can U.S. Environmental Historians Learn from Non-U.S. Environmental Historiography?" *Environmental History* 8 (January 2003); Lynda Jessup, "The Group of Seven and the Tourist Landscape in Western Canada, or The More Things Change…", *Journal of Canadian Studies* 37 (Spring 2002): 146–47; George Colpitts, *Game in the Garden: A Human History of Wildlife in Western Canada to 1940* (Vancouver: UBC Press, 2002).

25   I'm thinking of those original works by Harold Innis and W. A. Mackintosh, but also more recent work, as well. See Daniel Drache, ed., *Staples, Markets and Cultural Change: The Centenary Edition of Harold Innis' Collected Essays* (Montreal: McGill-Queen's University Press, 1995); Larry A. Glassford, "The Evolution of 'New Political History' in English-Canadian Historiography: From Cliometrics to Cliodiversity," *American Review of Canadian Studies* 32 (Autumn 2002): 349–50; Gary Burrill and Ian McKay, *People, Resources, and Power: Critical Perspectives on Underdevelopment and Primary Industries in the Atlantic Region* (Fredericton, NB: Gorsebrook Research Institute of Atlantic Canada Studies and Acadiensis Press, 1987); J. I. (Hans) Bakker, "Canadian Political Economy and Rural Sociology: Early History of Rural Studies in Canada," *Rural Sociologist* 7 (September 1987); Robert J. Brym and R. James Sacouman, eds., *Underdevelopment and Social Movements in Atlantic Canada* (Toronto: New Hogtown Press, 1979); John McCallum,

Unequal Beginnings: Agriculture and Economic Development in Quebec and Ontario Until 1870 (Toronto: University of Toronto Press, 1980).

26 For an in-depth investigation of this, see Caroline Desbiens, *Power from the North: Territory, Identity, and the Culture of Hydroelectricity in Quebec* (Vancouver: UBC Press, 2013); and "Producing North and South: A political geography of hydro development in Québec," *Canadian Geographer/Le Géographe canadien* 48, no. 2 (2004):101–18.

27 William Cronon, "The Uses of Environmental History," *Environmental History Review* 17 (Fall 1993); Richard White, "Afterword: Environmental History: Watching a Historical Field Mature," *Pacific Historical Review* 70 (February 2001); Alfred W. Crosby, "The Past and Present of Environmental History," *American Historical Review* 100 (October 1995).

28 Desbiens, *Power from the North*, 133.

29 Personal email, 4 November 2010.

30 For my interpretation of the beginnings of these negotiations, see Hans M. Carlson, "A Watershed of Words: Litigating and Negotiating Nature in Eastern James Bay, 1971–1975," in *Contemporary Quebec: Selected Readings and Commentaries*, ed. Matthew Hayday and Michael Behiels (Montreal: McGill-Queen's University Press, 2011).

31 Government of Quebec, *James Bay and Northern Québec Agreement* (Quebec City: Editeur officiel du Québec, 1976), 291.

32 Carlson, *Home Is the Hunter*, 287–333; *James Bay and Northern Québec Agreement*, 332, 359.

33 Roy MacGregor, *Chief: The Fearless Vision of Billy Diamond* (Middlesex, UK: Penguin, 1990), 123–25; Naila Clerici, "The Cree of James Bay and the Construction of Their Identity for the Media," *Canadian Issues-Themes Canadiens* 21(1999): 143–65.

34 Matthew Coon-Come, "A Reduction of Our World," in *Our People, Our Land: Perspectives on the Columbus Quincentenary*, ed. Kurt Russo (Bellingham, WA: Lummi Tribe and Kluckholn Center, 1992), 82; Glenn McRae, "Grassroots Transnationalism and Life Projects of Vermonters in the Great Whale Campaign," in *In the Way of Development*, ed. Mario Blaser, Harvey A. Fiet, and Glenn McRae (New York: Zed Books, 2004), 111–29.

35 Steve Turner and Todd Nachowitz, "The Damming of Native Lands," *The Nation*, 21 October 1991, 473–77; Jeff Jones, "Power Struggle: Will New York's Search for Clean Energy Destroy Cree and Inuit Homelands in Northern Quebec," *UpRiver/DownRiver*, March–April 1991, 24–28; Chris Busby, "Bourassa's Dream for Quebec: An Environmental Knockout Punch," *Borealis* 2, no. 4 (1991): 8–13; Sam Howe Verhovek, "Power Struggle: Flooding Quebec to Light New York," *New York Times Magazine*, 12 January 1992, 17–21; André Picard, "James Bay II," excerpted from the *Globe and Mail, Amicus Journal*, Fall 1990, 10–16; Harry Thurston, "Power in a Land of Remembrance," *Audubon*, November 1991, 52–59.

36 For more on logging issues and the Paix des Braves agreement, see Naomi C. Heindel, "The Cree and

the Crown: Management Stories From North America's Northern, Northern Woodlands," *Northern Woodlands* 75 (Winter 2012): 26–37; see also Hans M. Carlson, "Strangers Still and the Land Nearly Devoured," *Yellow Medicine Review* (Spring 2014): 50–61.

37 This was during the second administration of Grand Chief Ted Moses (1999–2005). The initiative was forwarded by his chief negotiator, Abel Bosum.

38 Carlson, *Home Is the Hunter*, 54–56.

39 Paul Robbins, *Political Ecology* (New York: Blackwell, 2004), 5.

40 See J. B. Greenberg and T. K. Park, "Political Ecology," *Journal of Political* Ecology 1 (1994): 1–12; R. Peet and M. Watts, "Liberation Ecology: Development, Sustainability, and Environment in an Age of Market Triumphalism," in *Liberation Ecologies: Environment, Development, Social Movements* (New York; Routledge, 1996), 1–45. For an example that touches on Solomon's statement, see Kristín Loftsdóttir, "Where My Cord Is Buried: *WoDaaBe* Use and Conceptualization of Land," *Journal of Political Ecology* 8 (2001): 1–24.

41 Christian Brannstrom, "What Kind of History for What Kind of Political Ecology?", *Historical Geography* 32 (2004): 71.

42 Karl H. Offen, "Historical Political Ecology: An Introduction," *Historical Geography* 32 (2004): 33.

43 Offen, "Historical Political Ecology," 21–22.

44 The use of the word "garden" was first recorded in Boyce Richardson, *Strangers Devour the Land* (New York: Knopf, 1976), in an interview with Job Bearskin. Whether he innovated this usage is not known, but he was always very careful to say that the land was "like" a garden, according to Richardson (personal communication).

45 For some of the pitfalls in making too much of this analogy, see Paul Nadasdy, "'We Don't *Harvest* Animals; We *Kill* Them': Agricultural Metaphors and the Politics of Wildlife Management in the Yukon," in *Knowing Nature: Conversations at the Intersection of Political Ecology and Science Studies* (Chicago: University of Chicago Press, 2011), 135–51.

46 Carlson, *Home Is The Hunter*, 4.

47 J. M. Cooper, "Is the Algonquian Family Hunting Ground System Pre-Columbian?", *American Anthropologist* 41 (1939): 66–90; Harvey Feit, "The Construction of Algonquian Hunting Territories," in *Colonial Situations: Essays on the Contextualization of Ethnographic Knowledge*, History of Anthropology, vol. 7, ed. G. Stocking (Madison: University of Wisconsin Press, 1991), 109–34. National Archives of Canada, Indian Affairs, RG 10, vol. 6752, file 420-10-1-3, 1943 Report for Indian Affairs Branch.

48 For a detailed description of the modern *Indoh-hoh* system, see "Eeyou Indoh-Hoh Weeshou Wehwun," http://creetrappers. ca/wp-content/uploads/2014/02/ CTA_EEYOU_HUNTING_LAW. pdf. This is a link off of the Cree Trapper's Association homepage (www.creetrappers.ca).

49 Carlson, *Home Is The Hunter*, 167–201.

50 This was before the formation of the Grand Council of the Cree in 1974.

51 My interpretation of the negotiations between *Kaanoowap-maakin* and the MNR and forestry companies comes from personal experience. These meetings are a fascinating mix of cordiality, resignation, and some outright hostility, depending on the age and personality of the *Kaanoowap-maakin*. The non-Natives involved are for the most part sympathetic to the Cree desire to protect land and culture, but there is an agenda driving these meetings written outside the region, which no one present is in a position to change. See also Heindel, n37.

10

Imposing Territoriality: First Nation Land Claims and the Transformation of Human-Environment Relations in the Yukon

Paul Nadasdy

Introduction

Environmental historians, anthropologists, and others have long argued that the imposition of Euro-American property regimes dramatically altered how Indigenous peoples could relate to the land, animals, and one another.[1] The effects of this imposition are very much in evidence across northern Canada,[2] where, over the past forty years, Indigenous people have engaged federal and territorial governments in negotiating comprehensive land claim and self-government agreements. These modern treaties grant northern Indigenous peoples real powers to govern themselves and manage their own lands and resources, but because they are conceived in the Euro-American language of property, they also necessarily transform how Indigenous people relate to that land and the animals upon it.[3] In this chapter, I expand the purview of my inquiry into the nature of these agreements beyond a narrow focus on property per se, and argue that they impose upon Indigenous people ideas and practices of *territoriality* more generally.

Geographer Robert Sack defines territoriality as "the attempt by an individual or group to affect, influence, or control people, phenomena, and relationships, by delimiting and asserting control over a geographic area."[4] He concedes that while there are many other, non-territorial ways of exercising power, "territoriality is the primary spatial form power takes."[5] Territoriality is thus a particular kind of political strategy, one that focuses on controlling people and processes through the demarcation and control of space. Sack (along with others) views the modern state as a fundamentally territorial form of political organization,[6] and he suggests that two state territorial strategies in particular have played a central role in the transformation of Indigenous people's relationship to land: the delimitation of property rights in land and the establishment of political jurisdiction.[7] As it happens, both forms of territorial practice are essential to the structure of comprehensive land claim and self-government agreements in Canada.[8]

These agreements are the principal contemporary means for incorporating northern Indigenous peoples into the Canadian state. They spell out the nature of government-to-government relations among the signatory governments, and grant northern First Nations real (if limited) powers of self-government, as well as a key role in the management of northern lands and resources. As a result, First Nation governments across the Canadian north have emerged as significant players in regional politics.[9] This is very different from the days, not so long ago, when they lived under the colonialist dictates of the federal Indian Act and had virtually no say either in their own governance or in the management of the lands and resources upon which they depend. As I have argued elsewhere, however, the price of this newfound authority has been high.[10] Northern Indigenous people have had to restructure their societies in dramatic ways just to gain a seat at the negotiating table. To be heard at all, they have had to frame their arguments in a language intelligible to lawyers, politicians, and other agents of the Canadian state. And, once in place, these agreements have a dramatic bureaucratizing effect, drawing Indigenous people off the land and into offices to become professional "managers."

In this chapter, I focus on the territorial (and territorializing) aspects of the Yukon land claim and self-government agreements. These agreements are fundamentally territorial; that is, it is primarily (though not solely) by demarcating space and assigning control over the resulting territories to various governments that the agreements constitute First Nations' authority in relation to other governments and their own citizens. The agreements create two distinct types of First Nation territory (in Sack's sense of the term). First, they carve the Yukon into fourteen distinct First Nation *traditional territories* (Fig. 10.1). Second, within each traditional territory the agreements also demarcate smaller areas of First Nation *settlement land* (Fig. 10.2). As we shall see, traditional territories and settlement lands are integral to the structure of both final and self-government agreements. It is tempting to view settlement lands as a form of property (which, in fact, they are) and traditional territories as being more about jurisdiction. There is some truth to this characterization, but things are actually considerably more complex. In fact, each of these territorial forms has both proprietary and jurisdictional aspects. Having considered the proprietary aspects of the Yukon agreements elsewhere, I focus in this chapter on their jurisdictional dimensions.[11]

Although many people—Indigenous and Euro-Canadian alike—assume that traditional territories reflect "traditional" patterns of land-use and occupancy, Indigenous society in the Yukon was not, in fact, composed of distinct political entities, each with jurisdiction over its own territory; such entities are actually a recent phenomenon in the Yukon. The new agreements, then, are not simply formalizing jurisdictional boundaries among pre-existing First Nation polities; they are mechanisms for *creating* the legal and administrative systems that bring those polities into being. In fact, the agreements are premised on the assumption that First Nation governments must be discrete politico-territorial entities if they are to qualify as governments at all.

Thus, although the Yukon agreements do grant First Nations some very real powers of governance, those powers come in the peculiarly territorial currency of the modern state. Not only does this implicitly devalue Indigenous forms of socio-political organization (which are not territorial), it is also transforming Indigenous society in radical and often unintended ways. My focus in this chapter is on how the land claim agreements and the territorial assumptions underlying them are transforming how Yukon Indian people, particularly those of the Kluane First Nation (KFN) and its neighbours in the southwest Yukon, relate to the land and

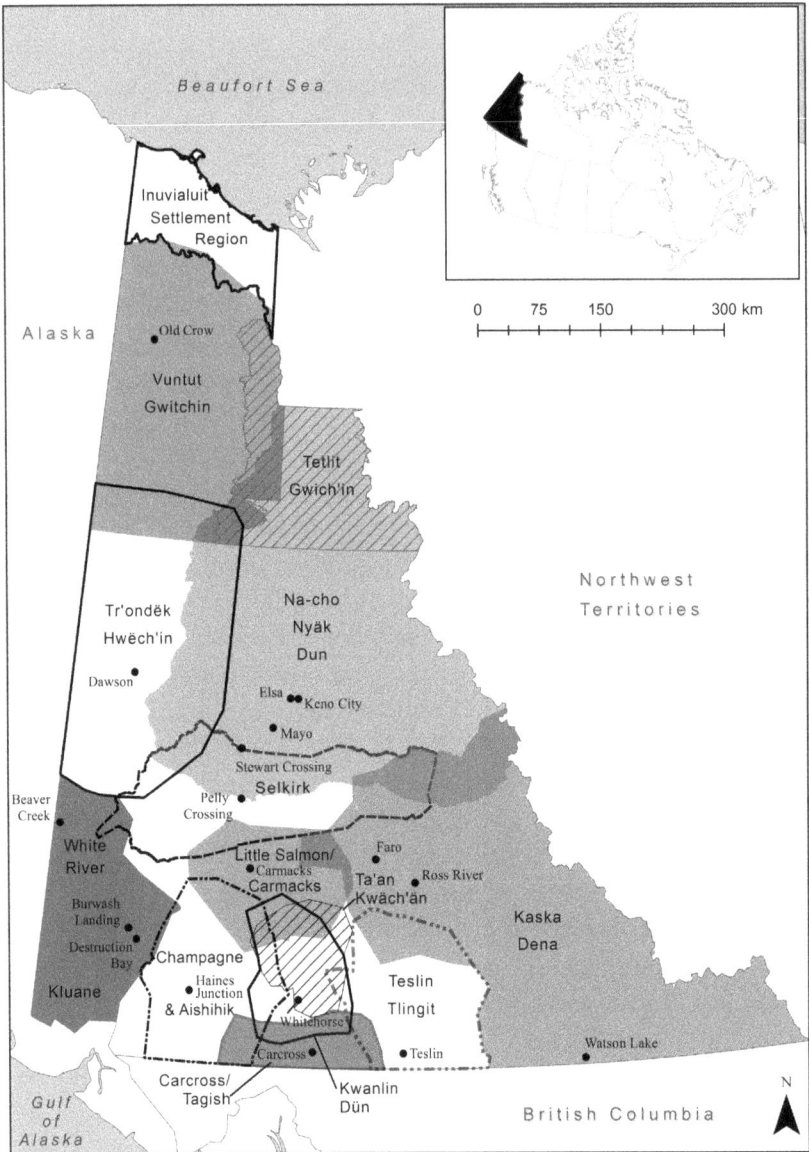

Fig. 10.1: Map of Yukon First Nation Traditional Territories by Tracy Sallaway (Maps, Data and Government Information Centre, Trent University). Contains information licensed under the Open Government Licence – Canada (http://open.canada.ca/en/open-government-licence-canada); all other data – Geomatics, Department of Environment, Government of Yukon.

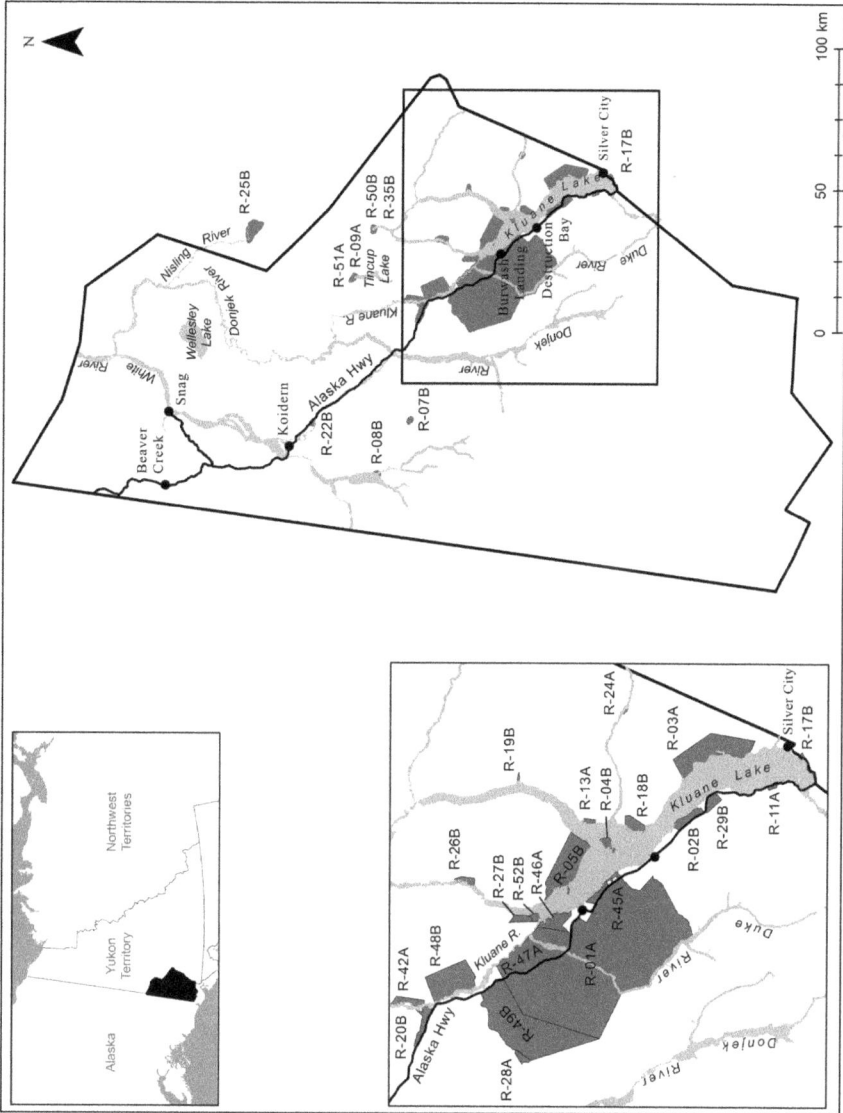

Fig. 10.2: Map of Kluane First Nation Traditional Territory and Settlement Lands by Tracy Sallaway (Maps, Data and Government Information Centre, Trent University). All data – Geomatics, Department of Environment, Government of Yukon.

animals—as well as to one another with respect to the land and animals. Before doing so, however, it is necessary to say a bit more about the territorial assumptions underlying the Yukon agreements.

Territoriality and the State

Over the past two decades, political geographers, anthropologists, and others have analyzed the territorial dimensions of the modern state. Rather than simply taking for granted its territorial boundedness, as scholars did previously, they have begun systematically to examine the role of territory—and of territoriality itself—in the constitution of the state as a particular socio-historical phenomenon. Building on Sack's political conception of territoriality, contemporary political geographers now agree that modern state power is largely—though by no means exclusively—an exercise in territoriality; that is, states seek to control people and resources principally through the demarcation and control of space. Indeed, the demarcation and control over territory, within which the state supposedly exercises exclusive sovereign power, has long been viewed as an essential aspect of the modern state.[12] The establishment and maintenance of clear boundaries both among and within states is essential to the concept of the modern state and the exercise of state power. Maps have played a particularly important role in the creation and consolidation of state power. In his study of the birth of the Thai nation, Thongchai Winichakul argues that early government-generated maps of the country were not simply abstract representations of reality. The situation was rather the reverse: "A map anticipated spatial reality, not vice versa. In other words, a map was a model for, rather than a model of, what it purported to represent."[13] He shows that maps not only helped constitute Thailand as a territorial state vis-à-vis other such states; they were also essential for developing essential *internal* administrative mechanisms.[14]

Sack observes that the modern state is territorial not only externally (i.e., vis-à-vis other states), but also internally: it produces a hierarchy of nested territories right down to the level of individual factories and households, which are themselves internally territorialized.[15] This internal territoriality involves not only the creation of jurisdictional boundaries within states but also the formalization and protection of property rights in land. These processes of internal territorialization produce administrative differences among citizens and carve the landscape into different geographical sub-units. This is essential because, although the creation of a unified state requires the erasure of certain kinds of difference, other kinds of differences among people and places are essential for the practice of governance. Internal boundaries (whether jurisdictional or proprietary) allow for the delegation of authority and the rationalization of jurisdiction among different levels of government, the coordination and delivery of government services, the management of people and resources, and so on.

Boundary making within the territorial state is far from politically neutral. Malcolm Anderson points out that boundaries within states, though "often presented as technical adjustments to promote efficiency of administration, are never independent of changes in power relationships."[16] Indeed, as Vandergeest and Peluso show, processes of internal territorialization have played a key role in efforts to expand and consolidate state control over peoples and resources.[17] In settler states, such processes have resulted in the (often violent) reworking of Indigenous social relations, transforming in fundamental ways how certain kinds of people can relate to one another as well as to the land and resources.[18] Yet, internal territorialization is not always a top-down process, nor are its socio-political consequences always those intended by either state officials or those who would resist them. There is a great deal to be said about this process of internal territorialization in the Yukon, but before considering its impact on human-environment relations there, a brief history of territoriality in the Yukon is in order.

Territoriality and Human-Environment Relations in the Yukon

Pre-Contact and Early Contact Yukon

Until the middle of the twentieth century, Yukon Indian people were nomadic, covering large distances in the course of their annual subsistence round; and for much of the year they lived in small hunting groups. These groups were extremely flexible; there were no formal rules for membership, and their composition was constantly changing as a result of seasonal and longer-term variations in the availability of resources, social tensions among group members, marriage, long-distance trading, and so on. Catharine McClellan characterized nineteenth-century Yukon socio-political organization: "A very sparse population was spread over a vast area, making a loosely linked social network with very few sharply defined linguistic and cultural boundaries. ... Cohesive political units did not exist—just widely scattered clusters of living groups whose composition and size changed throughout the year as people moved about in quest of food."[19] She noted that Aboriginal leaders "certainly never had clearly defined judicial or punitive powers over all persons living in a delimited territorial unit."[20] Social relations among Yukon Indian people were ordered by principles of kinship and reciprocity rather than territoriality. People drew on far-flung networks of bilateral kin to travel widely and exploit resources more or less where they pleased.[21] Although these kinship networks certainly existed in space, they were not defined by—nor did they define—specific territories, in Sack's sense of the term. This is not to say that Yukon Indian people never used territorial strategies of the sort Sack described. It seems clear, for example, that some important fishing sites were "owned" by particular moieties, the members of which regulated access to them. In practice, though, because the moieties are exogamous, everyone had close relatives from the opposite moiety on whom they could prevail for access.[22]

Yukon Indian people were not organized into distinct "tribes" with control over fixed territories. Nor did tribal categories organize their kinship practices, trade, or political relations. In fact, it is questionable whether they had "tribal" categories at all. In her ethnographic survey of the southern Yukon, McClellan did divide up Yukon Indians geographically on the basis of language and referred to these divisions as "tribes," but she stressed that this was purely for the convenience of the ethnographer and warned that "it is not the kind of classification which the Indians themselves are likely to stress, or perhaps even recognize." "Tribe" was a particularly problematic term, she suggested, because "usually it implies a sense of political unity which the Yukon natives ... never had."[23]

Although Yukon Indian people did not regard themselves as divided into distinct territorially organized "peoples," they did, of course, have ways of classifying one another. McClellan sketched the outline of a very complex Indigenous system for classifying people, which, she maintained, "is highly relative, depending on the particular vantage points in time and space of both the classifier and the classified. Also, various modes of classification cross-cut each other. Finally, the Yukon Indians prefer to think in terms of selected individuals rather than of total geographically bounded groups."[24] So, although social differences certainly existed in the Yukon in pre-contact times, those differences were not used to mark off the territories of distinct political units, nor were they used for regulating people's access to resources.[25] It seems that Yukon Indian people did not organize themselves into distinct politico-territorial units until well after European contact; and, as we shall see, it was as a result of their contact with the Canadian state that they first began to do so.[26]

The Expanding State, Territoriality, and First Nation Land Claims

Although Yukon Indian political organization did change in response to the fur trade, the pace of change accelerated markedly in the 1940s and 1950s, when federal officials began asserting control over the lands and peoples of the Canadian north. To this end, they divided the nomadic Indigenous population into distinct administrative "bands," each with its own elected chief and council. These bands, created under the federal Indian Act, had no relation to any existing political units; rather, they were composed of different families who had in many cases very different patterns of seasonal movement and who had settled—or sometimes had been coerced to settle—in a number of central locations.[27] The enforced settlement of nomadic populations in easily accessible locations is a strategy that has been used by expanding states the world over; it enables officials to assert control over these peoples and to provide them with government programs and services. The bands themselves, despite having elected chiefs and councils, had little real self-government authority, and acted instead as bureaucratic intermediaries between the federal government and local populations, helping to administer government programs such as the provision of social assistance, medical care, and housing.[28] So the geographical division of the Yukon Indian population into separate administrative bands had more to do with federal administrative objectives than with any cultural or linguistic factors. In fact, the federal government occasionally amalgamated and otherwise reorganized previously distinct bands for purely administrative reasons, principally to streamline and decrease the cost of service delivery.[29]

In contrast to the situation in much of southern Canada and the United States, the Canadian government never negotiated any land cession treaties in the Yukon.[30] Even so, it claimed to control all lands in the Yukon and maintained the position that Yukon Indian people had no legal entitlement to the land, except that with which the government had explicitly provided them, of which there was very little.[31] Yukon Indian people made no unified effort to assert their rights to land until 1973. In that year, Elijah Smith, president and co-founder of the newly formed Yukon Native Brotherhood, a political organization representing status Indians throughout the Yukon, presented the federal government with a document entitled *Together Today for Our Children Tomorrow*.[32] This was the first comprehensive land claim formally accepted for negotiation by the government of Canada.

Despite this, a Yukon land claim agreement was still a long way off. Negotiations dragged on for twenty years, until, in 1993, representatives of the federal and territorial governments, along with the Council for Yukon Indians, the Yukon Native Brotherhood's successor organization, signed the Yukon Umbrella Final Agreement (UFA). This is not in itself a land claim agreement; rather, it is a framework for the negotiation of specific Final Agreements between each of the fourteen individual Yukon First Nations and the federal and territorial governments. It is a complex document of twenty-eight chapters that deals not only with land and financial compensation, but also with self-government, taxation, renewable and non-renewable resources, heritage, economic development, and more. The UFA contains many general provisions that apply to the entire Yukon, and also identifies the areas in which individual First Nations may negotiate provisions specific to their own needs. Each individual Yukon First Nation was to negotiate its own specific Final Agreement within the framework of the UFA. The complex structure of the Yukon agreement was the result of compromise between Yukon First Nations, who were generally wary of a single Yukon-wide agreement, preferring instead multiple agreements that would be more sensitive to local First Nation needs, and the federal government, which preferred a Yukon-wide agreement so as to avoid the administrative nightmare of having fourteen completely distinct treaties in the Yukon.

Territorial Dimensions of the Yukon Land Claim Agreements

With the signing of land claim and self-government agreements, self-governing First Nations have replaced Indian Act bands throughout much of the territory. Although the transformation from band to First Nation has led to some important changes in their demographic composition, there is a great deal of continuity between these new, self-governing First Nations and their Indian Act predecessors. To some extent, this was probably inevitable. As we saw, Indian people were loath to enter into a Yukon-wide agreement, preferring instead a series of individual First Nation Final Agreements that would allow them to address local issues and concerns. Since popularly elected band governments already existed throughout the territory when land claims negotiations began in the 1970s, they naturally played an important role in the political organization of Yukon Indian people. It was individual bands that entered into negotiations with the federal and territorial governments, and ultimately became signatories to the agreements. In fact, people regularly referred to "Band Final Agreements" (rather than the official "First Nation Final Agreements") through the mid-1990s.

The political continuity between Indian Act bands and self-governing First Nations is also evident in the fact that First Nations inherited responsibility for the delivery of programs and services that had previously been administered by bands (and, in fact, funding levels for First Nation self-government were based directly on the bands' historical spending levels). Although First Nations have also assumed responsibility for additional programs and services that were not administered by bands, there is nevertheless an important sense in which self-governing First Nations evolved from the Indian Act bands that preceded them. The current configuration of First Nations in the Yukon, then, reflects quite closely the legacy of the Department of Indian Affairs' administration of Indian people in the territory.

There are, however, some very important differences between Indian Act bands and the self-governing First Nations that are succeeding them. Primary among these is the fact that First Nations, unlike bands, are political entities whose powers and authorities are territorially constituted. As noted above, the two principal forms of First Nation territory created by the agreements are traditional territories and settlement lands. Both are defined and mapped in First Nation final agreements, and, as we shall see, they are integral to the structure of both final and self-government agreements. They play a particularly important role in the new regime for managing wildlife and other natural resources.

Traditional Territories, Overlap, and Black Holes

First Nations do not "own" their traditional territories; but they do retain some rights on these lands that can be viewed as proprietary. Most important for the purposes of this chapter, First Nation citizens retain the right to hunt and fish throughout their entire traditional territory.[33] When they ratified their agreements, they exchanged their Aboriginal right to hunt anywhere in Canada for the more limited right spelled out in their agreements.[34] In addition to granting Yukon Indian people residual use-rights of this sort, the agreements also grant First Nations a prominent role in the management of wildlife, heritage, and other resources throughout their traditional territories—primarily through participation in formal co-management processes created under the agreements. Of central importance to First Nations is the process for co-managing fish, wildlife, timber, and other renewable resources. Each agreement establishes a Renewable Resources Council (RRC) as the "principal instrument for renewable resource management" throughout a First Nation's traditional territory.[35] Composed of members appointed by the Yukon and First Nation governments, RRCs make management recommendations directly to the relevant Yukon minister (usually the Minister of the Environment) and/or First Nation.[36] Thus, traditional territories have become significant administrative units for the management of renewable resources throughout the Yukon.

The federal and Yukon governments did not play a major role in the original creation of First Nation traditional territorial boundaries; instead, they left it up to the bands to work these out among themselves,

FIG. 10.3: Entering Kluane First Nation Territory. Photo by Paul Nadasdy.

presumably based on patterns of historic land use. Several officials (both federal and territorial) told me their governments had been loath to get involved in disputes over territory among the different bands. Yet, as we have seen, administrative bands were themselves recent and fairly arbitrary amalgamations of different families and individuals, each with their own historically distinct patterns of land use and residency. In some cases, members of the same immediate family—with very similar land-use practices—became members of different administrative bands. Intermarriage among members of different bands was also common. These factors, along with increased individual mobility, made extremely problematic any attempt to map a band's traditional territory based on the historical use and occupancy of its members and their ancestors.[37] What is more, there was little coordination among First Nations in mapping their territorial boundaries, and First Nations seem to have pursued different strategies when confronted with the task. Some took an inclusive approach, drawing their boundaries as widely as possible to capture the historic land use of all their members; others were more conservative, giving up their claim to certain areas in an apparent effort to minimize potential overlap with other First Nations.

There are three direct results of this ad hoc and uncoordinated process for drawing up First Nation traditional territory boundaries. First, there is a great deal of overlap among First Nation traditional territories in the Yukon (see Fig. 10.4). Some First Nations are in a situation where well over half their traditional territories overlap with those of their neighbours. Second, there are a few areas in the Yukon that do not fall within any First Nation's traditional territory. The largest of these is located in the southwest Yukon. Dubbed "the black hole," it includes a large portion of the Nisling River valley northeast of Kluane First Nation territory (see Fig. 10.4).[38] Third, some Yukon Indian people feel that they have been cheated, because they have lost all say in the management of—and, in the case of the black hole, the right to hunt as Indian people in—certain areas of special importance to them personally. Today, the question of traditional territory boundaries is quite contentious in some parts of the Yukon.

Although the federal and Yukon governments played a minimal role in the creation of traditional territory boundaries per se, they have generally been adamant about the need to minimize and even eliminate overlap. Because traditional territories are essentially administrative boundaries that determine the jurisdiction of various management boards and councils set up under the land claim agreements, any territorial overlap necessarily creates jurisdictional conflict. In anticipation of this problem, First Nation Final Agreements require First Nations to "resolve" any overlap by negotiating an "Overlap Resolution Boundary," a contiguous line that in effect eliminates the conflict by specifying where one board's jurisdiction begins and another's ends.[39] Until overlap is resolved in this way, overlap areas exist in a jurisdictional void, and several very important provisions of the Final Agreements do not apply within them. This poses a particularly acute problem in the arena of resource management, because Renewable Resources Councils have no jurisdiction at all in overlap areas. Although this way of handling overlap does prevent jurisdictional conflicts between Renewable Resources Councils, it also means that First Nations and their citizens have virtually no say over the management of fish, wildlife, and timber in overlap areas (except on settlement lands, to be discussed below).[40]

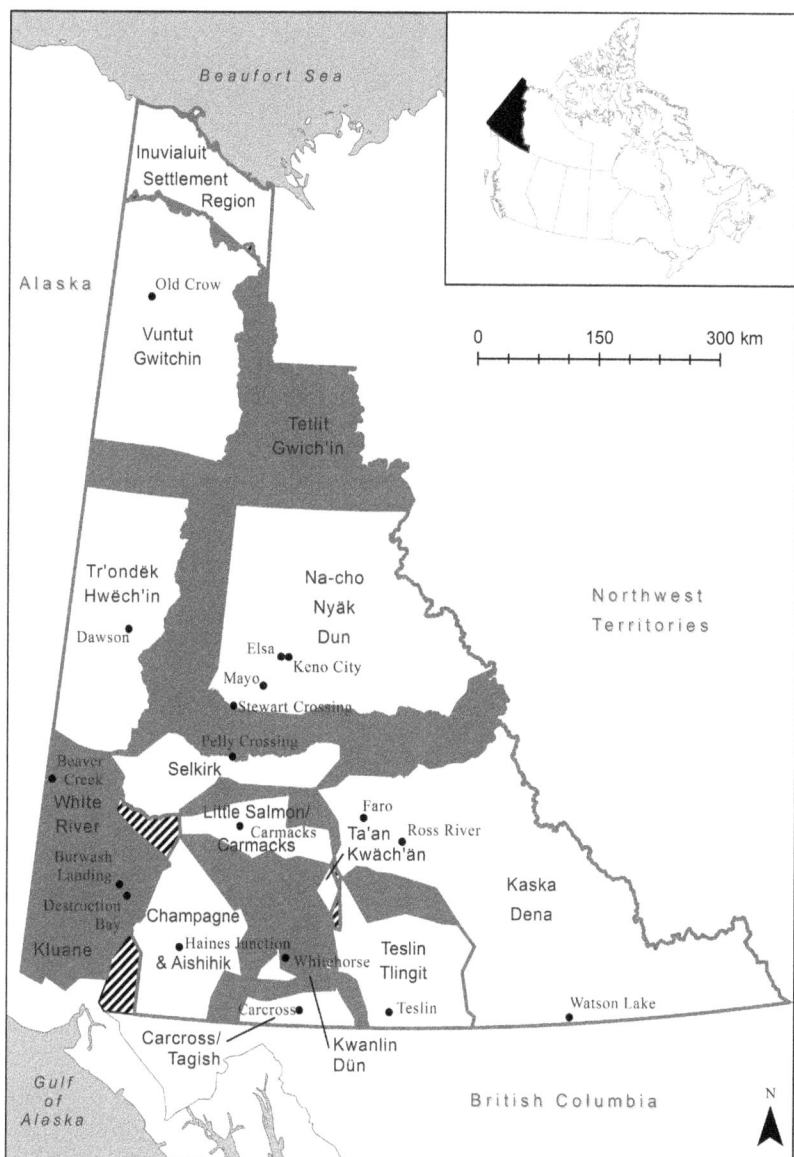

FIG. 10.4: First Nation Traditional Territory Overlap in the Yukon by Tracy Sallaway (Maps, Data and Government Information Centre, Trent University). Shading denotes overlap; cross-hatching indicates areas not included in any First Nation traditional territory. Contains information licensed under the Open Government Licence – Canada (http://open.canada.ca/en/open-government-licence-canada); all other data – Geomatics, Department of Environment, Government of Yukon.

In theory, overlap resolution boundaries would be used only to establish the jurisdiction of a few co-management boards and to bring a handful of other provisions (mostly regarding economic development) into effect in the overlap areas. For most other things, most importantly the exercise of hunting rights, the First Nations could continue to "share" the overlap areas (i.e., citizens of both First Nations could hunt there). In fact, there has been some interest in and activity around the negotiation of "sharing accords" among First Nations. These are reciprocal agreements that go beyond the shared use of overlap areas by extending to one another's citizens certain rights, particularly hunting rights, throughout the signatory First Nations' *entire* traditional territories.[41] In practice, however, the negotiation of overlap resolution boundaries has often been contentious and difficult. This is not surprising, since it requires Yukon Indian people to do something they have never done before: construct firm political boundaries between themselves and their neighbours (who are, often enough, close kin). The notion of a contiguous line separating "us" from "them" flies in the face of important cultural values of kinship and reciprocity, which continue to structure social relations among Yukon Indian people.[42]

Technically, First Nations could continue to share most everything in the overlap area, since an overlap resolution boundary applies only to certain jurisdictional issues, but in my experience this is often poorly understood. The very term "traditional territory," with its invocation of "tradition," seems to imply a link to historical use and occupancy, and Yukon Indian people (along with government negotiators) by and large think of them in this way. Yet, any well-defined territorial boundary between First Nations must necessarily be cross-cut by kinship relations and inconsistent with historical and contemporary patterns of use and occupancy. What is more, in the minds of many Yukon Indian people, traditional territories have come to be emblematic of self-governing First Nations' history and sovereignty.[43] As a result, there is often great reluctance to "give up" land to neighbouring First Nations through the "resolution" of overlap, because to many this seems tantamount to denying their affective ties to the land derived from historical and contemporary use.[44]

Those areas, such as the black hole, which fall outside the bounds of any traditional territory, present First Nations with a different set of issues. Although such lands comprise but a tiny percentage of the total area of the

Yukon, their exceptional status illustrates just how important traditional territories are for structuring First Nation hunting and management rights in the post–land claims Yukon. Falling as they do outside all traditional territories, areas like the black hole do not come under the jurisdiction of any Renewable Resources Council at all—in effect giving the Yukon government a disproportionate role in wildlife management there.[45] The issues posed by these administrative anomalies are even more significant with respect to hunting rights. By ratifying their agreements, citizens of self-governing First Nations have lost their Aboriginal right to hunt in the black hole without having gained any treaty-based hunting right to replace it.[46] If they wish to hunt on such lands legally, they now have no choice but to obtain a Yukon hunting license and abide by Yukon hunting regulations like any non–Indian resident hunter.[47]

In practical terms, obtaining a Yukon hunting license is not a problem for Yukon Indian people, but to do so is to submit to the authority of the Yukon Fish and Wildlife Branch. As I have detailed elsewhere, fish-and-game laws were among the principle mechanisms used by federal and territorial officials to establish and maintain control over Yukon Indian people, and the consequences of their imposition were especially dire in the southwest Yukon, where in the 1940s they threatened Indigenous peoples' very survival.[48] Indeed, opposition to Yukon fish-and-game laws was one of the prime factors motivating Yukon Indian people to organize politically and push for the settlement of land claims in the 1970s. They had been staunch defenders of what they saw as their Aboriginal hunting rights and would never have agreed to a treaty that subjected them to Yukon hunting regulations.[49] That they must now do so in the black hole and a few other such areas not only highlights how important traditional territories are for structuring contemporary Yukon First Nation hunting rights, it also shows just how unanticipated and problematic anomalous areas of this sort are for the territorial regime established by the agreements.[50]

Settlement Lands: First Nation Property in a Sea of Crown Land

In contrast to traditional territories, where First Nations retain only some residual use and management rights, settlement land is a form of property. Much smaller than the traditional territories within which they are located, the parcels that comprise settlement land are owned by First Nations, who are deemed to possess in them "the rights, obligations and liabilities equivalent to fee simple"[51] (see Fig. 10.2).

But settlement lands are not simply First Nation property; they also define the territorial limits of a self-governing First Nation's full political, legal, and administrative jurisdiction. It is on their settlement lands—and only on their settlement lands—that Yukon First Nations exercise full self-government authority, including significant powers of governance, taxation, management, regulation, and administration.[52] It is for this reason that the federal government was reluctant to agree to a First Nation's selection of settlement land outside its traditional territory—and then only with the consent of the First Nation within whose traditional territory the selection lay.[53] Consent was deemed necessary because these parcels of settlement land become embassy-like pockets of jurisdictional exception within another First Nation's traditional territory. Consent from another First Nation was also generally required when a First Nation chose a parcel of settlement land in an overlap area *within* its own traditional territory, because, depending on the location of any overlap resolution boundary, jurisdictional issues might be nearly as complex in overlap areas as outside the traditional territory altogether.

Especially significant for the purposes of this chapter is the fact that First Nations have full jurisdiction over hunting and many other forms of land use on settlement lands (Category A settlement lands in particular).[54] First Nation jurisdiction on these lands interacts in complex ways with the Yukon government's jurisdiction over wildlife on non-settlement land. While a Yukon First Nation has the power to regulate its own citizens' hunting throughout its traditional territory, it is the Yukon government that has final jurisdiction over non–First Nation hunting on non-settlement lands.[55] All non–First Nation citizens must obtain a Yukon hunting license to hunt in the Yukon, which entitles them to hunt anywhere in the Yukon except on First Nation Category A Settlement lands, which are owned and exclusively managed by First Nations. Hunting on all lands is subject to regulation by the government that has jurisdiction over wildlife management on the land in question—First Nation governments on settlement land, the Yukon government on non-settlement land. First Nation governments may permit non–First Nation citizens to hunt on their lands, and may charge them a fee to do so, but those hunters are then subject to First Nation wildlife laws, as well as general Yukon hunting laws.[56]

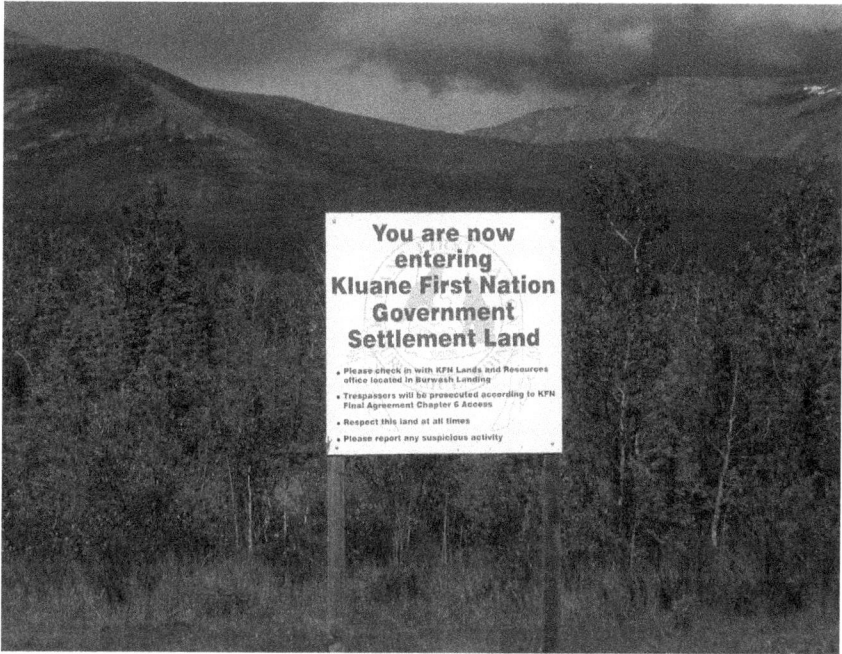

FIG. 10.5: Entering Kluane First Nation Government Settlement Land. Photo by Paul Nadasdy.

Imposing Territoriality in the Southwest Yukon

The new territorial divisions created by the Yukon agreements were initially devoid of social meaning for most Yukon Indian people. Over time, however, this has begun to change. Like Thongchai's maps of Siam, the maps attached to the Yukon land claim agreements were created as models for a new territorially ordered form of governance, as mechanisms for the creation of a particular system of legal, administrative, and jurisdictional relations. To the extent that they now undergird a complex system of political and legal relations, these maps and the territories they create have gradually assumed significance not only in the realm of Indigenous-state relations, but among Yukon Indian people, as well. One of the most important aspects of this is the rise of ethno-territorial nationalism among First Nations, about which I have written elsewhere.[57] Closely associated with the rise of First Nation nationalism, however, is what Vandergeest and Peluso refer to as the "territorialization of resource control."[58]

They argue that, "all modern states divide their territories into complex and overlapping political and economic zones, rearrange people and resources within these units, and create regulations delineating how and by whom these areas can be used. These zones are administered by agencies whose jurisdictions are territorial as well as functional." And they note further that the maps created by these bureaucratic agencies play "a central role in the implementation and legitimation of territorial rule."[59] In Canada, as elsewhere, the imposition of state regimes for the management of wildlife and other renewable resources has played an important role in this territorializing process, which has taken place on multiple scales. The federal government long ago devolved the management of fish and wildlife to the provinces and territories, which generally have jurisdiction over these matters within their respective borders.[60] But the provinces/ territories are themselves internally territorialized. The Yukon government, for example, has subdivided the entire Yukon into game management zones and sub-zones (Fig. 10.6). At the same time, other government maps subdivide the Yukon into conservation officer districts, outfitting concessions, and/or trapping concessions (Figs. 10.7–10.9). Although they overlap and otherwise fail to correspond to one another, each of these internally territorializing maps is essential for administering some aspect of human-wildlife (and human-human) relations in the Yukon. As models *for* wildlife management, they (or some earlier version of them) were crucial for developing the administrative mechanisms of the Yukon government's Fish and Wildlife Branch, and they remain vital tools for implementing and enforcing hunting, fishing, and trapping regulations.

Arbitrary as they are, the internal territories created by these maps have come to structure people's actual experiences on the land. Wildlife biologists, for example, tend to bound their wildlife surveys by management sub-zone, and outfitters and trappers, because they are only legally permitted to guide hunters or trap within their respective concessions, tend to focus their activities there. As a result, these maps now play a critical role in shaping how wildlife biologists, outfitters, trappers, and others actually come to know and think about the land and animals. In this way, the internal territories of wildlife management serve to structure

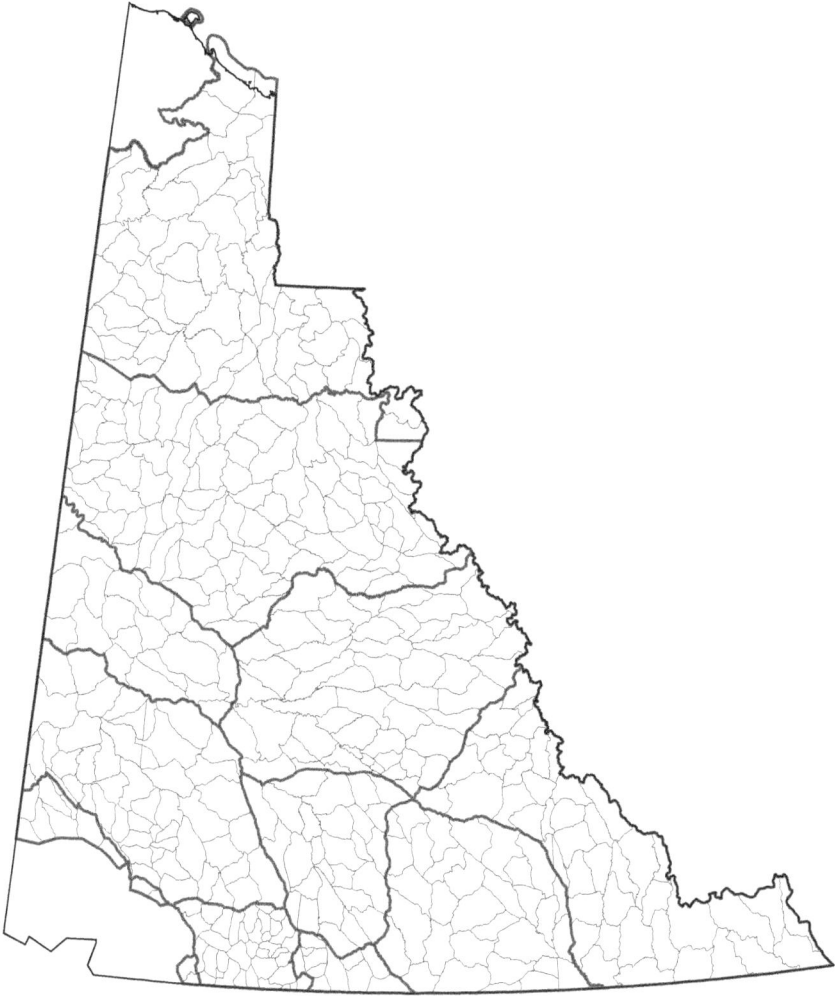

Fig. 10.6: Game Management Areas, © Yukon Department of Environment, Information Management and Technology Branch, 2016.

Fig. 10.7: Conservation Officer Districts, © Yukon Department of Environment, Information Management and Technology Branch, 2016.

Fig. 10.8: Outfitting Concessions, © Yukon Department of Environment, Information Management and Technology Branch, 2016.

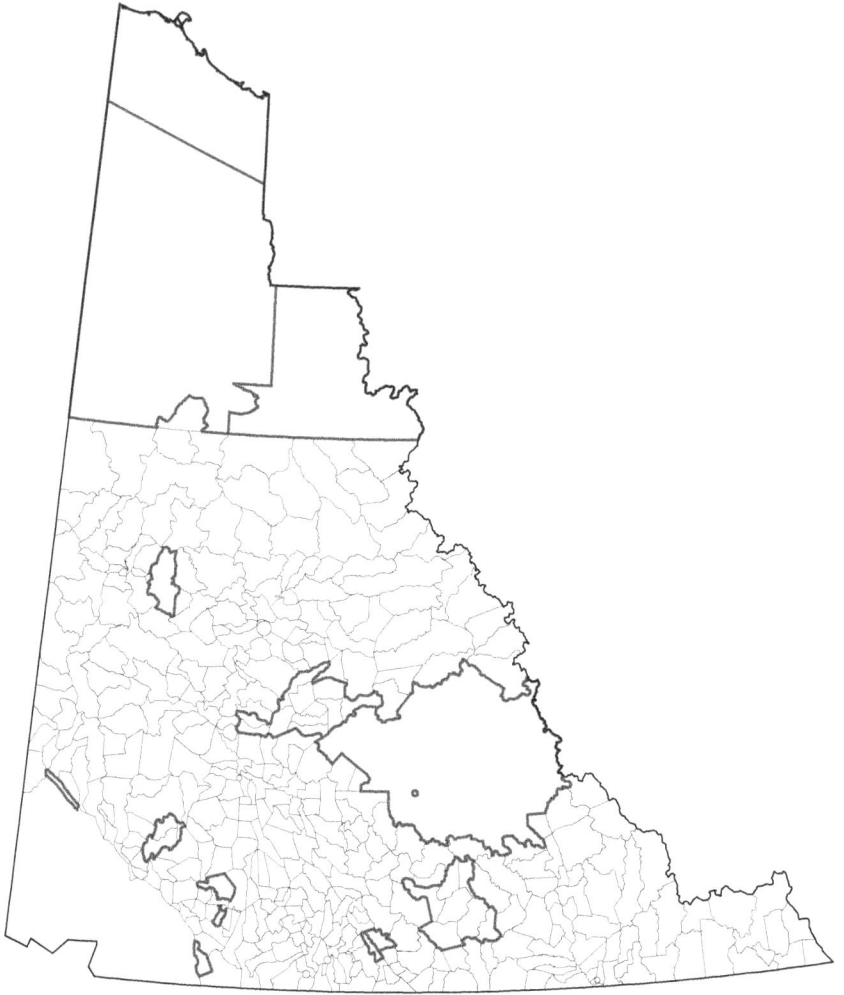

Fɪɢ. 10.9: Trapping Concessions, © Yukon Department of Environment, Information Management and Technology Branch, 2016.

the kinds of management interventions that are possible and even *thinkable* by various actors.[61]

Yukon First Nation final agreements do not disrupt the territorial order of state wildlife management; on the contrary, they merely superimpose a new layer of internal management-relevant territories upon the old. Maps of First Nation traditional territories and settlement lands have now joined those showing game management zones and trapping concessions as important tools for delineating how and by whom Yukon lands and resources can be used.[62] Just as the older administrative maps were models *for* the complex of human-environment relations they helped bring into being, the new First Nation Final Agreement maps also envision a new set of relations among humans, land, and animals; and they take for granted a set of far-reaching social and institutional changes whose intent is to transform that vision into reality. Principal among these changes has been the institutionalization of First Nation "management" through the creation of First Nation bureaucracies modeled on those of the Yukon government.

Because they now have legal jurisdiction over fish and wildlife (and other renewable resources) on their settlement lands and over First Nation hunting throughout their entire traditional territories, self-governing Yukon First Nation governments have had to create their own bureaucratic mechanisms for managing these resources. As in the Yukon government, First Nation resource officers are generally housed within departments that manage all aspects of land and resource use. Kluane First Nation's resource officer, for example, is in charge of managing all renewable resources, and answers to the director of KFN's department of lands, resources, and heritage (hereafter Lands Department). The creation of First Nation management bureaucracies is having a significant effect on how First Nation citizens can relate to one another and to the land and animals. One of these effects, predicted by Sack, is that "by classifying at least in part by area rather than by kind or type, territoriality helps make relationships *impersonal*."[63] As the following story illustrates, the replacement of personal by impersonal social relations is an important aspect of the territorialization of resource control in the Yukon.

In the summer of 2006, as we were about to depart for several days of moose hunting around Fourth of July Creek just across the KFN-CAFN (Champagne and Aishihik First Nations) border, Joe Johnson surprised me when he said that in preparation for the hunt he had obtained a Yukon hunting license. Recall that under the terms of KFN's Final Agreement, Joe, a KFN citizen, was free to hunt anywhere in KFN's traditional territory without a Yukon hunting license. To hunt within another First Nation's territory (such as along Fourth of July Creek), however, he had a choice: he could either get permission to hunt there from that First Nation (in this case, CAFN), or he could obtain a Yukon hunting license in the same fashion as any non–Indian Yukon resident and be subject to Yukon—rather than First Nation—hunting regulations. I was completely taken aback to learn that Joe had done the latter. As noted above, it was no more difficult for him to obtain a Yukon hunting license than to get permission to hunt from CAFN, but to do so was politically and culturally unpalatable, because it required him to submit to the authority of the Yukon Fish and Wildlife Branch. Indeed, years before, Joe himself had flouted Yukon hunting regulations and risked imprisonment by "illegally" hunting Dall sheep for his father's funeral potlatch, without which, he felt, it would have been impossible to carry out a proper ceremony.[64]

Relations between KFN and CAFN are friendly, and there is considerable day-to-day social interaction between Kluane and Champagne/Aishihik citizens, nearly all of whom, including Joe, have very close relatives (spouses, parents, children, siblings, or cousins) in the other First Nation.[65] Hence my surprise when I learned about his choice to obtain a Yukon hunting license rather than go through the CAFN lands department. When I asked him why he had done so, he explained that the summer before, the last time he and I had hunted together up Fourth of July Creek, he had asked CAFN for permission to hunt. In response, they sent him a letter granting him permission to hunt there for one specific week only. After telling me this, he looked at me for a while, his eyes smouldering. Then he went on to say that if a CAFN citizen asks for permission to hunt in Kluane territory, "we give them as much time as they want." He said he had decided to get a Yukon license so he could hunt wherever he wanted, and would not have to put up with such insulting treatment anymore.

Two decades before, it would also have been unthinkable for a Yukon Indian person to deny or limit another's right to hunt anywhere he or she wanted. In light of the continuing cultural and economic importance of Indigenous hunting in the Yukon, and the colonial legacy of wildlife management in the territory, it is hard even now to imagine one Yukon Indian person denying another's right to hunt. In the case of Joe's application to CAFN the previous year, wildlife officials were able to restrict his ability to hunt in the way they did only because of the prior creation of a bureaucratic apparatus for managing fish and wildlife, a development that was itself wholly dependent upon the creation of jurisdictional boundaries between the two First Nations. CAFN's agreements have been in place since 1995, so in 2006 they were much further along in the process of establishing their own government institutions than was KFN, which had been self-governing only since 2004. CAFN had by then established a formal permitting process to deal with requests like Joe's. As a result, the limitation on Joe's ability to hunt was issued not by a specific person, but by the First Nation's bureaucratic apparatus. Thus, Joe's anger was directed not at a specific person acting in a culturally inappropriate manner, but at "them"—CAFN's impersonal management bureaucracy.

Joe's impression that KFN regularly and routinely granted citizens from other First Nations permission to hunt in their territory was rooted in his many years of experience in the KFN office, where he served in various capacities, including several terms as chief, until his retirement in 1996. During those years, KFN dealt with requests for permission to hunt in an informal, ad hoc, and personalized manner. From my own observations in the KFN office between 1995 and 2003, I, too, got the impression that KFN routinely granted such requests from citizens of other First Nations. The issuance of letters of permission—mostly to CAFN hunters—had seemed to me little more than a formality. But things began to change after KFN's agreement came into effect on 2 February 2004. On 15 April of that year, the director of KFN's lands department told me that she was regularly fielding calls from CAFN citizens requesting permission to hunt in KFN territory. Since the newly self-governing KFN as yet had no policy in place for dealing with these requests, she said, she had no choice but to deny them permission. She admitted this was a difficult situation, because many of them got quite angry when she told them they could not hunt. She said she hoped KFN and CAFN would soon sign a sharing accord to deal with the problem. In fact, KFN and CAFN, along with the Ta'an Kwäch'än Council (the Southern Tutchone Tribal Council's third member), signed a sharing accord back in 1997, but KFN did not ratify it because of concerns about whether they had the legal authority to do so, given their one-hundred-percent overlap with White River First Nation (WRFN).[66] The legal implications of the overlap issue did not disappear when KFN became self-governing, with the result that there was still no sharing accord in place in 2006, when Joe Johnson wanted to hunt on Fourth of July Creek.

Overlap with WRFN was not the only obstacle to formalizing reciprocal hunting rights between CAFN and KFN citizens. By 2006 there was a growing concern, especially among some younger KFN citizens, about possible over-hunting of moose in KFN's territory owing to encroachment by CAFN citizens. Several of them independently expressed to me their concerns about the number of CAFN citizens who had recently been hunting in the Donjek River valley, deep in Kluane territory. One man—the same, incidentally, who also orchestrated the raising of signs on the Alaska Highway to mark the boundaries of Kluane territory (see Fig. 10.3)—told me it was imperative that KFN get its government up and running as soon as possible, so that Kluane people could control hunting by citizens of other First Nations and thereby protect their animal populations. The implication of his statement was clear: without an impersonal bureaucratic screen of laws and regulatory processes in place, it would remain very difficult for any KFN citizen to deny—or provide only limited approval to—requests by citizens of other First Nations to hunt in Kluane territory.

As of the summer of 2011, the Southern Tutchone First Nations had yet to ratify a Sharing Accord. In fact, much of what I describe above had become institutionalized; as a matter of policy, CAFN was issuing permits that were valid for two weeks only and were also species-specific. CAFN is a bit unusual because of its proximity to Whitehorse; other First Nations were generally responding to requests for permission to hunt by issuing permits that were good for the whole season. By contrast, and in a break with the recent past, KFN was not granting citizens of other First Nations permission to hunt in their territory at all. One Kluane citizen who worked for a time in KFN's lands department acknowledged that this causes anger among those whose requests are denied, but defended the policy as necessary for protecting Kluane animal populations.

The Instability of the Territorial System

It is becoming more common to hear Yukon Indian people invoke the language of territoriality to assert their First Nation's exclusive rights to control the resources within its traditional territory. But territorial sentiments of this sort do not always take precedence over cross-cutting relations of kinship and reciprocity. Non-territorial principles governing social relations remain strong among Yukon Indian people, and provide the basis for a trenchant moral critique of the new territorial regime. Vandergeest and Peluso note that territorialized resource management regimes are often unstable due to factors ranging from the state's inability to enforce them in the face of local resistance to conflict among government agencies with competing territorial mandates.[67] Because of "the lack of fit between lived space and abstract space," they argue, a government's territorial strategy for controlling resources is "often a utopian fiction unachievable in practice because of how it ignores and contradicts peoples' lived social relationships and the histories of their interactions with the land."[68] The territorialized resource management system in the Yukon is unstable for some of the same reasons.

One source of the new territorial regime's instability is its own complexity. Even a cursory comparison of Figs. 10.1, 10.2, and 10.4–10.7 reveals inconsistencies across and among the internal territories created for wildlife management in the Yukon; they cross-cut, overlap, and otherwise fail to correspond to one another. The result is a territorial regime of such complexity that even officials in the Yukon Department of Environment can sometimes get confused. In 2007, for example, a member of the Dän Keyi Renewable Resources Council (DKRRC)—the Renewable Resources Council established pursuant to KFN's final agreement—told me that Yukon government officials had recently contacted the council to inform them that they were planning to conduct fisheries research in two lakes within the council's jurisdiction, Tincup Lake and Wellesley Lake. He told me that the council had thanked them for the heads-up, but politely informed them that Wellesley Lake was actually well outside the council's jurisdiction. Shaking his head, he told me that the government biologists had seemed completely unaware of the terms of KFN's agreement and the jurisdictional implications it has for wildlife management in the region.[69] Similarly, in 2011, I was told that a CAFN citizen had recently shot a moose within the no-hunting corridor along the Alaska Highway near the Duke River on KFN settlement land. KFN officials wanted the Yukon government to charge him for hunting illegally, but the regional conservation officer had refused, claiming he had no authority to do so because KFN and CAFN were both signatories of the Southern Tutchone Sharing Accord.[70] The conservation officer had then been surprised to learn that KFN had never ratified the accord (back in 1997), and promised that in the future he would bear that in mind.[71]

If even Yukon government officials—whose job it is to know these things—can sometimes be unclear on the details of the new resource management regime, it should hardly come as a surprise that First Nation hunters and the general public, too, sometimes have trouble keeping it all straight. This general ignorance of a complex regime contributes to its instability; yet, for many First Nation hunters, something more than mere ignorance is at work. Unlike Yukon government officials, who accept—even take for granted—the underlying territorial premise of the new management regime, the same cannot be said for many First Nation hunters.

Yukon Indian people, like Indigenous peoples elsewhere, long chafed under hunting and trapping regulations imposed upon them by state wildlife managers. It was not merely the specific regulations that they resented (although some of these were particularly onerous), but the very idea of management itself. The idea of wildlife management, rooted as it is in the political and economic context of capitalist resource extraction and based on an agricultural metaphor, sits uneasily with Yukon Indian people's ideas about proper human-human and human-animal relations.[72] For them, to hunt is to participate in a complex web of reciprocal social relations among human and other-than-human persons.[73] To be sure, these relationships can at times be difficult and fraught with danger, so hunters must manage them with considerable care. But this kind of "management" is an intensely personal affair, involving introspection and self-control rather than the coercion of others. The idea that some outsider—who is by definition ignorant of the delicate and complex web of social relations in which the hunter is enmeshed—should dictate the terms of the hunt is anathema to the maintenance of proper interpersonal relations among persons. And the idea that regulations, along with the authority to make them, should vary according to arbitrarily defined geographical areas—rather than the particularities of those interpersonal relations—verges on the nonsensical. Finally, I would note that the very act of regulating the behaviour of others by telling them directly what they are permitted to do and where is itself deeply inconsistent with the norms of proper social interaction among Athapaskan peoples.[74]

Thus, although the new agreements do grant First Nation people significant new powers to manage and co-manage renewable resources throughout their territories, the very exercise of those powers threatens to transform the ways in which they relate to animals, and, perhaps even more significantly, to one another with respect to animals. The fact that it is now First Nation—rather than Yukon—government officials who claim the authority to interfere with the interpersonal relations of the hunt by dictating to hunters what they may do and where does not seem like much of an improvement to many First Nation hunters. As a result, some—even among those who may be otherwise supportive of the new agreements— find ways around the territorial regime of resource management or else refuse to comply with it altogether.

A case in point: one of the first acts of the newly self-governing Kluane First Nation in 2004 was to institute a permit system for the cutting of firewood on its settlement lands. Nearly five years earlier, a fire had raged through the area, nearly burning down the village and leaving a twelve-kilometre-long swath of easily accessible standing dead timber. The new permit system was primarily intended to control the harvest of firewood by commercial cutters from as far away as Haines Junction and even Whitehorse. Permits to cut firewood for personal use were issued free to KFN citizens and other local residents, although they did specify where the permit holders were allowed to cut. For non–First Nation citizens, in-cluding commercial cutters, the new system was but a minor change; it simply meant getting a permit from KFN rather than from the federal government, as they had previously been required to do. For KFN citizens, however, who until then had had the right cut firewood anywhere they wanted without a permit, it meant submitting to the territorial authority of a government, in this case KFN. Even though permits were free to KFN citizens, and KFN's Lands Department granted permits to cut wherever its citizens requested, there was a great deal of grumbling in the village about the new policy. Some KFN citizens refused outright to obtain permits on the grounds that no one had the right to tell them when, where, or how much firewood they could cut. Indeed, it seems that Yukon Indian people now increasingly view their own First Nations' management bureaucracies—along with co-management bodies like the Renewable Resources Councils—with considerable suspicion, simply because they have assumed the role of "managers."[75]

Even when First Nation people do abide by the new rules, they are sometimes able to subvert the impersonal bureaucratic control imposed by the new territorializing agreements. Indeed, the agreements may even be giving rise to new kinds of social relationships that cross-cut traditional territory boundaries. For example, because First Nation hunters can effectively avoid the need to obtain a hunting permit from another First Nation if they are accompanied on the hunt by a citizen of that First Nation (who can claim to have done the shooting), at least some hunters are cultivating a Yukon-wide network of "hunting buddies" upon whom they can prevail whenever they wish to hunt outside their own First Nation's territory. One First Nation hunter told me in 2011 that he has hunting buddies all over the Yukon. The first thing he does if he wants to hunt in another First Nation's territory is to phone one of his buddies in that First Nation. Only if none of them is available to hunt with him does he formally request permission to hunt from the First Nation government concerned.

In light of all this, it would be inaccurate to claim that Yukon Indian people have internalized the new management regime and its territorial assumptions. Nor can one say that the new agreements have completely transformed Yukon Indian people's relationship to the land and resources. But there can be little doubt that the new territorializing maps and boundaries function as *models for* a powerful new system of legal and administrative relations. While this system has not replaced Indigenous ways of relating to one another and the land, it does undermine them, and poses a serious obstacle to those First Nation citizens who would continue to relate to the land and animals—and to one another—as their grandparents did. Those who choose to do so often transgress not only Yukon law, but now also their own First Nation's laws, risking fines and possibly even imprisonment in the process. Even when First Nation citizens find creative ways to subvert the new territorial system (through the cultivation of hunting buddies, for example), they must grapple with and adjust their practice to accommodate it. Thus, the new territorial system must be seen for what it is: a powerful engine for social-ecological change.

Conclusion

Although I maintain that land claim and self-government agreements in the Yukon are imposing a new territorial political order in the Yukon, I do not mean to imply that it is being imposed—in some straightforward, top-down fashion—on Indigenous people by the Canadian government. Rather, the agreements have resulted from decades of struggle against colonial policies; and they grant Yukon First Nations significant powers to govern their peoples and resources. Those powers, however, come in the currency of territorial sovereignty, and to wield them Yukon Indian people have had to alter their forms of social and political organization in dramatic and often unforeseen ways. One of the most important dimensions of this territorializing process has been the rise of First Nation resource management bureaucracies, which compel Indian people—bureaucrats and citizens alike—to relate to the land and animals in new ways (though not always in the ways resource managers intend). The fact that the agreements are having such a transformative effect on human-animal-land relations is significant because the preservation of hunting practices, and the social relations they entail, was one of the principal goals motivating Indian people to enter into land claim negotiations in the first place. There is, then, a certain political ambiguity in the territorially ordered political system currently emerging in the Yukon. Rooted as it is in colonial administrative practices, the new configuration of territorially constituted First Nations must be viewed as a legacy of colonial rule, of federal efforts to incorporate Yukon Indian peoples more firmly into the Canadian state. At the same time, however, this territorial system is also the product of Indigenous resistance to colonial incorporation, a result of forty years of struggle and compromise. In other words, the new territorially ordered system must be viewed as both an assertion of territorial sovereignty by Yukon First Nations (and recognition of that sovereignty by Canada), and, simultaneously, as a process of internal territorialization that is creating new territorial units (and identities) within the Canadian state.[76]

It is hard to imagine how it could have been otherwise. The territorial assumptions underlying First Nation final and self-government agreements in the Yukon are so fundamental to contemporary political thought that non-territorial forms of governance are not even recognized as potential alternatives to colonial rule: First Nation self-government must be territorially ordered or it will not qualify as "government" at all.[77] Michael Asch has argued convincingly that the Canadian government's claim to sovereign jurisdiction over Canadian territory is itself rooted firmly in the underlying assumption that contact-era Indigenous people were too primitive to have governments.[78] A central challenge for Yukon First Nations in the land claim process has been to convince the federal government that they are capable of self-government (i.e., that they are no longer too "primitive" to govern themselves), and the only permissible evidence of this capacity is the ability to establish and run a European-style, territorially ordered government. Kluane negotiators were regularly forced to counter veiled suggestions that they were not yet ready for self-government with assurances that they would indeed develop Euro-Canadian style laws and political institutions.[79] This same paternalistic subtext is evident in calls, by now taken for granted, in Canada's self-government discourse for First Nations to "build capacity"—a euphemism for Euro-Canadian-style training that enables them to serve as the bureaucratic functionaries increasingly required by land claim and self-government agreements—as though they lacked the "capacity" to govern themselves before the arrival of Euro-Canadians.[80] Yukon First Nation negotiators thus found themselves in a grimly ironic position: the only way they could convince federal negotiators that they were politically mature enough to handle "self-government" was to agree to the establishment of a socio-political system that was not their own. In an important sense, then, the Yukon agreements can be viewed as part of an ongoing process of internal territorialization in Canada. Because this process compels Indigenous people to adopt Euro-Canadian forms of governance, it serves to extend the colonial project even as the agreements grant newly emerging First Nation polities a measure of power within the state context.

Notes

1 Although based in part on an article previously published in *Comparative Studies in Society and History*, this chapter has been significantly rewritten and reframed for inclusion in this volume. Thanks to all the participants at the authors' meeting in Peterborough for comments and discussion, which inspired me to develop this chapter in a new direction. Thanks especially to Stephen Bocking and Brad Martin for their insightful comments on my draft manuscript and for all their editorial work, as well as to Gerry Perrier at Environment Canada for help with the maps. I would also like to acknowledge the National Science Foundation and the Wenner-Gren Foundation for funding the research on which this article is based. As always, I wish to thank the people of Burwash Landing for their friendship and hospitality, particularly the late Joe Johnson, Luke Johnson, Robin Bradasch, and Gerald Dickson, who, along with Jim Bishop, Ron Sumanik, and other government negotiators, helped me understand many of the processes I discuss in this article. Any errors are mine alone.

Thomas Biolsi, "The Birth of the Reservation: Making the Modern Individual among the Lakota," *American Ethnologist* 22, no. 1 (1995): 28–53; Eric Cheyfitz, *The Poetics of Imperialism: Translation and Colonization from The Tempest to Tarzan* (New York: Oxford University Press, 1991); William Cronon, *Changes in the Land: Indians, Colonists, and the Ecology of New England* (New York: Hill and Wang, 1983); Carolyn Merchant, *Ecological Revolutions: Nature, Gender, and Science in New England* (Chapel Hill: University of North Carolina Press, 1989); James Tully, "Rediscovering America: The Two Treatises and Aboriginal Rights," in *An Approach to Political Philosophy: Locke in Contexts* (New York: Cambridge University Press, 1993), 137–76; Nancy M. Williams, *The Yolngu and Their Land: A System of Land Tenure and the Fight for its Recognition* (Stanford: Stanford University Press, 1986).

2 Notwithstanding the long and contentious debate over whether the family hunting territories of the eastern North American subarctic constituted an Aboriginal form of private property. See Paul Nadasdy, "'Property' and Aboriginal Land Claims in the Canadian Subarctic: Some Theoretical Considerations," *American Anthropologist* 104, no. 1 (2002): 247–61 for an overview of that debate.

3 Nadasdy, "'Property' and Aboriginal Land Claims."

4 Robert Sack, *Human Territoriality: Its Theory and History* (New York: Cambridge University Press, 1986), 19.

5 Sack, *Human Territoriality*, 26.

6 Although Sack and others view the modern state as the quintessentially territorial macro-political form, territorial strategies are utilized in all societies and at all scales. According to Sack, the appeal of territoriality lies in its efficiency. If a parent wants to prevent his or her very young children from playing with dangerous items in the kitchen, for example, it is more efficient simply to exclude them from the kitchen (a territorial strategy) than to try to explain and enforce

a distinction between those items with which they may and may not play. Sack, *Human Territoriality*, 22.

7 Sack, *Human Territoriality*, 15.

8 Although often treated as distinct, property and jurisdiction are so deeply implicated in one another that they might better be viewed as different aspects of state territoriality than as distinct forms of territoriality in themselves. Indeed, assertions of state jurisdiction are based on the assumption that the sovereign state possesses "underlying title" to all the land within its territorial boundaries. For some implications of this in the realm of Indigenous-state relations in Canada, see Michael Asch, "First Nations and the Derivation of Canada's Underlying Title: Comparing Perspectives on Legal Ideology," in *Aboriginal Rights and Self-government: The Canadian and Mexican Experience in North American Perspective*, ed. C. Cook and J. D. Lindau (Montreal: McGill-Queen's University Press, 2000), 148–67.

9 Some, like the Grand Council of the Cree, have become prominent—and sometimes effective—players on the international stage, as well (see the chapter by Carlson in this volume).

10 Paul Nadasdy, *Hunters and Bureaucrats: Power, Knowledge, and Aboriginal-State Relations in the Southwest Yukon* (Vancouver: UBC Press, 2003).

11 Nadasdy, "'Property' and Aboriginal Land Claims."

12 For example, Anthony Giddens, *The Nation-State and Violence: Volume Two of a Contemporary Critique of Historical Materialism* (Berkeley: University of California Press, 1981); Max Weber, "Politics as a Vocation," in *From Max Weber: Essays in Sociology*, ed. H. H. Gerth and C. W. Mills (New York: Oxford University Press, 1946), 78.

13 Thongchai Winichakul, *Siam Mapped: A History of the Geo-body of a Nation* (Honolulu: University of Hawaii Press, 1994), 130.

14 For more on the way in which maps construct rather than merely represent the world, see Denis Wood, *The Power of Maps* (New York: Guilford Press, 1992).

15 Sack, *Human Territoriality*, 53.

16 Malcolm Anderson, *Frontiers: Territory and State Formation in the Modern World* (Malden, MA: Polity Press, 1997), 107.

17 Peter Vandergeest and Nancy Peluso, "Territorialization and State Power in Thailand," *Theory and Society* 24, no. 3 (1995): 385–426.

18 Cronon, *Changes in the Land*.

19 Catharine McClellan, *My Old People Say: An Ethnographic Survey of Southern Yukon Territory*, 2 vols. (Ottawa: National Museum of Man, 1975), 14.

20 McClellan, *My Old People Say*, 481.

21 See McClellan, *My Old People Say*, 95–105; and Catharine McClellan and Glenda Denniston, "Environment and Culture in the Cordillera," in *Handbook of North American Indians. Vol. 6: Subarctic*, ed. J. Helm (Washington, DC: Smithsonian Institution Press, 1981), 372–86, for a description of the annual subsistence round in the southern Yukon and the entire Cordillera, respectively. For good descriptions of the social dynamics in Athapaskan hunting bands, see Henry Sharp, *The Transformation of Bigfoot: Maleness, Power, and Belief among the Chipewyan*

(Washington, DC: Smithsonian Institution Press, 1988); and Roger McDonnell, *Kasini Society: Some Aspects of the Social Organization of an Athapaskan Culture between 1900–1950* (Ph.D. diss., University of British Columbia, 1975).

22   McClellan, *My Old People Say*, 483. Dominique Legros paints a very different picture of Yukon Indian society in the nineteenth century, one in which family control over critical fishing sites led to the emergence of mafia-like corporate kin groups and extreme social inequality. Dominique Legros, "Wealth, Poverty, and Slavery among 19th-Century Tutchone Athapaskans," *Research in Economic Anthropology* 7 (1985): 37–64. Despite their differences, however, neither Legros nor McClellan ever suggests the existence of distinct polities with jurisdiction over their own bounded territories.

23   McLellan, *My Old People Say*, 13. In the Yukon today there are several "tribal councils," mid-level political entities situated between individual First Nation governments and the Council for Yukon First Nations. These are recent constructions that correspond roughly to linguistic boundaries laid out by early ethnographers in the region; see Barbra A. Meek, *We Are Our Language: An Ethnography of Language Revitalization in a Northern Athabascan Community* (Tucson: University of Arizona Press, 2010), 128–30.

24   McClellan writes that among the more important Indigenous criteria for classifying people were: "the kinds of technology, the specific food staples or the natural environment most closely associated with the group, the group's distance and direction from the speaker, the name of the particular place where families are congregated at a given point in time, or the histories and kin ties of important persons within a group" (McClellan, *My Old People Say*, 14). Notably absent from her list is language/dialect. Although she notes that Yukon Indian people "easily distinguish dialectical variations and are quick to comment that an individual speaks either 'the same as' they do, or 'a little bit different,' or that they 'can't hear [understand] him at all,'" these linguistic differences were cross-cut by other, often more salient kinds of difference, so that in practice, "Yukon Indians rarely single out language as a primary guide in circumscribing or naming a geographical group" (13–14; but see her entire discussion: 13–16). The low social salience of linguistic difference in the nineteenth century may be attributable in part to the fact that many Yukon Indian people spoke multiple Indigenous languages/dialects (McClellan, *My Old People Say*, 14); Kluane elders confirmed to me the multi-lingual abilities of their grandparents' generation.

25   Catharine McClellan, "Before Boundaries: People of Yukon/Alaska," in *Borderlands: A Conference on the Alaska-Yukon Border, Whitehorse, Yukon, 2–4 June 1989* (Whitehorse: Yukon Historical and Museums Association, 1992), 8–34. This is not to say that Yukon Indian people had no way of distinguishing legitimate from illegitimate land use. It is simply to say that their criteria of legitimacy were derived from interpersonal rather than territorial relations. As Colin Scott put it for the Cree, "this

system [of human-animal-land relations] entails specific criteria for inclusion within the network of human beings who practice it. Cree, in their own view, legitimately exercise and maintain their rights as against alien claimants who fail to conform to criteria of sharing and stewardship. Historically, when white men have apparently conformed to tenets of reciprocity, and contributed to stewardship of resources, they have been accorded a measure of legitimate participation in the Cree system. Thus when white men fail these standards, evasion or opposition is deemed legitimate by Cree." Colin Scott, "Property, Practice, and Aboriginal Rights Among Quebec Cree Hunters," in *Hunters and Gatherers: Property, Power, and Ideology*, ed. T. Ingold, D. Riches, and J. Woodburn (New York: St. Martin's Press, 1988), 40. See also Hans Carlson, *Home is the Hunter: The James Bay Cree and Their Land* (Vancouver: UBC Press, 2008) for a detailed historical account of European efforts to conform to—and later reject and undermine—these Cree social standards.

26 Long ago, Morton Fried argued that "tribes," in this sense, far from being a primordial political form, are instead what he called a "secondary phenomenon" because they have emerged almost everywhere as a result of—and in complex relation with—states. Morton Fried, *The Notion of Tribe* (Menlo Park, CA: Cummings, 1975). Recently, James Scott has elaborated on Fried's argument; see Scott's *The Art of Not Being Governed: An Anarchist History of Upland Southeast Asia* (New Haven: Yale University Press, 2009), especially ch. 7.

27 This is not to say that Indian Act Bands had no relationship whatsoever to the geographical distribution of Yukon Indian peoples at the time of their creation; I simply suggest that the situation was too complex to allow for any straightforward one-to-one mapping along those lines. McClellan characterizes the situation as follows: "Most of the modern government 'bands' have probably been organized on the basis of aboriginal territorial groupings, in the sense that those individuals who most often came together in the past probably segregated into the particular local groups that first built their cabins around a particular trading post, mining centre, church or school. Yet each such centre also attracted other families whose ties to the nuclear group were loose or non-existent" (*My Old People Say*, 481). To this one could add a further complication: families that came together at the same trading post at certain times of the year—and so ended up in the same band—often ranged over very different country at other times of the year. For example, some families that ended up in the Burwash band spent much of their year along the Donjek and White rivers, while others were oriented more in the direction of Kloo Lake and the Alsek drainage. The important point here is that Indian Act Bands did not correspond to any preexisting politico-territorial entities, since such entities did not exist.

28 Kenneth Coates, *Best Left as Indians: Native-White Relations in the Yukon Territories, 1840–1973* (Montreal: McGill-Queen's University Press, 1991), 233.

29 See Paul Nadasdy, "Boundaries among Kin: Sovereignty, the Modern Treaty Process, and the Rise of Ethno-Territorial Nationalism among Yukon First Nations," *Comparative Studies in Society and History* 54, no. 3 (2012): 499–532.

30 The only potential exception is a small part of the southeast Yukon, which was included in Treaty 11, signed in 1921. Even there, however, it is doubtful that Yukon Indians willingly ceded their lands. Although the Canadian government regards Treaty 11 as a land cession treaty, there is ample ethno-historical evidence to show that the Indians who signed it did not view it that way at all, believing instead that they were signing a treaty of peace and friendship. See Michael Asch, "On the Land Cession Provisions in Treaty 11," *Ethnohistory* 60, no. 3 (2013): 451–67.

31 It was estimated in 1962 that only 4,800 acres of land in the entire Yukon had been allocated for the use of Yukon Indian people. This included a few very small reserves. Most Indian villages in the territory, however, were not legally designated reserves but were set aside for Yukon Indian people's use by Orders in Council (Coates, *Best Left as Indians*, 235).

32 Yukon Native Brotherhood, *Together Today for Our Children Tomorrow* (Whitehorse, YT: Yukon Native Brotherhood, 1973).

33 The right to hunt can be viewed as a residual usufructuary right; see Nadasdy, "'Property' and Aboriginal Land Claims." The agreements also provide First Nations and their citizens with some opportunities for preferential employment and other forms of economic development within their traditional territories.

34 For example: Kluane First Nation, *Kluane First Nation Final Agreement among the Government of Canada and Kluane First Nation and the Government of the Yukon* (Ottawa: Public Works and Government Services Canada, 2003), 236.

35 Kluane First Nation, *Kluane First Nation Final Agreement*, 241.

36 Who gets the recommendation depends on which government (First Nation or Yukon) and which department within a government has jurisdiction over the matter in question.

37 See Brian Thom, "The Paradox of Boundaries in Coast Salish Territories," *Cultural Geographies* 16, no. 2 (2009): 179–205 for discussion of a similar situation among the Coast Salish on Vancouver Island.

38 The only other sizeable area in the Yukon not included within any First Nation traditional territory lies between the Kluane and Champagne/Aishihik territories. It poses less of a problem for First Nations, however, because it is composed of uninhabited and virtually inaccessible mountains and ice fields.

39 For example: Kluane First Nation, *Kluane First Nation Final Agreement*, 37.

40 For a complete list of provisions not in effect in overlap areas, see Kluane First Nation, *Kluane First Nation Final Agreement*, 35, 39–40, 44.

41 The three Northern Tutchone First Nations of the central Yukon signed a Sharing Accord in 1995.

42　Nadasdy, *Hunters and Bureaucrats*, ch. 2; see also Norman Easton, "King George Got Diarrhea: The Yukon-Alaska Boundary Survey, Bill Rupe, and the Scotty Creek Dineh," *Alaska Journal of Anthropology* 5, no. 1 (2007): 95–118; and Thom, *"Paradox of Boundaries."* Carlson also vividly describes Cree hunters' discomfort with drawing territorial boundaries among themselves (Carlson, *Home is the Hunter*, 215–16).

43　In our very first meeting, for example, Joe Johnson, then Chief of Kluane First Nation, told me that when driving north on the Alaska Highway, he feels that he has "come home" when Kluane Lake comes into view, and that when driving south he gets that same feeling once he crosses the White River. I did not know it at the time, but this corresponds to the traditional territory boundaries he and other KFN delegates were advocating at overlap negotiations with the White River First Nation; see Nadasdy, "Boundaries among Kin."

44　This is precisely what happened in failed efforts to resolve overlap between the Kluane and White River First Nation territories; see Nadasdy, "Boundaries among Kin."

45　These areas do still fall under the jurisdiction of the Yukon Fish and Wildlife Management Board, a co-management board established under the Yukon Umbrella Final Agreement with jurisdiction over the entire Yukon. Given the much broader scope of the board's authority, however, it is unlikely its members could devote as much focused attention to issues in the black hole as more "local" members of an RRC.

46　The members of First Nations that have not ratified a final agreement, such as WRFN, retain their aboriginal right to hunt in the black hole.

47　First Nation citizens, however, do have a treaty-right to hunt on settlement lands within the black hole. Kluane First Nation, for example, has a 18.9-square-kilometre parcel of settlement land at the junction of Onion Creek and the Nisling River, well inside the black hole; see Kluane First Nation, *Kluane First Nation Final Agreement*, 238.

48　Nadasdy, *Hunters and Bureaucrats*, 38–41.

49　Nadasdy, *Hunters and Bureaucrats*, 55.

50　That no First Nation of the central or southwest Yukon claimed the black hole as part of its traditional territory does not mean it was historically unimportant to them. On the contrary, the Nisling River valley has long been a vital hunting and (especially) fishing area for people throughout the region. McClellan, *My Old People Say*, 30–31, 483; see also Yukon Archives, Kluane First Nation Collection, SR234-(1), 1977 interview with Sophie Watt, trans. Billy Joe. Several Kluane people regularly hunted and trapped in the valley (mostly in winter) throughout the period of my fieldwork and continue to do so into the present. Prior to the construction of the Alaska Highway, the Nisling valley was also at the nexus of a set of important trade and travel routes. It was on the "Carmacks Trail," which linked the Yukon River to the upper White River, with branches connecting to Kluane Lake, Aishihik, and the Donjek/White River country. Frederick Johnson and Hugh Raup, "Appendix III: Notes on the Ethnology of the Kluane Lake Region," in *Investigations in Southwest Yukon: Geobotanical and Archaeological Reconnaissance,* 159–98, Papers of the Robert S. Peabody Foundation for Archaeology (Andover, MA: Peabody Foundation, 1964), 196. A couple of kilometres above the Nisling's confluence with the Donjek lie the remains of Lynx City, a key settlement in the region until its abandonment in about 1940. The historical use and occupancy of Lynx City epitomizes the cosmopolitanism of pre-1950s Indigenous society. It was a meeting place for Southern Tutchone speakers from Kluane, Aishihik, and Kloo Lakes; Upper Tanana people from the Yukon/Alaska borderlands; and Northern Tutchone people from along the Yukon River—all of whom were connected to one another by ties marriage and kinship. Gotthardt et al. note that, "at one time, many large potlatches were held [at Lynx City] and it was said that approximately 16 people were buried [there]. Many people lived [there] year round, and when they left, they dispersed to Snag, Burwash or Beaver Creek." The well-known Copper family, whose descendants are now citizens of half a dozen First Nations in the central and southwest Yukon, was based in the Nisling and lower Donjek valleys. Ruth Gotthardt, Sally Robinson, and Greg Skuce, *Historic Sites in the White River/ Donjek River Area* (Whitehorse, YT: Heritage Resources, Government of Yukon, 2004), 6; Catharine McClellan, Lucie Birckel, Robert Bringhurst, James Fall, Carol McCarthy, and Janice Sheppard, *Part of the Land, Part of the Water:*

A History of the Yukon Indians (Vancouver, Toronto: Douglas and McIntyre, 1987), 301. There are still elders alive today—in various contemporary First Nations—who were born at Lynx City and/or spent many of their younger years in and around the Nisling valley. Although Lynx City itself falls just within several First Nations' traditional territories (Kluane and White River First Nations of the southwest Yukon, and the Selkirk First Nation of the central Yukon; see Figure 10.1), the fact that most of the Nisling valley is not part of any traditional territory at all must be attributed to the ad hoc and uncoordinated nature of the boundary-making process. Prior to ratification of its final agreement, Kluane First Nation attempted to rectify this problem by amending its traditional territory to include the black hole, but it was unable to obtain the required consent of all Yukon First Nations. The chief stumbling block was overlap with White River First Nation; see Nadasdy, "Boundaries among Kin."

51 Kluane First Nation, *Kluane First Nation Final Agreement,* 67. The reason for this somewhat strange phrasing is that under their final agreements First Nations retain aboriginal rights and title to most of their settlement lands (Kluane First Nation, *Kluane First Nation Final Agreement,* 18, 65). The exact nature of aboriginal title remains undefined, but it is deemed for purposes of the agreements to be at least equivalent to fee simple. To complicate matters further, the agreements actually define three different categories of settlement land—category A, category B, and fee simple—and the exact nature of

First Nation rights differs for each. Ch. 5 of the Yukon agreements (Kluane First Nation, *Kluane First Nation Final Agreement*, 65–79) lays out these rights in some detail.

52 Kluane First Nation, *Kluane First Nation Final Agreement*, 68; Kluane First Nation, *The Kluane First Nation Self-Government Agreement among Kluane First Nation and Her Majesty the Queen in Right of Canada and the Government of the Yukon* (Ottawa: Public Works and Government Services Canada, 2003). It should be noted that First Nations also have a set of powers, particularly relating to governance and the delivery of services to their own citizens, that are not confined to settlement land (Kluane First Nation, *Kluane First Nation Self-Government Agreement*, 13–14).

53 For a discussion of the settlement land-selection process, see Paul Nadasdy, "The Antithesis of Restitution? A Note on the Dynamics of Land Negotiations in the Yukon, Canada," in *The Rights and Wrongs of Land Restitution: Restoring What was Ours*, ed. D. Fay and D. James (London: Routledge, 2008), 85–97. Kluane First Nation, for example, has five parcels of settlement land that lie outside its traditional territory. By far the largest of these (18.9 square kilometres) is located in the black hole (see n44). The other four are small, site-specific selections within CAFN's territory. The largest of these is a fifteen-hectare parcel near the village of Haines Junction, approximately 120 kilometres south of Burwash Landing. Since there is no high school in Burwash, KFN students (along with their parents) often relocate to Haines Junction for several years so

they can attend school there. KFN selected land in Haines Junction so it could provide housing for its citizens residing there.

54 Kluane First Nation, *Kluane First Nation Self-Government Agreement*, 14.

55 First Nations also have the power to regulate the hunting of other First Nations' citizens in areas that lie "within the geographical jurisdiction of the [Renewable Resources] Council established for that First Nation's Traditional Territory" (Kluane First Nation, *Kluane First Nation Final Agreement*, 238–39). This means that in overlap areas where there is no overlap resolution boundary, First Nations can only regulate hunting by their own citizens; citizens of another First Nation may be subject to different regulations on the same land; see also Kluane First Nation, *Kluane First Nation Final Agreement*, 35, 40, 44.

56 See, for example, Kluane First Nation, *Kluane First Nation Final Agreement*, provisions 16.5.0 and 16.12.0.

57 Nadasdy, "Boundaries among Kin."

58 Vandergeest and Peluso, "Territorialization and State Power," 389.

59 Vandergeest and Peluso, "Territorialization and State Power," 387.

60 In fact, however, provincial/territorial jurisdiction over wildlife is complicated by a number of factors. First, the federal government retains jurisdiction over some species, such as salmon and other anadromous fish, which fall under the purview of the federal Department of Fisheries and Oceans. Second, the federal government can pass legislation, such as the Species

at Risk Act, and enter into international agreements, such as the Migratory Bird Treaty of 1916 and the *1973 International Agreement on the Conservation of Polar Bears*, which impinge on provincial/territorial jurisdiction over fish and wildlife. Third, because animals do not respect political boundaries, their effective management often requires provinces/territories to work collaboratively with other jurisdictions; a prominent example of this kind of inter-jurisdictional cooperation in the Yukon is the co-management of the Porcupine Caribou herd.

61 For analysis of a specific case, see Nadasdy, *Hunters and Bureaucrats*, 193–96.

62 It is perhaps worth pointing out that Figures 1–2, which depict First Nation traditional territories and settlement lands, were created by Environment Yukon, the same government agency that produced Figures 6–9, and all are available from the same government website: http://www.environmentyukon.ca/maps/.

63 Sack, *Human Territoriality*, 33.

64 I put "illegally" in quotation marks because he killed them in the Kluane Game Sanctuary; had he been caught, he would have been charged under Yukon law. The Yukon Court of Appeal, however, subsequently ruled in *R. v. Michel and Johnson* (1983) that status Indians did in fact have an aboriginal right to hunt in the sanctuary.

65 Some of Joe's closest relatives were citizens of CAFN, including his paternal uncles, who helped raise him and taught him to hunt.

66 Kluane and White River First Nations emerged as distinct political entities from amidst the breakup of the Kluane Tribal Council in 1990 (KTC itself had resulted from the prior amalgamation in 1961 of the Kluane and White River bands). Because traditional territorial boundaries had by then already been agreed to by the parties to the UFA, Kluane and White River found themselves with hundred-percent territorial overlap. For an extended discussion of this overlap and the complications to which it has given rise, see Nadasdy, "Boundaries among Kin."

67 Vandergeest and Peluso, "Territorialization and State Power," 389–91.

68 Vandergeest and Peluso, "Territorialization and State Power," 389.

69 The situation here is particularly complex because it involves special provisions in KFN's agreement designed to deal with the fact that the Kluane and White River First Nations have hundred-percent overlap. See Nadasdy, "Boundaries among Kin," for a detailed discussion of the issue and a map from KFN's agreement that shows Wellesley Lake well outside the area over which the DKRRC has jurisdiction.

70 Unless and until KFN negotiates terms by which it will assume powers of law enforcement and the administration of justice (see Kluane First Nation, *The Kluane First Nation Self-Government Agreement*, 17–19), it is Yukon— rather than KFN—officers who are charged with enforcing both KFN and Yukon laws.

71 There are different ways in which one might interpret these incidents of official ignorance. Andrew Mathews shows that the "ignorance" of state officials is

sometimes the product of delicate negotiations with their political superiors, on one hand, and their "clients" and/or the local population, on the other. Andrew Mathews, "State Making, Knowledge, and Ignorance: Translation and Concealment in Mexican Forestry Institutions," *American Anthropologist* 110, no. 4 (2008): 484–94. See also Nadasdy, *Hunters and Bureaucrats*, 56–59. I do not have sufficient information about these incidents to offer an analysis of this sort, but the possibility exists that something other than simple ignorance was involved.

72   Paul Nadasdy, "Adaptive Co-Management and the Gospel of Resilience," in *Adaptive Co-Management: Collaboration, Learning, and Multilevel Governance*, ed. D. Armitage, F. Berkes, and N. Doubleday (Vancouver: UBC Press, 2007), 208–27; Paul Nadasdy, "'We Don't *Harvest* Animals, We *Kill* Them': Agricultural Metaphors and the Politics of Wildlife Management in the Yukon," in *Knowing Nature: Conversations at the Intersection of Political Ecology and Science Studies*, ed. M. Goldman, P. Nadasdy, and M. Turner (Chicago: University of Chicago Press, 2011), 135–51.

73   Nadasdy, *Hunters and Bureaucrats*, ch. 2; Paul Nadasdy, "The Gift in the Animal: The Ontology of Hunting and Human-Animal Sociality," *American Ethnologist* 34, no. 1 (2007): 25–43.

74   Jean-Guy Goulet, *Ways of Knowing: Experience, Knowledge, and Power among the Dene Tha* (Lincoln: University of Nebraska Press, 1998); Nadasdy, *Hunters and Bureaucrats*, 102–8.

75   In 2007, the executive director of the DKRRC and two of its members independently complained to me that local residents seemed to regard the RRC as "the enemy"— even though its members were all also local residents.

76   It seems to me that Carlson (his chapter in this volume) makes a very similar argument regarding developments in *Eeyou Istchee* since the signing of the James Bay Agreement. For a discussion of this ambiguous political space, see also Kevin Bruyneel, *The Third Space of Sovereignty: The Postcolonial Politics of U.S.-Indigenous Relations* (Minneapolis: University of Minnesota Press, 2007).

77   This resonates with Liza Piper's observation (see her chapter in this volume) that Canadian government officials who were interested in improving northern Native diets often did not regard country food as food at all.

78   Michael Asch, "From Calder to Van der Peet: Aboriginal Rights and Canadian Law, 1973–96," in *Indigenous People's Rights in Australia, Canada, and New Zealand*, ed. P. Havemann (New York: Oxford University Press, 1999), 428–46.

79   Nadasdy, *Hunters and Bureaucrats*, 250–51.

80   Stephanie Irlbacher-Fox, *Finding Dahshaa: Self-government, Social Suffering, and Aboriginal Policy in Canada* (Vancouver: UBC Press, 2009). In her chapter in this volume, Tina Loo shows that the Canadian government's patronizing preoccupation with the need to "build capacity" among northern Native peoples long predates the modern land claim era.

11

Ghost Towns and Zombie Mines: The Historical Dimensions of Mine Abandonment, Reclamation, and Redevelopment in the Canadian North

Arn Keeling and John Sandlos

In the past two decades a new approach to mining history has emerged to ask, in effect, what happens after the gold rush. Authors such as Richard V. Francaviglia, Ben Marsh, William Wyckoff, and more recently David Robertson have all extended their narratives beyond the demise of mining towns to question what they consider to be the "mining imaginary," the idea that the historical end-point for mining activity is inevitably community collapse and ecological destruction. They provide valuable case studies where communities have survived past the end of mining, diversifying their economies through industrial activity or the development of tourism. Historical memory often provides a sense of continuity for these communities, as mining heritage landscapes and museums become touchstones of tourist activity, and ecological restoration activities reveal a deep sense of attachment to the mining landscape. For this loosely defined community resilience school of mining history, mining is not an ephemeral economic activity but offers communities a long-term sense of deep intimacy with their history of labour within the local landscape.[1]

Without a doubt, this newer approach to mining history provides a powerful corrective to environmental histories that position mineral exploitation as a physical and symbolic marker of environmental decline. Some authors, such as Jared Diamond, have gone so far as to evoke mining as a metaphor for the ecological collapse of civilization.[2] However important it may be to critique such overblown, declensionist narratives, much of the published work of the community resilience school suffers from its own limitations. Many of these scholars extrapolate their theories of community renewal and survival from individual case studies rather than regional, national, or international studies, and most of the case studies are situated in or close to relatively well populated areas of Britain, the United States, and southern Canada. In these regions, communities are likely to have better access to markets and infrastructure to support economic diversification. In general, the impact of mines on individual communities varies across time and space according to several regional factors. The sociologist Lisa Wilson has argued, for example, that the consequences and impacts of mining on local communities are conditioned by the nature of the resource being exploited, the type of mining technology employed, the status of labour–management relations, fluctuations in international commodity prices, and varying levels of regional economic dependence on mining.[3] Similarly, reacting to simplistic models of resource-based development versus underdevelopment, rural sociologists Scott Frickel and William Freudenberg have called for detailed historical-geographical studies of "the *ways* in which the relationships between resource extraction and regional development have changed over time."[4]

Drawing on these ideas of regional heterogeneity, this paper will argue that themes such as continuity and local attachment to place often have very little relevance in hinterland regions such as Canada's remote territorial north, where geographically isolated communities and mining operations sit at the economic margins of an international commodity trade, and where historical experiences of collapse and continuity are governed more by mineral price fluctuations than the ties that bind a community to its local place.[5] Abandoned mines and ghost towns, or at least severely depressed former mining settlements, remain a prominent feature of the northern Canadian landscape. Another characteristic of the northern

Canadian context is the presence of a significant Aboriginal population, who may interpret historical mining developments undertaken by settler society as a colonial appropriation of their local environments and a threat to their local attachment to place, impacts symbolized by the physical legacies of abandoned mine sites in their traditional territories.[6]

As the literature on the "mining imaginary" suggests, however, mine abandonment does not constitute an end to the material and social relations that mining generates. Many former mine sites in northern Canada (and around the globe) are seeing renewed mineral exploration and development activity.[7] The redevelopment of previously worked mineral deposits may be driven by rising commodity prices, changing extraction technologies, and improvements in transportation and access that permit formerly uneconomic deposits to become attractive to investors and mining companies. In some cases, this redevelopment has the prospect of renewing long-closed, once-profitable mines and reanimating depressed or even moribund mining communities. This phenomenon intersects with ongoing reclamation, rehabilitation and/or re-use of abandoned mine landscapes; in some cases, redevelopment may take place simultaneously with post-closure cleanup and remediation activities.

The redevelopment of formerly abandoned mines raises both theoretical and practical questions for understanding the impact of mineral development on resource-dependent regions and the environment. The classical image of the "natural history" of a mine, outlined by Homer Aschmann, suggests a linear (and inevitable, based on the finitude of the resource) process of mine development and exhaustion, closure and/or abandonment.[8] Similarly, the idea of the mining cycle is widely promoted by the industry as a model for understanding mining's purportedly transient impacts on local environments, and is even linked to the notion of "sustainable" mining.[9] This notion of a mining cycle may, to some extent, accommodate the "repass" phenomenon, whereby changing technology or market and regulatory conditions make possible the renewed exploitation of formerly profitable mineral deposits (though often under very different scales of operation, capitalization, labour arrangements, etc.).[10] But this cyclical, yet ultimately terminal model fails to account for the afterlife of closed mines, in both physical and human terms.

Our research suggests that the environmental and social conflicts surrounding mining do not dissipate with closure and abandonment; rather, historical discord over mining developments is frequently revisited and re-engaged during the redevelopment or remediation phases, if under different circumstances. Renewed activities at former mine sites take place within the context of the ongoing environmental implications of previous mining, including acid mine drainage, waste piles, long-term landscape disturbances, etc. The advent of redevelopment and/or reclamation also raises economic and cultural questions surrounding both current and former mineral development, including: the costs and human hazards associated with environmental liabilities from abandoned mines and their cleanup; the status of local mining-dependent communities and infrastructure; alternative economic development prospects; and public perception (positive or negative) of previous rounds of mining activity.[11] This phenomenon, then, challenges the notion of the "finality" of the mining cycle and demands a reconceptualization of these reanimated, or not-quite-dead places that takes into account both their histories and the contemporary challenges posed by mineral redevelopment in formerly active territories. If the collapsed communities left in the wake of mine closure are known as "ghost" towns, we suggest these sites may be thought of as "zombie" mines.[12]

It is difficult to determine the precise number of potential zombie mines in northern Canada, given the sometimes disaggregated nature of public information on abandoned mines. In the north, remediation of mines that have reverted to public control slowly became a priority for Indigenous and Northern Affairs Canada (INAC; formerly Aboriginal Affairs and Northern Development Canada and Indian and Northern Affairs) as part of the broader Northern Contaminants Program established in 1991 to address a broad spectrum of contaminants such as persistent organic pollutants, radionuclides, and heavy metals that have concentrated in the region from sources further south. In keeping with the mandate of the NCP, subsequent remediation projects have tended to revolve around technical engineering issues associated with persistent toxins rather than the historical social and political conflicts that attended the development of the mines.[13] In terms of toxic sites requiring remediation, the Commissioner of the Environment in the federal Auditor General's office identified the issue of abandoned mines as a significant environmental and financial liability in northern Canada. In a 2002 report, the Commissioner highlighted the fact that INAC had identified thirty priority sites, seventeen of which required urgent action in terms of remediation. The department also identified twenty-nine additional sites that are suspected of being contaminated. The Commissioner's report estimated the total clean-up bill for the remaining abandoned mines at $555 million (a severe underestimate), all of it public money because, prior to the diamond mining era in the late 1990s, no financial security was collected from mining companies to cover remediation costs.[14] INAC or territorial government authorities are currently overseeing remediation, assessment, and/or monitoring activities at sixteen of these contaminated mine sites in Nunavut, Yukon, and the Northwest Territories. In addition, private companies have attempted to redevelop at least six mine sites spread throughout the territorial north that had previously closed, with four of these sites also simultaneously undergoing remediation (Table 11.1).

This chapter examines two of the largest zombie mines in the Great Slave Lake Region: the Pine Point lead-zinc mine east of Hay River and the notorious Giant gold mine near Yellowknife. The key issues at these two abandoned sites are very different: at Giant, the most prominent environmental concern is the containment of arsenic stored on site, while at Pine Point, massive landscape change is the mine's most important lasting legacy. Yet in both cases, nearby Aboriginal communities (Fort Resolution and K'atl'odeeche First Nation near Pine Point; Dettah and Ndilo adjacent to Giant Mine) frame their historical understanding of these mines in terms of colonial land appropriation and environmental degradation, historical injustices that today shape their responses to the reanimation of these mines. In contrast, the non-Native communities associated with the mines have promoted preservation of a proud mining history reminiscent of other local mining heritage campaigns in North America. In an online forum and memorial, former residents of Pine Point celebrate the history of the town as a suburban paradise, an exercise in nostalgia and memory that is captured brilliantly in the National Film Board multi-media documentary *Welcome to Pine Point*.[15] Similarly, the Northwest Territories Mining Heritage Society promotes an idealized historical understanding of gold mining at Yellowknife as the progenitor of modernity and civilization in the north, and the organization is very active in trying to preserve the material culture (equipment, headframes, buildings, etc.) and memories of the "good old days" of early gold mining.[16] The boundaries between these contrasting Native and non-Native

TABLE 11.1: Mines Undergoing Remediation and/or Redevelopment in Canada's Territorial North

Site	Territory	Issues	Operational period	Current Status
Cantung	NWT/Yukon	Heavy metals; acid mine drainage potential	1962-1986; reopened 2003-04; 2005-09; 2010-present)	Currently undergoing redevelopment and remediation
Colomac Mine	NWT	Cyanide; hydrocarbons	Operations 1989-1997	Remediation 2007-2011; monitoring
Con Mine	NWT	Arsenic in tailings; landscape impacts	Operations 1938-2003	Remediation since 2007
Contact Lake Mine	NWT	Arsenic and uranium	Silver in the 1930s, uranium 1949-50, and irregular operations to 1980	Currently undergoing remediation and redevelopment
Discovery	NWT	Mercury contamination	Operations 1950-1969	Remediation 1998-2001; ongoing monitoring; redevelopment of area as Yellowknife Gold Project after purchase by Tyhee in 2010
El Bonanza/ Bonanza	NWT	Hydrocarbons; diesel drums; waste rock in Silver Lake	Silver mining 1934-40; irregular activity 1965-1984;	Remediation commenced 2009
Faro Mine	Yukon	Acid mine drainage; heavy metal contamination	Operations 1965-98	Remediation planning ongoing
Giant Mine	NWT	Arsenic stored in tailings and underground	Operations 1948- 2004	Environmental assessment of remediation plan ongoing
Indore/ Beaverlodge (Hottah Lake)	NWT	Radioactive tailings; asbestos	Intermittent operations 1950-56	Undergoing assessment with remediation pending

Site	Territory	Issues	Operational period	Current Status
Keno Hill	Yukon	Zinc in water	Operations 1914-1989	Remediation and redevelopment
Lupin	NWT	Cyanide	Operations 1982-2005	Test drilling for further development commenced 2011
Nanisivik	Nunavut	Heavy metal contamination (zinc, lead, cadmium)	Operations 1976-2002	Remediation since 2002 and ongoing monitoring
Mt. Nansen	Yukon	Cyanide and heavy metals in tailings	Operation 1968-69, 1975-76, 1996-98;	Remediation and monitoring ongoing since 1999
North Inca	NWT	Fuel Tanks; asbestos	Exploration mining 1945-49	Remediation commenced 2009
Port Radium	NWT	Uranium and copper; elevated radiation;	Operations 1932-1982	Remediation completed, 2005-2008; ongoing monitoring
Pine Point	NWT	Landscape impacts	Operations 1964-1988	Redevelopment currently proposed
Rayrock	NWT	Radioactive tailings	Operations 1957-1959	Remediation completed 1997 and ongoing monitoring
Silver Bear Properties	NWT	Heavy metals (cadmium, lead, mercury, uranium, zinc and arsenic)	Operations early 1960s to 1985 at four mines	Remediation commenced 2009

(Sources: AANDC, Contaminants and Remediation Directorate, "Contaminated Sites Remediation: What's Happening in the Sahtu?," March 2009 http://publications.gc.ca/collections/collection_2010/ainc-inac/R1-27-2009-eng.pdf; AANDC, Major Mineral Projects and Deposits North of 60 Degrees, accessed March 3, 2012, http://www.aadnc-aandc.gc.ca/DAM/DAM-INTER-HQ/STAGING/texte-text/mm_mmpd-ld_1333034932925_eng.html; AANDC, Northern Contaminants Program, last updated August 15, 2012, http://www.aadnc-aandc.gc.ca/eng/1100100035611).

attitudes to mining in the past and present are not always neatly drawn. The Aboriginal communities adjacent to these mines often share a sense of pride in the mining past, particularly their own involvement in the mining workforce or experience with life in an idealized modern town such as Pine Point, while the non-Native community may share concerns about long term environmental problems at the mines, as is evident in the recent involvement of the Yellowknife city council and the environmental NGO Alternatives North in the environmental assessment of the federal government's arsenic remediation plan at Giant Mine.[17] Despite this fluid nature of various community engagements with reanimated mines, we argue that remediation and redevelopment projects force Aboriginal communities in particular to revisit their uniquely conflictual rather than consensual local mining histories, negotiating the reanimation of zombie mines though the historical lens of colonial dispossession and the environmental injustices associated with the original mining development.

Mining the Great Slave Region

The history of mining in Canada's territorial north properly begins with the Yukon Gold Rush in 1898, but the first large-scale industrial mines— what we would today call mega-projects, complete with capital-intensive industrial development and transportation infrastructure—did not appear until the early twentieth century. Prior to the Second World War, the number of major developments remained small, limited to the Keno Hill silver mines (1913), the Port Radium radium and uranium mine (1932), and three closely related mines on the northern shore of Great Slave Lake: Con (1938), Negus (1939), and Giant (1948, first staked in 1935). In the 1950s and 1960s, at the height of the post-war economic boom, mining companies proceeded with significant exploration and development activities at such sites as the lead-zinc deposits at Pine Point, NWT, lead deposits at Faro, Yukon (the Cyprus-Anvil Mine), and nickel at Rankin Inlet, NWT (Fig. 1). The pattern of development was haphazard, a constant logistical tug-of-war among factors such as the ore grade and the size of the deposit, its proximity to viable transportation routes, and prevailing commodity prices. Nevertheless, after the Second World War the Canadian government actively promoted northern mineral development

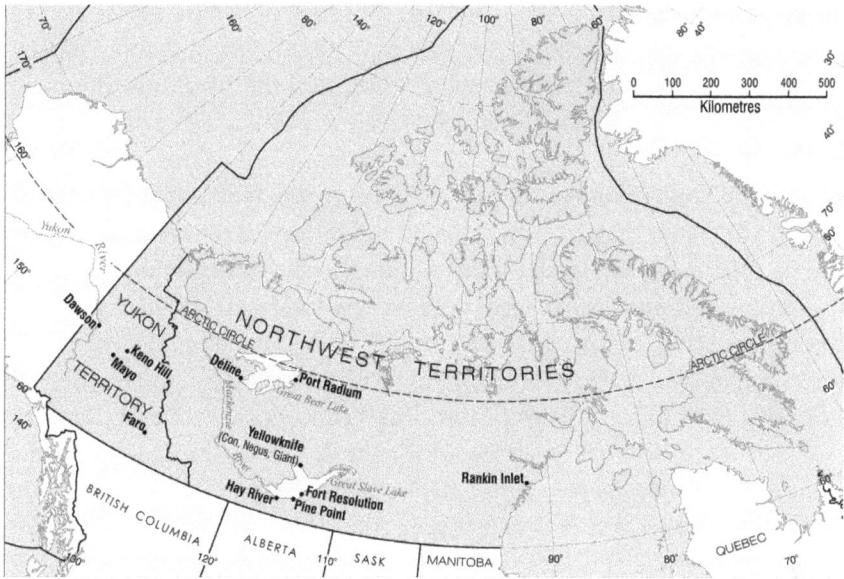

FIG. 11.1: Major mining developments and communities in the Canadian north in the early to mid-twentieth century. Map by Charlie Conway.

as a means of promoting settlement, modernizing the region, and pulling local Native inhabitants out of their anachronistic hunting and trapping economy. The government took an active role promoting a mineral-led opening of the north, funding transportation and hydroelectric infrastructure to facilitate rapid development of minerals. Certainly the enthusiasm of post-war government boosters was abetted by the fact that the initial gold rush at Yellowknife in the 1930s, and the longevity of the Con, Negus, and Giant mines on the north shore of the lake, had provided the economic base for a 1930s frontier outpost to grow into a modern town by the 1950s.[18]

Prospectors and mining promoters had been attracted to the mineral-rich region around Great Slave Lake as early as the late 1890s, when the first claims were staked at Pine Point on the lake's southern shore by prospectors initially interested in silver and gold.[19] From 1935 to the early 2000s, the land around Great Slave Lake within a 250-kilometre radius of Yellowknife, particularly to the north, became pock-marked with sixty-five mines, mostly devoted to gold, but also tungsten, uranium, cobalt, nickel, lead, zinc, rare earth metals such tantalum and lithium, and, more recently, diamonds. As of 2006, only one of these sites, the Snap Lake diamond mine, remained operational, though the large Ekati and Diavik diamond mines just over three hundred kilometres from Yellowknife remain the core of the contemporary mining industry in the Northwest Territories. Such an intensive regional concentration of mining activity has no historical parallel in the territorial north other than the much smaller complex of uranium, radium, silver, and gold mines developed at the east end of Great Bear Lake. Many of the sites in the Yellowknife region were small-scale operations, never proceeding beyond the advanced exploration or bulk-sampling stage of development, and often collapsing after only one or two years of unpromising samples. But others, such as the Discovery gold mine and town eighty-four kilometres north of Yellowknife that operated from 1950 to 1969, or the Colomac gold mine that operated intermittently in the 1990s, were full-scale industrial mining operations carved out of the northern edge of the boreal forest. By 2005, seventy years of mineral extraction had left a legacy of sixty-four abandoned mines in the Yellowknife region, though only fourteen of the sites could be considered significant producers. At only eleven of these sites have the mining companies or governments conducted any remediation or clean-up activities. In one case, a small company, Slave Lake Gold Mines Ltd., was in such a hurry in 1942 to abandon its increasingly unpromising and short-lived gold and tungsten mine on Outpost Island on the East Arm of Great Slave Lake, it neglected even to provide for the removal of the miners and their families, forcing them to build a barge with lumber from the mine buildings.[20]

Many of these abandoned mine sites left behind much more than industrial relics such as old buildings, headframes, and scattered mining equipment. At almost all stages of a mine's development, the social and environmental impacts on local Native communities (in this region, the Dene) could be severe. At the exploration phase, activities such as cutting seismic lines, road development, burning forest to access prospecting sites, and digging test pits and trenches could impact fish and wildlife populations and interfere with the hunting and trapping economy. At larger operations, full-scale development often produced extensive landscape changes due to the digging of pits and piling of waste rock. The construction of instant mining towns brought a sudden influx of outside workers, with the attendant introduction of alcohol, store-bought food, and competition for local wildlife resources (particularly in the pre-Second World War mining camps, where miners relied more on wild game for food).[21] Prior to the 1970s, almost no northern mines had developed a Native employment policy (the Rankin Inlet nickel mine being a lone exception; see Tina Loo's chapter in this volume).[22] Hence few employment or other economic benefits accrued to Native communities, other than seasonal line-cutting, hauling, road-building, or wood-cutting work, despite the impact on the hunting and trapping economy and the fact that much of the mining took place on land where claims arising out of Treaty settlements remained unresolved.[23] The mining developments around Great Slave Lake thus represented the leading edge of industrial colonization in the Northwest Territories. Although Native people were never swept aside into reserves as in the southern provinces, resource development proceeded with almost no meaningful attempt to integrate the Dene into the new mining economy and little regard for its impact on the mixed subsistence and small-scale trade economy of the Dene.

Perhaps the most severe and persistent legacy arising from mining activity near northern Aboriginal communities was the emission of toxins into water, soil, and air at many sites, particularly when ore was processed directly at the mine. The processing of gold in particular can produce toxic byproducts such as arsenic and cyanide: the former a naturally occurring element that is emitted into the air when ore containing arsenic is roasted or into water and soil when released with slag and tailings; the latter, used to extract gold from ore, may subsequently leach into soil and water as it is disposed with mine tailings. Heavy metals such as lead or zinc may also leach into water and soil depending on the mineral composition of mine wastes. Acid mine drainage, or low-pH mine water produced through the oxidization of sulphides, may also severely impact aquatic ecosystems depending on local conditions.[24] All of these toxins persist well after the closure of a mine, presenting environmental risks and liabilities that have become a critical public policy issue in northern Canada in recent years.[25] Most immediately, the environmental legacies of abandoned mines present grave concerns for residents of the region, particularly Aboriginal communities adjacent to the mines where human health and hunting economies may be impacted by landscape-scale ecological changes or toxins accumulating in water, fish, and game animals.

Arsenic Emissions at Giant Mine

The abandoned Giant Yellowknife gold mine represents one of the most high-profile cases of mine contamination in northern Canada. Although gold was discovered at the site as part of the early gold rush that had led to the development of the Con and Negus mines in the 1930s, the company Giant Yellowknife Mines Ltd. did not pour its first gold brick until 1948, as markets for precious metals recovered after the Second World War. Because the ore at Giant was contained within arsenopyrite rock formations, the company had to build a roasting facility to burn off sulfur that prevented separation of gold using the standard cyanide chemical leaching method. The roasting process also produced large volumes of highly toxic arsenic trioxide in the form of a fine airborne dust, which spread from a smokestack broadly over the Yellowknife area. By 1949, airborne emissions of arsenic trioxide from the roasting facility at Giant Mine had reached 7,500 kilograms per day.[26] During this early period, the company had no pollution abatement installed on the roaster stack, though emissions control technology, particularly the Cottrell electrostatic precipitator, had been available since 1908.[27]

Very quickly the area surrounding Yellowknife became an example of what several environmental historians have described more generally as "landscapes of exposure"—toxic hot spots that present enormous public health and ecological concerns due to the heavy loading of industrial pollutants.[28] The situation was particularly dire for the nearby Yellowknives Dene First Nation communities now known as Ndilo (on Latham Island adjacent to Yellowknife) and Dettah (on the east side of Yellowknife Bay across from Con Mine). These two villages were often downwind of the pollution from the roaster stack, as the breeze funnelled down from the mine site at the head of Yellowknife's Back Bay to the town sites located in the front part of Yellowknife Bay. The potential for arsenic poisoning in these communities was high due to the use of local berries in the diet and the practice of melting snow for drinking water. The Giant Mine site was, in fact, a very important area for the Yellowknives' berry-picking and hunting activities, an area which they eventually abandoned as the dangers of arsenic became known in the community.[29] In 1951, arsenic pollution resulted in tragedy for the Yellowknives, as a two-year-old Native child died due to arsenic poisoning, most likely from drinking contaminated snowmelt. The company offered the family $750 in compensation for the loss of their child.[30] Yellowknives Dene members have suggested that arsenic-related sickness such as skin irritation or hair loss were particularly acute during this period, while one local physician recalled at least one case of arsenic-related skin conditions such as keratosis (thickening of the skin), hyperpigmentation (black spots on the skin), and paresthesias (a burning sensation in the extremities) in a middle-aged patient from Dettah in 1954.[31]

The impacts of the mine were not limited to humans and were not confined to the issue of air pollution. Environmental studies from the 1970s suggested that arsenic emissions from mine tailings and settled dust had destroyed all biota in Baker Creek, a stream running through the Giant site that had once been a significant fishing area for the Yellowknives. Emissions of sludge from Giant, Con, and Negus created a zone of influence that extended three kilometres into Yellowknife Bay, where phytoplankton had disappeared.[32] Members of the Yellowknives Dene First Nation remember stories of sled dogs, cattle, and chickens dying from drinking or swimming in tailings water.[33] Local testimony from Yellowknife's non-Native residents recalls that several horses had died in 1949 due to drinking melting snow water in spring, and the entire cattle herd from the fledging Bevan farm in the area died from arsenic poisoning close to 1951.[34]

In response to pressure from local public health officials and federal bureaucrats within the departments of Indian Affairs, Health and Welfare, and Resources and Development, the mining company installed a Cottrell electrostatic precipitator in 1951.[35] This technology uses an electrical current to knock solid particles out of air emissions, similar to how static electricity on an older, pre-digital television screen collects dust.[36] To provide more refined air filtration, the company installed a baghouse in 1958, essentially a second-stage air filter that further blocked small arsenic trioxide particles produced in the roasting facility. In 1957 the company also began partly to remove arsenic from mine tailings in order to mitigate water pollution in Yellowknife's Back Bay. The installation of pollution control equipment did dramatically reduce the spread of airborne arsenic trioxide, as emissions dropped to between two hundred and three hundred kilograms per day by 1959.[37]

This focus on a technological fix to air pollution problems at Giant Mine was entirely in keeping with prevailing approaches to smelter pollution problems throughout North America, where governments imposed the principle that mitigation rather than outright elimination of pollution was acceptable so long as the polluter paid the monetary cost. At the same time, public health authorities were steeped in the idea that that it was possible to determine scientifically a safe level of human exposure for every toxic substance, a "dose makes the poison" approach that ignored the potential impacts of long-term arsenic exposure. The safe threshold level for Yellowknife drinking water, for example, was set at five times the current safe level of 0.01 parts per million for drinking water, levels we now know can produce serious long-term cancer risk. As long as nobody was suffering from acute arsenic poisoning, however, and as long as the company was at least attempting to control emissions, Giant Yellowknife Mines was allowed to operate full tilt. Certainly in the early to mid-twentieth century, there was no appetite among government regulators in North America to impose shutdowns on smelting and roasting facilities; technological mitigation, dispersal, and dilution remained the dominant public response to pollution and public health threats throughout this period.[38]

But as mentioned above, pollution control technology did not prevent all arsenic from entering air or water sources in the Yellowknife region after the 1950s, particularly as production increases boosted total emissions from the stack. In the wake of Rachel Carson's *Silent Spring* (1962), broad social acceptance of the environmental costs of industrial growth in North America began shifting to concern about the presence of toxic chemicals in the human environment, and, slowly, government regulators began to respond.[39] As Stephen Bocking shows in his chapter in this volume, these concerns extended to the contamination of northern environments by both distant pollution sources and local industrial developments. In 1965, federal public health officials became concerned when testing showed persistently high arsenic levels in vegetables and drinking water in the Yellowknife area.[40] In response, the federal government organized a comprehensive study of the issue under the direction of Dr. A. J. DeVilliers of the Department of Health and Welfare. Although the testing of water bodies, local vegetables, and individuals was highly visible in Yellowknife during the study period from 1966 to 1969, results of the survey were impossible to obtain, despite repeated requests from local municipal government, public health officials, and Native groups. DeVilliers blamed staffing issues on his failure to produce a report, but the lack of public information about the adverse health impacts of arsenic pollution—indeed, the perception that the government might be hiding information—created heightened public anxiety and controversy in Yellowknife.[41]

In 1975, the issue reached the national media when CBC Radio's *As it Happens* suggested the results of the DeVillers study had been suppressed, particularly sections pointing to high rates of lung cancer in Yellowknife, possibly due to long-term exposure to arsenic in the air.[42] Public health officials in the federal government remained skeptical of the link between arsenic and cancer in the Yellowknife area, arguing that many cancer patients in the local hospital were Inuit from the arctic coast, while Native populations from non-arsenic polluted regions such as Whitehorse in the Yukon exhibited similar rates of lung cancer.[43] Nonetheless, in response to local concerns the federal government did initiate new public health studies in 1975. These surveys found elevated levels of arsenic in hair and urine samples taken from mill workers at the mine, but not in the general population of Yellowknife and adjacent Native communities.[44] Two years later, however, the National Indian Brotherhood released its own hair-sample study showing high rates of arsenic in twenty-five Native children.[45] As a further response, the federal government contracted an independent body, the Canadian Public Health Association (CPHA), to conduct a study of local arsenic contamination in humans and the environment. Much of the CPHA's work focused on urine samples from workers and local Native people. After extensive testing, the CPHA concluded once again that the impacts were largely confined to the workplace and levels in the general population remained below threshold safety levels. The CPHA final report recommended ongoing monitoring of arsenic levels, careful washing of vegetables and berries, and the trucking of water to Ndilo and Dettah in winter, with warnings to locals not to use snow as a source of drinking water as studies still indicated high levels of concentration in this source.[46] Further upgrades to the Cottrell precipitator and baghouse equipment reduced air emissions from 76.6 mg/Nm^3 (milligrams per normal cubic metre) in 1975 (when testing began) to 14.07 mg/Nm^3 in 1981, part of an effort to meeting impending regulations limiting emissions to 20 mg/Nm^3 under the federal government's Clean Air Act.[47] The result was an eighty-percent drop in arsenic trioxide in snow by the mid-1980s.[48]

Despite such improvements in environmental conditions, the poisoning of marginalized Native communities and their local environments in the late 1940s and early 1950s through the uncontrolled spread of industrial air pollution clearly resonates with other instances of environmental injustice in North America.[49] Obviously the mine was situated according to the presence of gold rather than along the path of least political resistance associated with many toxic waste facilities. Nonetheless, the growth of an industrial frontier on Dene land with no consultation afforded to affected communities, the expansion of mining development and settlement on some of the most important Yellowknives Dene hunting and gathering territories, and finally the spread of toxic arsenic into basic sources of water, food, and air, suggest the colonial appropriation of land and resources for outside corporate and state interests and the dispossession of Native people from significant sources of local subsistence.[50]

Moreover, as Tim LeCain has argued for other mining sites, the technological fix at Giant Mine did not resolve but merely deferred the problem. Removing arsenic from the air does not destroy the substance, but concentrates it as a fine dust.[51] In 1951, Giant Yellowknife Mines began to pump the material captured by the Cottrell into old mine chambers, sealing off fourteen underground storage compartments containing 237,000 tonnes of arsenic trioxide by the time the mine closed in 2004. Such a massive amount of toxic materials presents an enormous contemporary and long-term toxic threat should the storage chambers leak, the arsenic mix with groundwater, or a seismic event disturb the underground environment. As we discuss below, this problem constitutes a major perpetual environmental and health risk for all people living in the Great Slave Lake region.

Changing Landscapes: The Pine Point Mine

The Pine Point lead-zinc mine was located almost directly south of Giant Mine on the opposite shore of Great Slave Lake. The site was first staked by the trader Ed Nagle in 1898, providing a major impetus for the government to seek a surrender of land from local Dene groups through the signing of Treaty 8 in 1899.[52] After a brief period of exploration in the 1920s and a more sustained and intensive effort in the 1950s, Pine Point Mines

Ltd., a subsidiary of mining giant Cominco, officially opened the mine in 1964, and it became one of the largest and most profitable base metal mines in Canadian history. The company also established the open town of Pine Point adjacent to the mine, a collection of trailers and bunkhouses that quickly became a settlement of two thousand with modern services and facilities such as a shopping centre, a bank, and a hotel. Although annual production rates varied, Pine Point Mines shipped roughly 9,628,000 US tons of concentrated ore dug from forty-seven open pits and two shafts over twenty-five years, most of it for further processing at the Cominco smelter located in Trail, British Columbia. By 1989 it all came to an end, as low mineral prices and high costs associated with the mine prompted the company to shut down the mine, and the territorial government to remove the town.[53]

The mine and townsite were located just over sixty kilometres west of the Chipewyan (a linguistic group of the Dene) community of Fort Resolution, and a hundred kilometres east of the K'atl'odeeche (Dene) First Nation near Hay River. Although Pine Point never presented the same issues of acute air and water pollution as at Giant Mine, the mine did impact the local environment of these two Native communities. Even in the earlier exploration phase, the cutting of seismic lines and roads destroyed traplines in the region, while in the full development phase mining activity closed access to former hunting areas and fishing creeks.[54] Environmental studies from the 1970s suggested that frequent overflows from the tailings pond and the pumping of water from the mine had deposited heavy metals and sulfuric acid into local streams, potentially affecting fish populations.[55] A post-closure study from 1998 indicated, however, that heavy metal deposition in the water presented no significant risk to human health, though oral interviews conducted by the authors revealed that local people remain concerned about water quality.[56] Perhaps the greatest concern for local First Nations is the extensive "mess" the mining company left at the site. Abandonment and reclamation activities for the extensive property included the complete removal of the town, mine and mill infrastructure, the closure of the extensive network of haul roads, and the blocking of access to open pits by berms. The 570-hectare tailings area was covered with loose gravel and its waters are subject to ongoing treatment to reduce high levels of zinc before discharge to the surrounding

FIG. 11.2: Tailings area at the former Pine Point Mine, where overflow water is treated with lime to precipitate zinc. Photo by Arn Keeling.

muskeg, but the area remains a large, moon-like landscape where almost no vegetation grows two decades later (Fig. 11.2).[57]

The mine's stark and sudden end contrasted sharply with the unbridled optimism and boosterism that accompanied its early development. Echoing the Stikine Railway proposals discussed by Jonathan Peyton in his chapter in this volume, federal government officials regarded the development of Pine Point and its associated rail link as a critical gateway development, spurring the exploitation of other mineral deposits in the Great Slave Lake region and speeding the emergence of a modern industrial economy in the region.[58] In 1955, R. Gordon Robertson, Deputy Minister of Northern Affairs and Commissioner of the Northwest Territories, testified to the Royal Commission on Canada's Economic Prospects that

a railway to Great Slave Lake will not be just another railway. It is not a railway to a lake, or to open a mine or to serve a community. A railway to Great Slave Lake will be one of the great development railroads of the country. It will not bring population to the Northwest Territories to the same extent that the western railroads brought it to the prairies, but it may well bring in the years ahead a comparable increase in the wealth of Canada. This railway is quite different from most of the branch lines constructed in recent years which were destined to serve one mine, or a group of mines; its purpose is to open up a whole new region. The fact that there happens to be a potential mine of great value at its northern terminus is a piece of great good fortune, for it will enable this railroad to be built without the long wait for reasonable returns which so often has been the lot of a pioneer railroad.[59]

The federal government's enthusiasm for the project was so great that it provided $100 million in subsidies for the railway and associated infrastructure, including a small hydro-electric project and a highway extension to the site. For many government officials, this was a small price to pay to kick-start an ambitious program of northern development, which would also contribute to the growth of a national economy that had gone into recession in the late 1950s.[60]

In her chapter in this volume, Tina Loo describes the efforts of government officials to promote modernization of Inuit in the Keewatin Region of the Northwest Territories (now Nunavut), in part through industrial wage labour opportunities in the Rankin Inlet Mine. Similarly, the government promoted the Pine Point Mine as a means to resolve the pressing economic issues facing the Native population of the Great Slave Region. For many government officials, the new mining economy would replace the moribund fur trade (which had experienced low prices during the recession of the late 1950s and early 1960s), keeping Native people off relief by pulling them into the modern industrial wage economy.[61] But as the marquee project of the new social and economic development strategy, Pine Point largely failed to fulfill the promise of local employment. In the mine's early stages, there were very few Aboriginal participants in the labour force, largely because the government did not extend the highway to Fort Resolution until 1972. In 1970, Aboriginal employment peaked at 17.1 percent of the mining labour force, and then declined to a steady rate of seven to nine percent through the rest of the decade.[62] How many of these labourers were local to the area is unknown, though some leaders in Fort Resolution complained that Cominco double-counted workers who had left and been rehired in the same year, and that most of the Aboriginal labour force was made up of itinerant workers from the south. Indeed, roughly one-third of the Aboriginal labour force of 78 was from outside the region between 1973 and 1976, while in 1978 the figure was 25 of 52 workers.[63] The turnover rate for Aboriginal workers was also high: between 1963 and 1978, Cominco hired 78 residents of Fort Resolution 125 times, but only 10 stayed for more than a year.[64] The reasons for the lack of local employment are complex, but there is no doubt that the government invested far more in the infrastructure supporting resource extraction than it did on roads or training programs to maximize Aboriginal participation in the mining industry.[65] Company officials complained that they had tried to reach out to Fort Resolution, even providing a bus for commuting, but high turnover kept the numbers of permanent workers low.[66] Ironically, however, employment numbers from the early 1970s indicate higher rates of transience among the non-Aboriginal workforce.[67] By the early 1980s, several federal government reports suggested that a mineral-led human development strategy had largely failed in the Northwest Territories, with Aboriginal employment rates hovering around five percent throughout the region in the late 1970s, and with few spinoff benefits for local communities.[68]

In the eyes of many Fort Resolution residents in particular, bitterness over the lack of economic benefits associated with the mine is compounded by the legacy of environmental damage left in its wake. Although many enjoyed working in the mine, noting in particular the good wages and high quality of life in a modern town such as Pine Point, they also resent the fact that very little was done to remediate the site and that no compensation has been forthcoming for the environmental damage and resources extracted in what is considered a traditional use area. In 1993, then Chief Bernadette Unka testified to the Royal Commission on Aboriginal Peoples about her community's experience with mining development:

> Pine Points [sic] Mines nor Canada have never compensated the Dene people that used those areas in their hunting, fishing and trapping. They have never been compensated for their

loss or for the land devastation. When I say land devastation, if you are to fly over Pine Point Mines you would look down and you would look down and think you were flying over the moon with the craters and open pits that are left open. The people have never been compensated for the hardships and the heartaches induced by mineral development. While the company creamed the crop at $53 million during their peak years, we got very little jobs and what we did get were very low-paying jobs.[69]

In an oral interview conducted in 2010, George Balsillie used remarkably similar language when he lamented the impact of the mine on the land and wildlife:

Oh, it made a mess out there. It's all just like a moon, you know, like when you go on a plane or something … they can only fill those holes up partly. Now they have put signs up there and gates. You can't even drive in there. Before you used to drive to go hunting and then in the fall time to go for moose or something. Now you can't, because it has all been dug up, and you can't drive on that road and you can't drive on this road. I suppose they made a … well, I'll say it, they made a mess. They should have just, you know, sure it was a mine put gates around it … it's all fenced in, but people are not crazy enough to go in there anyways, but you want to go hunting, you gotta see if there's other routes open. That's the only thing.[70]

Although he noted some recent recovery in vegetation and wildlife populations, Gord Beaulieu recorded similar themes as he recollected his first impressions of the mine after living for several years in Yellowknife:

But a few years later I moved back on this side of the lake, and to me it was just like a wasteland. There was nothing. There was no life, nothing. Even the trees were dead. It was just like a wasteland. Roads all over the place, you know? And no life. Like now, it's been twenty-five years now, and the life is coming back to the land over there. We see moose tracks and we see other wildlife around there. And everything turns green now, in the summer. But back then, when the mine first shut down, it was just like a wasteland. And it reminded me of those Mad Max movies.[71]

Some K'atl'ochdeeche First Nation elders, whose reserve lies west of Pine Point along the Hay River, shared memories of how the mine brought work, especially during the construction phase. But they also decried how the company abandoned the town and landscape, and, as Fred Tambour noted, "just leave it laying around like that lookin' like a real ghost territory, to me."[72] These conflicting memories—the "good life" associated with wage work and a modern town, the shattered landscape of the mine, and the possible recovery of the land despite the unfinished remediation of the mine—have all come to the fore in recent years, as the community considers a new proposal to mine the Pine Point area.

From discovery to development, and eventual closure and abandonment, then, the advent of mining around Great Slave Lake transformed environment and society in the region. On the one hand, mineral development brought unprecedented settlement and infrastructure to this remote frontier and generated considerable wealth; on the other, mining left a legacy of environmental degradation and economic marginalization that disproportionately affected the region's Indigenous inhabitants. This history—a source of contestation between settler and Native communities in the region—echoes the experience of other historic mining frontiers, such as the US west, where minerals have attracted outsiders and generated conflicts over land, labour, and livelihood.[73] However, as we have found at Giant and Pine Point, this history is more than past business; it remains a potent source of conflict over both the physical and cultural legacies of mining, especially as these "abandoned" sites become the focus of renewed mining-related activity.

Giant and Pine Point as Zombie Mines

At Giant Mine in Yellowknife, the "reanimation" of the abandoned mine site was instigated by the toxic deposits left behind following decades of gold extraction and arsenic collection. Although current activities are not aimed at redevelopment for minerals, the remediation of the mine site entails complex technical questions and large capital investments that have prompted intense regulatory and community scrutiny. When mining ceased at Giant, most of the surface works were abandoned intact, including the headframe, mill, and ore-roasting plant, and a small abandoned town site was left adjacent to the mine works. Millions of tonnes of arsenic-contaminated tailings filled several impoundments near the mine. Of greatest concern, however, were the 237,000 tonnes of toxic arsenic trioxide dust stored in sealed underground chambers. This odourless, white-powder form of arsenic is considered a human carcinogen and subject to dispersion into the environment through the atmosphere, in solution, or, if heated, in gaseous form. The concentration of arsenic wastes at Giant makes it one of the largest and most expensive toxic waste sites in Canada.[74]

Underground storage of the arsenic initially aimed to take advantage of the presumed permafrost conditions as a permanent storage facility. However, the extensive transformation of the underground environment by decades of mining meant that permafrost has never re-established in the mined-out chambers, and the combination of water infiltration and a rising water table threatens to mobilize the arsenic through groundwater and into the environment. As a result, water has to be continuously pumped from the mine and treated to remove arsenic (again). The mine's operators and, subsequently, INAC have sought various solutions to the problem.[75] Schemes to extract the stored arsenic from Yellowknife-area mines (including Giant and Con), process it, and sell it for use in wood preservatives had been considered since the 1980s.[76] But concerns over spills and the lack of a viable market meant this alternative for disposal of the remaining arsenic dust never gained traction.

When the mine's owners, Royal Oak Mines, went into receivership in 1999, INAC assumed responsibility for the mine's liabilities, while another company, Miramar, operated the mine. Since the closure for good of Giant in 2004, INAC has proposed a comprehensive remediation of the

FIG. 11.3: Thermosyphon test plot, Giant Mine Remediation Project. Photo by John Sandlos.

mine, including a novel method of securing the arsenic underground. The government's plan, estimated to cost over $600 million, calls for the drilling and installation of thermosyphons (passive heat-exchange convection pipes) surrounding the underground arsenic chambers, allowing for the re-freezing of the rock using refrigerants. Thereafter, the freezing would be maintained through the circulation of carbon dioxide in the pipes, which exchanges heat from the warming ground to the cooler air (Fig. 3). According to proponents, the frozen chambers will require maintenance "in perpetuity." The remediation plan also calls for the removal of contaminated soils (filling open pits), revegetation of tailings, removal of buildings, and rehabilitation of a polluted creek running through the site. Virtually all the surface works would be removed.[77]

Although the project aims to mitigate a toxic hazard EA process is over, these plans have met with controversy. At a meeting in May 2010 with the Yellowknives Dene First Nation at Ndilo, sceptical community members posed tough questions to INAC consultants and staff about the long-term effects of arsenic in the environment. Many doubted they would ever again be able to gather berries or consume fish from the mine area.[78] In debates over the remediation plan, history and memory also assert themselves. Some at the community meeting recalled being tested as children for arsenic exposure in the studies mentioned above, and the community continues to fear poisoned water and fish in Back Bay, which they had been previously warned not to consume. The Yellowknives Dene First Nation presentation to an earlier environmental assessment hearing on the remediation outlined the Dene people's long-term occupancy of the mine site and asserted that they were never consulted about the original development. After outlining the history of exposure and exclusion at the site, the First Nation requested that the environmental assessment process investigate these "legacy" issues and the full scope of the development's historical and contemporary impacts on the community and its land. For its part, the territorial regulator, the Mackenzie Valley Environmental Impact Review Board, decided to restrict its environmental assessment to the site remediation plan itself, excluding questions of historical impacts and any environmental effects beyond the remediation of the mine site itself.[79]

First Nations people and critics in the town of Yellowknife also question who bears the ultimate burden of exposure and care for this toxic site. In oral interviews, elders and community members from the Yellowknives Dene linked their history of exposure to their concerns about the long-term hazards left behind in their traditional territory. Some questioned the ability of the government and its technological solutions to protect the community and its land for generations to come.[80] The requirement for perpetual care of the frozen arsenic chambers, located right on the doorstep of Yellowknife (and a regional population of about twenty thousand) raised significant concerns from civic officials and local community members. In fact, the City of Yellowknife (rather than territorial regulators) triggered the environmental assessment of the project in 2008, responding to public concerns (and those of the Yellowknives Dene First Nation) over the supposed permanence of the arsenic freezing solution (Fig. 4). Alternatives North, a local environmental group, has prepared a submission to this ongoing assessment examining the concept and practice of "perpetual care" of contaminated sites.[81] When the Mackenzie Valley Environmental Impact Review Board released its decision on the Giant Mine Remediation

FIG. 11.4: In spite of its slogan, "Moving Forward Together," the federal Giant Mine Remediation Project has generated considerable controversy in Yellowknife. Photo by John Sandlos.

Project in June 2013, it represented a major victory for Alternatives North, the Yellowknives Dene, and other concerned citizens. Although the report covered many issues, the board was severely critical of the government's "leave it and freeze it" approach. It recommended a reduced time frame for the project (to one hundred years, with investment and research on a more permanent solution), periodic reviews of new approaches every twenty years, and the creation of an independent oversight committee to monitor key environmental issues during the lengthy remediation process. Although the fate of these recommendations rests with the federal cabinet, the complexity and intractability of the environmental problems at Giant Mine guarantee that the site will remain reanimated and in the public eye for decades, if not centuries to come.[82]

FIG. 11.5: Open pit at Pine Point, slowly filling with water. Note berms around edge in background. Photo by John Sandlos.

The reanimation of the Pine Point site, in contrast, involves both renewed mining activity and a proposed "brownfield" mineral processing site. By the turn of this century, the scarred landscape of the mine began to show signs of "healing." Slowly, vegetation began to recolonize disturbed areas, and local residents noted the presence of moose, lynx, and even woodland caribou in the area, and worried that these animals might fall into unfenced open pits or become sick from drinking pit water (Fig. 11.5). Some hunting and trapping resumed in the area, and residents of both Hay River and Fort Resolution, both Aboriginal and non-Aboriginal, began to use the extensive road network and cleared spaces of the mine area and former town as an informal recreation area for camping, all-terrain vehicles, and other activities.

The resurgence of mineral markets brought renewed geological interest to the area, and in 2004 a junior mining company, Tamerlane Ventures, began to acquire rights to unmined, historic lead-zinc deposits near the former Cominco mine. Tamerlane commenced an extensive drilling program that included activities at the former mine site, and, in 2008, applied to operate a one-million-tonne "test mine" to produce high-grade ores from its holdings just west of the old mine. Tamerlane exploration crews are regularly encountered in the maze of haul roads around Pine Point, as well as further to the west where its R-190 deposit is staked. More recently, the company announced its delineation of millions more tonnes of ore at the Pine Point site, which it had hoped to bring into production by 2013, "just when the demand for lead and zinc begins to peak," according to CEO Mike Willett.[83]

Tamerlane's plans to revive mining at Pine Point stirred the ghosts in the region. During environmental assessment hearings, residents from Fort Resolution expressed their concerns about the legacies of past mining and their fears that the benefits of development would, once again, pass them by. Like the interviewees cited above, participants at the public hearing expressed their anger that the previous Cominco operation had left a degraded landscape, and challenged Tamerlane's contention that there would be no cumulative impacts at the site. The Deninu Ku'e First Nation, of Fort Resolution, wrote: "This is a sensitive land in the process of healing, why is [the developer's biophysical assessment consultant] saying there will be no cumulative impacts in this area?"[84] For its part, the company declared that it "wanted to avoid the mistakes of past mining activities."[85] But for many Fort Resolution residents, the current development and past experiences cannot be neatly severed. The social, economic, and environmental effects of the former Pine Point mine continue to resonate in the community, whether concerns about the ongoing impact of historic mining on traditional harvesting activities, resentment at the low levels of Aboriginal employment at the Pine Point mine, or memories of the disruptive influence of alcohol and outsiders brought by the extension of the road from Pine Point to Fort Resolution.[86] Memories of the former mine and town are not universally negative—at hearings and in oral history interviews conducted by the authors, some residents proudly recalled working and living in Pine Point—but these memories are also tinged with regret at the hardships caused by the closure of the mine and the disappearance of the town.[87]

Many in the Northwest Territories are uncertain of the company's environmental commitments, given Tamerlane's leadership. The company's executive chairman and CFO is Margaret Kent (formerly Peggy Witte), the former CEO of Royal Oak Mines Ltd., a company notorious in the north for its poor labour and environmental record. Royal Oak owned Giant Mine in Yellowknife during the period of the vicious 1991–92 strike, which was tragically punctuated by the murder of nine replacement workers by a disgruntled striker who planted a bomb in the mine—more ghosts.[88] In addition to overseeing the massive underground accumulation of arsenic at Giant, the company was also responsible for several abandoned mines in the region, including the toxic Colomac Mine, but managed to evade financial liability for these sites by declaring bankruptcy. Kent's return to the territory has sparked considerable comment and concern that Royal Oak's dismal environmental and labour record might accompany her.[89]

More recently still, the former Pine Point mill site has attracted re-development interest due to its status as a brownfield. The growing global demand for rare earth element (REE) metals (used in high-tech devices such as computers, hybrid cars, and flat-screen monitors) launched the rapid development of an REE deposit on the north shore of Great Slave Lake. The Nechalacho deposit, being developed by Avalon Rare Metals, is the second-largest in the world, and drilling and development activities since 2005, along with rising prices, have brought it close to production feasibility.[90] As part of its planning, the company initially discussed siting a hydrometallurgical processing facility at the former Pine Point mill site, just across the lake from the mine, where it could take advantage of the existing power and transportation infrastructure, as well as the existing tailings containment facility. Throughout the planning stage, this project generated considerable discussion and interest at Fort Resolution and Hay River for the potential of both employment and cash benefits from both the mine and the production facility. In July 2013, the mine and hydromet project obtained regulatory approval, though uncertainty remains the watchword for this project, as Avalon warned that they still had the "daunting" task of raising $1.5 billion in capital investments, and as the company has not fully committed to siting the processing at Pine Point.[91]

Ironically, one potential hitch in the redevelopment plan emerged in Avalon's pre-feasibility study, released in June 2010: the cost of trans-porting chemical reagents to the hydromet facility at Pine Point led the company to consider other sites in southern Canada. In response, the for-mer head of the NWT & Nunavut Chamber of Mines called on the federal government to re-establish the rail line to Pine Point, removed after the closure of the former mine. "They took out the existing line for no real reason at all, and it would have made a huge difference to the viability of a hydromet facility at Pine Point," he said.[92] In a cruel twist, the trackless Great Slave Lake Railway spur line—that symbol of failed northern modernization and development plans—now stands as an obstacle to redevelopment, even in its moribund state.

Conclusion

In the Canadian north, redevelopment and reclamation activities not only reawaken local conflicts over the impacts of past mining activities, but also reflect a renewed discourse of minerals-based modernization and development in the region. Rising prices in global mineral markets have, until the recent commodities bust, spurred massive investments in mineral exploration and development in Canada's three northern territories (Nunavut, Canada's eastern arctic territory, separated from the NWT in 1998). Mineral exploration and deposit appraisal expenditures throughout Canada reached a record $3.3 billion in 2008, and, while slowed by the global economic downturn, rebounded strongly to new record levels prior to very recent declines in commodity prices.[93] This investment, and the new mines it spawns or old ones it reopens, is welcomed by many in the territories, whose economies remain strongly dependent upon extractive industries. Echoing the development vision of past promoters, contemporary government and industry figures forecast extensive mineral developments as drivers of the northern economy for years to come. For instance, in 2009 the NWT & Nunavut Chamber of Mines released an ambitious infrastructure proposal calling for the construction of railways, roads, air routes, an arctic port, and new power developments (including "pocket" nuclear plants) to drive northern mineral extraction.[94]

Our research suggests that such grandiose visions of minerals-based arctic industrialization would benefit from some critical historical-geographical perspective. The history of mineral development and abandonment in the Great Slave Lake region reveals that while mining brought settlement and prosperity to parts of the region, it also acted to advance the colonial objectives of the Canadian state in a hitherto lightly settled, predominantly Aboriginal territory. In the north, where extreme climatic conditions, poor soils, and distance from markets long restricted Euro-Canadian economic interests to the fur trade, mineral development in the twentieth century held the key to the final "industrial assimilation" of these far-flung territories into the orbit of Canadian state and capital.[95] The geography of minerals, in turn, shaped the pattern of this process: discontinuous, nodal developments centred around economically viable deposits of precious metals, and, later, other high-value industrial minerals.

The government sought to enrol Aboriginal inhabitants into the modern mineral economy of the north, but by and large the results of this process were displacement, marginalization, and the creation of unstable, "cyclonic" economies prone to sudden collapse.[96] In a region without a highly developed agricultural or urban economy, the classic mining boom-bust pattern has proven particularly devastating, although uneven, as the contrasting fates of Yellowknife and Pine Point illustrate. Aboriginal communities were also disproportionately affected by the environmental changes associated with mineral development. At Giant Mine, the Dene communities of N'dilo and Dettah found their waters, lands, food, and bodies contaminated with arsenic from mine wastes, even as they struggled to engage with the sudden arrival of modern settlement life and state control. In the South Slave region, Dene people found intermittent work at the Pine Point mine (particularly in its construction phase), but also struggled with the impacts of large-scale landscape transformations on traditional livelihoods and social arrangements.

While the closure of mining communities may leave behind "ghost towns," zombie mines emerge where renewed activity at former mine sites threatens to reawaken or reproduce the negative experiences and outcomes of previous mining operations. At both Pine Point and Giant—as at numerous other sites in the Canadian north where former mines are being brought back to life—historical conflicts over the impacts and benefits of mining are being revisited through the reanimation of the mines themselves. Certainly there is some truth to the argument, often advanced by mining interests, that contemporary mine development, redevelopment, and remediation takes place under very different historical circumstances today than it did in the 1950s and 1960s, with the industry-led Whitehorse Mining Initiative of the early 1990s marking an increased attentiveness to Aboriginal political and economic priorities and environmental performance issues.[97] Both in Canada and internationally, the mining industry's turn toward "sustainability" has included a reckoning with the negative legacies of mining, including the problem of abandoned mines and the links between mining and the dispossession of Indigenous peoples.[98] Nevertheless, those proposing redevelopment or remediation, whether governments or mining companies, often fail to recognize how their activities can stir deeply held feelings about historic mining. For local First Nations communities, old mines are not simply historical artefacts; their legacies persist in landscapes encountered through daily activities and memories of work, life changes, or other experiences, positive or negative. As Ginger Gibson wrote in relation to diamond mining developments in Dene territory in the 1990s, modern miners "may seek to enter the political geography of the north without acknowledging the past, [but] this relational view of history reveals they will arrive with the shadows of ghost-mines behind them."[99] When these "ghosts" reside at the same site as the original mine, we argue, these sites are more properly considered zombie mines. The global trend toward accelerated mineral development and the renewed interest in abandoned mine sites for both remediation and redevelopment suggests an important role for mining historians in highlighting the importance of this history in contemporary debates over the industry and its impacts.

Notes

1   The authors are grateful to the Social Sciences and Humanities Research Council of Canada (SSHRC)
and the Rachel Carson Center for
Environment and Society for their
generous support of this research.
They also appreciate the feedback
on earlier drafts from the editors
and participants in the northern
environmental history workshop
held in Peterborough in 2011.

Richard V. Francaviglia, *Hard
Places: Reading the Landscape of
America's Historic Mining Districts*
(Iowa City: University of Iowa
Press, 1991); Ben Marsh, "Continu-
ity and Decline in the Anthracite

Towns of Pennsylvania," *Annals
of the Association of American
Geographers* 77, no. 3 (1987): 337–52;
William Wyckoff, "Postindustrial
Butte," *Geographical Review* 85, no.
4 (1995): 478–96; David Robertson,
*Hard as the Rock Itself: Place and
Identity in the American Mining
Town* (Boulder: University Press of
Colorado, 2006). See also Peter Goin
and C. Elizabeth Raymond, *Chang-
ing Mines in America* (Santa Fe, NM:
Center for American Places, 2004);
Peter Goin and Elizabeth Raymond,
"Living in Anthracite: Mining
Landscape and Sense of Place in Wy-
oming Valley, Pennsylvania," *Public
Historian* 23, no. 2 (2001): 29–45;

and John Harner, "Place Identity and Copper Mining in Sonora, Mexico," *Annals of the Association of American Geographers* 91, no. 4 (2001): 660–80. For northern Canada, see Liza Piper, "Subterranean Bodies: Mining the Large Lakes of North-west Canada, 1921–1960," *Environment and History* 13, no. 2 (2007): 155–86; and Liza Piper, *The Industrial Transformation of Subarctic Canada* (Vancouver: UBC Press, 2009).

2   Jared Diamond, *Collapse: How Societies Choose or Fail to Succeed* (New York: Viking, 2005).

3   Lisa J. Wilson, "Riding the Resource Roller Coaster: Understanding Socioeconomic Differences between Mining Communities," *Rural Sociology* 69, no. 2 (2009): 261–81.

4   Scott Frickel and William R. Freudenburg, "Mining the Past: Historical Context and the Changing Implications of Natural Resource Extraction," *Social Problems* 43, no. 4 (1996): 448.

5   Trevor Barnes, "Borderline Communities: Canadian Single Industry Towns, Staples, and Harold Innis," in *B/ordering Space*, ed. Henk Van Houtum, Olivier Kramsch, and Wolfgang Zierhofer (Burlington, VT: Ashgate Publishing: 2005), 109–22; J. H. Bradbury, "Towards an Alternative Theory of Resource-Based Town Development in Canada," *Economic Geography* 55, no. 2 (April 1979): 147–66. J. H. Bradbury, "The Rise and Fall of the 'Fourth Empire of the St. Lawrence': The Quebec-Labrador Iron Ore Mining Region," *Cahiers de Géographie du Québec* 29, no. 78 (1985): 351–64; Arn Keeling, "'Born in an Atomic Test Tube': Landscapes of Cyclonic Development at Uranium City, Saskatchewan," *Canadian Geographer* 54, no. 2 (Summer 2010): 228–52.

6   Saleem H. Ali, *Mining, the Environment, and Indigenous Development Conflicts* (Tucson: University of Arizona Press, 2003); Ginger Gibson and Jason Klinck, "Canada's Resilient North: The Impact of Mining on Aboriginal Communities," *Pimatisiwin* 3, no. 1 (2004): 115–41; Arn Keeling and John Sandlos, "Environmental Justice Goes Underground? Historical Notes from Canada's Mining Frontier," *Environmental Justice* 2, no. 3 (2009): 117–25.

7   It is difficult to locate data on the precise number of mines reopening globally, or even nationally. But data provided by the Minerals Economics Group to the *Globe and Mail* newspaper suggests that, since the upturn in the commodities markets in early 2009, dozens of mines have reopened globally (although it is unclear how long any one of them was closed). See Brenda Bouw and David Ebner, "The commodity cycle speeds up," *Globe and Mail*, 14 January 2011, http://www.theglobeandmail.com/report-on-business/industry-news/energy-and-resources/the-commodity-cycle-speeds-up/article1871363/ (accessed 14 May 2012).

8   Homer Aschmann, "The Natural History of a Mine," *Economic Geography* 46, no. 2 (1970): 172–89.

9   S. Durukan, et. al., "Mining Life Cycle Modelling: A Cradle-to-Gate Approach to Environmental Management in the Minerals Industry," *Journal of Cleaner Production* 14, nos. 12–13 (2006): 1057–70; G.

M. Mudd, "Sustainable Mining: An Evaluation of Changing Ore Grades and Waste Volumes," paper presented to the International Conference on Sustainability Engineering and Science, Aukland, New Zealand, July 2004; *Breaking New Ground: The Report of the Mining, Minerals and Sustainable Development Project* (London: Earthscan, 2002).

10 Daviken Studnicki-Gizbert, "Exhausting the Sierra Madre: Long-term trends in the environmental impacts of mining in Mexico," paper presented to the American Society for Environmental History Conference, Tallahassee, FL, 2009.

11 Some of these implications are discussed from a policy perspective in Rhys Creswell Worrall, David T. Neil, David J. Brereton, and David R. Mulligan, "Towards a Sustainability Criteria and Indicators Framework for Legacy Mine Land," *Journal of Cleaner Production* 17, no. 16 (2009), 1426–34; Patricia Nelson Limerick, Joseph N. Ryan, Timothy R. Brown, and T. Allan Comp, *Cleaning up Abandoned Hardrock Mines in the West: Prospecting for a Better Future* (Boulder, CO: Center of the American West, 2005); Mining, Minerals and Sustainable Development Project (MMSD), "Mining for the Future, Appendix C: Abandoned Mines Working Paper," International Institute for Environment and Development, April 2002; Chilean Copper Commission and United Nations Environment Program, "Abandoned Mines: Problems, Issues and Policy Challenges for Decision Makers," workshop summary report, June 2001.

12 Since we began thinking about zombie mines, others have adopted the term independently of our work. See, for example, Traci Brynne Voyles, *Wastelanding: Legacies of Uranium in Navajo Country* (Minneapolis: University of Minnesota Press, 2015).

13 Indian and Northern Affairs Canada, *The Big Picture: Contaminated Sites in the NWT* (Ottawa: Indian and Northern Affairs Canada, 2008).

14 Office of the Auditor General of Canada, *Report of the Commissioner of the Environment and Sustainable Development, 2002* (Ottawa: Public Works and Services, 2002), ch. 2–3. See Indigenous and Northern Affairs Canada (INAC), "Northern Contaminated Sites Program," last modified 15 August 2010, http://www.aadnc-aandc.gc.ca/eng/1100100035301.

15 For the Web memorial, see "Pine Point Revisited," http://pinepointrevisited.homestead.com/Pine_Point.html (accessed 7 May 2012). For the multi-media documentary, see *Welcome to Pine Point*, dir. Paul Shoebridge and Michael Simons (National Film Board, 2011).

16 See NWT Mining Heritage Society, "Mining History," https://www.nwtminingheritage.com/mining-history (accessed 7 June 2016).

17 For oral history recollections of Pine Point, see John Sandlos, "A Mix of the Good and the Bad: Community Memory and the Pine Point Mine," in *Mining and Communities in Northern Canada: History, Politics and Memory*, ed. Arn Keeling and John Sandlos (Calgary: University of Calgary Press, 2015), 137–65. For an overview of the environmental assessment of the Giant Mine Remediation Project,

see Kevin O'Reilly, "Liability, Legacy, and Perpetual Care: Government Ownership and Management of the Giant Mine, 1999–2015," in *Mining and Communities in Northern Canada: History, Politics and Memory*, ed. Arn Keeling and John Sandlos (Calgary: University of Calgary Press, 2015), 341–76.

18   For overviews of northern development see, K. J. Rea, *The Political Economy of the Canadian North* (Toronto: University of Toronto Press: 1968); Morris Zaslow, *The Northward Expansion of Canada, 1914–1967* (Toronto: McClelland and Stewart, 1988); and Piper, *The Industrialization of Sub-Arctic Canada.*

19   Ted Nagle and Jordan Zinovich, *The Prospector North of Sixty* (Edmonton: Lone Pine Publishing, 1989), 18–19.

20   The data on mining in the Yellowknife region was culled from Ryan Silke, "The Operational History of Mines in the Northwest Territories, Canada" (unpublished report, 2009).

21   For overviews of the social and environmental impacts of northern mining, see Lisa Sumi and Sandra Thomsen, *Mining in Remote Areas: Issues and Impacts* (Ottawa: MiningWatch Canada, 2001); Claudia Notzke, *Aboriginal Peoples and Natural Resources in Canada* (North York, ON: Captus University Publications, 1994), 216–17; Janet E. Macpherson, "The Pine Point Mine" and "The Cyprus Anvil Mine," in *Northern Transitions, Volume I*, ed. Everett B. Peterson and Janet B. Wright (Ottawa: Canadian Arctic Resources Committee, 1978), 65–148. For the impact of prospecting on barren-ground caribou, see Anthony G. Gulig, "'Determined to Burn off the Entire

Country': Prospectors, Caribou, and the Denesuliné in Northern Saskatchewan, 1900–1940," *American Indian Quarterly* 26, no. 3 (Summer 2002): 335–59.

22   Frances Abele, *Gathering Strength* (Calgary: Arctic Institute of North America, 1989); NWT Archives, Prince of Wales Northern Heritage Centre (hereafter cited as PWNHC), G 2002-004, box 23, S. Collymore, "Native Labour in the Northern Mining Industry," unpublished draft paper, Department of Indian Affairs and Northern Development; Paul Deprez, *The Pine Point Mine and the Development of the Area South of Great Slave Lake* (Winnipeg: Centre for Settlement Studies, 1973).

23   René Fumoleau, *As Long As This Land Shall Last: A History of Treaty 8 and Treaty 11, 1870–1939* (Calgary: University of Calgary Press, 2004).

24   Bernd Lottermoser, *Mine Wastes: Characterization, Treatment and Environmental Impacts* (Berlin: Springer, 2003). A 1994 report on active and abandoned mines in the Northwest Territories listed seven mines with the potential for acid generation: see G. Feasby and R. K. Jones, "Report of Results of a Workshop on Mine Reclamation," MEND Report 5.8e, August 1994, http://mend-nedem.org/wp-content/uploads/2013/01/5.8e.pdf (accessed 14 May 2012).

25   Auditor General of Canada, *Report of the Commissioner of the Environment and Sustainable Development*; Brian Bowman and Doug Baker, *Mine Reclamation Planning in the Canadian North* (Canadian Arctic Resources Committee, 1998).

26   For early emissions data and the unregulated nature of emissions

from 1949 to 1951, see INAC, "Historical Timeline: Giant Mine Remediation Project," last modified 15 August 2010, http://www.aadnc-aandc.gc.ca/eng/1100100023233. Con Mine also roasted ore during this period from its own mine and also from Negus, but because only twenty percent of the gold from these two mines was contained in arsenopyrite formations, the contribution to the arsenic pollution problem was initially limited to about twenty-five percent of the total. The mining company Cominco abandoned roasting in favour of flotation methods in 1970 and remediated arsenic stored in tailings sludge to less toxic forms in 1982. "Cominco's Arsenic Recovery Process Solves Environmental Problem in Yellowknife, NWT," *Canadian Institute of Mining, Metallurgy and Petroleum (CIM) Reporter* 8, no. 2 (October 1982): 1.

27 Timothy LeCain, "The Limits of 'Eco-Efficiency': Arsenic Pollution and the Cottrell Electrical Precipitator in the U.S. Copper Smelting Industry," *Environmental History* 5, no. 3 (July 2000): 336–51.

28 The term was adopted from the title of the following collection of essays on the history of public health: Gregg Mitman, Michelle Murphy, and Christopher Sellers, eds., *Landscapes of Exposure: Knowledge and Illness in Modern Environments*, Osiris 19 (Chicago: University of Chicago Press, 2004). See also the discussion by Jody A. Roberts and Nancy Langston, "Toxic Bodies/Toxic Environments: An Interdisciplinary Forum," *Environmental History* 13, no. 4 (2008): 629–35.

29 Yellowknives Dene First Nation presentation to Mackenzie Valley Environmental Impact Review Board Scoping Session, 23 July 2008, http://www.reviewboard.ca/upload/project_document/1219099111_15606YKDF-NUndertaking10.pdf (accessed 20 May 2011).

30 Reference to the death and settlement are contained in Library and Archives Canada (hereafter cited as LAC), RG85, vol. 40, file 139-7, pt. 1, J. K. Muir, General Manager, Giant Yellowknife Gold Mines Limited, to G. B. Sinclair, Director, Northern Administration and Lands Branch, Department of Resources and Development, 10 August 1951.

31 Reference to hair loss was found in Yellowknives Dene First Nation, *Weledeh Yellowknives Dene: A History* (Dettah: Yellowknives Dene First Nation Council, 1997). The specific references to skin conditions were found in PWNHC, G-2008-028, box 9, file 16, Memo from Dr. O. Schaeffer to Dr. B. Wheatley, Environmental Contaminant Program, Medical Services Branch, Health and Welfare Canada, 1 May 1978.

32 The condition of Baker Creek and Yellowknife Bay was discussed in PWNHC, G-2008-028, box 9, file 16, Minutes of the Standing Committee on Arsenic Meeting in Yellowknife, 31 January 1978.

33 Yellowknives Dene First Nation, *Weledeh Yellowknives Dene: A History*, 54–55.

34 The information on the horses comes from the testimony of Laurie Cinnamon, *Yellowknife, NWT: An Illustrated History*, ed. Susan Jackson (Yellowknife: Nor'West

Publishing, 1990), 85. Helen Kilkenny became a farmhand on Bevan's farm in 1947, and stated that the cattle died from arsenic poisoning "four years later." Kilkenny mentions this incident after she describes how cattle were watered in local ponds and fed at the side of the road, implying that the arsenic was ingested through ground and water contamination. See Helen Kilkenny, "Bevan's Farmhand," Yellowknife, NWT, 114–15. Barbary Bromley claimed that the source of the arsenic that killed the Bevan cattle was Con Mine, and that a court case resulted. Considering the high levels of emissions from Giant, the exact source of the arsenic would likely have been hard to pinpoint. See testimony of Barbary Bromley, Yellowknife Tales: Sixty Years of Stories from Yellowknife (Yellowknife: Outcrop, 2003), 97–98.

35 See LAC, RG 29, vol. 2977, file 851-5-2, pt. 1, Minutes of Meeting held to Discuss the Death of Indian Boy, Latham Island, 1 June 1951.

36 LeCain, "The Limits of Eco-Efficiency."

37 AAND, "Historical Timeline: Giant Mine Remediation Project."

38 For an overview of mining pollution, smelter disputes, and regulation, see John D. Wirth, Smelter Smoke in North America: The Politics of Transborder Pollution (Lawrence: University of Kansas Press, 2000); and Duane A. Smith, Mining America: The Industry and the Environment (Lawrence: University of Kansas Press, 1987). For the formation of regulatory approaches to toxins, particularly the "dose makes the poison approach," see Nancy Langston, Toxic Bodies:

Hormone Disruptors and the Legacy of DES (New Haven: Yale University Press, 2010); Nash, Inescapable Ecologies; Linda Nash, "Purity and Danger: Historical Reflections on the Regulation of Environmental Pollutants," Environmental History 13, no. 4 (October 2008): 651–58; Sarah A. Vogel, "From 'the Dose Makes the Poison' to 'the Timing Makes the Poison': Conceptualizing Risk in the Synthetic Age," Environmental History 13, no. 4 (October 2008): 667–73.

39 On the rise of pollution and toxics as an environmental issue in the 1960s, see Samuel Hays, Beauty, Health and Permanence: Environmental Politics in the United States, 1955–1985 (New York: Cambridge University Press, 1987); for the Canadian context, see Arn Keeling, "The Effluent Society: Water Pollution and Environmental Politics in British Columbia, 1889–1980" (PhD diss., University of British Columbia, 2004); Jennifer Read, "'Let us heed the Voice of Youth': Laundry Detergents, Phosphates and the Emergence of the Environmental Movement in Ontario," Journal of the Canadian Historical Association 7, no. 1 (1996), 221–50.

40 For vegetables, see LAC, RG 29, vol. 2977, file 851-5-2, pt. 1, M17 to Dr. Procter, 10 December 1965. The issue of water contamination remained irresolvable until intake for the city of Yellowknife was moved upstream from Back Bay to mouth of the Yellowknife River in 1969. See Canadian Public Health Association, Task Force on Arsenic—Final Report, Yellowknife Northwest Territories (Ottawa: CPHA, 1977), 49–51.

41 LAC, RG 29, vol. 2977, file 851-5-2, pt. 1, Dr. A. J. de Villers to E. A. Watkinson, Health Services Branch, 26 August 1969. For an overview of this third period of heightened public concern over arsenic, see PWNHC, G-2008-028, box 9, file 17, F. J. Colvill, Senior Advisor, NWT Region, "Arsenic in Yellowknife—A Perspective," 25 September 1979.

42 See Colvill, "Arsenic in Yellowknife—A Perspective."

43 PWNHC, G-2008-028, box 9, file 16, J. A. Hildes, Co-Director, Northern Medical Unit, to C. J. G. Mackenzie, Head, Department of Health Care and Epidemiology, University of British Columbia, 16 February 1978. The table attached to this letter provides detailed statistical surveys of comparative cancer rates in Whitehorse and Yellowknife.

44 PWNHC, G-2008-028, box 9, file 17, R. D. P. Eaton, "Analysis of Hair Arsenic Results, Yellowknife, 1975."

45 An overview of the water issue and the various studies in the 1970s was found in PWNHC, G-2008-028, box 9, file 17, F. J. Colvill, Senior Advisor, NWT Region, "Arsenic in Yellowknife—A Perspective," 25 September 1979.

46 Canadian Public Health Association, *Task Force on Arsenic—Final Report, Yellowknife Northwest Territories* (Ottawa: CPHA, 1977).

47 PWNHC, G-1993-006, box 29, file 13 408 023, R. J. Kent, "Environmental Protection Services, A Report on Arsenic Emissions During August 1981 at Giant Yellowknife Mines (January 1982)."

48 PWNHC, G-1993-006, box 19, file 13 408 024, Lorne C. James, Pollution Control Officer, Department of Renewable Resources, Government of the Northwest Territories, to Ranjit Soniassy, Northern Affairs Program, 23 June 1986.

49 Luke Cole, *From the Ground Up: Environmental Racism and the Rise of the Environmental Justice Movement* (New York: New York University Press, 2000). For the application of environmental justice as a frame for global Indigenous conflicts with mining companies, see Nicholas Low and Brendan Gleeson, "Situating Justice in the Environment: The Case of BHP at the Ok Tedi Copper Mine," *Antipode* 30, no. 3 (1998): 201–26; Joan Martinez-Alier, "Mining Conflicts, Environmental Justice, and Valuation," *Journal of Hazardous Materials* 86, nos. 1–3 (2001), 153–70.

50 For a theoretical elaboration on the relationship between colonialism, mining, and environmental justice in northern Canada, see Keeling and Sandlos, "Environmental Justice Goes Underground?"

51 Tim LeCain, *Mass Destruction: The Men and Giant Mines That Wired America and Scarred the Planet* (New Jersey: Rutgers University Press, 2009).

52 Nagle, *Prospector North of Sixty*; Fumoleau, *As Long as This Land Shall Last.*

53 For a summary of the mine's history, see Macpherson, "The Pine Point Mine."

54 For oral testimony on this issue, see Mackenzie Valley Pipeline Inquiry Transcripts, vols. 30 (Pine Point), 31, and 32 (Fort Resolution), October 1975, http://www.allwestbc.com/MVP/MVP_AllTranscripts.

html (accessed 30 March 2010). In his final report to the inquiry, Justice Thomas Berger highlighted Pine Point as an example of the "social, economic and geographic dislocations" of Native people. Thomas Berger, *Northern Frontier, Northern Homeland: The Report of the Mackenzie Valley Pipeline Inquiry,* vol. 1 (Ottawa: Supply and Services Canada, 1977), 123–24.

55 J. N. Stein and M. R. Miller, *An Investigation into the Effects of a Lead-Zinc Mine on the Aquatic Environment of Great Slave Lake* (Winnipeg: Resource Development Branch, Fisheries Service, Department of Environment, 1972); Yves Berube, et al., *An Engineering Assessment of Waste Water Handling Procedures at the Cominco Pine Point Mine* (Ottawa: Department of Indian Affairs and Northern Development, 1972).

56 M. S. Evans, L. Lockhart, and J. Klaverkamp, *Metal Studies of Water, Sediments and Fish from the Resolution Bay Area of Great Slave Lake: Studies related to the decommissioned Pine Point Mine* (Burlington and Saskatoon: Environment Canada, National Water Research Institute, NWRI Contribution No. 98-87, 1998). Persistent concerns about water pollution and the unremediated landscape changes were voiced during oral interviews that were conducted in Fort Resolution in May 2010 by the authors, with the assistance of Frances Mandeville and Catherine Boucher.

57 On the abandonment and reclamation activities undertaken, see Pine Point Mines Limited, "Abandonment & Reclamation Plan, Pit Water License N1L3-0034,"

Submission to the NWT Water Board, 1 July 1987, Mackenzie Valley Land and Water Board Public Registry, http://www.mvlwb.ca/Boards/mv/SitePages/registry.aspx (accessed May 1 2012); Environment Canada, *The State of Canada's Environment 1996* (Ottawa: Government of Canada, 1996), 8–31; D. L. Johnston, "Pine Point, NWT: Closing a Mine and Removing the Whole Townsite," *Hazardous Materials Management* 4 (December 1992): 21–22.

58 These ambitions are discussed in greater detail in John Sandlos and Arn Keeling, "Claiming the New North: Development and Colonialism at the Pine Point Mine, Northwest Territories, Canada," *Environment and History* 18, no. 1 (2012): 5–34.

59 LAC, RG 33, ser. 41, vol. 7, R. Gordon Robertson, *The Northwest Territories: its Economic Prospects, a Brief Presented to the Royal Commission on Canada's Economic Prospects* (Ottawa: Queen's Printer, 1955), 29.

60 Paul Deprez, *The Pine Point Mine and the Development of the Area South of Great Slave Lake.*

61 LAC, RG 85, vol. 1512, file 1000-181, pt. 4, Extract from the Minutes of the Thirty-first Meeting of the Advisory Committee on Northern Development, 6 February 1956. LAC, RG 85, vol. 1512, file 1000-181, pt. 4, R. Gordon Robertson, Internal Memo, "The Economic Crisis of the Resident Population in the North," 1 February 1956.

62 Employment data from 1967 to 1978 was culled from the following sources: Deprez, *The Pine Point Mine and the Development of the Area South of Great Slave Lake,*

67–70; PWNHC, G-2002-004, box 38, "Northern Mineral Strategy: Discussion Paper," Draft, 5 June 1978; PWNHC G-2002-004, box 38, R. P. Douglas, Assistant Vice President, Cominco, to A. D. Hunt, Assistant Deputy Minister, Department of Indian Affairs and Northern Development, 11 September 1975; PWNHC, G-2002-004, box 38, G. D. Tikkanen, General Manager, Cominco, to S. M. Hodgson, Commissioner, Government of the Northwest Territories, 9 February 1977; PWNHC, G-2002-004, box 38, R. P. Douglas, Vice-President, Cominco, to Frederick J. Joyce, Northern Operations Branch, 16 March 1977; PWNHC, G-2002-004, box 35, Pine Point Mines, Ltd., Annual Reports, 1979–82; R. P. Douglas, "Utilization of Human Resources North of Sixty," *CIM Forum*, April 1978, 13–17.

63   Macpherson, "The Pine Point Mine," 89. For a statement about most Native employees coming from outside the region, see Testimony of Anvid Osing, Mayor, Pine Point, Mackenzie Valley Pipeline Inquiry, Proceedings at a Community Hearing (transcripts), vol. 30, Pine Point, NT, 6 October 1975), 3000–01, http://www.pwnhc.ca/extras/berger/report/NT%20Pine%20Pt%20Berger%20V30.pdf.

64   PWNHC, G-2002-004, box 38, "Fort Resolution Residents Employed by Pine Point Mines Limited," document sent by G. D. Tikkanen, General Manager, Cominco, to S. M. Hodgson, Commissioner, Government of the Northwest Territories, 9 February 1977.

65   Deprez, *The Pine Point Mine and the Development of the Area South of Great Slave Lake*.

66   PWNHC, G-2002-004, box 38, R. P. Douglas to Frederick Joyce, Director, Northern Operations Branch, 16 March 1977; PWNHC, G-2002-004, box 38, W. H. R. Gibney, Manager, Pine Point Operations to D. P. Mersereau, Regional Director, Government of the Northwest Territories, 8 July 1975; R. P. Douglas, "Utilization of Human Resources North of Sixty," 13–17.

67   For the issue of Native turnover versus non-Native turnover, see Deprez, *The Pine Point Mine*, 107.

68   PWNHC, G-2002-004, box 38, "Northern Mineral Strategy: Discussion Paper," Draft, 5 June 1978; PWNHC, G-2002-004, box 23, "Native Labour in the Northern Mining Industry," Draft, sent by Robin Bricel, Resource Economist, Mineral and Petroleum Development Section, Department of Economic Development and Tourism, Government of the Northwest Territories, to Chief, Manpower Development Division, Department of Economic Development and Tourism, Government of the Northwest Territories, 3 February 1982.

69   University of Saskatchewan Archives, Native Law Centre Fonds, RCAP vol. 167, box 26, Presentation by Chief Bernadette Unka of the Deninu Kue First Nation to the Royal Commission on Aboriginal Peoples, 17 June 1993.

70   George Balsillie, interview by John Sandlos, 20 May 2010.

71   Gord Beaulieu, interview by Arn Keeling and John Sandlos, 19 May 2010.

72   Fred Tambour, interview by Arn Keeling, 21 May 2010.

73   Thomas G. Andrews, *Killing for Coal: America's Deadliest Labor War* (Cambridge, MA: Harvard University Press, 2008); Gray Brechin, *Imperial San Francisco: Urban Power, Earthly Ruin* (Berkeley: University of California Press, 1999); Kathleen A. Brosnan, *Uniting Mountain and Plain: Cities, Law and Environment along the Front Range* (Albuquerque: University of New Mexico Press, 2002); William Cronon, "Kennecott Journey: The Paths out of Town," in *Under an Open Sky: Rethinking America's Western Past*, ed. William Cronon, George Miles, and Jay Gitlin (New York: W. W. Norton, 1992); Kathryn Morse, *The Nature of Gold: An Environmental History of the Klondike Gold Rush* (Seattle: University of Washington Press, 2003).

74   Auditor General of Canada, "Abandoned Mines in the Canadian North," in *Report of the Commissioner of the Environment and Sustainable Development* (Ottawa: Office of the Auditor General of Canada, 2002). Giant Mine was highlighted in another recent report of the federal environment commissioner; see Auditor General of Canada, "Federal Contaminated Sites and Their Impacts," in *Spring Report of the Commissioner of the Environment and Sustainable Development* (Ottawa: Office of the Auditor General of Canada, 2012).

75   Auditor General of Canada, *Report of the Commissioner of the Environment and Sustainable Development*, 9–10.

76   See correspondence in PWNHC, G1993-006, box 46, file 165 022.

Some limited bulk transfers of arsenic had taken place earlier in that decade.

77   An overview of the project can be reviewed at INAC, "Giant Mine Remediation Project" (last modified 17 July 2015), http://www.aadnc-aandc.gc.ca/eng/1100100027364. Maps and engineering specifications are available via the Mackenzie Valley Review Board Public Registry, http://www.reviewboard.ca/registry/. See also Tim Edwards, "Underground chill on arsenic," *NWT News/North*, 10 August 2009, 1.

78   This meeting was attended by the authors.

79   Yellowknives Dene First Nation presentation to MVEIRB Scoping Session, 23 July 2008; Mackenzie Valley Review Board, "Reasons for Decision on Scope," Public Registry, 19 December 2008, http://www.reviewboard.ca/registry/ (accessed May 1, 2012).

80   The authors conducted interviews in May 2011 with nine Dene people as part of a collaborative project with the Yellowknives Dene First Nation to document the history of the mine and its impact on the community. As this work is ongoing, we are not yet at liberty to quote directly from interviews.

81   Alternatives North, *From Despair to Wisdom: Perpetual Care and the Future of Giant Mine*, Report on a Community Workshop, 26–27 September 2011, http://aged.alternativesnorth.ca/pdf/Perpetual%20Care%20Workshop%20Full%20Report%20(lo-res%20revised).pdf (accessed 16 March 2012); Jess McDiarmid, "Forever is a long time," *Northern News Services*, 21 March

2008, http://www.reviewboard.
ca/upload/project_document/
EA0809-001_Collection_of_Gi-
ant_Mine_Media_Coverage.pdf
(accessed 7 June 2016).

82 Mackenzie Valley Environmental
Impact Review Board, "Report of
Environmental Assessment and
Reasons for Decision," 20 June
2013, http://www.reviewboard.
ca/upload/project_document/
EA0809-001_Giant_Report_of_
Environmental_Assessment_
June_20_2013.pdf (accessed 17
August 2013).

83 "Production start for Pine Point
a perfect time, says Tamerlane
Ventures," *Resource Intelligence*
(last updated 21 March 2011),
http://www.tamerlaneventures.
com/images/pdf/Resource_Intelli-
gence_February_2011.pdf. See also
Margaret Kent's pitch for financing
at the 2012 Asia Mining Congress,
available as a video and slide show
at http://fnncompanypresenta-
tions.s3.amazonaws.com/Tamer-
lane_Venture_Inc/index.html
(accessed 14 May 2012). In spite
of this optimism, as of the time of
writing the Pine Point project had
not commenced and Tamerlane
was struggling to remain solvent.

84 Mackenzie Valley Environmental
Impact Review Board, "Report of
Environmental Assessment and
Reasons for Decision on Tamerlane
Ventures Inc.'s Pine Point Pilot
Project, EA-0607-002," 22 Febru-
ary 2008, p. 91, available on the
MVEIRB Public Registry, http://
www.reviewboard.ca/registry/
(accessed 1 May 2012).

85 "Meeting Report from the Fort
Resolution Scoping Session,"
compiled by the Mackenzie Valley
Environmental Impact Review
Board, 17 August 2006, p. 6,
available on the MVEIRB Public
Registry, http://www.reviewboard.
ca/registry/ (accessed April 2011).

86 "Meeting Report from the Fort
Resolution Scoping Session," 17
August 2006; DKFN Library, Fort
Resolution, NT, Deninu Ku'e First
Nation, "Tamarlane Ventures Inc.
Pine Point Pilot Project Written
Sub-mission," August 2006. These
concerns were reflected in the final
report and approval of the propos-
al, which recommended further
study of the cumulative effects of
development in the area, especial-
ly if a full-scale mine were to be
developed, "to help bring closure
to the historic Pine Point Mine."
Mackenzie Valley Environmental
Impact Review Board, "Report of
Environmental Assessment and
Reasons for Decision on Tamerlane
Ventures Inc.'s Pine Point Pilot
Project," 98.

87 Sandlos, "A Mix of the Good and
the Bad."

88 Lee Selleck and Francis Thompson,
*Dying for Gold: The True Story of
the Giant Mine Murders* (Toronto:
HarperCollins, 1997).

89 Matthew McClearn, "Mining:
Sh*t Happens But You Move On,"
Canadian Business, 27 May 2009,
http://liquidbriefing.com/twiki/
pub/BusinessDetox/Environment-
Detox/090427_CDNBIZ_shithap-
pensbutmoveon.pdf (accessed 7
June 2016).

90 See description on the company
webpage, "Nechelano Overview,"
http://avalonadvancedmaterials.
com/nechalacho/nechalacho_over-
view/ (accessed 7 June 2016).

91 Bob Weber, "Miner Seeks $1.5 bil-
lion for Canada's First Rare Earths
Project," *Globe and Mail*, 29 July

2013, http://www.theglobeandmail.com/report-on-business/industry-news/energy-and-resources/rare-earth-metals-miner-seeks-capital-for-northern-project/article13482908/#dashboard/follows/ (accessed 17 August 2013); "Avalon Rare Metals may not build facility at Pine Point, NWT," *CBC News*, 9 August 2013, http://www.cbc.ca/news/canada/north/story/2013/08/09/north-nechalacho-pine-point-location.html (accessed 17 August 2013).

92 Guy Quenneville, "Nechalacho study casts doubt on Northern plant location," *Northern News Services*, 23 June 2010, http://www.nnsl.com/frames/newspapers/2010-06/jun23_10re.html (accessed 14 May 2012).

93 Current and some historical mineral exploration statistics may be found at Natural Resources Canada, "Mineral Exploration" (tables from the federal-provincial-territorial Survey of Mineral Exploration, Deposit Appraisal and Mine Complex Development Expenditures), http://sead.nrcan.gc.ca/expl-expl/sta-sta-eng.aspx (accessed 14 May 2012).

94 NWT and Nunavut Chamber of Mines, "Proposed Infrastructure North of 60," Map, 2009, http://www.miningnorth.com/docs/Infrastructure.jpg (accessed January 2009); G. Quenneville, "Chamber of Mines drafts ambitious infrastructure plan," *NWT/Nunavut Mining 2009*, 16 November 2009. This call for infrastructure development was echoed by conservative commentator Diane Francis in her blog entry, "Mining Marshall Plan for Canada now," 7 March 2009, http://www.republicofmining.com/2011/06/25/mining-marshall-plan-for-canada-now-by-diane-francis-national-post-march-8-2009/ (accessed 8 June 2016). More recently, a Conference Board of Canada report series touts minerals-based economic development for the north. See, for instance, Gilles Rhéaume and Margaret Caron-Vuotari, "The Future of Mining in Canada's North" (Conference Board of Canada, 2013).

95 This phrase is from Piper, *The Industrial Transformation of Subarctic Canada*.

96 Sandlos and Keeling, "Claiming the New North"; Arn Keeling, "'Born in an Atomic Test Tube.'"

97 Patricia Fitzpatrick, Alberto Fonseca, and Mary Louise McAllister, "From the Whitehorse Mining Initiative Towards Sustainable Mining: Lessons Learned," *Journal of Cleaner Production* 19 (2011): 376–84; Mary-Louise McAllister, "Shifting Foundations in a Mature Staples Industry: A Political Economic History of Canadian Mineral Policy," *Canadian Political Science Review* 1, no. 1 (June 2007): 73–90.

98 The global mining industry's embrace of sustainability and "social licensing" issues is best captured in the International Council on Mining and Metals' program report "Mining, Minerals and Sustainable Development," in *Breaking New Ground* (London: Earthscan, 2002), which included chapters on Indigenous peoples, displacement, and abandoned mines. For a skeptical view of these initiatives, see Andy Whitmore, "The Emperor's New Clothes: Sustainable Mining?", *Journal of Cleaner Production* 14,

nos. 3–4 (2006), 309–14. The new
focus on the policy challenges
surrounding abandoned mines
is exemplified in Canada by the
creation of the National Orphaned
and Abandoned Mines Initiative in 2001, a national advisory
committee composed of mines
ministers, industry representatives, and public stakeholders. See

National Orphaned Abandoned
Mines Initiative, http://www.aban-
doned-mines.org/en/ (accessed 8
June 2016).

99   Virginia Valerie Gibson, "Negotiated Spaces: Work, home and
relationships in the Dene diamond
economy" (PhD diss., University of
British Columbia, 2008), 81–82.

12

Toxic Surprises: Contaminants and Knowledge in the Northern Environment

Stephen Bocking

In 2003, Sheila Watt-Cloutier, chair of the Inuit Circumpolar Conference, described her people's reaction to the discovery of elevated levels of contaminants in their bodies: "Imagine the shock, confusion, and rage that we initially felt when evidence of high levels of persistent organic pollutants was discovered in our cord blood and nursing milk in the mid-1980s. ... We were being poisoned—not of our doing but from afar." Inuit shock and outrage would eventually energize negotiations toward a global convention restricting these pollutants.[1]

Scientists were also surprised. In 1987, Eric Dewailly, an environmental health researcher, found contaminants in breast milk from women in Nunavik (northern Quebec); this, he recalled, "belied all logic."[2] The following year, a study concluding that Inuit of Broughton Island (then in the Northwest Territories, now in Nunavut) were exposed to contaminants in their food generated intense media coverage and urgent responses from government officials.[3]

These episodes were pivotal events in the history of northern contaminants. They provoked intensive research on their distribution and effects, new environmental and health policies and practices within Canada and in the circumpolar north, and global negotiations. Along the way, the conduct of science in the north was reconsidered and reshaped, as was the relationship between Aboriginal peoples, experts, and governments.

This history echoes several themes in the history of northern Canada. The presence of contaminants exemplified increasing human impacts on the northern environment, and the reality that the region is not isolated from the global environment. Scientists surveyed contaminants in the atmosphere, ecosystems, and species, and sketched their implications for environmental and human health—extending their historical role as interpreters of the north for audiences elsewhere. Official responses epitomized the influence of government administration on relations between humans and the northern environment. Aboriginal communities and institutions asserted their own perspectives on contaminants and food—extending into a new realm the assertion of their right to self-determination.

As we have seen, the discovery in the 1980s of contaminants provoked surprise, implying that this was a novelty. Yet there had been numerous previous instances of contaminants being discovered in the northern environment. Pilots in the 1950s, biologists and toxicologists in the late 1960s and early 1970s, atmospheric scientists in the mid-1970s—all were surprised to find substances that did not "belong" in the region. Contaminants were discovered several times, by separate communities of scientists, often where they did not expect them: in the atmosphere, in ecosystems, and in human bodies. Looking back, these surprises puzzled them: they wondered why, for example, it took so long to connect the presence of contaminants in arctic animals to the risk they may pose to Inuit who eat them.[4] These surprises can also tell us much about northern science. They are hints of what Michelle Murphy has called "regimes of perceptibility"—the combinations of scientific and social phenomena that determine which hazards will be visible, and which will remain invisible.[5] For historians seeking to probe these regimes, contaminants are a useful analytical tool: just as ecologists track the movement of substances to understand the structure of ecosystems, historians can track the movement (and lack of movement) of contaminants knowledge to describe the evolving structure of the scientific community and its relations with other communities, including governments and Aboriginal peoples.[6] Tracking knowledge reveals its uneven distribution among scientific and professional disciplines;

just because some scientists know something does not imply that all do. Instead, distinct strands of knowledge proliferated, linked to evolving disciplines and environmental circumstances.

This history of surprises can also tell us about the influence of ideas about the north. Foremost amongst these ideas is that of a pristine northern environment, protected by distance—a perspective inspired by the historical notion of the north as remote, unknown, and unspoiled.[7] This notion has been remarkably durable; even after a century of incidents of northern contamination, at mines, DEW Line sites, and other locations, it still persisted among scientists, expressed even amidst discussions of arcane technical topics, such as the chemistry of organic compounds or the details of atmospheric dynamics. As one scientist noted in the 1990s, "we are accustomed to regarding the Arctic and Antarctic as remote, unpolluted, and undisturbed areas of the world."[8] Conversely, once northern contaminants became evident, it became seemingly obligatory for scientists to remind their readers that the region was "no longer pristine."[9] Indeed, their presence in the north now carries rhetorical force. In his foreword to *Our Stolen Future*, a 1996 book that presented the dangers of global pollutants, Al Gore emphasized how humans "in such remote locations as Canada's far northern Baffin Island now carry traces of persistent synthetic chemicals in their bodies." Theo Colborn and her coauthors also discussed arctic contaminants in a chapter titled "To the ends of the Earth"—their presence demonstrating that "there is no safe, uncontaminated place."[10] To observers elsewhere, contaminants in the remote north have global implications: if they are there, they must be everywhere.

Contaminants exist at the most intimate scale: in the relations between people and food; and the most expansive: across the circumpolar region, and throughout the planet.[11] They include radionuclides (still present decades after the end of atmospheric nuclear testing), metals such as mercury, lead, and cadmium, and to a lesser extent arsenic and selenium, and persistent organic pollutants (POPs), such as PCBs, DDT, and many other synthetic compounds. Their sources are scattered across the landscape: mines (as John Sandlos and Arn Keeling discuss in their chapter), DEW Line sites, and, in the Russian arctic, discarded nuclear reactors and other relics of the Cold War. A few toxic substances, including metals such as mercury, cadmium, and arsenic, are present in local geological formations, and so are considered "natural."[12]

My focus, however, is on contaminants that originate in distant places, and are transported to the north through the atmosphere. Their ubiquity and extreme mobility make it difficult to place boundaries around them: they do not create contaminated sites that can be avoided, but entire "landscapes of exposure."[13] It is similarly difficult to limit the environmental history of northern contaminants. Like contaminants elsewhere, their presence is the result of diverse causal factors that expand outward from the substances themselves to encompass the global distribution of modern industry and agriculture—from electrical transformers that leak PCBs to farmers that use insecticides. Their history could even extend to Monsanto's marketing department, which after 1929 facilitated the global distribution of PCBs as a useful but toxic industrial chemical, or to Paul Hermann Müller's laboratory in Basel, where in 1939 he demonstrated DDT's insecticidal properties. The presence of these and other substances in the north underlines the role of the political economy of modern industry and agriculture in making their use a seemingly rational choice.

Those who study contaminants have defined their topic in several ways: in terms of international relations, foreign and circumpolar policy, or public health and environmental justice. Contaminants are not only physical matter, but social, political, and cultural phenomena. They raise interesting questions regarding scientific expertise and Aboriginal knowledge and their application to policy development and international negotiations, the definition of acceptable levels of exposure, risk, and uncertainty, and issues of equity, choice, trust, and power.[14]

Insights into contaminants elsewhere can be applied (albeit with care) to the north. From historians of science, we can learn about the place of science in political and regulatory contexts, its evolving disciplinary structure, and the links between scientific knowledge and other ways of thinking about the world. Work on the history of the field sciences is particularly relevant, including studies of the production of reliable knowledge in complex environments, and the relations between science in the field and in the laboratory. In addition, the construction by field scientists of a vertical dimension of the environment—whether downward, in oceanographic or geological research, or upward, in mountain research—presents interesting parallels with scientists' inclusion of the atmosphere as part of the northern environment.[15]

The insights generated by environmental historians can help us interpret the links between knowledge, peoples' actions, and non-human actors, including the atmosphere, animals, and contaminants, while reminding us that however contaminants are understood—as poisons, waste, or pollution—they are historically situated; that is, they are the product of particular ways humans have of organizing the world. Among these ways are the various scales that humans apply when defining a problem, whether as a local, national, or global issue. Each of these scales has political implications. Working with medical historians and historians of science, environmental historians have also considered how to include the human body within the history of the environment. Finally, environmental historians remind us—as Watt-Cloutier did—that these substances have moral implications. They demonstrate the fallacy of assuming that modern industry can be kept separate from the rest of nature, or that we can isolate our own bodies from the changes we impose on the rest of the world.[16]

Research by geographers is also relevant. For example, recent work on the geography of air can illuminate the intersections between institutions, economic activity, and the movement of atmospheric matter. Studies in the historical geography of science can demonstrate the significance of place and movement to the production and application of contaminants knowledge; like contaminants themselves, knowledge about them is located in specific places, and can move.[17] In summary, by applying all these perspectives to the history of contaminants we can achieve a better understanding of two essential themes in northern history: the dynamics of knowledge, both scientific and Indigenous; and the relationships—material, cognitive, and political—between the north and the rest of the globe.

Northern contaminants must also be examined in the context of the region's political and ecological history: the extension of government authority, development of resource industries, emergence of public health and environmental concerns, evolving scientific knowledge, assertion of Aboriginal rights and self-determination, and negotiation of regional and global treaties.[18] Northern contaminants have attracted the attention of many specialists—atmospheric chemists, wildlife biologists, toxicologists, and health scientists—who defined certain features of the north as of particular interest: the atmosphere, feeding relationships between species, and the relationship between Aboriginal peoples and country foods. This history demonstrates the power of scientific disciplines, institutions, and ideologies to shape perceptions of the north. Indigenous people have also developed their own interpretations, relating contaminants to how they understand the landscape, food, and health. The history of northern contaminants thus links with numerous themes in northern environmental history, including Aboriginal perceptions of landscape, as discussed in this volume by Hans Carlson and Paul Nadasdy, and the importance of food, as Liza Piper explains in her chapter. The history of northern contaminants is thus a history of diverse approaches to making sense of the world.

The history of northern contaminants knowledge also exhibits a series of striking transformations—in how they were defined and studied, how their consequences were understood, and whose knowledge about them was considered trustworthy. Contaminants often contradicted expectations. In doing so, they forced scientists, officials, and Aboriginal peoples to reconsider how they understood the northern environment and its relations to the rest of the world. Contaminants provide an opportunity to consider how the north itself is defined: as a place that is distinctive, yet embedded within political and environmental systems that extend far beyond its boundaries. Like climate change (as Emilie Cameron explains in her chapter in this volume), they require historians to consider how to write the environmental history of a globalized Arctic.

Northern Contaminants: First Observations

There were early hints of Arctic contamination. Norwegian explorer Fridtjof Nansen was among the first observers: during his Fram Polar Expedition of 1892–96, he noted dark stains on the ice—possibly, he thought, traces of air pollution.[19] In 1933, Charles and Anne Lindbergh collected samples of spores and pollen during flights over Labrador, Baffin Island, Greenland, and adjacent waters, demonstrating that winds could carry microorganisms (and presumably other particles) as far as the Arctic.[20] However, the first sustained observations of the arctic atmosphere came in the 1950s. During the Cold War, the Arctic became subject to aerial surveillance, and in 1956 J. Murray Mitchell, a climatologist, recorded the observations of US Air Force pilots flying weather reconnaissance missions.

They were surprised to encounter patches of haze hundreds, sometimes thousands of kilometres wide. According to Mitchell's summary, they saw a "grey-blue hue in antisolar directions, and a reddish-brown hue in the direction of the sun"—a visual account illustrating how the haze was of interest solely in relation to flying, navigation, and reconnaissance. He termed it "Arctic haze," signalling that the phenomenon was specific to this region.[21]

As Ken Wilkening has noted, the Inuit word "poo-jok" refers to "mist or haze," indicating, he suggests, an awareness of arctic haze.[22] Northerners have considered haze a familiar phenomenon: "People understand very well how things travel in air ... we've always known. In the summer some days the sky gets very hazy in a certain way. It's quite distinctive and elders will comment that there must be a fire in the south. Sometimes we can smell the smoke, last year the smoke from a fire in Northern Manitoba travelled straight up here—we could smell it for days on the wind."[23] However, Mitchell also stressed that arctic haze was only visible from the air, not the ground—a view consistent with the assumption during this era of the superiority of airborne over ground-based observations.[24] Thus, even if Inuit had already perceived this haze, it was only because of the post-war extension of aviation throughout the High Arctic, and the Air Force's concerns regarding pilot vision, that it became "visible"—that is, a phenomenon worth noting in official records.

This was also the era of above-ground nuclear weapons testing by the United States, the Soviet Union, and Great Britain. During the 1950s, awareness grew that, contrary to official reassurances, radioactive fallout could travel long distances. This awareness stemmed from both tragic accidents (such as radiation poisoning suffered by the crew of the *Lucky Dragon*, a Japanese fishing boat, near the March 1954 Bikini hydrogen bomb test), and from observations that strontium-90 from nuclear tests had circled the globe.[25] In 1953, the US Atomic Energy Commission (AEC) launched "Operation Sunshine," a secret effort to track the global distribution of strontium-90. This included the Arctic, where it was found that fallout could descend to earth within a few months, long before decay would have rendered it less radioactive.[26] This and other studies exemplified the expansion of the physical environmental sciences in response to Cold War imperatives, particularly in the strategically crucial Arctic region.[27]

Fallout data, once declassified, indicated that the north received less than did temperate regions. But evidence also accumulated that fallout was not only a global phenomenon; local ecological conditions also determined its consequences. The surprising discovery was made—first, apparently, in Norway—that concentrations in caribou and reindeer of Strontium-90 were higher than in grazing animals elsewhere, even those closer to the sources of these radionuclides. In addition, Eville Gorham, a British botanist, noted the peculiar capacity of lichen—a favoured food of caribou—to accumulate fallout. As he explained, "the chief practical conclusion to be drawn from this work is that animals feeding on mosses and lichens may well exhibit high intakes of radioactive fall-out on this account. In this connection a few reindeer bones from Norway have been shown to contain markedly greater concentrations of radioactive strontium-90 than sheep bones from the same country."[28] By the early 1960s, these observations were indicating the distinctive vulnerability of northern ecosystems and people to radioactive fallout: lichen accumulate fallout, caribou eat lichen, and many northerners eat caribou. Bill Pruitt, a wildlife biologist working on environmental studies associated with Project Chariot, the AEC plan to test the feasibility of excavating a new harbour in Alaska using "peaceful" nuclear explosions, helped publicize these conclusions, and elevated levels of cesium-137 were detected in numerous northern peoples, in Sweden, Finland, and Alaska.[29] Concerns regarding fallout were sharpened in September 1961, when the Soviet Union (followed by the United States) ended a three-year moratorium on atmospheric nuclear testing. In 1963 a study of the presence of cesium-137 in Canadian Inuit was initiated by the Radiation Protection Division of the Department of National Health and Welfare, to determine where Canada sat in relation to studies of other peoples in the circumpolar region.[30] This research concluded that it was below the maximum permissible body burden, and that there was no need to restrict consumption of caribou.

In 1963, with the signing of the Limited Test Ban Treaty, nuclear fallout concerns began to diminish (except for briefly renewed concern in the aftermath of the Chernobyl accident in 1986).[31] What remained was the awareness that contaminants could travel long distances—a lesson that echoed powerfully in Rachel Carson's *Silent Spring*.[32] By analogy with nuclear fallout, it was thought that organic contaminants (such as pesticides) could also be distributed as "fallout," and prominent scientists like

George Woodwell made explicit the parallel between radioactive material and pesticides such as DDT.[33] This perspective provoked global studies of the movement and distribution of pesticides. Among those pursuing this research, Alan Holden, a British scientist, was apparently the first to note the presence of PCBs and DDT in seals in arctic Canada, in the course of a study in Britain, Norway, and northern and southern Canada.[34] He and other scientists considered seals and other marine mammals of particular interest, because they accumulated contaminants in their fat, thereby serving as indicators of environmental contamination. The Canadian Wildlife Service (CWS) also began research: in 1967 biologists working under contract for the CWS in northern Quebec and the Northwest Territories measured DDE (a derivative of DDT) in the fat and eggs of peregrines and thinning of their egg shells, demonstrating that even in the north, this species was affected by pesticides.[35] Fisheries Research Board scientists also became involved, measuring DDT residues in beluga whales from the Mackenzie Delta.[36] Both studies reflected these agencies' interest in expanding their research beyond their traditional focus on resource management, thereby demonstrating their relevance to the federal government's new environmental responsibilities.[37] Canada's role in international arctic science was also a consideration. Gerald Bowes and Charles Jonkel of the CWS measured PCBs in arctic char, seals, and polar bears, confirming that they were found throughout the north, and in increasing concentrations as one ascends the food chain.[38] Their research was a contribution to the Polar Bear Specialist Group of the International Union for the Conservation of Nature—an early effort in circumpolar science and conservation.[39]

In summary, between the 1950s and the early 1970s several northern contaminants issues had attracted attention, including exposure to radioactive fallout and the presence of contaminants in peregrines and polar bears. Changes in how the north was known and experienced also influenced perceptions: the presence of aviators rendered arctic haze visible, fallout studies were provoked by global Cold War concerns, and research by the CWS and other agencies on contaminants signalled the extension of federal authority into arctic environmental affairs. The Arctic also gained a new status: while no longer pristine, it, like a few other remote places, could now indicate the global background level of contamination. The Arctic became a "baseline reference area"—as clean a place as one could find on a now-polluted planet, and a necessary station in international monitoring networks.[40]

Each of these phenomena: haze, fallout (especially its concentration in caribou and reindeer), and DDT, surprised those who had assumed the Arctic was pristine. But eventually, the concerns provoked by each dispersed. Haze seemingly had no implications for the environment or human health, and so remained only a scientific curiosity. The Limited Test Ban Treaty and restrictions on DDT eliminated any sense of urgency regarding fallout and pesticides. While contaminants in the Arctic (and in the Antarctic, and other remote places) remained, they were present in lower concentrations than elsewhere. Environmentalists and other observers turned their attention to more immediate northern issues, including resource development and proposals for pipelines and oil tankers. A new generation of Aboriginal leaders focused on land claims and authority over wildlife and renewable resources. Research and regulation shifted accordingly.

The Atmospheric Arctic

In 1972, Glenn Shaw, a scientist at the University of Alaska, was surprised to observe that the supposedly pristine air above Barrow was less clear than expected; in the language of atmospheric physics, he recorded high "atmospheric turbidity." Subsequent observations during flights over the pack ice north of Barrow confirmed that it took the form of distinct layers of brownish-yellow haze—just as, Shaw noted, Mitchell had recorded nearly two decades before.[41]

Shaw reinterpreted Mitchell's observations in terms of his own discipline. To an atmospheric physicist, turbidity didn't mean impaired flying conditions, but the presence of aerosols—tiny suspended particles. Scientists had been studying these for a long time, even before the era of environmentalism: tracking dust swept aloft from deserts, ejected from volcanoes (like Krakatoa in 1883), and blown away in the American Dust Bowl of the 1930s.[42] During the early Cold War, these studies were sometimes linked to strategic concerns: for example, Harry Wexler of the United States Weather Bureau published his studies of volcanic dust even while pursuing classified research on the dust swept aloft by nuclear explosions.[43]

Wexler's research exemplifies how atmospheric research done for strategic purposes, including meteorological studies of particle transport and distribution, would eventually have implications for understanding northern contaminants. However, much of the study of the movement of material in the atmosphere remained focused on "natural" sources—as reflected, for example, in the discovery in the 1960s that desert dust from Africa could cross the Atlantic.[44]

Shaw accordingly evaluated arctic haze in the context of studies of the intercontinental movement of dust, applying techniques to determine its origins that were similar to those of his colleagues elsewhere. These included meteorological maps of the movements of continental air masses, and chemical analysis of the material itself. A particle has a chemical "signature" (a distinctive elemental composition) that can indicate whether it is, say, desert dust, or the product of combustion—that is, pollution. At first, he and his collaborator, Kenneth Rahn of the University of Rhode Island, interpreted arctic haze as a natural phenomenon—it was dust from Asian deserts. But then it turned out that this conclusion was the product of a chance occurrence: they had collected samples after a storm in Asia had blown unusual quantities of dust into the atmosphere.[45] In 1977, new samples indicated vanadium, manganese, aluminum, and sulfates, suggesting industrial sources.

Shaw and Rahn also realized that what they were describing was not simply a northern instance of a global phenomenon. Arctic haze had properties distinct from those of haze elsewhere; to scientists, it seemed to "break all the rules."[46] For one thing, it was a complex mixture, formed predominantly from sulfates, as well as graphitic carbon, organic compounds, several metals, and carbon dioxide, methane, and carbon monoxide.[47] It also had a distinctive seasonal pattern, occurring only in winter and spring, not summer. And finally, arctic haze particles were transported at lower elevations than was typical in the south.[48]

This material distinctiveness found a parallel in a distinctive research community that emerged during the 1970s and 1980s. Besides Shaw and Rahn, an early member was Len Barrie of the Atmospheric Environment Service at Environment Canada. An informal Arctic Chemical Network formed, which, like arctic haze itself, covered most of the circumpolar region, including the United States, Canada, and Scandinavia (but not the Soviet Union). They organized a series of conferences on arctic air chemistry, as well as cooperative research, including three Arctic Gas and Aerosol Sampling Programs (in 1983, 1986, and 1989), which tracked the movement of the aerosols that constituted haze from Eurasia across the Arctic to Canada and Alaska.[49] While this was an interdisciplinary community—mainly meteorologists, atmospheric chemists, and physicists—they focused almost entirely on the atmosphere. And while they often noted that the ecological consequences of arctic haze were worthy of study, in practice these received very little attention.[50] This reflected the power of disciplinary boundaries: these consequences only became evident when contaminants left the atmosphere—at which point they were of less interest to atmospheric scientists.

With its focus on the atmosphere, the arctic haze research community eventually moved out of step with developments elsewhere. By the late 1970s, atmospheric contaminants, reconceived as Long-Range Trans-boundary Air Pollution (LRTAP), had emerged as a major international concern thanks to the newly acquired notoriety of acid rain, as well as an emerging awareness that the atmosphere was a source of contaminants affecting the Great Lakes and other ecosystems. Arctic researchers shared in the resources that now became available for atmospheric science: they tied their studies to work in the Great Lakes (the most active region for Canadian contaminants research), and the Canadian Network for Sampling Precipitation collected samples of snow and surface water on Ellesmere Island. But acid rain was not really an issue in the north, and neither were other prominent sources of pollution: metal smelters, coal plants, the Alberta tar sands, or motor vehicles. For arctic researchers, particulates remained a scientific matter—intriguing, and a way of understanding the movement of continental air masses, but remote from the environmental and health issues motivating scientists elsewhere. As scientists later recalled, in the 1970s the detection of contaminants in the north "was generally regarded as little more than a curiosity."[51] This was also reflected in the focus on elements like vanadium—not significant in terms of toxicity, but relevant to scientific questions, such as the origins of contaminants and the mechanisms by which they travel to the Arctic. It is not surprising, then, that arctic haze research did not lead to significant political initiatives, such as an international agreement; this was consistent with the nature of the phenomenon and of the scientific effort devoted to it.[52]

Persistent Pollutants

In the late 1980s a new phase in arctic contaminants research began. Instead of examining the stew of disparate substances that together formed arctic haze, research focused on a single category of synthetic chemicals: persistent organic pollutants (POPs, also referred to as organochlorines). This research was tied to developments elsewhere: the global political economy of chemicals, and environmental and health concerns in affected regions, such as the Great Lakes. Their presence in both the environment and in political affairs testified to the status of POPs as a category defined in terms of both science and policy.[53]

The environmental history of POPs is the product of both their intended characteristics (including their persistence) and their unintended behaviour once released. They can be classified in terms of purpose into three main categories (see Table 12.1). PCBs were among the first, introduced in 1929, and used in electrical equipment virtually everywhere. They were followed by thousands of other synthetic chemicals. Many were pesticides, with DDT only the most notorious. Other substances, including dioxins and furans, are waste products of combustion and industrial processes. Given enough time, POPs can travel everywhere, and do not depend on the meteorological processes that had been of interest to those studying arctic haze. By the mid-1960s, they were found everywhere on the planet. Eventually some became subject to bans or stringent regulation: the insecticide hexachlorobenzene in 1965, DDT in numerous countries in the early 1970s, aldrin in the United States in 1975, PCBs in Europe and the United States in 1976 (1977 in Canada), and dieldrin, another pesticide, in the United States in 1984. POPs continue to be used in many countries; some pesticides still evaporate from the soil in areas where they were once, but are no longer, used; and novel chemicals continue to be invented and released (deliberately or accidentally) into the environment.

By the mid-1980s, these substances had become a focus of concern regarding their often-insidious consequences for health, even in minute concentrations. Graphic images—deformed fish in the Great Lakes, and abnormal sexual development in amphibians, for example—captured attention, as did the arguments of those who urged action, including Theo Colborn, co-author of *Our Stolen Future*.[54] These concerns were reinforced by the emerging view, expressed in Carson's *Silent Spring*, that the health

TABLE 12.1: Categories of Persistent Organic Pollutants (POPs).

Category of Use	Examples
Compounds for industrial applications	PBBs (polybrominated diphenylethers), PCBs (polychlorinated biphenyls), PCP (pentachlorophenol)
Compounds for agricultural applications (e.g. pesticides)	Aldrin, chlordane, DDT, dieldrin, endrin, heptachlor, HCB (hexachlorobenzene)
Unwanted by-products of chemical production processes (e.g. incineration)	Polychlorinated dioxins, polychlorinated furans, PAHs (polycyclic aromatic hydrocarbons)

Source: R. Kallenborn, "Persistent Organic Pollutants (POPs)" in: *Encyclopedia of the Arctic*, ed. M. Nuttall, (New York: Routledge, 2005), Vol. 3, 1622-1624.

of bodies is tied to the health of their surrounding environments, undermining the modernist view of health as strictly a matter of protecting individual bodies from external pathogens.[55]

Concerns about chemicals in the industrial heartland eventually reached the Arctic. Evidence of their presence was, in part, the product of sampling networks mainly based in the south, but some studies also addressed specific northern concerns. Surveys of their distribution across the north helped demonstrate, in combination with other evidence, that they did not come from local sources.[56] By the late 1980s, there had been extensive surveys of contaminants across the arctic environment: in the atmosphere, snow, plants, and animals. Favoured species received particular attention: in particular, scientists examined polar bears (killed by Inuit hunters) throughout the Northwest Territories in an effort to identify effects on their health.[57]

But these surveys did not explain the larger puzzle: these chemicals seemed to be everywhere in the Arctic, in quantities that suggested there was some sort of mechanism—almost a conveyor belt, it seemed—facilitating their movement north from industrial centres. This notion had, in fact, been mentioned before. In 1973, Max Dunbar, the McGill University oceanographer, had noted the "rule of the cold wall"—that volatile substances such as pesticides evaporate, and then condense and concentrate in cold places. And in 1975, E. D. Goldberg coined the term "global distillation" to describe this process of evaporation and condensation, re-conceiving the global atmosphere as a laboratory experiment.[58] The result, as the Norwegian scientist Brynjulf Ottar noted in 1981, was a "systematic transfer of the more persistent compounds from warmer to colder regions," so that "the Arctic is a region where a general accumulation of partly volatile pollutants may be expected."[59]

In the late 1980s, a few scientists, including Don Mackay of the University of Toronto, applied these ideas to the presence of chemicals in the Arctic. Since their movement in the atmosphere could not be observed directly, Mackay and his colleagues constructed models of how they expected them to behave. These models were simplified descriptions of nature: they represented the atmosphere as a few compartments, each described in terms of a few characteristics, such as temperature. By combining these models with knowledge of the chemicals' properties (such as their tendency to evaporate at various temperatures), scientists could predict how and where they should move. These predictions could then be compared with their actual distribution. When predictions and data matched, scientists could be confident that the models were describing the chemicals' actual behaviour.

These models exemplified how contaminants research was changing; they were a way of bringing together previously distinct forms of knowledge—about global air movements, and the behaviour and distribution of chemical molecules—to form a synthetic explanation of their presence in the Arctic. Thus, by combining in a novel way what was already known about these substances, new knowledge was created. New disciplines also now became defined as relevant to understanding contaminants: meteorologists and atmospheric chemists were joined by modellers, toxicologists, marine ecologists, and wildlife biologists, while new techniques—and knowledge of additional aspects of the Arctic environment, such as marine ecology and caribou biology—were called upon to help make sense of their presence, movement, and ecological consequences.[60] In all these ways: techniques, disciplines, and the phenomena themselves, the study of POPs broke from studies of arctic haze. These substances also displaced arctic haze as the focus of arctic contaminants research.

As these models demonstrated, POPs demonstrate a distinctive behaviour in the atmosphere, reflecting characteristics of the substances themselves. They evaporate when it is warm, and condense on surfaces when it is cold. They may repeat this cycle several times, as winds carry them toward the north, finally being deposited where it is too cold to evaporate again—in the Arctic. The more volatile a substance, the more readily it travels north. Scientists referred to these cycles of evaporation and deposition as the "grasshopper effect," and deduced that they could account for both the observed delay between the release of chemicals in the south and their appearance in the north, and their substantial presence in the region.[61]

The grasshopper effect meant that the north was peculiarly vulnerable to these chemicals. While they could travel anywhere, they tended to condense and concentrate in the north. The region was now a "sink," actively attracting contaminants. Once there, the cold, diminished biological activity, and lack of winter sunlight allowed them to persist longer without breaking down. Arctic animals were also distinctively vulnerable: large marine mammals, near or at the tops of food chains, with ample body fat, readily accumulated these contaminants. (And, looking to the future, climate change may, for a variety of reasons, magnify these impacts.)[62] In effect, therefore, the Arctic became not just a passive receptacle for contaminants but an active agent in their environmental history, adding thereby a regional complication to their global distribution. However, they would only become a political priority to the extent that they had implications for humans.

The Unhealthy Arctic

A striking feature of this research was its inattention to humans. In hindsight, this is puzzling—as scientists themselves later noted. Contaminants were known to concentrate as they moved up the food chain. Inuit were involved in the research; they often provided the wildlife samples used for analysis; and their position at the top of the food chain as hunters was obvious. Given the history of synthetic contaminants production, it can be assumed that northern people have been exposed to them since at least the 1960s. However, only in the 1980s were these connections made. Why was there such a delay? Evidently, the scientific disciplines and policy agencies involved in contaminants research shaped a particular regime of perceptibility that drew attention away from risks to humans—in effect rendering invisible some of the implications of the use of country foods.

A new regime of perceptibility began to emerge once official agencies and their scientific experts began to consider Aboriginal use of country foods on its own terms, and not simply as a vestige of a disappearing traditional lifestyle. As Liza Piper discusses in her chapter in this volume, federal and territorial governments have long included food and nutrition as part of their responsibility for Aboriginal health and wellbeing; from this stemmed their initial actions regarding food and contaminants. Extensive nutrition surveys in the 1960s and 1970s provided the foundation for harvest surveys in the 1980s, which showed that Aboriginal communities still relied on wildlife for food. Information about the contaminant content of northern food species had also been assembled. There was clearly a potential risk. But to fulfill their administrative responsibility, a formal risk assessment had to be done. This required specific information: how much contaminants were being consumed, in what foods, and with what consequences in terms of the presence of contaminants in human bodies. With this information, officials and health scientists could compare exposures to generally accepted allowable limits, and advise people accordingly.

Because this was novel territory for northern medical experts, a pilot study was deemed necessary. In 1985, the community of Broughton Island was chosen, as harvest data had indicated that it had the highest per-capita consumption of country foods in the Baffin Island region. After conducting dietary surveys and analyzing samples of the food types reported in these surveys, as well as samples of blood and breast milk, David Kinloch, the regional medical officer, and Harriet Kuhnlein, an experienced scholar of Aboriginal food and nutrition, were surprised to find that a significant fraction of the population consumed more than the acceptable daily intake of PCBs, or had levels in their blood above "tolerable levels."[63] These PCBs came mainly from eating narwhal.

In 1985, meanwhile, a province-wide survey of PCBs and other contaminants in breast milk was getting under way in Quebec. It was a response to concerns in industrial centres, not the north; in fact, no northern communities were included, as it was assumed that they were not affected. However, Eric Dewailly, a health scientist based in Quebec City, was given an opportunity to include samples from a community in Nunavik. He thought they would serve as useful "blank" controls—that is, samples without contaminants—but to his surprise they indicated levels of PCBs five times those recorded in southern Canada. At first, he suspected the samples had been contaminated in the lab.[64] This reaction stemmed from distance—both geographic (the community was far from obvious sources of contaminants), and intellectual (he was unaware that scientists were already studying contaminants in the Arctic). To make sense of this discovery, he would need to shorten this intellectual distance by reaching beyond his medical training and studying the ecological literature.

As I noted at the start of this paper, these surprises attracted a great deal of attention. Kinloch and Kuhnlein's study received front-page coverage in the national media.[65] Several factors—some specific to the north, others of more general significance—had converted a scientific result into a public health crisis. Inuit themselves felt shock (as Watt-Cloutier emphasized), and because of anxiety, some reduced their consumption of country foods, with unfortunate consequences.[66] The special vulnerability of women and infants made it an issue of gender and environmental justice. Contamination of northern wildlife also contradicted the federal government's legal obligations to Aboriginal peoples, including their rights to hunt and to consume country foods.[67] The then-recent Krever Inquiry, instigated in response to a scandal regarding contaminated blood, had sensitized the public to federal responsibilities regarding the protection of the purity of the nation's blood supply; it was noted that the federal government had an analogous responsibility to Aboriginal people to ensure their safe access to uncontaminated country foods, with failure to do so implying potential legal liability and a requirement for compensation.[68]

Beyond the north, contaminants had, of course, become a matter of general concern, and news of their presence in the Arctic—still widely considered to be pristine—accentuated perceptions that no place was immune. The Canadian Arctic Resources Committee, together with environmental organizations active in the Great Lakes and other contaminant-rich regions, helped focus attention on the issue.[69] Acid rain and ozone-layer depletion made tangible the idea that pollution was a long-distance and not simply a local issue (just as nuclear fallout had at the time of *Silent Spring*). And finally, the public and governments were receptive to an issue that involved the environment and Aboriginal people. The Brundtland Commission in 1987 and the run-up to the 1992 Earth Summit made the environment a political priority; for various reasons, Aboriginal issues had also become prominent in Canada. Thus, through scientific and political developments, in both the north and elsewhere, northern contaminants were "discovered"—that is, they became not just a northern, but a national issue.[70] Knowledge previously held within the north, and within a limited professional community, now flowed beyond the region.

The federal government responded by redefining the contaminants issue in terms of its administrative structures. This had begun even before it became an issue in the media. In 1985, a Technical Committee on Contaminants in Ecosystems and Native Diets was assembled, composed of federal scientists and science managers from four federal departments and the Northwest Territories government. This committee was eventually expanded to include representatives from five northern Aboriginal groups. Much of its attention focused on Kinloch and Kuhnlein's results from Broughton Island.

More research was also underway—reflecting, as did the formation of the technical committee, an effort to define the issue as a scientific matter. Studies sought a link between country foods and exposure to contaminants, as evaluated by comparing an individual's diet with his or her contaminant burden. During the late 1980s, surveys in Inuit communities across much of the north accumulated evidence of high levels of contaminants.[71] Scientists also found contaminants throughout the northern environment—in the atmosphere, surface water, and living organisms, evidence that the atmosphere was a pathway by which these substances entered northern food chains.[72] Aboriginal peoples initiated research: reflecting a broadening of concern beyond Inuit, Dene and Métis communities requested studies of contaminants in their food, including fish from the Mackenzie River downstream of Norman Wells.[73]

By 1990, research on contaminants had shifted from studies of their presence in various arctic species to a focus on how people encountered them, particularly through food—their most immediate link with their environment. The meaning of northern contaminants was shifting from being a chemical and ecological, to a human health and cultural issue. A view of the north as a region distinctively vulnerable to contaminants was also emerging. However, this view was not uniform across the north. Marine mammals—specifically, narwhal, beluga, and walrus—emerged as the main sources of contaminants. These are a more important part of the diet in the eastern than the western Arctic. This implied a new geography of exposure, introducing a human dimension to the mapping of northern contaminants.

By 1989, preparations were underway to build on this emerging view of northern contaminants. The Department of Indian and Northern Affairs organized two workshops to synthesize information and plan a more integrated research strategy; this would eventually become the basis for the Northern Contaminants Program (NCP).[74] Funding became available through the Green Plan (an initiative intended to demonstrate the federal government's environmental commitment). The Arctic Environmental Strategy was one component of the Green Plan, and it, in turn, provided funding for the NCP. It ran from 1991 to 1997, with funding of $5 million per year. Renewed in 1997, its second phase continued until 2003.[75] The program then began a third phase, at a lower level of support.

The NCP represented an effort to redefine a politically difficult issue in terms of both administrative and scientific priorities. It would do so by replacing individual research efforts with a more systematic approach that could provide an overview of contaminant movements from the atmosphere, through ecosystems, to people, as well as specific advice regarding food and health. This involved several challenges, some of which were inherent in interdisciplinary research, or in community-based research, or stemmed from diverse views of food, hazards, and knowledge. To understand how the NCP developed, we can begin by examining these challenges.

Studies involving humans required a combination of toxicological, medical, and social expertise and sensitivity entirely unlike that required by studies of contaminants in wildlife. It required novel areas of research, including the social, cultural, and nutritional importance of country foods. There were also challenges encountered in communicating between scientists and non-scientists, including translating technical terms and working out protocols and expectations for community-based research.

Other challenges were more specific to the north, its history, environment, and communities. Scientists and northerners had distinctly different ideas regarding nature, health, and knowledge. The scientific understanding of the effects of contaminants—as subtle, long-term, and invisible—challenged Aboriginal perspectives on food safety, which did not involve these characteristics. Other differences were apparent in views of how knowledge of contaminants should be applied to decisions regarding food and health. Scientists trained in the south were guided by ideas about risks, which, it became evident, were not appropriate in northern communities. Conventional risk analysis (as was employed on Broughton Island in the 1980s) involves several steps: determining the toxicity of the contaminant and possible pathways of exposure, evaluating potential risks, and then managing these risks through consumption advisories (advice on what foods should or should not be eaten, by whom, and in what quantities). This process defines contaminants as a biophysical phenomenon, to be understood in terms of scientific descriptions of substances and hazards, with advice formulated by experts and based on rational analysis of risks and benefits. It relies on several assumptions: that the boundary between bodies and the environment is clearly defined, that individuals can act autonomously in response to advice, and that knowledge of health consequences will be the determining factor in individual actions.

None of these assumptions were valid in the north, given the importance of country foods, the lack of alternatives, and controversies over risk communication. Individual and community health are considered inseparable from country foods—encouraging a reluctance to believe they could be unhealthy; as one Inuk explained, "Country food is preventing you from diseases. Therefore it is a medicine. When you are sick and you are trying to gain back your strength, you eat country food. It's your medicine."[76] The importance of country foods—nutritionally, economically (given the expense of imported foods), and socially (through their contribution to community relationships), as well as the ethical ties between people and wildlife—contradicts the view of health as a characteristic of individual bodies. Disruption of traditional community lifestyles has its own consequences, illustrating how the effects of contaminants are not limited to direct toxicity. And although Aboriginal people had no experience with substances such as PCBs or DDT, hunters able to draw on their own experiences and their community's traditional knowledge are intimately aware of where their food comes from, and pay close attention to its quality. As one hunter explained, "When you've been working with caribou all of your life you just know when it is healthy."[77] They rely on concrete, visual forms of evidence, such as spoiled food, garbage, or parasites.[78] This reflects a particular view of what counts as reliable knowledge; protection against unhealthy meat requires careful observation and checking (for example, for parasites), while monitoring the overall health of wildlife. This knowledge has several dimensions: cultural (ideas about health and how it relates to the environment), social (the importance of hunting and sharing food), political (defining who has authority to make decisions about food and animals), and epistemological (the reliability of sensory evidence).[79]

It is helpful to place in historical context this distinction between scientific and Aboriginal perspectives. As Linda Nash and other scholars have argued, roughly a century ago a shift occurred in ideas about the relationship between health, landscapes, and bodies. Health had been seen as a characteristic shared by a landscape and the people living in it.[80] A healthy place could ensure healthy people; accordingly, as Gregg Mitman has described, certain places—mountains or deserts, for example—became noted as health resorts.[81] Conversely, illness was linked to an unhealthy environment. But this view was eventually displaced by a modernist perspective that discounted local environments, focusing instead on individuals. Guided by germ theory, health became defined as the absence of disease, landscapes served only as neutral spaces occupied by harmful agents (germs, poisons, contaminated food), and the function of medicine was to protect bodies from these agents. Health expertise no longer implied an understanding of local landscapes, but a mastery of lab-based, universally applicable knowledge.

This modernist perspective on health was applied widely in the post-war north: through efforts to reduce Aboriginal peoples' reliance on country foods in favour of vegetables and other "southern" food, to apply the advice of nutritional science (as Piper discusses in her chapter in this volume), and to extend modern medical expertise into northern communities, backed up by transfers to southern hospitals. This represents, as Nash has noted, the project of modernity: erasure of local environments and their recreation as homogenous, controlled space.[82]

Given these contrasting views, a research program on contaminants satisfactory to all parties would require considerable negotiation. The challenge was exacerbated by a history of difficult relations between government officials, scientists, and northerners—the product of a colonial relationship and of scientists' failures to consult while attempting to manage Aboriginal relationships with northern wildlife.[83] One result of this history has been Inuit skepticism toward scientists' claims about contaminants—perhaps, some thought, this was merely another strategy to discourage hunting.[84] Distrust was further exacerbated when the

Department of Indian and Northern Affairs initially excluded Aboriginal leaders (as well as the media) from meetings about the Broughton Island situation.[85] Yet these difficult circumstances also made a socially responsible research program all the more essential; only then would those affected by contaminants accept advice from experts or government officials.

The Northern Contaminants Program

The NCP thus took form amidst distrust and uncertainty. Planning research that addressed both scientific and community priorities required extensive consultations and workshops. Researchers and communities gained considerable experience in collaborating. There were ample precedents to draw upon, including the work of the territorial science institutes, which licensed researchers and administered ethical guidelines for the conduct of research, including requirements relating to conduct, participation, and communication. In 1994, a new concept of practice was implemented known as "responsible research."[86] Responsible research became one aspect of the distinctive nature of northern contaminants research, exemplifying its new social relations.[87]

The NCP involved a wide range of research activities on a range of topics corresponding to the ecological and human systems relevant to the movement and effects of contaminants. However, the NCP also had objectives beyond research. These included building the capacity of Aboriginal organizations and communities to evaluate evidence and to make decisions about contaminants and other environmental factors. In addition, because it presents its own challenges in northern communities, communication became a discrete research area, with a focus on strategies to communicate effectively to various audiences. This work was guided by the view that scientific and Indigenous knowledge, applied through community-based research and consultation that considers local customs, cultures, and ways of life, could together provide the basis for advice regarding food consumption. As one research group explained, "it has been most effective for local public health authorities working in concert with the community at risk and experts from a variety of disciplines to develop risk reduction strategies that address the risks and benefits components of each specific concern."[88] This included study of the ways and means of dietary advice, such as the value of positive options: instead of banning consumption of one species, consumption of an alternative food could be encouraged. When exposure to contaminants is greater than recommended, people are not immediately advised to alter their diets, but instead are provided with the information required so that they themselves might evaluate the risks in relation to benefits of traditional foods.[89]

In the second phase of the program, beginning in 1997, research emphasized the human dimensions of contaminants, with substantial community involvement in setting priorities, conducting research, communicating, interpreting, and applying results. Northern contaminants research gained a distinctive regional character by acknowledging its social and cultural dimensions. This included enabling Aboriginal people to help determine research priorities and allocation of research funding—a feature that likely makes the NCP unique among federal research programs.[90] The political significance of these dimensions became evident in the fact that this research was the only part of the Arctic Environmental Strategy to continue after 1997, with human dimensions remaining central to the program throughout its third phase (2002–08).[91]

The NCP was also an effort of synthesis. In 1997, both the NCP and the Arctic Monitoring and Assessment Programme (to which the NCP contributed) published extensive assessments combining knowledge from within and outside the region, and linking the scientific and social dimensions of contaminants. But the NCP's synthesis function extended beyond assessment reports. Synthesis—bringing together knowledge of disparate parts of nature and society—was also evident in the program's organization. A complex management structure was constructed, far more elaborate than the model of the individual researcher, which brought together scientists of various disciplines, managers, and Aboriginal organizations.[92] Synthesis was also evident in the formation of institutions that enabled new combinations of expertise. The Centre for Indigenous Peoples' Nutrition and the Environment, established in 1992 with funding from the Arctic Environmental Strategy and the NCP, focused on diet-related research conducted through collaboration with Aboriginal peoples. This centre represented a novel approach to the organization of northern research—independent of government, and with leadership provided by Aboriginal people. During the 1990s, it conducted three large, participatory studies of the risks and benefits of contaminants and country foods in more than forty northern communities.[93] Another institution combining disciplines in a novel way was (and still is) the environmental health research group at Université Laval's Public Health Research Unit, which includes researchers in community medicine, epidemiology, toxicology, nutrition, psychology, and anthropology.[94]

However, synthesis and new ways of studying and acting on contaminants were most evident in the formation of new objects of research and practice. Periodically in the history of science a new object is constructed, which becomes a template for the organization of nature, research, and practice.[95] One such object in northern contaminants research was the pathways by which contaminants travel to humans. This included the mechanisms by which contaminants move from their origins, through the atmosphere and ecosystems into country food species, reaching humans through the harvesting, sharing, and consuming of food. This object implied a shift in research effort: less emphasis on ecology, and more on medical and toxicological science, and the community dimensions of food. This object also underlined the specifically northern character of NCP research: for example, it encompassed the specific mixtures of contaminants that enter these pathways in the north; and it also corresponded to the focus of the Centre for Indigenous Peoples' Nutrition and the Environment.

A second research object related even more directly to the dual biophysical and social dimensions of northern risk assessment. This object, the "Arctic Dilemma," emerged in the late 1990s as a way of referring to the fact that while country foods provide essential benefits, they are also the primary conduit of exposure to contaminants.[96] It thus expressed a regionally specific version of the challenge of risk assessment: determining the appropriate balance between benefits and hazards. When contaminants were seen as a strictly biophysical phenomenon, they could be understood in terms of global mechanisms of atmospheric transport and ecosystem behaviour. Once, however, attention focused on how humans exposed to contaminants actually think about them, they became a dilemma—a local issue, situated within Aboriginal communities and the spaces in which food is hunted and consumed. As such, the Arctic Dilemma is a hybrid concept: it combines environmental and social dimensions, scientific and Indigenous perspectives, and general and local knowledge. It is also an administrative concept, framed in terms of northern government jurisdiction. Resolving the Arctic Dilemma requires consideration not only of the type, amount, and nutritional value of foods, but of their social, cultural, economic, and spiritual benefits, and appropriate ways of communicating and developing options for communities.[97]

Shaping the NCP

The NCP was framed in terms of science and Aboriginal perspectives, and the specific character of each in the northern context. It was also a federal administrative initiative. To understand how and why the NCP took the form it did, we need to examine the influence of all three factors.

Consistent with its position in northern environmental affairs, science has been central to the NCP. One major role of scientists has been as a conduit for knowledge and experience from outside the region. By the late 1980s, health scientists elsewhere had accepted that effective research required that communities gain a sense of ownership over it, by participating at all stages, and with the research considering not just the biophysical, but all dimensions of the environment and health, including well-being and not merely absence of disease. This view developed through experience: a study in the late 1970s and early 1980s of mercury exposure in Canadian Aboriginal communities, as well as the Effects on Aboriginals from the Great Lakes Environment (EAGLE) project, had demonstrated the importance of the indirect health effects of contaminants, such as reluctance to consume traditional foods.[98] Thus, the NCP's distinctive northern approach to contaminant research can also be traced back to national studies.

Scientists also brought to the NCP ideas about how to combine knowledge from the north and elsewhere, and from diverse disciplines. Perspectives on health in northern communities have been based not only on observations in the north, but on information obtained elsewhere, such as from laboratory studies and accidental exposures.[99] However, this information was not necessarily directly applicable to specific northern conditions, thus necessitating studies in the region itself.[100] In several ways, the Arctic presents unique circumstances: a distinctive mixture of contaminants, chronic exposure to relatively low levels of contaminants, nutrients in arctic seafood that may affect its toxicity, human genetic variability, and diverse socioeconomic and lifestyle factors, such as seasonal patterns of food consumption specific to the Arctic. As Dewailly and Furgal noted, the "specificity of the Arctic situation raises the question: to what extent can results and conclusions from epidemiological studies conducted outside the Arctic apply to this region."[101] Knowledge of arctic contaminants therefore had to combine knowledge from elsewhere with knowledge from within the region, with scientists working out this combination. Evaluating the risks and benefits of country foods also means bringing together previously separate research communities: atmospheric chemists and ecologists studying the long-range movement of contaminants and their distribution in the environment; health scientists and toxicologists examining the implications of contaminants for humans; and social scientists involved in research in northern communities. This presented challenges, such as how to balance the uncertain risks described by toxicology against the certain benefits outlined by nutritional science.[102]

But beyond its scientific dimensions, the NCP was above all a federal administrative initiative—part of the long history of such initiatives in the north. It was also the federal government's chief political response to the northern contaminants controversy. It was designed in relation to federal jurisdictional mandates over the environment and human communities in the two northern territories (three after 1999, with the creation of Nunavut). It also defined "contaminants" quite specifically: as substances that had been transported long distances to the north. These were distinct from "waste": namely, unwanted substances originating from within the north itself. The NCP thus excluded local issues such as drinking water quality and cleanup of DEW Line sites.[103] Overall, as an effort to relate scientific activity to federal jurisdiction in the region, the NCP exemplified the co-production of the scientific and political dimensions of a research area.[104]

The NCP's status as a federal program was also apparent in its international role. Federal jurisdiction included all foreign, including circumpolar, affairs. By the late 1980s, contaminants were becoming a circumpolar issue thanks both to new scientific information and to improved prospects for arctic cooperation with the thawing of the Cold War and dissolution of the Soviet Union.[105] Circumpolar initiatives were reinforced by the view of the Arctic as a distinct place facing a distinctive challenge from contaminants.[106] Arctic Monitoring and Assessment Programme studies showed that Indigenous people, and particularly those who consume marine mammals, were among the most exposed to POPs of any people on earth. Other organizations, such as the Northern Aboriginal Peoples' Coordinating Committee on POPs, also urged recognition of the Arctic, and arctic people, as distinctively vulnerable to POPs.[107] Through such initiatives, POPs gained an identity not only as scientific entities, defined in terms of chemical characteristics and environmental behaviour, but as a political concept—the end-products of activism and negotiation through which the "dirty dozen" POPs were identified as a collective priority.[108] Through the Stockholm Convention, POPs also became both a global issue and one situated in a specific region, with the Convention mentioning only one region—the Arctic—as facing specific risks. In effect, a regional perspective on contaminants became globalized.[109]

The Canadian government both responded to and led international actions on contaminants. Canada participated in negotiations that led to the Declaration on the Protection of the Arctic Environment (signed by eight arctic nations in June 1991) and the creation of the Arctic Environmental Protection Strategy, and co-chaired with Sweden the United Nations ECE Task Force on Persistent Organic Pollutants. The regional Protocol on Persistent Organic Pollutants (the "Arhus Protocol") was adopted in 1998, followed by the global Stockholm Convention, which was adopted on 22 May 2001, with Canada becoming the first nation to ratify it the following day. Canada's international leadership was largely facilitated by the NCP, which gave it the necessary scientific capacity. The NCP also provided Canada's chief contribution to the Arctic Monitoring and Assessment Programme, a component of the Arctic Environmental Protection Strategy.[110]

In addition to scientists and the federal government, Aboriginal communities and organizations helped shape the NCP. One way they did this was by asserting alternative perspectives on contaminants, and thus orienting the NCP toward human health.[111] Their coalition, Canadian Arctic Indigenous Peoples Against POPs, also asserted the definition of contaminants as both a northern and an Aboriginal issue. And, as I noted at the beginning of this chapter, Aboriginal representatives, led by Watt-Cloutier, became a powerful voice urging application of NCP results to the pursuit of environmental justice at the Stockholm negotiations—engaging, in effect, in "scale jumping" by transforming a local concern into a global challenge.[112] Community interests and concerns have also influenced the priorities and practices of science; for example, by ensuring that tradition-al harvest areas and populations are the focus of research. Aboriginal per-ceptions regarding the relative importance of contaminants have also had an influence. For example, endocrine disruptors have received less atten-tion in the north because communities have not identified reproductive or fertility issues as significant concerns.[113]

We have seen how Aboriginal views of contaminants varied from those of scientists; but they also differed from how the federal government defined them. For example, communities expressed concerns regarding contaminants that were "of a local nature." While these were excluded from the federal definition, there was no other source of funding avail-able to deal with them. Accordingly, the NCP made provision to respond to these "Local Contaminants Concerns," thereby improving its relations with northerners.[114] This arrangement represented a negotiation between divergent definitions—in terms of government jurisdiction, or local per-ceptions—of contaminants. Aboriginal concerns, and scientific infor-mation, also compelled the federal government to expand its program beyond its territorial jurisdiction. Although Nunavik and Labrador are within provinces, not territories, and thus are not under direct federal jurisdiction, after 1997 the NCP was enlarged to include them, because people there also consume marine mammals and are thus exposed to contaminants.[115] Social (food consumption) and ecological (presence of contaminants) dimensions thus trumped federal jurisdictional limits in defining the geography of northern contaminants.

The NCP and Northern Research

The NCP represented a distinctively northern form of the challenge, often encountered in environmental affairs, of bringing together different ways of perceiving, experiencing, and knowing the environment. Knowledge from elsewhere about contaminants and their impacts on humans and other species was relevant to understanding them in the north. This knowledge then had to be combined with knowledge that was distinct-ive to the region, such as the special sensitivity of some northern species, and the distinctive ways in which northern people value and consume country foods. By forming new knowledge and practices, as well as new institutions and working relationships, the NCP had a variety of consequences for northern science and environmental history.

By providing an arena for the influence of regional factors, the NCP reinforced the distinctive character of northern contaminants research. Its status as a separate institution contributed to this—likely more than would have been the case if funding for contaminants research had been the responsibility of a national granting agency. The result was juxtapositions of research fields, such as atmospheric chemistry alongside community-based food studies—highly unlikely anywhere else. Explicit attention to community and ethical issues was also central to this regional character. The attention devoted to community decision-making and involvement in research and communication represented a new social contract for northern research. This encompassed an evolving view of the citizenship of northern Aboriginal peoples—from seeing them as objects of expert guidance to playing an active role in shaping expert knowledge and advice.

This distinctive character was evident to scientists from elsewhere. As one report noted, from "the perspective of a scientist trained in the South, the conduct of research in the North presents its own unique and unfamiliar challenges."[116] Contaminants research thus became a pathway by which scientists were encouraged to adopt specifically northern approaches to doing science. One scientist commented that first-hand experience of Dene culture had "changed the way that I implement southern standards in the North"—acknowledging that scientific standards could be situated geographically.[117] In effect, contaminants researchers and their community partners constructed a distinctively northern definition of "good science."[118] Their research thus evolved differently than other projects in the history of northern science that aimed to construct the north (often without success) as a "placeless" laboratory, producing knowledge valid everywhere.[119] In contrast, the NCP aimed to formulate contaminants advice that eschewed universal standards of risk and tolerable limits in favour of community-specific perceptions and values.

Contaminants research reinforced two other ideas about the north. One was quite obvious—that the region is linked to the rest of the world: physically, politically, intellectually. But even while this research drew attention to these connections, it also reinforced the view of the north as a distinctive space, as defined in terms of both scientific and Aboriginal perspectives on health, community, and the land. Among its distinctive characteristics was the practice of research in an inhabited landscape, with people that have a distinctive relationship with the environment. In this way, the NCP exemplified the more general evolution of northern environmental research. Although its practice of Aboriginal involvement in priority-setting and funding allocation has not been emulated in other research programs, the underlying social principle has: by 2010, and via Canadian involvement in the International Polar Year, northern environmental research had gained a strong social dimension focused on the wellbeing of northern peoples, integrated with northern communities, acknowledging local cultural beliefs and knowledge, drawing on concepts of vulnerability, adaptation, and resilience, and organized in ways that parallel how people relate to the environment. In areas ranging from climate change to caribou, as well as contaminants, environmental change was reinterpreted not only in terms of scientists' perspectives, but the perspectives of people and communities.[120]

Conclusion

One could imagine a linear history of northern contaminants in which these substances are produced and then deposited in the north, scientists describe their properties, and generate knowledge that is translated into action. But the surprises punctuating the actual history hint at a more complex tale. Repeatedly, the north was found not to be the pristine environment it was assumed to be. Instead, contaminants were discovered in places—the atmosphere, animals, human bodies—that confounded expectations. Their presence recalled the classic definition of pollution as misplaced matter—with "misplaced" being a category constructed on the basis of knowledge, experience, and expectations of what belongs in a place.[121]

Contaminant surprises stemmed in part from the persistent assumption that the Arctic was a pristine space, remote from sources of pollution. In reality, features of atmospheric transport mechanisms and of certain contaminants made the Arctic more contaminated than elsewhere. The complex and unpredictable behaviour of contaminants also contributed: in the Arctic, contaminants have tended to "break the rules" that scientists had formulated through research in more temperate regions. The structure of knowledge was also a factor. Disciplines influenced scientists' choices of where and what to study, and these choices rendered certain phenomena and places for contaminants visible, while obscuring others. In the early years of contaminants research, scientists focused on species that were ecologically significant or useful as indicators of environmental contamination; in doing so, species important as food were neglected. Scientists also produced discipline-specific descriptions of contaminants in the atmosphere or in ecosystems, and thereby limited their knowledge to what could be understood within that discipline, excluding other ways of knowing, including local knowledge.[122] Thus, scientists constructed a geography of knowing and unknowing in which partial knowledges of various kinds were distributed across disciplines and institutions.[123] Surprises resulted when observations failed to match the expectations formed by the regimes of perceptibility that these partial knowledges engendered.

These surprises had consequences. Scientists interpreted them as hints about how contaminants move and behave, and about the structure of nature itself. To align their research with this structure, they linked previously distinct bodies of knowledge: atmospheric chemistry and ecology; ecology and toxicology; toxicology and community health.[124] Scientists of different disciplines worked together, and also with non-scientists, including people in communities affected by contaminants. Spurred on by surprises, contaminants science exhibited a dynamic, evolving structure, with shifting regimes of perceptibility—an interdisciplinary research community that supported a view of the north as a single system linked to the rest of the planet.

Even as knowledge about contaminants evolved, so did the objects themselves: both contaminants, and ecological and human systems. As scientists, policymakers, public health experts, and Aboriginal people became involved, the actual substances they were talking about changed. In the 1950s, arctic haze was a visual phenomenon that could be described through observation. In the 1970s, it became an aerosol—a physical phenomenon describable in terms of chemical composition. Until the mid-1980s, organic contaminants were described in the context of an arctic ecosystem that apparently lacked humans. By the early 1990s, the food chain linking country foods and humans had become the focus of attention. These diverse and partial identities testify to the instability of contaminants as a category; it shifted and reshaped as different communities became involved in contaminant affairs, provoking surprise whenever its behaviour contradicted its assigned identity.[125]

Arctic contaminants were thus not simply "discovered." Their contemporary identity—as substances that link industrial regions to the north, and that have consequences for northern ecosystems, wildlife, and people—was constructed over several decades, and out of various forms of evidence and reasoning. Several communities were involved in constructing contaminants: scientific disciplines (each concerned with an aspect of contaminants: sources, behaviour in the atmosphere or in ecosystems, implications for human communities or bodies), forums for policy-making and international negotiations, and northern communities.

These complexities, formed by the interaction between the processes by which knowledge is formed and the materiality of the north, underline how northern contaminants present opportunities to consider the relations between the history of science (the evolution of scientific practices and knowledge) and environmental history (the production, distribution, and consequences of contaminants). Each shapes the other; our awareness of the presence of contaminants has depended on production of scientific knowledge, and this production has been, in part, in response to their presence. For several decades, observers have tracked the movement and transformation of contaminants into and within the north, their motivations ranging from Cold War strategic priorities, to scientific curiosity, to concerns regarding the health of arctic people and wildlife. Similarly, this paper has tracked the movement (and sometimes non-movement) and transformation of knowledge about contaminants, seeking clues as to how knowledge about the north is constructed and shared, and by whom. The shifting regimes of perceptibility produced through these activities resulted, at different times, in certain contaminants becoming highly visible, even as others did not—an outcome that implies a need for caution in using scientific evidence to reconstruct the material history of northern contaminants. The environmental history of contaminants science also illuminates how the places where science is done shape knowledge, and conversely, how doing science changes a place—through research, through the activities that accompany scientific work, and through the knowledge that results. This history therefore illustrates how science constitutes part of living in a place, and so has been central to the environmental and political history of northern Canada.

The presence of contaminants, and knowledge about them, has also influenced ideas about the north: its identity, and those of its inhabitants. Their presence eroded the identity of the Arctic as remote and pristine (although it remains oft invoked, notably when presenting the region to tourists). Instead, this knowledge helped form a distinctive identity of northern Canada as a landscape of exposure, in which contaminants are everywhere (albeit usually at very low concentrations). This identity was, in part, about vulnerability. The links between the region and the rest of the planet encouraged this view; as did features of arctic ecosystems: cold that slows breakdown of contaminants, and the fat-based metabolism of mammals that encourages their accumulation. Reliance on country foods was another factor, as was the longstanding view of the Arctic as a fragile environment. This identity was shared with the rest of the circumpolar region, becoming the basis for circumpolar institutions such as the Arctic Council, and for the notion of arctic citizenship.[126] Finally, contaminants reinforced perceptions of the intimate ties between Aboriginal people and their environments: they belonged there, because their wellbeing was tied to the wellbeing of their homeland.

Notes

1 Sheila Watt-Cloutier, "The Inuit Journey towards a POPs-Free World," in: David L. Downie and Terry Fenge, eds., *Northern Lights against POPs: Combatting Toxic Threats in the Arctic* (Montreal: McGill-Queen's University Press, 2003), 257; on the surprise felt in Inuit communities, see also Heather Myers and Chris Furgal, "Long-Range Transport of Information: Are Arctic Residents Getting the Message about Contaminants?", *Arctic* 59, no. 1 (2006): 49.

2 Marla Cone, *Silent Snow: The Slow Poisoning of the Arctic* (New York: Grove Press, 2005), 31.

3 David Kinloch and Harriet Kuhnlein, "Assessment of PCBs in Arctic Foods and Diets: A Pilot Study in Broughton Island, Northwest Territories, Canada," *Arctic Medical Research* 47, supp. 1 (1988): 159–62; Matthew Fisher, "Soviet, European pollution threatens health in Arctic," *Globe and Mail*, December 15, 1988.

4 Cone, *Silent Snow*, 39.

5 Michelle Murphy, *Sick Building Syndrome and the Problem of Uncertainty: Environmental Politics, Technoscience, and Women Workers* (Durham: Duke University Press, 2006).

6   Angela Creager applied a similar historiographical approach in her study of knowledge and practice relating to radioisotopes. Creager, *Life Atomic: A History of Radioisotopes in Science and Medicine* (Chicago: University of Chicago Press, 2013).

7   On the implications of perceptions of the north as remote and unspoiled, see Ken Coates, "The Discovery of the North: Towards a Conceptual Framework for the Study of Northern/Remote Regions," *Northern Review* 12–13 (1993–94): 15–43; and Sherrill E. Grace, *Canada and the Idea of North* (Montreal: McGill-Queen's University Press, 2002).

8   Mats Olsson, "Ecological Effects of Airborne Contaminants in Arctic Aquatic Ecosystems: A Discussion on Methodological Approaches," *Science of the Total Environment* 160–61 (1995): 619; see also Shawn G. Donaldson, et al., "Environmental Contaminants and Human Health in the Canadian Arctic," *Science of the Total Environment* 408 (2010): 5167.

9   See, for example, Robie W. Macdonald, et al., "Contaminants in the Canadian Arctic: 5 Years of Progress in Understanding Sources, Occurrence and Pathways," *Science of the Total Environment* 254 (2000): 94.

10   Theo Colborn, Dianne Dumanoski, and John P. Myers, *Our Stolen Future* (New York: Plume Books, 1996), 109.

11   David L. Downie and Terry Fenge, eds. *Northern Lights Against POPs: Combatting Toxic Threats in the Arctic* (Montreal: McGill-Queen's University Press, 2003).

12   Derek C. G. Muir, Rudy Wagemann, Barry T. Hargrave, David J. Thomas, David B. Peakall, and Ross J. Norstrom, "Arctic Marine Ecosystem Contamination," *Science of the Total Environment* 122 (1992): 5–134; Eva C. Voldner and Yi-Fan Li, "Global Usage of Selected Persistent Organochlorines," *Science of the Total Environment* 160–61 (1995): 201–10. For an analysis of contaminants that, unlike those that are the focus of this chapter, originate in sites within the Arctic (and, for comparison, the Antarctic), see John S. Poland, Martin J. Riddle, and Barbara A. Zeeb, "Contaminants in the Arctic and in the Antarctic: A Comparison of Sources, Impacts, and Remediation Options," *Polar Record* 39, no. 211 (2003): 369–83. The ambiguities of classifying some contaminants as "natural" is illustrated by the case of arsenic: as Keeling and Sandlos explain in their chapter, while it is found naturally in gold ore, in the Yellowknife region it became a significant problem only because it was released when the ore was roasted to release the gold.

13   Gregg Mitman, Michelle Murphy, and Chris Sellers, eds., *Landscapes of Exposure: Knowledge and Illness in Modern Environments*, Osiris 19 (Chicago: University of Chicago Press, 2004).

14   See, for example, Javier Auyero and Debora Swistun, "The Social Production of Toxic Uncertainty," *American Sociological Review* 73, no. 3 (2008): 357–79; Downie and Fenge, eds., *Northern Lights Against POPs*; Chris Furgal, Theresa D. Garvin, and Cynthia G. Jardine, "Trends in the Study of Aboriginal Health Risks in Canada," *International Journal*

of *Circumpolar Health* 69, no. 4 (2010): 322–32; Henrik Selin, *Global Governance of Hazardous Chemicals: Challenges of Multilevel Management* (Cambridge, MA: MIT Press, 2010); Charles Thrift, Ken Wilkening, Heather Myers, and Renata Raina, "The Influence of Science on Canada's Foreign Policy on Persistent Organic Pollutants," *Environmental Science and Policy* 12 (2009): 981–93.

15 Robert E. Kohler, "History of Field Science: Trends and Prospects," in *Knowing Global Environments*, ed. J. Vetter (New Brunswick, NJ: Rutgers University Press, 2010), 212–40; Jeremy Vetter, "Introduction," in *Knowing Global Environments*, 1–16. And as Marionne Cronin explains in her chapter in this volume, aviators also contributed to defining the atmosphere as part of the northern environment.

16 Craig E. Colten, "Waste and Pollution: Changing Views and Environmental Consequences," in *The Illusory Boundary: Environment and Technology in History*, ed. Martin Reuss and Stephen H. Cutcliffe (University of Virginia Press, 2010); Michael Egan, "Mercury's Web: Some Reflections on Following Nature across Time and Place," *Radical History Review* 107 (2010): 111–26; Toshihiro Higuchi, "Atmospheric Nuclear Weapons Testing and the Debate on Risk Knowledge in Cold War America, 1945–1963," in *Environmental Histories of the Cold War*, ed. John R. McNeill and Corinna R. Unger (Cambridge: Cambridge University Press, 2010): 301–22; Nancy Langston, *Toxic Bodies: Hormone Disruptors and the Legacy of DES* (New Haven: Yale University Press, 2010); Gregg Mitman, Michelle

Murphy, and Chris Sellers, "Introduction: A Cloud over History," in *Landscapes of Exposure*, 1–17; Jody A. Roberts and Nancy Langston, "Toxic Bodies/Toxic Environments: An Interdisciplinary Forum," *Environmental History* 13, no. 4 (2008): 629–35; Brett Walker, *Toxic Archipelago* (Seattle: University of Washington Press, 2010); Chris Sellers, "The Artificial Nature of Fluoridated Water: Between Nations, Knowledge, and Material Flows," in *Landscapes of Exposure*, 182–200; Tina Loo, "Disturbing the Peace: Environmental Change and the Scales of Justice on a Northern River," *Environmental History* 12, no. 4 (2007): 895–919; Chris Sellers and Joseph Melling, "Towards a Transnational Industrial-Hazard History: Charting the Circulation of Workplace Dangers, Debates and Expertise," *British Journal of the History of Science* 45, no. 3 (2012): 401–24.

17 Peter Adey, Ben Anderson, and Luis L. Guerrero, "An Ash Cloud, Airspace, and Environmental Threat," *Transactions of the Institute of British Geographers* 36 (2011): 338–43; Diarmid A. Finnegan, "The Spatial Turn: Geographical Approaches in the History of Science," *Journal of the History of Biology* 41 (2008): 369–88; David N. Livingstone, "Landscapes of Knowledge," in *Geographies of Science*, ed. Peter Meusburger, David N. Livingstone, and Heike Jöns (Springer, 2010), 3–22.

18 Ronald E. Doel, "Constituting the Postwar Earth Sciences: The Military's Influence on the Environmental Sciences in the USA after 1945," *Social Studies of Science* 33, no. 5 (2003): 635–66; P. Whitney Lackenbauer and Matthew Farish,

"The Cold War on Canadian Soil: Militarizing a Northern Environment," *Environmental History* 12, no. 4 (2007): 921–50; Stephen Bocking, "A Disciplined Geography: Aviation, Science, and the Cold War in Northern Canada, 1945–1960," *Technology and Culture* 50 (2009): 320–45.

19 Ken Wilkening, "Science and International Environmental Non-regimes: The Case of Arctic Haze," *Review of Policy Research* 28, no. 2 (2011): 125–48.

20 Fred C. Meier, with Charles A. Lindbergh, "Collecting Micro-Organisms from the Arctic Atmosphere," *Scientific Monthly*, January 1935, 5–20.

21 J. Murray Mitchell, "Visual Range in the Polar Regions with Particular Reference to the Alaskan Arctic," *Journal of Atmospheric and Terrestrial Physics* 57 (1956), special supp.: 195–211. These "weather reconnaissance missions" fulfilled a variety of military purposes relating to aircraft and missile performance; see Paul N. Edwards, "Meteorology as Infrastructural Globalism," in *Global Power Knowledge: Science and Technology in International Affairs*, Osiris 21, ed. John Krige and Kai-Henrik Barth (Chicago: University of Chicago Press, 2006): 243–44. It is also now known that at this time the USAF was actively involved in sampling the arctic atmosphere for traces of radioactivity as a means of detecting Soviet atomic tests; this may have been another purpose of these flights. Jacob Hamblin, *Arming Mother Nature: The Birth of Catastrophic Environmentalism* (Oxford University Press, 2013), 85–91.

22 Wilkening, "Case of Arctic Haze," 130–31.

23 Quoted in Peter J. Usher, Maureen Baikie, Marianne Demmer, Douglas Nakashima, Marc G. Stevenson, and Mark Stiles, *Communicating about Contaminants in Country Food: The Experience in Aboriginal Communities* (Ottawa: Inuit Tapirisat of Canada, 1995), 176.

24 Bocking, "Disciplined Geography."

25 Laura A. Bruno, "The Bequest of the Nuclear Battlefield: Science, Nature, and the Atom During the First Decade of the Cold War," *Historical Studies in the Physical Sciences* 33, no. 2 (2003): 237–60.

26 Higuchi, "Atmospheric Nuclear Weapons Testing"; see also Asker Aarkrog, "Radioactivity in Polar Regions—Main Sources," *Journal of Environmental Radioactivity* 25 (1994): 21–35.

27 Doel, "Constituting the Postwar Earth Sciences."

28 Eville Gorham, "A Comparison of Lower and Higher Plants as Accumulators of Radioactive Fall-Out," *Canadian Journal of Botany* 37 (1959): 329.

29 William O. Pruitt, "A New 'Caribou Problem,'" *The Beaver*, Winter 1962, 24–25. Project Chariot, analyzed and publicized by Barry Commoner and other environmentalists, also encouraged perceptions of nuclear contamination as not only a global, but a specifically local, northern issue, which could not be understood only in terms of knowledge produced elsewhere (such as at the Nevada nuclear testing site); see also Barry Commoner, M. W. Friedlander, and Eric Reiss, "Project Chariot," *Science* 134, no. 3477 (1961): 495–96, 499–500;

Harvey E. Palmer, Wayne C. Hanson, Bobby I. Griffin, and William C. Roesch, "Cesium-137 in Alaskan Eskimos," *Science* 142, no. 3588 (1963): 64–66; D. G. Watson, W. C. Hanson, and J. J. Davis, "Strontium-90 in Plants and Animals of Arctic Alaska, 1959–61," *Science* 144, no. 3621 (1964): 1005–9; Peter A. Coates, *The Trans-Alaska Pipeline Controversy: Technology, Conservation, and the Frontier* (University of Alaska Press, 1993): 111–33; and, especially, Dan O'Neill, *The Firecracker Boys* (New York: St. Martin's Griffin, 1994). Another link between radioactive materials and northern scientific and military affairs resulted from interest in human performance and adaptation to the arctic environment. Alaska and its Aboriginal inhabitants were together viewed as a "natural laboratory" for the study of these topics; accordingly, iodine-131 was administered to humans and used to trace physiological functions. By the 1990s, these and many other Cold War studies of humans that used radioactive materials had become highly controversial; Matthew Farish, "The Lab and the Land: Overcoming the Arctic in Cold War Alaska," *Isis* 104, no. 1 (2013): 1–29.

30 V. K. Mohindra, "Cesium 137 Burdens in the Canadian North," *Acta Radiologica Therapy Physics Biology* 6 (1967): 481–90. A follow-up study was conducted between 1967 and 1969; Bliss L. Tracy, Gary H. Kramer, Jan M. Zielinski, and H. Jiang, "Radiocesium Body Burdens in Residents of Northern Canada from 1963–1990," *Health Physics* 72, no. 3 (1997): 431–42.

31 Colin R. Macdonald, Brett T. Elkin, and Bliss L. Tracy, "Radiocesium in Caribou and Reindeer in Northern Canada, Alaska and Greenland from 1958 to 2000," *Journal of Environmental Radioactivity* 93 (2007): 1–25. However, fallout from this era, including cesium-137, persists in northern soils and vegetation, with traces released into the atmosphere each summer by forest fires; Arctic Monitoring and Assessment Programme, *AMAP Assessment 2009: Radioactivity in the Arctic* (Oslo, 2010), 27.

32 Ralph H. Lutts, "Chemical Fallout: Rachel Carson's Silent Spring, Radioactive Fallout, and the Environmental Movement," *Environmental Review* 9, no. 3 (1985): 210–25.

33 George M. Woodwell, "Toxic Substances and Ecological Cycles," *Scientific American* 216, no. 3 (1967): 24–31; Creager, *Life Atomic*, 392–93.

34 Alan V. Holden, "Monitoring Organochlorine Contamination of the Marine Environment by the Analysis of Residues in Seals," in *Marine Pollution and Sea Life*, ed. M. Ruivo (Surrey: Fishing News Books, 1970): 266–72.

35 Daniel D. Berger, Daniel W. Anderson, James D. Weaver, and Robert W. Risebrough, "Shell Thinning in Eggs of Ungava Peregrines," *Canadian Field-Naturalist* 84 (1970): 265–67.

36 Richard F. Addison and Paul F. Brodie, "Occurrence of DDT Residues in Beluga Whales (Delphinapterus leucas) from the Mackenzie Delta, N.W.T.," *Journal of the Fisheries Research Board of Canada* 30 (1973): 1733–36.

37 J. Alexander Burnett, *A Passion for Wildlife: The History of the Canadian Wildlife Service* (Vancouver: UBC Press, 2003).

38 Gerald W. Bowes and Charles J. Jonkel, "Presence and Distribution of Polychlorinated Biphenyls (PCB) in Arctic and Subarctic Marine Food Chains," *Journal of the Fisheries Research Board of Canada* 32 (1975): 2111–23.

39 Anne Fikkan, Gail Osherenko, and Alexander Arikainen, "Polar Bears: The Importance of Simplicity," in *Polar Politics: Creating International Environmental Regimes*, ed. Oran R. Young and Gail Osherenko (Ithaca: Cornell University Press, 1993), 96–151.

40 Jost Heintzenberg, "Arctic Haze: Air Pollution in Polar Regions," *Ambio* 18, no. 1 (1989): 50–55; see also Wilkening, "Case of Arctic Haze."

41 Kenneth A. Rahn and Glenn E. Shaw, "Sources and Transport of Arctic Pollution Aerosol: A Chronicle of Six Years of ONR Research," *NR Reviews* (1982): 5.

42 John E. Stout, Andrew Warren, and Thomas E. Gill, "Publication Trends in Aeolian Research: An Analysis of the Bibliography of Aeolian Research," *Geomorphology* 105 (2009): 6–17; Ken Wilkening, "Intercontinental Transport of Dust: Science and Policy, pre-1800s to 1967," *Environment and History* 17 (2011): 313–39.

43 Matthias Dörries, "The Politics of Atmospheric Sciences: 'Nuclear Winter' and Global Climate Change," in *Klima*, Osiris 26, ed. James R. Fleming and Vladimir Jankovic (Chicago: University of Chicago Press, 2011), 202.

44 Wilkening, "Intercontinental Dust."

45 Rahn and Shaw, "Sources and Transport."

46 Rahn and Shaw, "Sources and Transport," 21.

47 Marvin S. Soroos, "The Odyssey of Arctic Haze: Toward a Global Atmospheric Regime," *Environment* 34, no. 10 (1992): 6–11, 25–27.

48 Wolfgang E. Raatz, "Meteorological Conditions Over Eurasia and the Arctic Contributing to the March 1983 Arctic Haze Episode," *Atmospheric Environment* 19, no. 12 (1985): 2121–26; Wilkening, "Case of Arctic Haze," 134–35.

49 While haze was evidently a circumpolar phenomenon, the research describing it was not; the distribution of air-sampling stations, with good coverage of North America and Scandinavia, but no sites in the Soviet half of the Arctic, demonstrated the continuing influence of the Cold War on contaminants research. See map in Brynjulf Ottar, "The Transfer of Airborne Pollutants to the Arctic Regions," *Atmospheric Environment* 15, no. 8 (1981): 1442.

50 See, for example, Bernard Stonehouse, ed., *Arctic Air Pollution* (Cambridge: Cambridge University Press, 1986); Heintzenberg, "Arctic Haze." One exception, however, was apparently the study of the consequences of nuclear winter scenarios—a prominent issue in the 1980s; see Russell C. Schnell, "The International Arctic Gas and Aerosol Sampling Program," in Stonehouse, ed., *Arctic Air Pollution*, 135–42; and Dorries, "Transfer of Airborne Pollutants."

51 David W. Schindler, Karen A. Kidd, Derek C. G. Muir, and W. Lyle Lockhart, "The Effects of Ecosystem Characteristics on Contaminant Distribution in Northern Freshwater Lakes,"

Science of the Total Environment 160–61 (1995): 1.

52   Wilkening, "Case of Arctic Haze." Future action on haze is most likely to be motivated by its relevance to other, larger issues, especially climate change; see Tracey Holloway, Arlene Fiore, and Meredith G. Hastings, "Intercontinental Transport of Air Pollution: Will Emerging Science Lead to a New Hemispheric Treaty?", *Environmental Science and Technology* 37 (2003): 4535–42.

53   Henrik Selin, "From Regional to Global Information: Assessment of Persistent Organic Pollutants," in *Global Environmental Assessments: Information and Influence*, ed. Ronald B. Mitchell, William C. Clark, David W. Cash, and Nancy M. Dickson (Cambridge, MA: MIT Press, 2006).

54   Colborn, et al., *Our Stolen Future*; Langston, *Toxic Bodies*.

55   Linda Nash, *Inescapable Ecologies: A History of Environment, Disease, and Knowledge* (Berkeley: University of California Press, 2006).

56   Michael Oehme and Sobia Mano, "The Long-Range Transport of Organic Pollutants to the Arctic," *Fresenius' Journal of Analytical Chemistry* 319 (1984): 141–46.

57   Ross J. Norstrom, Mary Simon, Derek C. G. Muir, and Ray E. Schweinsburg, "Organochlorine Contaminants in Arctic Marine Food Chains: Identification, Geographical Distribution, and Temporal Trends in Polar Bears," *Environmental Science and Technology* 22 (1988): 1063–71.

58   Max J. Dunbar, "Stability and Fragility in Arctic Ecosystems," *Arctic* 26, no. 3 (1973): 179–85; Edward D.

Goldberg, "Synthetic Organohalides in the Sea," *Proceedings of the Royal Society of London*, ser. B 189 (1975): 277–89.

59   Ottar, "The Transport of Airborne Pollutants," 1444, 1445.

60   Vera Alexander, "The Influence of the Structure and Function of the Marine Food Web on the Dynamics of Contaminants in Arctic Ocean Ecosystems," *Science of the Total Environment* 160–61 (1995): 593–603; Brett T. Elkin and Ray W. Bethke, "Environmental Contaminants in Caribou in the Northwest Territories, Canada," *Science of the Total Environment* 160–61 (1995): 307–21.

61   Frank Wania and Donald Mackay, "Global Fractionation and Cold Condensation of Low Volatility Organochlorine Compounds in Polar Regions," *Ambio* 22, no. 1 (1993): 10–18; Frank Wania and Donald Mackay, "A Global Distribution Model for Persistent Organic Chemicals," *Science of the Total Environment* 160–61 (1995): 211–32; Donald Mackay and Frank Wania, "Transport of Contaminants to the Arctic: Partitioning, Processes and Models," *Science of the Total Environment* 160–61 (1995): 25–38.

62   Arctic Monitoring and Assessment Programme, *Arctic Pollution 2009* (Oslo: AMAP, 2009): 2–4.

63   Kinloch and Kuhnlein, "Assessment of PCBs."

64   Eric Dewailly, Albert Nantel, Jean-P. Weber, and François Meyer, "High Levels of PCBs in Breast Milk of Inuit Women from Arctic Quebec," *Bulletin of Environmental Contamination and Toxicology* 43 (1989): 641–46; his reaction to this result is described in Cone, *Silent Snow*, 29–32.

65   Fisher, "Soviet, European Pol-
      lution"; Usher et al., provide a
      detailed narrative and discussion
      of implications of the Broughton
      Island and Nunavik episodes in
      Communicating About Contami-
      nants in Country Food, 61–70.

66   Chris M. Furgal, Stephanie Powell,
      and Heather Myers, "Digesting the
      Message about Contaminants and
      Country Foods in the Canadian
      North: A Review and Recommen-
      dations for Future Research and
      Action," *Arctic* 58, no. 2 (2005):
      104. See also Bruce E. Johansen,
      "The Inuit's Struggle with Dioxins
      and Other Organic Pollutants,"
      American Indian Quarterly 26,
      no. 3 (2002): 479–90; Joanna
      Kafarowski, "Contaminants in the
      Circumpolar North: The Nexus
      Between Indigenous Reproductive
      Health, Gender and Environmen-
      tal Justice," *Pimatisiwin* 2, no. 2
      (2004): 39–52.

67   Terry Fenge, "POPs and Inuit:
      Influencing the Global Agenda,"
      in *Northern Lights against POPs:*
      Combatting Toxic Threats in the
      Arctic, ed. David L. Downie and
      Terry Fenge (Montreal: Mc-
      Gill-Queen's University Press,
      2003), 200.

68   Terry Fenge, personal
      communication.

69   Ken Wilkening and Charles
      Thrift, "Canada's Foreign Policy
      on Persistent Organic Pollutants:
      The Making of an Environmental
      Leader," in *Environmental Change*
      and Foreign Policy: Theory and
      Practice, ed. Paul Harris (New
      York: Routledge, 2009), 136–52.

70   Kathryn Harrison alludes to
      this phenomenon of a northern
      issue becoming a national issue
      when she notes that the presence

      of contaminants in breast milk
      received national attention in
      Canada only with reference to Inuit
      women, not the general population;
      Kathryn Harrison, "Too Close to
      Home: Dioxin Contamination of
      Breast Milk and the Political Agen-
      da," in *This Elusive Land: Women*
      and the Canadian Environment, ed.
      Melody Hessing, Rebecca Raglon
      and Catriona Sandilands (Vancou-
      ver: UBC Press, 2005), 213–42.

71   Pierre Ayotte, et al., "Arctic Air
      Pollution and Human Health:
      What Effects Should be Expected?",
      Science of the Total Environment
      160–61 (1995): 529–37.

72   Terry F. Bidleman, G. W. Patton,
      M. D. Walla, Barry T. Hargrave, W.
      Peter Vass, Paul Erickson, Brian
      Fowler, Valerie Scott, and Dennis
      J. Gregor, "Toxaphene and Other
      Organochlorines in Arctic Ocean
      Fauna: Evidence for Atmospheric
      Delivery," *Arctic* 42, no. 4 (1989):
      307–13.

73   Harriet Kuhnlein, Laurie H. M.
      Chan, Grace Egeland, and Olivier
      Receveur, "Canadian Arctic Indig-
      enous Peoples, Traditional Food
      Systems, and POPs," in *Northern*
      Lights against POPs: Combatting
      Toxic Threats in the Arctic, ed.
      David L. Downie and Terry Fenge
      (Montreal: McGill-Queen's Univer-
      sity Press, 2003), 22–40.

74   Russel Shearer and Siu-Ling Han,
      "Canadian Research and POPs: The
      Northern Contaminants Program,"
      in *Northern Lights against POPs:*
      Combatting Toxic Threats in the
      Arctic, ed. David L. Downie and
      Terry Fenge (Montreal: McGill-
      Queen's University Press, 2003),
      41–59; Thrift et al., "Influence of
      Science," 2009; Wilkening and
      Thrift, "Canada's Foreign Policy."

75 Shearer and Han, "Canadian Research and POPs" provides a useful overview of the NCP.

76 John D. O'Neil, Brenda Elias, and Annalee Yassi, "Poisoned Food: Cultural Resistance to the Contaminants Discourse in Nunavik," *Arctic Anthropology* 34, no. 1 (1997): 29–40; quote on p. 32. See also Joslyn Cassady, "A Tundra of Sickness: the Uneasy Relationship between Toxic Waste, TEK, and Cultural Survival," *Arctic Anthropology* 44, no. 1 (2007): 87–98.

77 Quoted in Usher et al., *Communicating About Contaminants in Country Food*, 160.

78 Myers and Furgal, "Long-Range Transport"; O'Neil et al., "Poisoned Food."

79 Usher, et al., *Communicating About Contaminants in Country Food*, provides a detailed overview of views regarding contaminants and northern communities as they had developed by the mid-1990s.

80 Nash, *Inescapable Ecologies*.

81 Gregg Mitman, *Breathing Space: How Allergies Shape Our Lives and Landscapes* (New Haven: Yale University Press, 2006).

82 Nash, *Inescapable Ecologies*, 123.

83 Peter Kulchyski and Frank J. Tester, *Kiumajut (Talking Back): Game Management and Inuit Rights, 1900–70* (Vancouver: UBC Press, 2007); John Sandlos, *Hunters at the Margin: Native People and Wildlife Conservation in the Northwest Territories* (Vancouver: UBC Press, 2007).

84 O'Neil et al., "Poisoned Food."

85 Furgal et al., "Digesting the Message."

86 Northern Contaminants Program, *Guidelines for Responsible Research.*

87 Northern Contaminants Program, *Canadian Arctic Contaminants Assessment Report II: Knowledge in Action* (Ottawa, 2003).

88 Jens C. Hansen, Lars-Otto Reiersen, and Simon Wilson, "Arctic Monitoring and Assessment Programme (AMAP); Strategy and Results with Focus on the Human Health Assessment under the Second Phase of AMAP, 1998–2003," *International Journal of Public Health* (2002): 316.

89 Furgal et al., "Digesting the Message"; Myers and Furgal, "Long-Range Transport."

90 Terry Fenge, personal communication, 23 April 2014.

91 For an overview of NCP research principles and results, see Donaldson et al., "Environmental Contaminants and Human Health."

92 This was evident in its organizational chart; see Shearer and Han, "Canadian Research and POPs," 53.

93 Kuhnlein et al., "Canadian Arctic Indigenous Peoples."

94 Eric Dewailly and Chris Furgal, "POPs, the Environment, and Public Health," in *Northern Lights against POPs: Combatting Toxic Threats in the Arctic*, ed. David L. Downie and Terry Fenge (Montreal: McGill-Queen's University Press, 2003), 3–21.

95 On the creation by research communities of new research objects, see, for example, Sheila Jasanoff, "Ordering Knowledge, Ordering Society," in *States of Knowledge: The Co-Production of Science and*

Social Order, ed. Sheila Jasanoff (New York: Routledge, 2004), 13–45.

96 The term appeared in the Arctic Monitoring and Assessment Programme's report *Arctic Pollution Issues: A State of the Arctic Environment Report* (Oslo: AMAP, 1997), 177; see also Eric Dewailly, "Canadian Inuit and the Arctic Dilemma," *Oceanography* 19, no. 2 (2006): 88–89.

97 David Kinloch et al., "Inuit Foods and Diet: A Preliminary Assessment of Benefits and Risks," *Science of the Total Environment* 122 (1992): 247–78; Furgal et al., "Digesting the Message."

98 Brian Wheatley, "A New Approach to Assessing the Effects of Environmental Contaminants on Aboriginal Peoples," Paper presented to 9th International Congress on Circumpolar Health, Reykjavik, Iceland, 20–25 June 1993; Peter J. Usher, "Socio-Economic Effects of Elevated Mercury Levels in Fish on Sub-Arctic Native Communities," in *Contaminants in the Marine Environment of Nunavik*, Proceedings of the Conference, Montreal, 12–14 September 1990 (Quebec City: Université Laval, 1992), 45–50; Usher et al., *Communicating about Contaminants in Country Food*, 30–53.

99 NCP, *Knowledge in Action*, 3.

100 Dewailly and Furgal, "POPs, the Environment, and Public Health."

101 Dewailly and Furgal, "POPs, the Environment, and Public Health," 13.

102 Kuhnlein et al., "Canadian Arctic Indigenous Peoples," 37.

103 Indian and Northern Affairs Canada, *Arctic Environmental Strategy* (Ottawa, 1991).

104 Jasanoff, "Ordering Knowledge."

105 On the emergence of post-Cold War Arctic cooperation, see Terry Fenge, "The Arctic Council: Past, Present, and Future Prospects with Canada in the Chair From 2013 to 2015," *Northern Review*, Fall 2013, 37.

106 Fenge, "POPs and Inuit," 194.

107 Fenge, "POPs and Inuit," 199; Noelle Eckley, "Traveling Toxics: The Science, Policy, and Management of Persistent Organic Pollutants," *Environment* 43, no. 7 (2001): 24–36.

108 David L. Downie, "Global POPs Policy: The 2001 Stockholm Convention on Persistent Organic Pollutants," in *Northern Lights against POPs: Combatting Toxic Threats in the Arctic*, ed. David L. Downie and Terry Fenge (Montreal: McGill-Queen's University Press, 2003): 133–59.

109 Selin, "From Regional to Global Information."

110 Thrift et al., "Influence of Science."

111 Shearer and Han, "Canadian Research and POPs."

112 On scale jumping, see Loo, "Disturbing the Peace," 898.

113 Clive Tesar, "Confirming the Effects of Contaminants on Inuit Children: An Interview with Eric Dewailly," *Northern Perspectives* 26, no. 1 (2000): 18.

114 NCP, *Knowledge in Action*, 32–33.

115 Shearer and Han, "Canadian Research and POPs," 50.

116 NCP, *Knowledge in Action*.

117 NCP, *Knowledge in Action*, 35.

118 On the construction of "good science" in another arena of toxicological research, see Kim Fortun and Mike Fortun, "Scientific Imaginaries and Ethical Plateaus in Contemporary U.S. Toxicology," *American Anthropologist* 107, no. 1 (2005): 43–54.

119   Richard C. Powell, "'The Rigours of an Arctic Experiment': The Precarious Authority of Field Practices in the Canadian High Arctic, 1958–1970," *Environment and Planning A* 39 (2007): 1794–811; on the notion of the "placeless" laboratory in the field, see Robert E. Kohler, *Landscapes and Labscapes* (Chicago: University of Chicago Press, 2002).

120   Brenda Parlee and Chris Furgal, "Well-Being and Environmental Change in the Arctic: A Synthesis of Selected Research for Canada's International Polar Year Program," *Climatic Change* 115, no. 1 (2012): 13–34; Marybeth Long Martello, "Global Change Science and the Arctic Citizen," *Science and Public Policy* 31, no. 2 (2004): 107–15.

121   Mary Douglas, *Purity and Danger: An Analysis of Concepts of Pollution and Taboo* (London: Routledge and Kegan Paul, 1966).

122   O'Neil et al., "Poisoned Food."

123   On the disciplinary geography of scientific knowledge, see, for example, Scott Kirsch, "Harold Knapp and the Geography of Normal Controversy: Radioiodine in the Historical Environment," in *Landscapes of Exposure*, 167–81.

124   Len A. Barrie, Dennis Gregor, Barry Hargrave, R. Lake, Derek Muir, Russel Shearer, B. Tracey, and T. Bidleman, "Arctic Contaminants: Sources, Occurrence and Pathways," *Science of the Total Environment* 122 (1992): 1–74.

125   In this discussion, I am drawing on the notion of "historical ontology"—that things emerge and become stabilized in relation to knowledge about those things; see Mitman et al., "Introduction"; Sellers, "Artificial Nature."

126   Martello, "Global Change Science."

13

Climate Anti-Politics: Scale, Locality, and Arctic Climate Change

Emilie Cameron

Introduction

Although the study of arctic climate change over the last several decades has been predominantly associated with the natural sciences, the relations between climate, people, and arctic environments have also long preoccupied scholars in the social sciences and humanities, including environmental historians and geographers. And while it is largely the dramatic physical evidence of global warming at the poles that has made the Arctic synonymous with climate change, it is difficult to identify a historical moment when the Arctic has *not* been associated (among non-residents) with its supposedly extreme climatic features and their impacts on human life and livelihoods. As Liza Piper observes, summarizing a large body of historical, anthropological, and archaeological scholarship, "the most significant arrival of people in the Circumpolar North and their movements across the region historically correspond to climatic changes."[1] Piper's account is far from deterministic; she traces the movements of Norse,

Thule, and Inuit, but also Basque, Siberian, French, English, and Russian peoples (among others) across arctic lands and seas over many centuries as they were articulated not just with climatic changes, but also with social, political, and economic shifts. While generations of anthropological and archaeological scholarship has focused on the capacity for northern Indigenous peoples to adapt to and thrive within changing arctic environmental conditions, Piper also identifies links between climate and the geographies of arctic whaling, fur trading, infectious disease, agriculture, and non-renewable resource extraction. Making a living in the Arctic, she suggests, whether in subsistence terms or in industrial, capitalist, or otherwise extractive terms, is deeply tied to climate.

This longer and more complex history of political, economic, social, cultural, and environmental change in the Arctic offers important insights into contemporary struggles with climatic change, but has had limited impact on arctic climate change research and policy development.[2] Although there is a great deal of climate-related research being undertaken in arctic communities, keen interest in understanding and mitigating the human impacts of climate change in the region, and significant research investigating Indigenous and traditional knowledges of climate history and contemporary climatic changes, the orientation of most (although not all) climate research is toward an anticipated future.[3] That is, although past experience with environmental change informs assessments of the "human dimensions of climate change" in the region, it tends to do so only insofar as past experiences aid in identifying current and future impacts of climatic change on the region's largely Indigenous population, and in developing policy that might facilitate adaptation to a rapidly shifting set of environmental conditions.[4]

This focus on present and future impacts responds, in part, to the ways in which human impacts are framed in the Arctic Climate Impact Assessment (ACIA) and the Intergovernmental Panel on Climate Change (IPCC) reports on climatic change, but also to decades of Inuit political mobilization, particularly the work of the Inuit Circumpolar Council, demanding that the international community take seriously the ways in which climate change impacts Inuit lives and livelihoods. Efforts to safeguard the "right to be cold," as Sheila Watt-Cloutier frames it, are at the heart of such political mobilizations. As scholars and policymakers have taken up the call to attend to the risks climate change poses to Inuit, however, there has been considerably less focus on pan-Inuit rights than on localized capacities to adapt to localized changes, changes that are already palpable. While many Inuit leaders and organizations have supported this pragmatic turn toward adaptation, they have done so alongside continuous calls both to address the historical and ongoing causes of climatic change and to attend to the interconnections between climate change and geopolitical struggles to control access to arctic resources and transportation routes.[5] But in spite of these calls, the translocal, historical, deeply political dimensions of climate change have been less prominent in the literature than assessments of local understandings of, and vulnerabilities to it. It is not only particular temporal scales that dominate arctic climate research, then, but also particular geographic scales, notably a particular understanding of the "local."

Interest in the local characterizes the climate change literature more broadly, where Brace and Geoghegan identify a turn toward a more "grounded and localized understanding of climate change" and a kind of consensus "that environmental knowledges, including those surrounding climate change, need to be understood on a local scale."[6] Indeed, scholars working in diverse traditions and with a range of objectives have increasingly turned to the local as a site for understanding climatic change. Duerden highlights the importance of understanding locally specific articulations of Arctic climatic change, noting that "human activity is highly localized, and impacts and responses will be conditioned by local geography and a range of endogenous factors, including demographic trends, economic complexity, and experience with 'change' in a broad sense."[7] Ford and colleagues similarly note the importance of in-depth, place-based, case study research for characterizing climate exposure and vulnerability in specific northern communities.[8] Riedlinger and Berkes point to the importance of locally specific, traditional knowledge and local expertise for understanding climate history, and for developing research questions, adaptive strategies, and monitoring plans.[9] And Hulme emphasizes the importance of understanding "the multiple meanings of climate change in diverse cultures" so as to "create new entry points for policy innovation," and highlights the role of the interpretative social sciences, arts, and humanities in complementing positivist scientific engagements with climatic change.[10]

Whether emphasizing its methodological, ontological, or epistemological merits, it would seem that the local has become a privileged site for understanding and responding to climate change in the region. But as geographers have long argued, scale is itself a social and deeply political frame through which to understand any issue.[11] That is, any attribution of scale makes apparent certain processes and relations and renders others illegible. To speak of climate change as a "global" issue with "local" dimensions (as has become the norm among arctic scholars and policymakers) is thus to make climate change knowable, analyzable, and amenable to intervention in certain ways and not others. Similarly, as a number of climate historians have argued, to speak of climate change as a distinctly contemporary phenomenon is to overlook the varied and complex histories of human relations with climate around the world, and the lessons they offer us in the present.[12]

How, then, might critical geographers and historians intervene in the study of arctic climate change? I argue here that a significant contribution can be made not just through study of past relations with climate (which has tended to preoccupy historical geographers and environmental historians interested in climatic change), but also by bringing key theoretical and conceptual resources to bear on climate scholarship and questioning some of the temporalities and spatialities underpinning contemporary research. My focus is on querying the intellectual and political implications of scaling arctic climate change. Drawing on critical interrogations of scale in geography and beyond, I consider the limitations of engaging "local knowledge" at the local scale in efforts to understand the effects of—and appropriate responses to—a "global" phenomenon like climate change. Rather than make a case against the local per se, however, I attempt to tease out some of the multiple locals at play in various approaches to climate change, and trace some of the unintended effects that can flow from engaging the local as a self-evident site, level, or method of analysis. Against tendencies to take the local for granted, I argue that environmental historians and geographers are well placed to clarify and expand the kinds of locals that are at stake in climate research and the histories informing their production and legibility. Next, I explore some alternative engagements with locality and their implications for understanding arctic climate change. I consider three potential ways in which climate scholars might engage the local differently: a) tracing the genealogies of academic, government, and corporate engagement with "local knowledge"; b) developing a more critical understanding of the ways in which Indigeneity, locality, and tradition become politically consequential; and c) localizing the geographies of the "global" dimensions of climatic change. Alert to the capacity for academic research to buttress—rather than dismantle—relations of exploitation and domination, these lines of intervention are offered not as a straightforward research agenda, but rather as possible sites for rethinking and rematerializing the interweaving of knowledge and power in the contemporary north.

Scale and Climate Change

Scale has long preoccupied human geographers. One of the most important insights to emerge through the 1980s and 1990s, however, was that scale is social, relational, and political; there is nothing natural or inevitable about the scale at which different issues are analyzed, managed, or known.[13] Rather than assume that invocations of the "local," "national," and "global" correspond with material reality, geographers began to turn their attention toward the ways in which scale produces and naturalizes social, economic, and political difference.

There has been a range of work undertaken within this broad literature. Although some understand scale to be wholly socially constructed, others conceive of scale as social and political in the sense that struggles to advance various social and political causes are inevitably shaped by the scale at which they are lived, and the scale at which they become politically legible.[14] For the former, scale is approached "not as an ontological structure which 'exists,' but as an epistemological one—a way of knowing and apprehending,"[15] and understandings of scale as a "nested hierarchy of differentially sized and bounded spaces"[16] are rejected on both intellectual and political grounds. As J. K. Gibson-Graham argues, "the global and the local are not things in themselves, nor are 'globalness' and 'localness' inherent qualities of an object. They are interpretive frames ... inherently empty of content."[17] For the latter group, scale is understood to be an abstraction but a necessary one, and rather than wholly reject understandings of scale as nested hierarchies or levels, or the attribution of particular bodies and events to particular scales, the emancipatory potential of scale is emphasized. Thus, for Neil Smith, "geographic scale is political precisely because it is the technology according to which events and people are, quite literally, 'contained in space.' Alternatively, scale demarcates the space or spaces people 'take up' or make for themselves. In scale, therefore, are distilled the oppressive and emancipatory possibilities of space, its deadness but also its life."[18]

Although they are informed by and imply slightly different political and intellectual commitments, both perspectives share a concern with the ways in which the local is mobilized to make sense of various people, practices, and places. For both, the local is neither self-evident nor neutral, and the framing of particular issues or struggles as predominantly "local" demands critical interrogation. Both also conceive of the local as having political potential, but in different ways and on different terms. Smith, for example, emphasizes the importance of connecting diverse local struggles into broader political movements, and thereby "jumping scales."[19] Implicit in such a formulation is an understanding of the local as more small, isolated, and politically limited than "higher" scales of activity. By contrast, those who reject hierarchical conceptualizations of scale do so, in part, as part of a political commitment to the local and a refusal of the notion that struggles must be scaled up to be effective. Marston et al. argue, for example, that, "the local-to-global conceptual architecture intrinsic to hierarchical scale carries with it presuppositions that can delimit entry points into politics—and the openness of the political—by pre-assigning to it a cordoned register for resistance."[20] They argue, in effect, that hierarchical understandings of scale and political struggle shore up an understanding of power as sweeping, pervasive, and hegemonic—a form of power that can only be effectively countered by equally expansive, "globalized" forms of resistance—and in so doing the very real political possibilities of the local and the specific are overlooked.

I will return to the implications of these slightly different conceptualizations of the politics of scale for understanding climatic change, but it is important to note that most (although not all) arctic climate research engages the local as a self-evident, contained site of analysis, not as a product of social and political struggle. A large body of scholarship considers the ways in which local landscapes, infrastructure, health, and other dimensions of wellbeing in the Arctic are affected by climatic change, and how local knowledges might be brought to bear on these localized changes.

This line of scholarship has identified important dimensions of climatic change in the region. But as I have argued elsewhere,[21] it has also tended to obscure some other, very pressing dimensions of climatic change. Insofar as the local is equated with the Indigenous, and the Indigenous, in turn, is equated with an externally defined understanding of the "traditional," localized studies of the human dimensions of climate change have tended to emphasize hunting, land travel, and traditional knowledge over other "local" dimensions of climate change, particularly climate-related transformations in resource extraction, shipping, and sovereignty. These latter transformations are of pressing concern to Indigenous northerners, but tend not to register as local dimensions of climate change in the literature, or as issues to which local or traditional knowledges might be brought to bear. Within the large body of literature considering the human dimensions of climate change in the Arctic, there has to date been no study focused on how, for example, climate-related transformations in the resource and shipping sectors will impact northern Indigenous peoples, in spite of the rapid and extensive changes afoot.

Approaching the local as a contained and self-evident scale of analysis has also tended to obscure the historical and political constitution of "local communities" and "local knowledge" in the Arctic. Indeed, most contemporary climate research overlooks the importance of imperial and colonial histories in shaping contemporary research objects, subjects, and research relations. Julie Cruikshank notes, for example, the cruel irony in contemporary efforts to gather traditional knowledge from Indigenous peoples about lands from which they have been forcibly removed.[22] The establishment of arctic settlements is itself deeply contested and an outcome of prior colonial interventions in the north; the very constitution of a local community to which contemporary researchers might travel is a legacy of prior efforts to manage and know northern Indigenous peoples. Similarly, contemporary efforts to "help" arctic Indigenous peoples adapt to a rapidly changing world (including academic and government interest in maintaining Inuit "in place") are themselves informed by long histories of helpful intervention in the north. There are lines of affiliation to be drawn between contemporary efforts to help Indigenous northerners in the face of climatic change, and what Tania Murray Li describes as a colonial "will to improve," a will that ultimately entrenches colonial authority and interests.[23] In other words, academic engagement with local communities and local knowledge in the Arctic is historically and politically informed, and is invariably shaped by inherited structures of knowledge and practice.

While there are clearly reasons to be cautious about the mobilization of the local in contemporary research, to pursue these arguments is not to suggest that scholars should abandon analysis of the localized geographies of climatic change. Rather, it is to insist on a critical, careful assessment of the ways in which the local is deployed in climate research, policy, and politics; to attend to the ways in which attributions of locality are affiliated with colonial systems of knowledge and practice; to challenge the presumption that the "local" knowledges and concerns made legible through academic and bureaucratic knowledge production fully reflect what arctic Indigenous peoples know about and care about with respect to climatic change; and to redirect attention to a series of other locals that are central to the constitution of climatic change. In so doing, some of the intellectual and political effects of scaling climate change might be both traced and challenged.

Toward Different Locals

What other lines of engagement with the local might we pursue, then, as part of a critical and careful engagement with arctic climatic change? I consider three lines of inquiry below that might be thought of as points of critique, lines of existing inquiry, and lines of potential inquiry through which studies of climatic change in the Arctic might be both interrogated and refocused. All three take scale to be socially and politically constructed—there is no self-evident "local" at stake here—and all three redirect attention from non-Indigenous study of distant, Indigenous "locals" toward a critical interrogation of the range of locations, peoples, and practices that produce, sustain, and profit from climatic change.

1. Genealogies of "Local Knowledge"

The locality of knowledge is firmly established in environmental history, geography, and anthropology; few would dispute Cruikshank's claim that "all knowledge is incontrovertibly local."[24] Such an insight emerges from geographic attention to the spatiality of knowledge and power, but also from the work of a broader network of scholars drawing on feminist, poststructuralist, postcolonial, actor-network, and science and technology studies (STS) writings. There is now a large body of literature attending to the very specific geographies of scientific, imperial, and other knowledges, a literature that insists on the importance of acknowledging the geographical and historical specificity of the universal or global, and the processes by which hegemonic knowledges are made "true."[25]

If all knowledges are "local," however, it does not follow that all knowledges enjoy equivalent mobility, legibility, and legitimacy. Indeed, there is an important distinction to be made between the philosophical premise that all knowledge is local (in the sense of being produced by particular people in particular places) and the broader discursive contexts within which particular knowledges become associated with the "local," "global," "universal," or "true." There is no doubt that "all knowledge is located and geographically and historically bounded, and ... the local conditions of its manufacture affect substantively the nature of the knowledge produced,"[26] but certain knowledges continue to be more firmly associated with the local scale. Such associations are politically consequential. As Nygren observes, "local knowledge" continues to be framed as "an out-of-the-way other, contrasted with progressive representatives of the expert world," and part of the "romantic past," or, alternatively, framed as "a panacea for dealing with the most pressing environmental problems" and a "critical component of a cultural alternative to modernization." In either case, "local knowledge has been represented as something in opposition to modern knowledge."[27] Indeed, the turn toward local knowledge in climate change research inherits, and in many ways remains informed by, a long tradition of associating the local with the small, powerless, antimodern, isolated, racialized, and feminized.[28]

At a time when gathering, documenting, and integrating local, traditional, and Indigenous knowledges has become orthodox in assessments of climatic change,[29] tracing the geographies of the rise of local knowledge in climate change research—situating and making explicit the "local conditions of its manufacture" as an object of research—might reveal some of the intellectual and political stakes of the mobilization of the local in contemporary climate discourse. Bocking contributes to such a project in his assessment of the evolution of the status of Indigenous knowledge in northern science. He traces the conditions and terms upon which "the status of Indigenous people and their knowledge has been radically revised" among northern scientists, from dismissal and denigration to acceptance and integration, and argues that such shifts must be understood in relation to dynamics of race and whiteness.[30] Cruikshank has long troubled the ways in which northern Indigenous knowledges are taken up by scholars and bureaucrats, and has situated recent enthusiasm for "local" climate knowledge in relation to a much longer timeline of encounter between scientists, explorers, and northern Indigenous peoples.[31] Forbes and Stammler also help situate the rise of local knowledge in climate change research by noting the awkwardness of researching both "traditional ecological knowledge" (TEK) and "climate change" in arctic Russia, where neither concept has the same purchase as it does in North America.[32] There is no necessary correspondence, they argue, between what becomes framed as "traditional" knowledge and the knowledges that emerge from intimate relations with locality, particularly when externally defined research agendas and questions are imposed on circumpolar Indigenous peoples.[33] These varied works draw on a larger body of scholarship unpacking the place of the local in environmental research, governance, and politics.

A genealogy of the ways in which "local knowledge" has been engaged (and ignored) in northern research would thus make important links with past efforts to engage or dismiss local knowledge, including the utter dismissal and denigration of local understandings of wildlife through the caribou crises of the 1960s,[34] the failure to account for local concerns and interests in post-war arctic mineral development and militarization,[35] the denigration of local relationships with sled dogs through the sled dog shootings of the 1960s,[36] the assumption that Indigenous knowledges of lands in one part of the Arctic could be seamlessly transferred to other regions as a rationalization for relocations,[37] the impacts of bureaucratized knowledge production, translation, and decision-making on Indigenous wildlife co-management (as discussed by Paul Nadasdy in his chapter in this volume),[38] and the failure of local knowledges about the importance of seal hunting and commercial trapping to effectively counter European and North American movements to ban seal and other fur trades.[39] Each of these interventions had devastating effects on "local communities," and is symptomatic of the ways in which strategic understanding and misunderstanding of local knowledges interweaves with broader power relations. Given that each intervention was also framed as helpful, well-meaning, and in the best interests of northerners, how, we might ask, do contemporary valuations of local knowledge risk extending such legacies?

Furthermore, a genealogy of local knowledge might assist in de-centering academic knowledge production and the false notion that local, Indigenous, or traditional knowledges must be documented by academics and policymakers in order to be meaningful or effective—or, in more extreme formulations, in order to survive at all. Indigenous peoples have always nurtured their own knowledges, for their own purposes, and continue to do so beyond the confines of academic or institutional documentation. These geographies of engagement with "local knowledge" are necessarily and importantly *not* known or knowable in academic or bureaucratic terms, and it is by no means clear that it is possible or desirable to translate such knowledges and practices into institutional spheres.[40] But it is crucial to acknowledge that, whether or not such processes are documented in academic or bureaucratic terms, Indigenous knowledges are continually engaged by Indigenous peoples, for their own purposes, and scholars must take seriously the necessary and inevitable gap between the institutionalized documentation, mobilization, and application of local knowledge, and the relationships and practices that sustain knowing and being in Indigenous communities. This is not a gap to fill or bridge (much less by non-Indigenous academics), but rather to respect and remain attentive to. Indeed, initiatives like the Digital Indigenous Democracy (DID) project recently launched in Nunavut challenge the notion that "local" Inuit knowledges and concerns can be meaningfully engaged by institutions, and underscore the necessity of sustaining and sharing knowledge outside of, alongside, and in opposition to institutional parameters. Emphasizing oral Inuktitut, DID uses Internet, community radio, local TV, and social media

> to amplify Inuit traditional decision-making skills at a moment of crisis and opportunity as Inuit face Environmental Review of the $6 billion Baffinland Iron Mine (BIM) on north Baffin Island. Through centuries of experience Inuit learned that *deciding together,* called *angiqatigiingniq* … is the smartest, safest way to go forward in a dangerous environment. Through DID, Inuit adapt *deciding together* to modern transnational development—to get needed information in language

they understand, talk about their concerns publicly and reach collective decisions with the power of consensus.[41]

Not only does Digital Indigenous Democracy initiative challenge the delimitation of local knowledge to a narrow understanding of the traditional, antimodern, or ecological (see the following section), it also displaces academic and institutional structures—and written English—as the most appropriate, effective, and meaningful venues within which such knowledges might be mobilized and through which decisions should be made. A genealogy of engagement with "local knowledge" must account for such practices, not in an attempt to make them knowable in academic or institutional contexts, but rather as instances that de-centre dominant forms of knowledge production, and challenge the presumption of equivalence between academic engagements with local knowledge and knowledge itself. In sum, a genealogy of "local knowledge" as an object of arctic climate research would call into question not only the nature and scope of the knowledges in question, but also the longer histories and politics of academic engagement with Indigenous knowledges. It would situate the turn among scholars and policymakers to "collect" and "integrate" local, Indigenous, and traditional knowledge in relation to much longer histories of engagement, including histories of selective and strategic misunderstanding and disregard.

2. Indigeneity, Locality, and Tradition

There is a tendency to use the concepts of "local knowledge," "Indigenous knowledge," and "traditional knowledge" interchangeably in studies of arctic climate change, a tendency that takes its cue from much longer histories of associating Indigeneity with the local and traditional. Such associations have been consequential. Appadurai argues that, in Western knowledge systems, the very idea of Indigeneity or nativeness is underpinned by assumptions of "intellectual and spatial confinement" in which Indigenous peoples are assumed to be not only from certain places, "but they are also those who are somehow incarcerated, or confined, in those places. ... Natives are in one place, a place to which explorers, administrators, missionaries, and eventually anthropologists, come."[42] This tendency to assume an equivalence between the local and the Indigenous, and to assume that non-Western and Indigenous peoples are more local than others, emerges from distinctly racialized and colonial epistemologies. Indeed, a number of scholars have specifically problematized the association of Indigenous peoples with the local and of non-Indigenous peoples with mobility, translocality, and globality.[43] Within such framings, to be Indigenous is to not only be more explicitly tied to a local place, but also to have one's Indigeneity itself delimited to one's relations with that place. That is, one is understood to be Indigenous only insofar as one is located in a particular place and engaging in recognizably "Indigenous" practices. According to Appadurai, the emphasis placed by anthropologists on the traditional relations forged between Indigenous peoples and their environment must be understood as rendered "in a language of incarceration" that perpetuates a fictional and deeply political confinement of particular peoples to particular cultural, intellectual, and spatial locations.[44] Notably, he identified this confinement as not just physical, but also cultural and intellectual: in other words, it is not just Indigenous peoples, but also their knowledge, that is understood to be more local, more directly contingent on relationships with the local environment, its relevance restricted to local processes and concerns.

Much of the past several decades has been spent dismantling this assumption among anthropologists, environmental historians, and geographers, and considering its intellectual and political implications.[45] Critical work has challenged the assumption that local cultures are isomorphic with particular territories (that is, that they are more likely and appropriately found in particular, delimited places), and also that their very being is more tied to the land, more reliant upon nature, and more shaped by things like climate, natural resource use, and so on. The association of Indigeneity with locality has come to be understood as not only a political, intellectual, and cultural-historical construction, but also as a profoundly relational reading of Indigeneity that produces non-Indigenous, colonizing identities, cultures, and knowledges as much as Indigeneity itself. As Howitt argues, "relationships between Indigenous peoples, colonial powers, settler populations, and postcolonial government have always been spatialized by a complex politics of geographical scale,"[46] and problematizing the scalar dimensions of (neo)colonial formations has been a key site of critical inquiry into Indigenous/non-Indigenous relations.[47]

The point is not that Indigenous peoples' knowledges are somehow *not* local, but rather that attributions of locality can be used to undermine that knowledge, even (and perhaps especially) when aiming to document, integrate, and represent "local" understandings of an issue. Indeed, northern Indigenous peoples themselves insist that their distinctive ontologies and epistemologies emerge from intimate relations with place and land.[48] Contemporary efforts on the part of non-Indigenous scholars and governments to take seriously these distinctive, specific, and rich knowledges flow from the political mobilization of northerners themselves; the systematic dismissal and exclusion of Indigenous knowledges and claims from northern research, governance, and politics motivated calls for the establishment of new institutional structures that might account for these knowledges and claims.[49] Indigenous ways of knowing and being are not *limited* to the local, however. Not only can distinctively Indigenous and traditional forms of knowledge be brought to bear on translocal, complex, contemporary dilemmas (and they are), but also a number of Indigenous leaders and scholars explicitly reject the delimitation of Indigeneity to an externally defined understanding of the local, traditional, or Indigenous.[50] They reject, in other words, the move from *recognizing* the importance of the traditional and intimately local, to *limiting* Indigeneity itself to these spheres. It is that delimitation that has been used, time and again, to restrict the legibility and efficacy of Indigenous peoples' knowledges, practices, and claims.

Such a delimitation has been specifically problematized in recent years in relation to the growing interest in integrating Indigenous knowledges into academic and policy settings. While this move can be understood as a progressive response to calls by Indigenous peoples to take seriously their distinctive knowledges, it remains an undertaking that is fraught with political and intellectual challenges.[51] Nadasdy notes, for example, that the emphasis placed on the traditional dimensions of Indigenous knowledges "makes it easy for scientists and resource managers to disregard the possibility that aboriginal people might possess distinct cultural perspectives on modern industrial activities such as logging or mining,"[52] activities that are just as "local" as hunting or other practices associated with Indigeneity. Others raise concerns about whether Indigenous knowledges can be effectively, accurately, and appropriately integrated into academic studies, including issues around translation, representation, decontexualization, and power.[53] Bravo notes that while there is a great deal of enthusiasm for integrating Indigenous knowledges into science, there is almost no attention directed toward understanding how scientific research is received and debated in northern Indigenous communities.[54] It would seem that "local" knowledge can inform scientific research, but not the reverse. These slippages and critiques are instructive. Local, traditional, and Indigenous knowledge has never been as highly valued in academic, government, or political spheres as it is today, and yet it is by no means clear that what registers as "local" or "traditional" knowledge corresponds with what Indigenous peoples know about and care about, that Indigenous ways of knowing and being can be integrated into institutional parameters, and that documenting, translating, analyzing, and otherwise engaging local and Indigenous knowledges serves the interests of Indigenous peoples themselves.

At such a moment, it seems to crucial to ask, then, what counts as a local issue or local knowledge, the conditions under which these issues and knowledges come to matter, and for whom? Mobilizing explicitly local and traditional perspectives on climate change has been an important political strategy for Inuit; it is precisely by making people care about the localized effects of climate change on traditional practices that organizations like the Inuit Circumpolar Council (ICC) have advanced their political objectives.[55] As Duane Smith, president of ICC Canada recently noted, ICC's priorities are to ensure Canadian Inuit have strong ties to Inuit in Russia, Greenland, and Alaska, and to represent Inuit rights internationally: "we make connections abroad so that Canadian Inuit can benefit at home. This is especially important because … challenges in the Arctic very often need to be addressed abroad. And we often call upon our Inuit cousins in other countries to help us."[56] Reflecting on the settlement of the Stockholm Convention on Persistent Organic Pollutants (POPs) in 2001 (also discussed by Stephen Bocking in his chapter in this volume), Sheila Watt-Cloutier similarly positioned her international political work in relation to her personal history, her connections to family and to the land, and her accountability to Inuit in communities.[57] These reflections make clear both the importance of "scaling up" localized concerns to achieve particular political objectives, but also the futility of framing Inuit knowledges and practices as either local or global, traditional or modern.

Indeed, the coordination and internationalization of Inuit concerns about climate change seem to exemplify the emancipatory possibilities of scale; making the local matter at UN conferences or at the US Human Rights Tribunal has been an enormously effective political strategy. But Duane Smith has also observed that there are distinct limitations on the mobility of Inuit knowledges and concerns. Inuit participation is only selectively solicited in governance, research, and consultation settings, he argues, and Inuit are systematically "shut out" when matters of "resource development and other issues of great importance to Inuit" are decided.[58] The ICC's submission to the Arctic Marine Shipping Assessment,[59] a study carried out by the Arctic Council to assess increased shipping through the Arctic, made clear the limitations of engaging Inuit along wholly "traditional" and "local" lines. Inuit have repeatedly called for a comprehensive plan to address oil spills in Arctic waters, calls that have thus far met with only incremental progress by bodies like the Arctic Council, even while oil rigs establish themselves off the coast of Greenland and interest in opening the Northwest Passage to seasonal transcontinental shipping accelerates.[60] Clearly, the legibility and mobility of Indigenous knowledges is highly variable, and that variability is itself tied to larger political-economic relations and interests.

Why do some knowledges travel and not others, and on what terms? While Indigenous knowledges are actively solicited in the identification of local vulnerabilities to climate change, acute concerns and knowledges about pressing, translocal climate-related threats tend not to register in community-based studies. Further attention to the conditions under which local, traditional, and Indigenous knowledges become defined, mobilized, and politically consequential thus represents an important line of inquiry into the geographies of arctic climate change.[61] It is a line of inquiry to which geographers and environmental historians have much to offer; as Carey observes, the task facing climate historians is not simply reconstruction of past climates, or studies of how past societies have adapted to climatic changes, but also directing keen attention to how "social relations and power dynamics" shape the deeply political unfolding of human-climate relations.[62] Challenging the conflation of the local, Indigenous, and traditional in assessments of climatic change draws attention precisely to the social relations and power dynamics underpinning northern research and policy development.

3. Localizing the Global

The knowledges that are codified and represented as "local knowledge" in studies of climatic change are themselves relational productions; they emerge from encounters between researchers and community informants and are profoundly shaped by the contours of these encounters. Framing Indigenous knowledges as more local than other knowledges not only risks delimiting the knowledges and claims of Indigenous peoples, then, but also overlooks the locality and specificity of scientific, imperial, and other translocal knowledges.[63] If, indeed, all knowledge is local, what might be revealed by attending to the locality of the "global" dimensions of climate change, including not just globalized climate-related knowledges (such as climate science) but also climate change itself? That is, what if we turned our attention toward the specific geographies of international scientific networks, CO_2 emissions, commodity finance, and geopolitical struggles for arctic resources? How might localizing these "global" processes sharpen our understanding of climatic change and its articulations in the Arctic?

Efforts to localize climate science have already begun: Bocking has considered not only the ways in which climate science has been shaped by "the particular combination of disciplines, ideas about science, and political imperatives that have attended its development," but also the ways in which political and economic inequality have been defined within climate science as "issues beyond the scope of investigation."[64] His work indicates that there are very specific social, cultural, and political practices underpinning climate science that demand further interrogation and elucidation. Liverman identifies a series of narratives shaping climate research, discourse, and the mobilization of climate science, thus problematizing the ways in which climate science comes to make sense and be made effective.[65] Others have been localizing climate science (and other forms of northern research) in different ways: Carr recently quantified some of the economic dimensions of northern Canadian research, highlighting not so much the ways in which scientific ideas and understandings are generated, but rather the materiality of economic flows in the production of academic industries.[66] Carr argues that research is a significant industry in the territorial north, with output, GDP, and income impacts that are comparable to the commercial hunting, fishing, and trapping industries, and employment impacts that are similar to the arts and heritage, entertainment, and recreational industries. Abele and Dalseg Kennedy similarly aim to situate northern research over the last several decades in relation to broader political-economic formations, including the often unacknowledged importance of non-academic consulting research in the north, and its role in environmental impact assessment and co-management processes.[67] These lines of inquiry go a long way toward addressing James Secord's important point that scholars must do more than merely demonstrate that scientific knowledge is "local" in the sense of being specific to a cultural-geographical context, and instead consider its "connections with and possibilities for interaction with other settings."[68]

Indeed, what political-economic objectives are advanced in making certain kinds of knowledge mobile and legible in both academic and non-academic institutions? Who benefits from northern knowledge production? Almost twenty years ago, as research funding related to the Royal Commission on Aboriginal Peoples (RCAP) was being disbursed, Martha Flaherty observed that almost none of it went to Inuit, and that RCAP failed to establish research guidelines that would "support full participation of the people being studied in the identification of research needs, design of research methods, collection and analysis of data, and control over the results and use of the results."[69] Similar concerns continue to be expressed by northerners, in spite of important shifts in northern research practices.[70] Even while community knowledges are solicited, most research funding continues to flow to southern researchers and institutions. To localize climate science, then, is not simply to attend to the institutional cultures within which scientists operate, or the settings in which their findings become legible, but also to query arctic knowledge production as an industry underpinned by profound inequalities, tied to processes of accumulation, and affiliated with longer histories of extraction. Here, again, history is instructive; as Carey observes, a range of scholars working in different historical and geographic settings have shown that "the accrual of climate science in the hands of government bureaucracies or among the intellectuals and the ruling elite has resulted in the accumulation of power for those groups—the power to withhold weather data, to manipulate understandings, or to economically benefit certain groups over others."[71] This is not the first time, in other words, that knowledge about climate has been politically and economically consequential.

Tracing the locality of the broader political-economic dimensions of climate change would also challenge the erasures inherent in framing climate change as a "global" issue. While there is no doubt that the effects of anthropogenic climate change are being observed across the planet, not only are these effects highly differentiated socially and geographically, but also the specific geographies of greenhouse gas emissions—the social, political, and economic practices that have caused anthropogenic climate change—fall from view when climatic change is framed as a global issue. There is a geography, Chakrabarty notes, to those who are "retrospectively guilty" of inducing climate change and those who are "prospectively guilty," and these geographies are thoroughly interwoven with "histories of capitalism and modernization."[72] If climate change forces us to confront our collective capacity to act as geological agents of change on a planetary scale, in other words, we must also continually pay attention to the differentiation within that collectivity. Chakrabarty thus cautions against thinking of climatic change "by use of such all inclusive terms as *species* or *mankind* when the blame for the current crisis should be squarely laid at the door of the rich nations in the first place and of the richer classes in the poorer ones."[73] Relatedly, Hulme has recently highlighted the dangers of "climate reductionism"—"a form of analysis and prediction in which climate is first extracted from the matrix of interdependencies that shape human life within the physical world ... then elevated to the role of dominant predictor variable."[74] Climate change, in such scenarios, becomes an abstracted and depoliticized explanation for itself; it is climate, rather than specific human activities and relations, that comes to be understood as the primary agent of social, political, ecological, and economic crisis.

If "thinking global" about climate change risks depoliticizing the processes that cause climatic change itself, there is perhaps no better illustration than the move to frame uranium development in Nunavut as a contribution to climate change mitigation. Although they did little to cause climatic change, and suffer disproportionately from its effects,[75] Nunavummiut are being urged by industry and government to think of uranium mining as a means of doing their part to reduce global greenhouse gas emissions.[76] As French uranium company AREVA emphasizes in its proposal to develop a uranium mine near Qamani'tuaq (Baker Lake), Nunavut, "uranium from the Kiggavik Project would help to meet the future needs for nuclear power, which will help reduce, on a global scale, greenhouse gas emissions."[77] Nunavummiut are, of course, acutely concerned about climate change and keen to contribute to mitigation efforts. But as Nunavut resident Dodie Kayuk noted in a Nunavut Planning Commission meeting tasked with assessing uranium development in the territory, Nunavummiut will bear distinctly localized risks for such "global" benefits: "what will happen when there's a leakage? Who will help us? ... We are the ones that will be affected. People from down south and government do not drink our water, eat our animals and the fish, they don't breathe the air we're breathing."[78] Some of the material implications of "thinking globally and acting locally" on climate change are thus laid bare. Framing uranium development as a responsible contribution to a planetary crisis advances the interests of investors and shareholders, and places the lives and livelihoods of others at risk. How might scholars work to make these very different "locals" not only legible, but also identify the ways in which they are interconnected and shape each others' fortunes?

Indeed, what might arctic climate change research look like if we attended not only to its localized effects, but to the very specific geographies of greenhouse gas emissions, commodity finance, and fossil fuel extraction? What might result if we turned our attention to the geographies of what Johnson terms "accumulation by degradation,"[79] the extension of a new round of resource extraction in a region rendered more accessible by the biophysical effects of *previous* rounds of accumulation? Inuit have been tracing and naming these connections for decades.[80] As Ginsburg shows, there is no lack of "local knowledge" about what causes climate change and how it matters in communities: "Most Salluit residents do not characterize climate change as a threat to Inuit culture. Instead, they highlight the damaging impacts of globalization and internal colonialism as a more serious problem. This ... suggests that focusing narrowly on climate change can obscure the broader and more immediate challenges facing Inuit communities. Such a realization demonstrates the need for researchers to locate climate change within a matrix of non-climatic challenges in order to mitigate threats to indigenous cultures."[81] According to Ginsburg, the more pressing and primary threats facing Sallumiut emerge from the wage labour system, school system, and other colonial and capitalist interventions, interventions whose spatiality extends far beyond the local, that intimately shape Inuit relations with climatic change.

To come to terms with these dimensions of climate change is to come to terms with colonial and capitalist histories and presents. And yet, in the vast majority of arctic climate change research, neither colonialism nor capitalism is within the frame of reference.[82] The local as a category of analysis facilitates these exclusions, insofar as the local is engaged as a resolutely contemporary and contained site, a site whose ties with broader histories and geographies are occluded. Not only is it essential to refuse this delimitation of the local, it is also essential to localize and make legible the specific geographies of climate-related accumulation shaping the Arctic, including both academic and industrial production and extraction.

Conclusion: Climate Anti-Politics

> *With slyness and flattery*
> *you pretend*
> *it is us you are serving*
> *not yourselves*
> —Aqqaluk Lynge[83]

At the conclusion of his landmark book, *The Anti-Politics Machine*, James Ferguson responds to an anticipated question. After analyzing the political and economic consequences of development discourses, including the role of academic knowledge production in sustaining the development apparatus, Ferguson addresses the reader who might ask, "what, then, is to be done?" His response remains compelling, particularly at a moment when the "rule of experts"[84] appears to be intensifying in the Arctic. He begins by noting that the question, "'what is to be done?' demands first of all an answer to the question, 'by whom?'"[85] Ferguson challenges the notion that scholars and policymakers have the capacity to create meaningful change, as well as their self-appointed responsibility and jurisdiction to do so. Instead, he argues, "it seems clear that the most important transformations, the changes that really matter, are not simply 'introduced' by benevolent technocrats, but fought for and made through a complex process that involves not only states and their agents, but all those with something at stake, all the diverse categories of people who craft their everyday tactics of coping with, adapting to, and, in their various ways, resisting the established social order."[86] It is crucial to acknowledge the limitations and dangers of academic engagement in various northern "problems" and to disentangle intentions to "help local communities" from the actual effects of well-meaning, helpful intervention. Although there is clearly a need for pointed, political effort to address the perpetuation and differentiated impacts of climatic change, current academic knowledge production is not necessarily achieving this objective, and it actually risks retrenching the very systems that have dominated northern Indigenous relationships, governance systems, and wellbeing.[87]

This is not to say that scholars should remain silent on arctic climate change; far from it. Rather, it is to call attention to the ways in which academic knowledge production—often unknowingly and unintentionally—can be implicated in the validation and extension of unjust social, political, and economic relations. Indeed, we cannot know in advance where our work will lead, what political strategies will be effective, and how knowledge produced in one context will be transformed by its interactions with others. If this is so, then we cannot assume that academic investigations of climate change do what they aim to do and claim to do, and neither can we assume that filling "gaps" in climate research will have meaningful impacts on the social, political, economic, and environmental processes through which climate change is produced, perpetuated, and ignored. In such a condition, it may be that *not* knowing is as important a response as knowing. That is, it may be that paying attention to the inheritances that shape academic practices in the Arctic, questioning the categories and methods through which we come to make sense of climate change, and de-centering academic capacities to "know" the north will be as important as the advancement of conventional research objectives.

Ferguson goes on to underscore the importance of "political engagement in one's own society" as an alternative to studying and "helping" distant others, as well as the importance of engaging and supporting oppositional groups in other locations, although always with an acute understanding that there may be "no need for what we do [as scholars] among such actors."[88] Both lines of engagement, it seems to me, warrant the attention of contemporary northern scholars, and it is with these cautions and possibilities in mind that I have outlined some alternative approaches to the "local" dimensions of arctic climate change. For Ferguson, political engagement in one's own society includes actively shaping the discursive terrain within which issues like climate change come to make sense. In this regard, tracing the genealogies of past and present interest in "local knowledge," challenging the political and intellectual incarceration of Indigeneity to the "local" scale, and tracing the very specific geographies of various "global" dimensions of climate change (including greenhouse gas emissions, academic knowledge production, and the acceleration of resource extraction and shipping in the Arctic) might contribute to the ways in which climate change is understood, managed, and lived. This is both an historical and a geographic task. And whether we conceive of local struggles as transformative in and of themselves, or as necessarily articulated with other scales of political engagement, orienting academic skills and resources toward supporting the struggles of those who are experiencing the palpable, profound, material effects of climatic change will remain pressing. If anything, it seems clear that incarcerating arctic climate change to both the local scale and the temporal present and future renders climate change knowable in very specific terms, terms that demand challenge, revision, and re-imagination.

Notes

1   Liza Piper, "The Arctic and Subarctic in Global Environmental History." In *A Companion to Global Environmental History*, ed. J. R. McNeill and Erin Stewart Mauldin (Oxford: Wiley-Blackwell, 2012), 153.

2   Mark Carey suggests of climate history more generally that it "could contribute much more than it has to present-day discussions about global climate change knowledge, impacts, and responses." Mark Carey, "Climate and History: A Critical Review of Historical Climatology and Climate Change Historiography," *WIRE's Climate Change* 3 (2012): 233.

3   See, for example, Scot Nickels, et al., *Unikkaaqatigiit: Putting the Human Face on Climate Change* (Ottawa: Inuit Tapiriit Kanatami, Nasivvik Centre for Inuit Health and Changing Environments, and National Aboriginal Health Organization, 2006); Gita Laidler, "Inuit and Scientific Perspectives on the Relationship between Sea Ice and Climate Change: The Ideal Complement?", *Climatic Change* 78, no. 2 (2006); Timothy Leduc, "Sila Dialogues on Climate Change: Inuit Wisdom for a Cross-Cultural Interdisciplinarity," *Climatic Change* 85, no. 3 (2007); Dyanna Riedlinger and Fikret Berkes, "Contributions

of Traditional Knowledge to Understanding Climate Change in the Canadian Arctic," *Polar Record* 37, no. 203 (2001); Natasha L. Thorpe, "Contributions of Inuit Ecological Knowledge to Understanding the Impacts of Climate Change on the Bathurst Caribou Herd in the Kitikmeot Region, Nunavut" (Simon Fraser University, 2000); IPCC, "Climate Change 2007: Impacts, Adaptation and Vulnerability," Working Group II Contribution to the Fourth Assessment Report of the Intergovernmental Panel on Climate Change (Geneva: IPCC, 2007); ACIA, "Arctic Climate Impact Assessment Report: Impacts of a Warming Climate" (Cambridge: Cambridge University Press, 2004); Elizabeth Weatherhead, Shari Gearheard, and Roger Barry, "Changes in Weather Persistence: Insight from Inuit Knowledge," *Global Environmental Change* 20, no. 3 (2010); Tristan Pearce, et al., "Inuit Vulnerability and Adaptive Capacity to Climate Change in Ulukhaktok, Northwest Territories, Canada," *Polar Record* 46, no. 237 (2010); James D. Ford and Barry Smit, "A Framework for Assessing the Vulnerability of Communities in the Canadian Arctic to Risks Associated with Climate Change," *Arctic* 57, no. 4 (2004); James D. Ford, Barry Smit, and Johanna Wandel, "Vulnerability to Climate Change in the Arctic: A Case Study from Arctic Bay, Canada," *Global Environmental Change* 16 (2006); James D. Ford, et al., "Climate Change in the Arctic: Current and Future Vulnerability in Two Inuit Communities in Canada," *Geographical Journal* 174, no. 1 (2008); James D. Ford, et al., "Vulnerability to Climate Change in Igloolik,

Nunavut: What We Can Learn from the Past and Present," *Polar Record* 42, no. 221 (2006); and Gita Laidler, et al., "Travelling and Hunting in a Changing Arctic: Assessing Inuit Vulnerability to Sea Ice Change in Igloolik, Nunavut," *Climatic Change* 94, no. 363–97 (2009).

4   Emilie Cameron, "Securing Indigenous Politics: A Critique of the Vulnerability and Adaptation Approach to the Human Dimensions of Climate Change in the Canadian Arctic," *Global Environmental Change* 22, no. 1 (2012).

5   For example, Inuit Circumpolar Council, "Circumpolar Inuit to Global Leaders in Cancun: Strong Action on Arctic Climate Change Urgently Needed," Press Release, 1 December 2010 (Cancun, Mexico, 2010); Inuit Circumpolar Council, *A Circumpolar Inuit Declaration on Resource Development Principles in Inuit Nunaat* (Nuuk, Greenland, 2011); Aqqaluk Lynge, "Arctic Riches: From Knowledge to Action … The Inuit Perspective," in *International Polar Year Conference: From Knowledge to Action* (Montreal: Inuit Circumpolar Council, 2012).

6   Catherine Brace and Hilary Geoghegan, "Human Geographies of Climate Change: Landscape, Temporality, and Lay Knowledges," *Progress in Human Geography* 35, no. 3 (2011): 284, 286. See also Neil Adger, Nigel Arnell, and Emma Tompkins, "Successful Adaptation to Climate Change across Scales," *Global Environmental Change* 15, no. 2 (2005); and Georgina Endfield, "Reculturing and particularizing climate discourses: weather, identity, and the work of Gordon Manley," *Osiris* 26 (2011): 142–62.

7   Frank Duerden, "Translating Climate Change Impacts at the Community Level," *Arctic* 57, no. 2 (2004): 204.

8   James D. Ford, et al., "Case Study and Analogue Methodologies in Climate Change Vulnerability Research," *WIRE's Climate Change* 1, no. 4 (2010).

9   Riedlinger and Berkes, "Contributions of Traditional Knowledge."

10   Mike Hulme, "Meet the Humanities," *Nature Climate Change* 1, no. 4 (2011): 179. See also Georgina Endfield and Carol Morris, "Cultural Spaces of Climate," *Climatic Change* 113 (2012): 1–4.

11   J. K. Gibson-Graham, "Beyond Global vs. Local: Economic Politics Outside the Binary Frame," in *Geographies of Power: Placing Scale*, ed. Andrew Herod and Melissa Wright (Hoboken, NJ: Wiley-Blackwell, 2002); Richard Howitt, "Scale and the Other: Levinas and Geography," *Geoforum* 33, no. 3 (2002); Sallie Marston, "The Social Construction of Scale," *Progress in Human Geography* 24, no. 2 (2000); Eric Sheppard, "The Spaces and Times of Globalization: Place, Scale, Networks, and Positionality," *Economic Geography* 78, no. 3 (2002); Neil Smith, "Scale," in *Dictionary of Human Geography*, ed. Ron J. Johnston, et al. (Oxford: Blackwell, 2000); Erik Swyngedouw, "Neither Global nor Local: 'Glocalization' and the Politics of Scale," in *Spaces of Globalization*, ed. Kevin Cox (New York: Guildford, 1997); Neil Smith, *Uneven Development: Nature, Capital, and the Production of Space*, 3rd ed. (Athens, GA: University of Georgia Press, 2008 [1984]); Doreen Massey, "A Global Sense of Place,"

in *Reading Human Geography: The Poetics and Politics of Inquiry*, ed. Trevor Barnes and Derek Gregory (London: Arnold, 1997).

12   Carey, "Climate and History"; Sam White, "Climate Change in Global Environmental History," in *A Companion to Global Environmental History*, ed. J. R. McNeill and Erin Stewart Mauldin (Oxford: Wiley-Blackwell, 2012).

13   For example: Swyngedouw, "Neither Global nor Local"; Marston, "The Social Construction of Scale"; Smith, *Uneven Development*; Cindi Katz, "On the Grounds of Globalization: A Topography for Feminist Political Engagement," *Signs: Journal of Women and Culture* 26, no. 4 (2001); Massey, "A Global Sense of Place."

14   See Sallie A. Marston, John Paul Jones, and Keith Woodward, "Human Geography without Scale," *Transactions of the Institute of British Geographers* 30, no. 4 (2005); Smith, "Scale."

15   Katherine Jones, "Scale as Epistemology," *Political Geography* 17, no. 1 (1998): 28.

16   Marston, Jones, and Woodward, "Human Geography without Scale," 416–17.

17   Gibson-Graham, "Beyond Global vs. Local," 30–31.

18   Smith, *Uneven Development*, 230.

19   Smith, *Uneven Development*, 232.

20   Marston et al., "Human Geography Without Scale," 427.

21   Cameron, "Securing Indigenous Politics."

22   Cruikshank, *Do Glaciers Listen? Local Knowledge, Colonial Encounters, and Social Imagination* (Vancouver: UBC Press, 2005), 257.

23 Tania Murray Li, *The Will to Improve: Governmentality, Development, and the Practice of Politics* (Durham: Duke University Press, 2007). I detail some of the interventions I am referring to later in this chapter. I also develop this argument more fully in Cameron, "Securing Indigenous Politics."

24 Cruikshank, *Do Glaciers Listen?*, 10.

25 See, for example, Derek Gregory, *Geographical Imaginations* (Oxford: Blackwell, 1994); Steven Shapin, *A Social History of Truth: Civility and Science in Seventeenth-Century England* (Chicago: University of Chicago Press, 1995); Felix Driver, *Geography Militant: Cultures of Exploration and Empire* (Oxford: Blackwell, 2001); David Livingstone, *Putting Science in Its Place: Geographies of Scientific Knowledge* (Chicago: University of Chicago Press, 2003); David Livingstone, *The Geographical Tradition: Episodes in the History of a Contested Enterprise* (Oxford: Blackwell, 1992); Donna Haraway, "Situated Knowledge: The Science Question in Feminism and the Privilege of Partial Perspective," *Feminist Studies* 14 (1988); Donna Haraway, *Primate Visions: Gender, Race, and Nature in the World of Modern Science* (New York: Routledge, 1989); Cruikshank, *Do Glaciers Listen?*; Daniel Clayton, *Islands of Truth: The Imperial Fashioning of Vancouver Island* (Vancouver: UBC Press, 2000); Bruno Latour, *Science in Action: How to Follow Scientists and Engineers through Society* (Cambridge: Cambridge University Press, 1987); Bruno Latour, *The Pasteurization of France*, trans. Alan Sheridan and John Law (Cambridge, MA: Harvard University Press, 1988).

26 Trevor Barnes, "Local Knowledge," in *Dictionary of Human Geography*, ed. Ron J. Johnston, et al. (Oxford: Blackwell, 2000), 452.

27 Anja Nygren, "Local Knowledge in the Environment–Development Discourse: From Dichotomies to Situated Knowledges," *Critique of Anthropology* 19, no. 3 (1999): 271.

28 For example: Gibson-Graham, "Beyond Global vs. Local: Economic Politics Outside the Binary Frame"; Arturo Escobar, "Culture Sits in Places: Reflections on Globalism and Subaltern Strategies of Localization," *Political Geography* 20 (2001); Massey, "A Global Sense of Place"; Swyngedouw, "Neither Global nor Local"; Tim Cresswell, *In Place/out of Place: Geography, Ideology, and Transgression* (Minneapolis: University of Minnesota Press, 1996); Tim Cresswell, *Place: A Short Introduction*, ed. Geraldine Pratt and Nicholas Blomley, Short Introductions to Geography (Oxford: Blackwell, 2004); Simon Dalby, "Global Environment/ Local Culture: Metageographies of Post-Colonial Resistance," *Studies in Political Economy* 67 (2002); Cruikshank, *Do Glaciers Listen?*

29 Although not without critique; see Bethany Haalboom and David Natcher, "The Power and Peril of 'Vulnerability': Lending a Cautious Eye to Community Labels in Climate Change Research," *Arctic* (forthcoming); Cruikshank, *Do Glaciers Listen?*; Julie Cruikshank, "Glaciers and Climate Change: Perspectives from Oral Tradition," *Arctic* 54, no. 4 (2001); Cameron, "Securing Indigenous Politics"; Tom Thornton and Nadia Manasfi, "Adaptation—Genuine and Spurious: Demystifying Adaptation

Processes in Relation to Climate Change," *Environment and Society: Advances in Research* 1, no. 1 (2010).

30 Stephen Bocking, "Indigenous Knowledge and the History of Science, Race, and Colonial Authority in Northern Canada," in *Rethinking the Great White North*, ed. Andrew Baldwin, Laura Cameron, and Audrey Kobayashi (Vancouver, UBC Press, 2011), 41.

31 Cruikshank, *Do Glaciers Listen?*; Julie Cruikshank, "Uses and Abuses of 'Traditional Knowledge': Perspectives from the Yukon Territory," in *Cultivating Arctic Landscapes: Knowing and Managing Animals in the Circumpolar North*, ed. Mark Nuttall and David G. Anderson (Oxford: Bergahn Books, 2004); Cruikshank, "Glaciers and Climate Change."

32 Bruce Forbes and Florian Stammler, "Arctic Climate Change Discourse: The Contrasting Politics of Research Agendas in the West and Russia," *Polar Research* 28 (2009): 28–42.

33 See also Tim Ingold and Terhi Kurttilla, "Perceiving the Environment in Finnish Lapland," *Body & Society* 6, nos. 3–4 (2000).

34 Peter Kulchyski and Frank Tester, *Kiumajut (Talking Back): Game Management and Inuit Rights 1900–1970* (Vancouver: UBC Press, 2007); John Sandlos, "From the Outside Looking In: Aesthetics, Politics, and Wildlife Conservation in the Canadian North," *Environmental History* 6, no. 1 (2001); Stephen Bocking, "Indigenous Knowledge"; Tina Loo, *States of Nature: Conserving Canada's Wildlife in the Twentieth Century* (Vancouver: UBC Press, 2006). For discussion of the ways

in which local knowledges and concerns about wildlife continue to be undermined, see Paul Nadasdy, "Reevaluating the Co-Management Success Story," *Arctic* 56 (2003), 367–80.

35 P. W. Lackenbauer and Matthew Farish, "The Cold War on Canadian Soil: Militarizing a Northern Environment," *Environmental History* 12, no. 4 (2007); Liza Piper, *The Industrial Transformation of Subarctic Canada* (Vancouver: UBC Press, 2009); Arn Keeling and John Sandlos, "Environmental Justice Goes Underground? Historical Notes from Canada's Northern Mining Frontier," *Environmental Justice* 2, no. 3 (2009); John Sandlos and Arn Keeling, "Claiming the New North: Mining and Colonialism at the Pine Point Mine, Northwest Territories, Canada," *Environment and History* 18, no. 1 (2012); Emilie Cameron, "Copper Stories: Imaginative Geographies and Material Orderings of the Central Canadian Arctic," in *Rethinking the Great White North*, ed. Andrew Baldwin, Laura Cameron, and Audrey Kobayashi (Vancouver: UBC Press, 2011).

36 Lisa Stevenson, "Of Names, Dreams and Sled-Dogs: Forms of Care in the Canadian Arctic," Carleton University Department of Sociology and Anthropology Colloquium (Ottawa, 2011); Frank Tester, "Can the Sled Dog Sleep? Postcolonialism, Cultural Transformation and the Consumption of Inuit Culture," *New Proposals* 3, no. 3 (2010); Qikiqtani Truth Commission, "QTC Final Report: Achieving Saimaqatigiingniq" (Iqaluit, NU: Qikiqtani Inuit Association, 2010).

37 Frank Tester and Peter Kulchyski, *Tammarniit (Mistakes): Inuit Relocation in the Eastern Arctic, 1939–63* (Vancouver: UBC Press, 1994).

38 Paul Nadasdy, *Hunters and Bureaucrats: Power, Knowledge, and Aboriginal-State Relations in the Southwest Yukon* (Vancouver: UBC Press, 2003)

39 George Wenzel, *Animal Rights, Human Rights: Ecology, Economy, and Ideology in the Canadian Arctic* (Toronto: University of Toronto Press, 1991).

40 Paul Nadasdy, "The Politics of Tek: Power and the 'Integration' of Knowledge," *Arctic Anthropology* 36, no. 1–2 (1999); Paul Nadasdy, "Reevaluating the Co-Management Success Story"; Paul Nadasdy, *Hunters and Bureaucrats*; Jackie Price, "Tukisivallialiqtakka: The Things I Have Now Begun to Understand: Inuit Governance, Nunavut and the Kitchen Consultation Model" (master's thesis, University of Victoria, 2007); Janet Tamalik McGrath, "Isumaksaqsiurutigijakka: Conversations with Aupilaarjuk towards a Theory of Inuktitut Knowledge Renewal" (PhD diss., Carleton University, 2011); Cruikshank, "Uses and Abuses of 'Traditional Knowledge'"; Glen Coulthard, "Subjects of Empire: Indigenous Peoples and the 'Politics of Recognition' in Canada," *Contemporary Political Theory* 6 (2007): 437–60.

41 Isuma, "Fact Sheet: Angjqatigiingniq—Deciding Together—Digital Indigenous Democracy," http://s3.amazonaws.com/isuma.attachments/DIDOverview120504.pdf.

42 Arjun Appadurai, "Putting Hierarchy in Its Place," *Cultural Anthropology* 3, no. 1 (1988): 38, 37.

43 For example: Cruikshank, *Do Glaciers Listen?*; Escobar, "Culture Sits in Places"; Emilie Cameron, "'To Mourn': Emotional Geographies and Natural Histories of the Canadian Arctic," in *Emotion, Place, and Culture*, ed. Mick Smith, et al. (London: Ashgate, 2009); Michael Bravo, "Ethnographic Navigation and the Geographical Gift," in *Geography and Enlightenment*, ed. David Livingstone (Chicago: University of Chicago Press, 1999); and Howitt, "Scale and the Other."

44 Appadurai, "Putting Hierarchy," 37.

45 See, for example, James Clifford and George Marcus, eds., *Writing Culture: The Poetics and Politics of Ethnography* (Berkeley: University of California Press, 1992); Akhil Gupta and James Ferguson, "Beyond 'Culture': Space, Identity, and the Politics of Difference," *Cultural Anthropology* 7, no. 1 (1992); James Clifford, *Routes: Travel and Translation in the Late Twentieth Century* (Cambridge, MA: Harvard University Press, 1997); Mary Louise Pratt, *Imperial Eyes: Travel Writing and Transculturation* (London: Routledge, 1992); Cruikshank, *Do Glaciers Listen?*; Hugh Raffles, *In Amazonia: A Natural History* (Princeton: Princeton University Press, 2002); Timothy Mitchell, *Rule of Experts: Egypt, Techno-Politics, Modernity* (Berkeley: University of California Press, 2006); James Secord, "Knowledge in Transit," *Isis* 95 (2004).

46 Richard Howitt, "Getting the Scale Right? A Relational Scale Politics of Native Title in Australia," in *Leviathan Undone: Towards a Political Economy of Scale*, ed. Roger Keil and Rianne Mahon (Vancouver: UBC Press, 2009), 141.

47   For example: Caroline Desbiens, "Producing North and South: A Political Geography of Hydro Development in Québec," *Canadian Geographer* 48, no. 2 (2004); Howitt, "Scale and the Other"; Howitt, "Getting the Scale Right? A Relational Scale Politics of Native Title in Australia"; David Rossiter and Patricia Wood, "Fantastic Topographies: Neo-Liberal Responses to Aboriginal Land Claims in British Columbia," *Canadian Geographer* 49, no. 4 (2005); Sarah de Leeuw, "Intimate Colonialisms: The Material and Experienced Places of British Columbia's Residential Schools," *Canadian Geographer* 51, no. 3 (2007).

48   Sheila Watt-Cloutier, "Keynote Address," 2030 NORTH National Planning Conference (Ottawa, ON: Canadian Arctic Resources Committee, Inuit Tapiriit Kanatami, and the University of Calgary, 2009); John B. Zoe, "Gonaewo—Our Way of Life," in *Northern Exposure: Peoples, Powers, and Prospects in Canada's North*, ed. Frances Abele, et al. (Montreal: Institute for Research on Public Policy, 2009); Price, "Tukisivallialiqtakka"; John Bennett and Susan Riley, eds., *Uqalurait: An Oral History of Nunavut* (Montreal: McGill-Queen's University Press, 2004).

49   Julia Christensen and Miriam Grant, "How Political Change Paved the Way for Indigenous Knowledge: The Mackenzie Valley Resource Management Act," *Arctic* 60, no. 2 (2007).

50   Leanne Simpson, "Anticolonial Strategies for the Recovery and Maintenance of Indigenous Knowledge," *American Indian Quarterly*

28, nos. 3–4 (2004); Linda Tuhiwai Smith, *Decolonizing Methodologies* (London: Zed Press, 1999); Deborah McGregor, "Linking Traditional Knowledge and Environmental Practice in Ontario," *Journal of Canadian Studies* 43, no. 3 (2009); Winona Stevenson, "Indigenous Voices, Indigenous Histories. Part 3: The Social Relations of Oral History," *Saskatchewan History* 51, no. 2 (1999); Taiaiake Alfred and Jeff Corntassel, "Being Indigenous: Resurgences against Contemporary Colonialism," *Government and Opposition* 40, no. 4 (2005).

51   See McGregor, "Linking Traditional Knowledge and Environmental Practice in Ontario," for one assessment, and Laidler "Inuit and Scientific Perspectives," for an interpretation in Inuit contexts.

52   Nadasdy, "The Politics of TEK," 4.

53   See, for example, Cruikshank, "Uses and Abuses of 'Traditional Knowledge'"; Stephen Ellis, "Meaningful Consideration? A Review of Traditional Knowledge in Environmental Decision Making," *Arctic* 58, no. 1 (2005); McGregor, "Linking Traditional Knowledge and Environmental Practice in Ontario"; Nadasdy, "The Politics of TEK"; Simpson, "Anticolonial Strategies for the Recovery and Maintenance of Indigenous Knowledge"; Laidler, "Inuit and Scientific Perspectives"; Christensen and Grant, "How Political Change Paved the Way"; Evelyn Peters, "Views of Traditional Ecological Knowledge in Co-Management Bodies in Nunavik, Quebec," *Polar Record* 208 (2003); Marybeth Long Martello, "A Paradox of Virtue? 'Other' Knowledges and Environment-Development

Politics," *Global Environmental Politics* 1, no. 3 (2001).

54   Michael Bravo, "Voices from the Sea Ice: The Reception of Climate Impact Narratives," *Journal of Historical Geography* 35, no. 2 (2009).

55   For an assessment of a particularly effective mobilization along these lines, see Terry Fenge, "POPs and Inuit: Influencing the Global Agenda," in *Northern Lights against POPs: Combatting Toxic Threats in the Arctic*, ed. David Leonard Downie and Terry Fenge (Montreal: McGill-Queen's University Press for the Inuit Circumpolar Conference of Canada, 2003).

56   Smith cited in Jane George, "ICC Canada Ponders Its Future at Kuujjuaq Agm," *Nunatsiaq News*, 2012.

57   Sheila Watt-Cloutier, "The Inuit Journey towards a POPs-Free-World," in *Northern Lights against POPs: Combatting Toxic Threats in the Arctic*, ed. David Leonard Downie and Terry Fenge (Montreal: McGill-Queen's University Press for the Inuit Circumpolar Conference of Canada, 2003).

58   Inuit Circumpolar Council, "Include Canadian Inuit in Implementing Arctic Foreign Policy Statement, Says Inuit Circumpolar Leader," Press Release, 20 August 2010 (Inuvik, NT, 2010).

59   Inuit Circumpolar Council, "The Sea Ice Is Our Highway. An Inuit Perspective on Transportation in the Arctic: A Contribution to the Arctic Marine Shipping Assessment" (Ottawa: Inuit Circumpolar Council—Canada, 2008), iii.

60   Patricia Bell, "Arctic Council Leaders Sign Rescue Treaty," *CBC News*, http://www.cbc.ca/news/canada/north/story/2011/05/12/arctic-council-greenland.html (accessed 29 July 2013).

61   Although not directly focused on climate change, Jessica Shadian's history of the ICC provides some useful leads into tracing the ways in which Inuit knowledges have informed translocal circumpolar debates and policy. Jessica Shadian, "Remaking Arctic Governance: The Construction of an Arctic Inuit Polity," *Polar Record* 42 (2006): 249–59.

62   Mark Carey, "Climate and History," 243.

63   Cruikshank, *Do Glaciers Listen?*

64   Stephen Bocking, *Nature's Experts: Science, Politics and the Environment* (New Brunswick, NJ: Rutgers University Press, 2004), 115, 123.

65   Diana Liverman, "Conventions of Climate Change: Constructions of Danger and the Dispossession of the Atmosphere," *Journal of Historical Geography* 35, no. 2 (2009).

66   Katrina Carr, "Impact of Publicly Funded Research on the Canadian Territorial Economies" (master's thesis, University of Saskatchewan, 2012).

67   Frances Abele and Sheena Dalseg Kennedy, "Seeing Like a Community: Research in Northern Indigenous Communities in Historical Perspective," in *IPY 2012: From Knowledge to Action* (Montreal, 2012).

68   James Secord, "Knowledge in Transit," 659, 664.

69   Martha Flaherty, "Freedom of Expression or Freedom of Exploitation?", *Northern Review* 14 (1995): 181.

70   Jackie Price, "Living Inuit Governance in Nunavut," in *Lighting the Eighth Fire: The Liberation, Resurgence, and Protection of Indigenous*

Nations, ed. Leanne Simpson (Winnipeg: Arbieter Ring Press, 2008).

71 Carey, "Climate and History," 239.

72 Dipesh Chakrabarty, "The Climate of History: Four Theses," *Critical Inquiry* 35, no. 2 (2009): 218.

73 Chakrabarty, "The Climate of History," 216, 221. See also Joel Wainwright, "Climate Change, Capitalism, and the Challenge of Transdisciplinarity," *Annals of the Association of American Geographers* 100, no. 4 (2010).

74 Mike Hulme, "Reducing the Future to Climate: A Story of Climate Determinism and Reductionism," *Osiris* 26, no. 1 (2011): 247.

75 Inuit Circumpolar Council, "Petition to the Inter-American Commission on Human Rights Seeking Relief from Violations Resulting from Global Warming Caused by Acts and Omissions of the United States," ed. Inuit Circumpolar Council (Iqaluit, NU: Inuit Circumpolar Council, 2005).

76 Warren Bernauer, "The Uranium Controversy in Baker Lake," *Canadian Dimension* 46, no. 1 (2012); Nunavut Planning Commission, "Report of the Uranium Mining Workshop, Baker Lake NU," ed. Nunami Jacques Whitford (Cambridge Bay, NU: Nunavut Planning Commission, 2007); AREVA, "2nd Draft Kiggavik Eis Popular Summary," in *Submission to NIRB, April 2012* (Cambridge Bay, NU: Nunavut Impact Review Board, 2012).

77 AREVA, "2nd Draft Kiggavik Eis Popular Summary," vi.

78 Cited in Nunavut Planning Commission, "Report of the Uranium Mining Workshop, Baker Lake, NU," 3–7.

79 Leigh Johnson, "The Fearful Symmetry of Arctic Climate Change: Accumulation by Degradation," *Environment and Planning D: Society and Space* 28, no. 5 (2010).

80 Jose Kusugak, "Foreword: Where a Storm Is a Symphony and Land and Ice Are One," in *The Earth Is Faster Now: Indigenous Observations of Arctic Environmental Change*, ed. Igor Krupnik and Dyanna Jolly (Fairbanks, AK: Arctic Research Consortium of the United States, 2002); Inuit Circumpolar Council, "Petition."

81 Alexander Ginsburg, "Climate Change and Culture Change in Salluit, Quebec, Canada" (master's thesis, University of Oregon, 2011).

82 See Cameron, "Securing Indigenous Politics."

83 Lynge, "Arctic Riches."

84 Mitchell, *Rule of Experts.*

85 James Ferguson, *The Anti-Politics Machine: "Development," Depoliticization, and Bureaucratic Power in Lesotho* (Minneapolis: University of Minnesota Press, 1994), 280.

86 Ferguson, *The Anti-Politics Machine*, 281.

87 Price, "Tukisivallialiqtakka"; Flaherty, "Freedom of Expression."

88 Ferguson, *The Anti-Politics Machine*, 286, 287.

Conclusion

14

Encounters in Northern Environmental History

Stephen Bocking

These chapters have ranged across northern space and time, exploring episodes from a century of relations between people and nature, from the Klondike to global environmental change. They have also demonstrated how the natural and cultural features of northern Canada imply many distinctive questions and issues regarding its environmental history. In telling these stories, our authors have drawn on diverse forms of evidence: public documents, archives, physical traces, and the voices of individuals from north and south. But above all, they have shown how this place has been a terrain of encounter: between people (often from elsewhere) and the land, expressed through technology, knowledge, and other dimensions of human thought and activity. Among these encounters, that between Indigenous and newcomer societies has been of special significance, particularly because it has been marked not, as elsewhere in North America, by displacement, but by a long period of cultural mixing. From this encounter have flowed others—between contrasting economic systems and ways of relating to the environment, between different systems of knowledge, and between different political and cultural priorities relating to the northern environment. In this conclusion we will extend our authors' exploration of these encounters by returning to those themes that we considered in the first chapter.

Environmental Change

Change has been a pervasive feature of the northern environment, evident in its histories of climate, fire, and species. As these chapters explain, change has also been central to human history in the north: as the consequence of development or other interventions, or as a factor compelling adaptation. Mining's toxic residues, lands flooded by dams in Quebec and elsewhere, and the local impacts of DEW Line stations all testify to the capacity of industry and technology to transform ecosystems. Development projects have also had wider impacts, as roads or railways (even those that failed, as Jonathan Peyton explains) opened up regions: in the Northwest Territories, Quebec, and elsewhere, mines and dams have been only the starting point for transformation. Fire has often played a role in these transformations, accompanying, as Liza Piper notes, prospecting and other industrial activities. Our chapters have explored other interventions, as well, including efforts to domesticate reindeer (as Andrew Stuhl describes) and to establish gardens to support southern diets. The entire north has experienced the consequences of human impacts on the global environment, including contaminants and climate change, as my and Emilie Cameron's chapters discuss.

Yet these episodes also illustrate how the concept of "impacts," with its implication that humans are an external force, does not adequately describe historical interactions between humans and the northern environment. Environmental change has most often been experienced and understood in terms of people, and particularly Indigenous ways of life and wellbeing. As Piper explains, after 1870, excessive hunting by explorers, whalers, and miners was experienced alongside other pressures on Indigenous northerners' food supplies imposed by natural variability and hunting regulations. Similarly, the effects of infectious diseases were felt not only directly but through disruption of relations with nature, including an accelerated shift to southern foods. Further, as Arn Keeling and John Sandlos note, the toxic aftermath of mineral exploitation has been experienced not in terms of pristine ecosystems, but in relation to existing ties between humans and nature, including hunting. Contaminants and climate change must also be understood in terms of their consequences for humans, again reminding us that the north is not only a natural but a cultural landscape.

These chapters also demonstrated how northern nature has itself shaped the consequences of human activities. The effects of reindeer grazing were partly determined by the ecology of tundra, peculiar atmospheric and ecological conditions rendered the region vulnerable to radioactive fallout and persistent organic pollutants, and arctic ecosystems and wildlife have a special sensitivity to climate change. Thus, ecosystems and humans together define the effects of changes set in motion by human activities. These chapters therefore underline the value of understanding change in terms of new relations—of domestication, exploitation, or contamination—between humans and nature, in which species, ecosystems, and human activities together produce new environments.

The Indigenous North

Several chapters in this volume considered the evolution of Indigenous peoples' relations with nature, particularly in the context of environmental change, resource development, and the initiatives of governments and other actors. Piper considers the changing sources and meanings of food. Hans Carlson explores, with the Quebec Cree, their landscape after the James Bay hydro project. Paul Nadasdy examines peoples' relations with wildlife and one another in the Yukon territory. Keeling and Sandlos examine the consequences of mining developments for Indigenous health and livelihoods, explaining how remediation and redevelopment is forcing communities to revisit these difficult histories. Matt Farish and Whitney Lackenbauer note the implications of the DEW Line for Indigenous ways of life. Other chapters consider these relations in terms of reindeer herding, prospecting, or contaminants.

Throughout this history, people from elsewhere have perceived the relations between Indigenous people and nature in various ways. In the 1920s, prospectors and others believed that through contact with outsiders Indigenous people had lost much of their knowledge of the land. As Tina Adcock explains, this was for some a matter of regret because of nostalgia or because of concerns regarding traditional diets. Others considered this merely the inevitable consequence of assimilation and modernization, made necessary by the uncertain fur economy, shifting caribou migration routes, incidents of scarcity and starvation, and the conclusion that the northern environment remained beyond control, implying the need for an agriculture-based diet. This perspective had a variety of consequences. For a time, reindeer herding was seen as a path toward a more secure northern food economy—and an occupation for which, as Stuhl explains, Inuit were thought to be particularly suited. But as Piper notes, food also became central to more ambitious interventions, including relocation, economic development, and wildlife management, all consistent with the shift from traditional foods to food from the south that was being encouraged by Family Allowances, nutrition surveys, relief rations, and education. In addition, new settlements were often situated in places that responded to southern priorities (like access to natural resources or transport routes), but that were distant from local food supplies.

Interventions had other implications for Indigenous environmental relations. Indigenous people were often kept at a distance from state and industry initiatives, including resource development, aviation, and military projects. But other interventions focused directly on Indigenous communities. As Tina Loo describes, sustainable development meant encouraging more cohesive Inuit communities, in part through communication projects and applied anthropology, and efforts to enable Inuit to continue living on the land even while joining the wage economy. And as Nadasdy explains, the division of Indigenous people in the Yukon into bands and the imposition of property and territorial regimes has devalued non-territorial forms of organization, encouraging bureaucratization of their relations with animals and each other. The results of these interventions could often be ambiguous. Yukon co-management arrangements are the product of resistance to colonization, but are also rooted in colonial administration and state-based practices of territorial governance. And while the James Bay and Northern Quebec Agreement maintains a Cree presence on and understanding of the land, it also established a new language of environment, embodying Western ideas regarding the division between nature and culture. This accommodation reflects, as Carlson notes, how development co-exists with traditional ideas of the bush, sometimes producing compromises, like the 2002 "Paix des Braves" agreement, which reflect the challenges involved in navigating this complex landscape.

However, while describing these interventions, our authors have, as have other historians, portrayed Indigenous people as active agents in northern environmental history, acting on the basis of their knowledge and values, even amidst the colonizing agenda of political, military and economic actors. Resisting wildlife regulations, insisting on changes to government relief and rations, influencing research activities, negotiating land claims, and asserting their views regarding contaminants and climate—through these and other means, they have sought to redefine their relations with government, science, and nature.

The State

Our authors considered several aspects of the relations between the state and the northern environment, through its own activities, as well as its influence on how others—scientists, Indigenous peoples, industries—have understood, used, or transformed the environment. The state's role in encouraging resource development has been central to these relations, linking the northern environment to global markets. In this volume, this has been evident in resource surveys and support for aviation and the expansion of mining and hydroelectric development. Economic activities by the state have had, as we have seen, far-reaching impacts on the environment and on Indigenous ways of life. The state has also promoted or imposed a variety of other ways of relating to the environment. During the interwar era, the state attempted domestication of wildlife through the Canadian Reindeer Project; as Stuhl describes, this "experiment" in expanding the role of the state in the northern environment aimed to replace Inuit reliance on the fur trade and caribou with what was hoped to be a more stable and predictable pastoral economy. But as northern priorities shifted in the postwar era, so did state activities—to encompass, for example, the "geographical engineering" of the DEW Line—one of many ways in which the state, as Farish and Lackenbauer explain, fused science and security. These interventions also included relocation of settlements and expanded social programs (including sustainable development initiatives, which, as Loo explains, demonstrated the limits of state action), and land and wildlife management regimes based on property rights and territorial jurisdiction. Another role of the state became evident in the Northern Contaminants Program: contaminants were redefined in terms of federal jurisdiction, reframing an environmental and health issue to be consistent with administrative boundaries. These and other initiatives exhibited the extension of the state's authority and capacity to impose order, making the north administratively legible and amenable to regulation.

Several chapters also illustrated the significance of food in state roles and policies in northern environmental history. Changing views of subsistence have been part of this picture, with an early preference that Indigenous people continue to rely on hunting and fishing being eventually displaced by the assumption that an agriculture-based diet was superior. The result was "food colonialism"—evident in the reindeer project, relocation to places lacking adequate local food supplies, and a variety of other educational, nutritional, and social interventions. Food thus became the basis for northern administration—a means of managing the relations between people and nature, bringing the region into the Canadian mainstream, and making Aboriginal peoples full citizens of Canada. Food itself was redefined, becoming a matter of nutrition, not culture. Ultimately, however, food also became a basis for resistance to the state, as became evident in reactions to wildlife management, nutritional policies, and contaminants research.

Technology

Our authors have explored several aspects of the history of technology in the north, including its roles in creating new environments and new relations between nature and northern society. One aspect has been the sometimes unpredictable or unintended consequences of technology, as experienced, for example, at DEW Line stations, or with the pollution control equipment installed in mines—which only displaced many of the problems associated with toxic contaminants. Another aspect has been the influence of technology on how northern environments are understood. As Farish and Lackenbauer note, military technology implied a view of the northern environment as a technical space remote from politics, merely a set of problems to be solved or held at a distance. This influence has also been evident in the case of scientific technologies: analytical equipment detecting the presence of contaminants enabled perception of previously unknown aspects of the northern environment, while also sharpening distinctions between scientific and Indigenous perceptions of the environment.

These chapters also illustrate the reciprocal relationship between technology and the environment. This has included the impacts in the north of technology situated elsewhere, such as industrial and agricultural technology that has affected the global climate and the movement of contaminants. In such ways, technology has linked the north and the rest of the world. And as Cronin explained, airplanes, although built elsewhere, were then recreated in northern environments, with modifications made on the basis of experience with their performance. The experience of flying became incorporated into aviation technology and practices, as pilots dealt with novel problems: where to land on Great Slave Lake, how to deal with seasonal rhythms of ice and open water, how to maintain machinery in extreme cold. Experience with aspects of the northern environment that could once be ignored, such as particular patterns of ice cover and weather, thus became crucial to northern travel, and northern landscapes, waterscapes, and airscapes left their traces in or on the airplanes themselves. A similar relation between the environment and technology became evident when equipment for the DEW Line was brought into the north, and then had to be modified for work under local conditions. This conclusion suggests the need for a broader focus in the history of northern technology, which considers not just its impact on the environment, but the role of the environment in shaping technology.

Experience

The history of technology also illustrates some ways in which experience has been linked to trends in northern environmental history. As airplanes extended the reach of prospectors and other newcomers, they changed the experience of northern travel, eliminating most of the effort and time required. As Adcock describes, time itself seemed to speed up—at the expense, however, of experience with the intimate details of the land. As a result, northern travellers like Douglas and Blanchet were ambivalent about aviation and wary of how experience defined by distance and dependent on fossil fuels was privileged over that based on proximity and exertion. Technology thus reshaped how people experienced the north, reordering how evidence from their senses was interpreted, and how they perceived time and space. This would have practical consequences for how the north would be valued and transformed.

There are many other links, as well, between experience and northern environmental history. Some have been evident in interventions in Indigenous ways of life. As Piper explains, residential schools removed children from families and thus also from opportunities to learn about hunting and other subsistence practices, while instilling a taste for gardening and southern foods. And as Nadasdy describes, the consequences of co-management in the Yukon have become evident through the disconnect between how the land has traditionally been experienced—through hunting, kinship and reciprocity—and how it is now experienced within a system of resource management based on institutions, laws, and social relations consistent with state-based administration. In Quebec, the history of relations between the Cree and resource development is embodied in stories that make sense of experiences associated with places. As Carlson notes, they have continued hunting and trapping through the Income Security Program, thereby maintaining their experience with and understanding of the land.

Experiences and perceptions have also been essential to encounters between the north and people from elsewhere. Their arrival implied new ways of experiencing the north, such as the struggles of would-be Klondike prospectors, Blanchet's playful "Indian" engagement and Douglas' more controlled forays, and the careful observations by pilots and mechanics of the effects of flying and landing on their aircraft. These encounters also exhibited the influence of experience on perception, illustrating how memory has shaped how the north, time, and change are understood. As Adcock explains, old campsites, piles of firewood, and other traces coloured Douglas' and Blanchet's perceptions, imparting a sense of nostalgia, a desire for a more primitive, intense experience with nature, and dismay over the impacts of modern mining and surveys. How Western Electric employees on the DEW Line perceived the north was similarly shaped by their experience both there and in the south. More recently, Indigenous communities' responses to the revival of "zombie" mines have been influenced by memories of past experiences with the industry; and scientists' surprise when encountering the residues of global industrial activity was rooted in the cultural memory of a once-pristine north. And as Cameron notes, past experience can also help in understanding perceptions of climate change today.

Knowledge

These chapters have much to tell us regarding knowledge and the northern environment. Knowledge has been linked to changes in this environment, to how people live on or transform it, and to the pursuit of diverse political or economic goals. New ways of relating to or manipulating northern nature and people have implied new definitions of relevant knowledge. In the Stikine, plans for trails and railways required information about the landscape; as Peyton notes, even these failed schemes encouraged awareness of the region's economic potential. As Stuhl explains, the reindeer industry required a new form of botanical expertise and a new relationship between science and the state, based not simply on describing the range but on experimenting with it to increase productivity. Geological information about the Great Slave and Great Bear lakes region that companies collected in the 1920s (using government maps, surveys, and aerial photos) supported mining development. Nutritional science guided interventions in Indigenous communities, justifying a view of their diets as inherently insecure. Cold War technology, including the DEW Line, demanded knowledge of northern geography and local construction sites. Anthropology became an applied science, guiding economic development in northern communities. Contaminants rendered newly relevant knowledge about the physical environment, ecosystems, and human health.

The social relations of northern knowledge have been especially evident in how various ways of knowing have been defined as authoritative. While the 1919–20 Royal Commission on Possibilities of Reindeer and Musk-ox Industries in the Arctic and Sub-arctic Regions defined northern experience as the basis for expert advice, by the 1930s knowledge of botany and project management experience obtained in Alaska and the western United States was considered more reliable. Aviation privileged observations obtained at a distance over those based on more immediate experience. Status within institutions could determine the authority of knowledge, as pilots and engineers struggled to have their ground-level, experience-based knowledge recognized within their company. New kinds of experts emerged during the Cold War, including technicians able to manage complex systems. Contaminants and climate change have encouraged reconsideration of the relative authority of scientific and Indigenous knowledge, and awareness of the social implications of science has led to the concept of "responsible research," redefining relations between researchers and northern communities.

Throughout this history, new ways of arranging knowledge have accompanied novel relations with the environment. As these new relations emerged, science, because of how it was organized, was often able to provide at best only an incomplete understanding. This occurred with contaminants, knowledge of which was divided across several disciplines and institutions. The consequences included new combinations of knowledge cemented by new research institutions and by new objects of research and practice, such as the Arctic Dilemma, which combined, in a way distinctive to the north, the social and physical dimensions of contaminants. The history of northern knowledge exhibits other such objects: the variables describing tundra that ecologists used to indicate its ability to support reindeer; the complex technical systems (such as airplanes and radar) that enabled mobility and surveillance; the territorial units used to define areas of wildlife management. Each of these objects served to make the northern environment amenable to study and administration.

Several chapters have also examined where knowledge is formed, identifying the multiple "locals" of northern knowledge. One aspect of this is the distinction between knowledge tied to a specific location and knowledge of wider relevance and authority. In the 1920s, expertise in reindeer was not thought to be available locally; accordingly Saami, with experience in herding, were brought over from Scandinavia, and the Porsilds, with their botanical knowledge, were brought over from Denmark (and they then travelled elsewhere to learn more about herding). More recently, as Cameron noted, climate research has tended to consider local activities and adaptations as the sites of Indigenous knowledge, in contrast to activities like shipping, which take place on larger scales. This association of Indigenous knowledge with local places has been resisted in international contaminants negotiations.

Across a variety of areas of activity, distinctive northern knowledge has been formed by combining knowledge of the north with knowledge from elsewhere. Cold War installations merged laboratory knowledge from elsewhere with local field knowledge, illustrating how high modernist projects adhering to "universal" ideals of rationality and efficiency nevertheless also required local knowledge. Similarly, contaminants scientists found, sometimes to their surprise, that standard assumptions about substances, bodies, health, and risk were not valid in the north, and therefore distinctive research approaches were necessary. The emerging community-based character of northern contaminants research reinforced this distinctive character, dispensing with the notion of the north as a placeless laboratory.

Mobility

As our authors have demonstrated, northern environmental history is a history of movement—of people, animals, machines, materials, ideas. They have described efforts to enable mobility by flying, importing reindeer, or exporting resources; to understand it, by walking through Eeyou Istchee or through studies of animal migrations or the flows of contaminants; and sometimes to prevent it, through strategic surveillance. Efforts to build an all-Canadian route to the Klondike illustrated how mobility was considered an essential response to the discovery of gold, as it also was during the Great Bear Lake radium boom of the early 1930s, and in subsequent resource developments. Mobility has often embodied an ambition to transcend environmental constraints, including distance, climate, and landscape. Yet these still exerted their influence: pilots followed waterways, while paying careful attention to freezing, breakup, and weather.

Mobility transformed travellers' lives: what they saw, heard, and felt; their perceptions of time and space; and distinctions of gender, class, and race. Materials have also been mobilized, including minerals, petroleum, electricity, and other commodities exported to meet market demands. Different kinds of mobilities have combined: capital with people, technology, and commodities; knowledge with the mobile phenomena it described. Technology and mobility have intersected: an aviation network, initially along waterways, emerged out of the interaction between geography, environment, technology, and the market; and airplanes, as well as roads and railways, spurred export of resources. Resource booms and busts have demonstrated how signals from distant markets could produce or disrupt mobility. The movement of materials into the region, such as surveillance equipment, and food and other consumer goods, have demonstrated how policy could drive mobility. Some materials have moved themselves, including contaminants. Together, these mobilities have eroded the notion of the north as the last inaccessible place (although the continuing challenge of expensive imported food illustrates the limits of mobility).

Ideas have also been mobilized, taken from elsewhere to influence attitudes and policies in northern Canada. These have included antimodernist attitudes that framed personal encounters with and perceptions of change; ideas about economic progress through commodity exports; principles of scientific management that reordered relations between hunters and wildlife in terms of capitalism, agriculture, and territory; ideologies of apolitical rationality and high modernism, which guided development of the DEW Line and other projects; models of sustainable and regional development, including cooperatives; and expert ideas about food and nutrition. But these chapters also portray a north that is part of the world's conversation, not merely receiving but forming ideas, which in turn have become influential elsewhere. Knowledge has flowed out of the north, including accounts of the exploitation, management, or conservation of its landscapes, and the region's place in global ecosystems. Indigenous perspectives have been mobilized: by the Cree advocating in New York and elsewhere regarding the Great Whale Project, and by communities seeking a voice in global contaminant negotiations. Mobility has been particularly evident in the stitching-together of sites of production and application of knowledge. In the Canadian Reindeer Project, ecological information and advice was transferred between Alaska (and before that, the rangelands of the American west), the Yukon and Northwest Territories, Norway, and Ottawa. Networks of knowledge also formed during the design and construction of the DEW Line, combining plans, equipment, and technical expertise from the south with local information from the north. The movement of knowledge within an airline illustrated how power and bureaucracy could shape the flow of observations and technical advice even within one organization. Contaminants science formed yet another network: knowledge moved among scientists, between scientists and communities, and between north and south.

Northern Places

Throughout these chapters, the north has been identified as not only a physical but an imagined space, with diverse ideas about where it is, who and what belongs there, and what its future could be. As Peyton explained, a failed railway to the Klondike catalyzed perceptions of the Stikine as a place of opportunity. Carlson and Nadasdy explored in Quebec and the Yukon the north imagined as a homeland defined by Indigenous identity, kinship, and ways of life—a view transformed in recent decades, but still essential to northern history and life. Other ways of imagining the north have encompassed ambitions to tame this space. One way was by combining, as Stuhl and Piper described, science, the state, and species (reindeer and garden crops) in attempts to form a domesticated agricultural region. Another was by converting its geology into commodities, and the north into a region accessible to global markets. This redefinition of the north in terms of natural resources neglected other features of the northern landscape, including its social dimensions. The north also became a proving ground for the military, as Farish and Lackenbauer explain; and this once-pristine place became an environment newly vulnerable to contaminants, as my chapter shows. This imagined north has been a dynamic place, with shifting perceptions, interests, and attitudes.

Experience has been essential to these imagined norths. Aviators and airborne prospectors perceived an apparently empty but newly accessible and resource-rich landscape; the Cree walked, worked, and camped on the land. Economic interests have also been influential, such as those of private operators and state agencies imagining northern resource wealth. Other attitudes, too, have made their influence felt: the strategic concerns that led Cold War military strategists to imagine the north as a defensive bulwark; and ideas about food quality that led officials and nutritionists to imagine a new agricultural frontier. Throughout, physical realities have both inspired and constrained these ideas. Agricultural aspirations collided with the cold. The presence of gold and other minerals justified formation of a toxic "landscape of exposure." Northern geography and climate shaped aviators' ideas about the north as a technological frontier. Ecological and cultural circumstances encouraged a view of the north as distinctively vulnerable to contaminants.

One feature that these imagined norths share is that they have been framed in relation to other places. The transformation of the northern landscape into commodities has been the most obvious relation, but there have been others, as well. Antimodernist ideals led some to view the north in terms of what southern Canada was not: a relic of the past, and a place to retreat from the modern world. Aviators described the northern environment as exceptional, implying a need to adapt technology accordingly. Military strategists located the north as a strategic space between Cold War superpowers—geographically distinct yet potentially universal. Government officials interpreted economic challenges in the postwar north in terms of underdevelopment—a concept more often applied to what was referred to during that period as the Third World. More recently, the challenge has been to imagine a north on its own terms, responding to aspirations and opportunities framed within the region itself.

Doing Northern Environmental History

Finally, we can see throughout this volume how these encounters have encouraged distinctive ways of doing environmental history. This includes acknowledging the nature of northern environmental change, which, though part of a larger history of global change, has distinctive physical and social features. These features imply a need to understand northern historical relationships in terms of the experience and perspectives of northerners themselves, especially Indigenous people, drawing on distinctive approaches to place and history, nature and culture. One way is that of exploring, as Carlson describes, the ecology of stories that embodies the ties between personal history and the history of a place. This also implies distinctive challenges. Forming relationships with Indigenous communities can take a great deal of time, which can be difficult for scholars with limited time and budgets. Environmental historians must also be aware of Indigenous rights, identities, knowledge, and community realities, and consider critically their role in representing the stories of others.

However, these chapters also illustrate numerous points of contact between environmental history in the north and elsewhere. A comparative approach can challenge assumptions of northern exceptionalism by identifying features shared by the north and other regions, such as the specific structural relationships between remote regions and centres of power.[1] Useful comparisons can also be made with other regions where resource exploitation imposed from the outside has reshaped lives. Struggles in the north between dominant institutions and Indigenous people over knowledge and authority can be compared with similar struggles elsewhere in the circumpolar region, or with other places with colonial histories, or wherever Indigenous people have faced pressures to adapt to or participate in development. A history of contentious relations between colonial officials and Indigenous people over land and wildlife has been seen in many other countries, even as these have assumed a distinctive character in the north. The gendered dimensions of food and contaminants, and the relation between human health and the health of the northern environment, suggest other opportunities for comparison with histories elsewhere.[2]

This volume has also demonstrated the value of an interdisciplinary perspective on environmental history, drawing on the work of historians, geographers, anthropologists, and scientists, as well as on perspectives formed by scholars elsewhere. Knowledge of how technology embodies power places the DEW Line into perspective. Political ecology provides insights into the global context of local environmental changes, such as those caused by the James Bay dams and other developments. Our understanding of the history of northern science has been informed by work elsewhere on the history of the field sciences, including the relations between expert and vernacular knowledge and the practice of natural history surveys.[3] The "envirotechnical" perspective, combining environmental history and the history of technology, illuminates the evolving relations between airplanes and the northern environment. Concepts of territoriality—how people and their actions are administered through the control of space—have guided our exploration of co-management regimes. Regimes of perceptibility and evolving views of the linkages between the health of environments and of bodies have been applied to the history of northern contaminants. Northern environmental history also draws on our understanding of the social and political dimensions of food, and its role in the relationships between bodies and environments.

This book began with the ice blink, symbolizing the encounters between people and nature that have shaped northern history, and the challenge of navigating this region and its history as it undergoes rapid change. Today, northern change is often taken to mean climate change. This is understandable, given the capacity of warming to transform sea and land—a reality already apparent in record losses of sea ice, with consequences for people and polar bears alike. But as our authors have shown, change in the north has taken many forms, driven not just by environmental transformation, but through the agency of Indigenous peoples and newcomers using varied technologies and guided by their experience and knowledge. Through this richer understanding of the nature of change, northern environmental history can provide an essential perspective on contemporary northern issues, and on the future of the region.

Notes

1   Kenneth Coates, "The Discovery of the North: Towards a Conceptual Framework for the Study of Northern/Remote Regions," *Northern Review* 12–13 (1993–94): 15–43.

2   Gregg Mitman, *Breathing Space: How Allergies Shape Our Lives and Landscapes* (New Haven: Yale University Press, 2007); Linda Nash, *Inescapable Ecologies: A History of Environment, Disease, and Knowledge* (Berkeley: University of California Press, 2007).

3   See, for example, Robert E. Kohler, *All Creatures: Naturalists, Collectors, and Biodiversity, 1850–1950* (Princeton: Princeton University Press, 2006).

Contributors

TINA ADCOCK is an assistant professor of history at Simon Fraser University. She studies the cultural and environmental history of the modern Canadian North, particularly the experiences of southern sojourners involved in exploration, scientific fieldwork, travel, and resource exploitation. Her research has previously appeared in Swedish, Norwegian, and American scholarly collections. She is completing a book manuscript on cultures of northern Canadian exploration in the first half of the twentieth century.

STEPHEN BOCKING is a professor of environmental history and policy in the Trent School of the Environment, Trent University. His research interests include the environmental history of science, and the roles of science in environmental affairs in northern Canada and elsewhere.

EMILIE CAMERON is an associate professor in the Department of Geography and Environmental Studies at Carleton University. Her current research focuses on geographies of resource extraction, empire, and labour in the contemporary North.

HANS M. CARLSON is Executive Director of the Great Mountain Forest, in Norfolk, Connecticut. Great Mountain is a 6,300-acre working forest dedicated to sustainable forestry practice, research, and the integration of human communities and human history in the stewardship of their land. Carlson previously taught in the American Indian Studies Department at the University of Minnesota. His contribution to this volume is part of an ongoing writing project on the cultural and environmental impacts of global resource use in James Bay, Quebec.

MARIONNE CRONIN is an honorary research fellow in the Northern Colonialism program at the University of Aberdeen. Her research interests include the histories and historical-cultural geographies of science, technology, and exploration, with a special interest in the history of circumpolar aviation. Her current research examines the material practices and imagined geographies of northern colonialism.

MATTHEW FARISH is a historical geographer and an associate professor in the Department of Geography and Planning at the University of Toronto. His research focuses on the relationships between militarization and geographical knowledge. Among his current research projects is a comprehensive history of the Distant Early Warning (DEW) Line.

ARN KEELING is Associate Professor of Geography at Memorial University, conducting research on the environmental history and historical geography of western and northern Canada. With John Sandlos, he led the Abandoned Mines in Northern Canada project, with funding from the Social Sciences and Humanities Research Council and ArcticNet. They are currently pursuing research on industrial development and pollution in the north.

P. WHITNEY LACKENBAUER is a professor of history at St. Jerome's University. His research explores historical and contemporary arctic sovereignty and security issues, Native-newcomer relations, and civil-military relations. He acknowledges the support of the Social Sciences and Humanities Research Council and ArcticNet for facilitating his archival, oral history, and field research on Canada and the Cold War Arctic.

TINA LOO teaches Canadian and environmental history at the University of British Columbia. The work in this volume is part of a larger project examining forced relocations, development, and the welfare state in postwar Canada.

BRAD MARTIN is Dean of the Faculty of Education, Health, and Human Development at Capilano University.

PAUL NADASDY is Associate Professor of Anthropology and American Indian Studies at Cornell University. He is the author of *Hunters and Bureaucrats: Power, Knowledge, and Aboriginal Relations in the Southwest Yukon* and numerous scholarly articles. The present chapter is part of a larger study he is conducting on the cultural assumptions that underlie the Yukon land claim and self-government agreements and the effects their implementation is having on Indigenous ways of life.

JONATHAN PEYTON is an assistant professor in the Department of Environment and Geography at the University of Manitoba whose research on extractive economies lies at the intersection of environmental historical geography and political ecology. He is the author of *Unbuilt Environments: Tracing Postwar Development in Northwest British Columbia.*

LIZA PIPER is an associate professor of History at the University of Alberta. Her 2009 book, *The Industrial Transformation of Subarctic Canada*, examines the role of industrial resource exploitation and science in the twentieth century transformation of subarctic environments. Her current research considers how disease and climate have changed human relations to nature in the Subarctic and the Arctic since the nineteenth century.

JOHN SANDLOS is an associate professor of History at Memorial University, where he teaches environmental history and conducts research on wildlife conservation, parks, and mining in northern Canada. He was recently a fellow at the Rachel Carson Center for Environment and Society in Munich, Germany. With Arn Keeling, he led the Abandoned Mines in Northern Canada project, with funding from the Social Sciences and Humanities Research Council and ArcticNet. They are currently pursuing research on industrial development and pollution in the north.

ANDREW STUHL is an assistant professor of Environmental Humanities at Bucknell University. His 2016 book, *Unfreezing the Arctic: Science, Colonialism, and the Transformation of Inuit Lands*, examines scientific engagement with the western North American Arctic between 1881 and 1984 as colonial and environmental history. He is currently researching the environmental histories of northern comprehensive land claim agreements and offshore drilling in the Beaufort Sea.

Index

G

game management zones, 352, 353, 357
game reserves, 6, 15, 17
gangrene, 194–95
gardens (mission gardens), 198, 211
gender, 165, 271, 304, 438, 509, 513
Geological Survey of Canada, (GSC) 8, 20, 43, 46, 48, 56, 73, 91,
Giant Mine, 381, 384–85, 388–92, 393, 398, 399–403, 405, 407
Gibson, Roy, 196, 200, 204–5, 211
Glenora, British Columbia, 36–37, 40, 42–3, 45–48, 50–51, 53
globalization, 320, 484
Great Bear Lake, 2, 6, 26, 86, 122, 124, 133–34, 137, 140–41, 153–55, 157–59, 162, 165, 168–69, 186, 386, 507, 509
Great Slave Lake, 6, 111, 133–34, 137, 142, 149, 155–56, 158, 162, 165, 167–68, 190–91, 381, 384–87, 392, 394, 398, 405–6, 505, 507
growth pole, 241
Gwich'in, 183, 185

H

Hanbury, David, 185
Harper, Stephen, 308
Harrington, Richard, 196, 223, 225–27
Hay River, 192, 193, 237, 381, 393, 398, 403, 405
heavy metals, 380, 382–83, 387 393
Herschel Island, 186
high modernism, 9, 91, 266–68, 274, 281, 510
Hoare, W. H. B., 23
Hoffmann, Erich, 242
Hudson Bay, 195
Hudson's Bay Company, 5, 19, 23, 25, 47, 72–74, 187–88, 197–200, 204, 206, 211, 231–32
Hunt, L.A.C.O., 201, 204
hunting rights, 344, 348–49, 360
Hydro Quebec, 296, 314, 315
hydroelectric projects, 9–11, 15, 17

I

ice blinks, 3, 4, 513
Iglauer, Edith, 242, 248
Indian Act, 68, 334, 341, 343–44
Indigenous people(s), 3, 5–6, 8, 11–4, 16–18, 21, 25–26, 28–29, 40–41, 50, 142, 146–47, 149, 152, 168, 186–88, 195–96, 198, 200–206, 208–9, 211–13, 226, 234, 239, 242, 251, 281–82, 298, 312, 333–34, 349
Industrial Division, 239, 242, 246, 249, 362, 365–6, 408, 426, 444–45, 447–48, 466, 471–79, 481, 501–4, 512–14
industry, military role in, 266–70, 275
influenza, 16, 69, 189–91, 194
Interdepartmental Reindeer Committee, 69, 78, 89–91
Intergovernmental Panel on Climate Change (IPCC), 466
International Polar Year, 186, 451
Inuit, 5, 9–10, 17–18, 21, 23–25, 27, 64–65, 67–72, 77–79, 84–85, 87–92, 151, 168, 185–88, 190, 192–96, 200–202, 204–6, 213, 223, 225–27, 230–32, 234–40, 242–44, 247–52, 282, 312, 395, 421–22, 427–28, 434, 436, 438–39, 442, 466–67, 471, 475, 479–80, 482, 484, 502–3
Inuit, and climate change, 466, 484
Inuit Circumpolar Conference/Council (ICC), 421, 466, 479
Inuvialuit, 10, 69, 88, 90, 186, 203, 207, 209, 281–82

J

Jackson, Sheldon, 67
James Bay, 11, 17, 28, 196, 206, 295, 303, 306–7, 309–10, 312–15, 319, 322, 501–2, 513
James Bay and Northern Quebec Agreement, 11, 17, 310–312, 313, 315, 323, 502
Jenness, Diamond, 70
Jennings, Francis, 303
Johnson, Joe, 357–60
jurisdiction, political, 334–35, 339, 346, 350–51, 359, 366

V

Vandergeest, Peter, 339, 352, 361
Vermont, 314–15
vitamins, 195, 197–98, 203, 206, 210–11

W

walrus, 192, 194–95
Waswanipi, 324
watery geographies, 110–12
Watt-Cloutier, Sheila, 421, 425, 438, 448, 466, 479
Welcome to Pine Point (National Film Board documentary), 381
Wensley, Gordon, 244
Western Electric Company, 261–84
Whale Cove, 224, 242–43
whaling, 14, 26, 67–68, 192, 466
Whitehorse Mining Initiative, 407
White River First Nation (WRFN), 360
wildlife science, 22
"will to improve", 227, 230, 249, 471
Williamson, Robert G., 232–38, 244
Willis, J. S., 201
Wood Buffalo National Park, 5, 15, 186, 242
World Bank, 319
Wrangell, Alaska, 38, 42, 47, 49–51

X

X-ray, 207

Y

Yellowknife, 12, 15, 158, 161, 165, 168, 193, 236–37, 381–82, 384–86, 388–91, 397, 399, 401–2, 405, 407
Yellowknives Dene (First Nation), 388–89, 392, 400–402
Yukon-Canadian Railway, 44, 51, 53
Yukon River, 36, 40, 42
Yukon Territory, 501–2, 506, 510–11
Yukon Umbrella Final Agreement (UFA), 11, 342

Z

zinc, 111, 381, 383–84, 386–87, 392–94, 403–4
"zombie" mines, 12, 380–81, 384, 407–8, 506

www.ingramcontent.com/pod-product-compliance
Lightning Source LLC
Chambersburg PA
CBHW051440270326
41932CB00025B/3379